THE ELECTRICAL ENGINEERING AND APPLIED SIGNAL PROCESSING SERIES

Edited by Alexander Poularikas

The Advanced Signal Processing Handbook:
Theory and Implementation for Radar, Sonar,
and Medical Imaging Real-Time Systems
Stergios Stergiopoulos

The Transform and Data Compression Handbook
K.R. Rao and P.C. Yip

Handbook of Multisensor Data Fusion
David Hall and James Llinas

Handbook of Neural Network Signal Processing
Yu Hen Hu and Jenq-Neng Hwang

Handbook of Antennas in Wireless Communications
Lal Chand Godara

Noise Reduction in Speech Applications
Gillian M. Davis

Signal Processing Noise
Vyacheslav P. Tuzlukov

Digital Signal Processing with Examples in MATLAB®
Samuel Stearns

Applications in Time-Frequency Signal Processing
Antonia Papandreou-Suppappola

The Digital Color Imaging Handbook
Gaurav Sharma

Pattern Recognition in Speech and Language Processing
Wu Chou and Biing-Hwang Juang

Propagation Handbook for Wireless Communication System Design
Robert K. Crane

Nonlinear Signal and Image Processing: Theory, Methods, and Applications
Kenneth E. Barner and Gonzalo R. Arce

Mobile Internet: Enabling Technologies and Services
Apostolis K. Salkintzis

Forthcoming Titles

Smart Antennas
Lal Chand Godara

Soft Computing with MATLAB®
Ali Zilouchian

MOBILE INTERNET

Enabling Technologies and Services

Edited by
Apostolis K. Salkintzis

CRC PRESS

Boca Raton   London   New York   Washington, D.C.

Library of Congress Cataloging-in-Publication Data

Mobile Internet: enabling technologies and services / edited by Apostolis K. Salkintzis.
   p. cm. -- (Electrical engineering and applied signal processing series)
  Includes bibliographical references and index.
  ISBN 0-8493-1631-6 (alk. paper)
   1. Wireless Internet. I. Salkintzis, Apostolis K. II. Title. III. Series.

TK5103.4885.M63 2004
004.67'8--dc22

2003069582

Visit the CRC Press Web site at www.crcpress.com

Dedication

To my mother and father

Preface

The migration of the most common Internet services to a mobile environment has long been an evolving demand of both business and consumer markets. The ability to be connected to the Internet while on the go and to benefit from using such applications as e-mail, instant messaging, audio and video streaming, Web browsing, and e-commerce creates an exciting new lifestyle and sets the foundation for increased work efficiency and productivity. In addition, however, it introduces numerous technical challenges inherently associated with the user's mobility and wireless connectivity. This book addresses these technical challenges and provides a thorough examination of the most recent and up-to-date technologies designed to deal with them and enable the so-called *mobile Internet*. It is written by some of the most eminent academic and industry professionals in the areas of mobile and multimedia networking.

In more detail, this book first discusses the evolution toward the mobile Internet and the associated technological trends and visions. In this context, it illustrates the key features of this evolution, including the migration to all-IP networks, the enabling of advanced multimedia services, the provision of end-to-end quality of service (QoS), and the integration of heterogeneous access networks. In addition, we discuss the key players in the evolution toward the mobile Internet (e.g., the Third-Generation Partnership Project [3GPP], 3GPP2, the Internet Engineering Task Force [IETF], and the Institute of Electrical & Electronics Engineers [IEEE]) and summarize their corresponding activities. This allows the reader to obtain a coherent understanding of how the technology is being evolved to meet the requirement of the mobile Internet and who is involved in this evolution.

Subsequently, the most common mobile and wireless network technologies that can provide the means for mobile Internet connectivity are examined. In this context, Internet access over technologies such as wireless local area networks (WLANs), General Packet Radio Service (GPRS), and satellites is discussed in a comprehensive way in order for the reader to understand their respective characteristics, advantages, and disadvantages. To explain how well each technology can support Internet applications, we review their QoS and routing features. In addition, we present performance results and demonstrate how WLAN and cellular networks can be integrated in order to provide heterogeneous mobile data networks that are capable of ubiquitous data services with very high data rates in strategic locations, such as airports, hotels, shopping centers, and university campuses. Such heterogeneous networks are generally considered to be a key characteristic of fourth-generation (4G) mobile data networks.

A large part of this book concentrates on mobility-management techniques, which are necessary in a mobile IP environment where the user's point of network attachment can constantly change. Both macromobility (e.g., Mobile IP) and micromobility (e.g., Cellular IP) management protocols are extensively discussed. In addition, we examine the security concerns associated with mobility management and look at several methods that can provide secure mobility management. We further elaborate on the issue of security and tackle such topics as security in WLANs and third-generation (3G) cellular networks. In this context, we explain the most common security attacks, including the man-in-the-middle attack, denial of service, and identity theft.

Because many modern Internet applications, such as context broadcasting and videoconferencing, are based on multicasting, this book also reviews the key technical challenges of multicasting in mobile environments and the most promising technologies that can support it. In addition, QoS provisioning, considered a key aspect of the mobile Internet, is examined at length. The classical IP QoS techniques (both Differentiated Services [DiffServ] and Integrated Services [IntServ]) are reviewed, and we propose extensions for addressing their deficiencies in mobile IP networks.

Moreover, for enabling the efficient provision of Internet multimedia services in wireless networks, where the use of radio resources needs to be highly optimized, this book examines the characteristics of several header compression schemes designed specifically for wireless environments. Finally, it discusses the challenges of video streaming in wireless IP networks as well as techniques tailored to operate in such networks and significantly improve their performance.

In short, this original book provides a thorough study of all the key technologies that can enable the provision of ubiquitous mobile Internet services, focusing on mobility, QoS, security, and multicasting. Apart from reviewing the technical challenges and the prime technologies developed to deal with them, this book proposes some novel techniques and examines performance results.

Apostolis K. Salkintzis, Ph.D.

Editor

Apostolis K. Salkintzis, Ph.D., received his diploma in 1991 (honors) and his Ph.D. degree in 1997, both from the Department of Electrical and Computer Engineering, Democritus University of Thrace, Xanthi, Greece. From 1992 to 1997, he was a research engineer at Democritus University, studying mobile data networks and working on research projects dealing with the design and implementation of wireless data networks and protocols. In 1999 he was a sessional lecturer at the Department of Electrical and Computer Engineering, University of British Columbia, Vancouver, Canada, and from October 1998 to December 1999 he was also post-doctoral fellow in the same department. During 1999, Dr. Salkintzis was a visiting fellow at the Advanced Systems Institute of British Columbia, Vancouver; during 2000, he was with the Institute of Space Applications and Remote Sensing (ISARS) of the National Observatory of Athens, Greece, where he conducted research on digital satellite communication systems. Since September 1999, he has been with Motorola Inc., working on the design, evolution, and standardization of mobile telecommunication networks, focusing in particular on General Packet Radio Service (GPRS), Universal Mobile Telecommunications System (UMTS), Terrestrial Trunked Radio (TETRA), and wireless local area networks (WLANs).

Dr. Salkintzis has served as lead guest editor for a number of special articles, such as "Mobile Data Networks: Advanced Technologies and Services," in *Mobile Networks and Applications Journal*; "The Evolution of Mobile Data Networking," in *IEEE Personal Communications*; "Satellite-Based Internet Technology and Services," in *IEEE Communications*; "Multimedia Communications over Satellites," in *IEEE Personal Communications*; "IP Multimedia in Next Generation Mobile Networks: Services Protocols and Technologies," in *IEEE Wireless Communications*; and "The Evolution of Wireless LANs and PANs," in *IEEE Wireless Communications*, among others. He has served as a TPC member and referee for numerous international conferences, including Globecom03, ICC200, ICC2003, VTC03-Fall, VTC03-Spring, VTC00, VTC99, Globecom01, ICC01, and INFOCOM03. He has organized many technical sessions and symposia in various conferences and has chaired most of them.

Dr. Salkintzis has published more than 40 papers in peer-reviewed journals and international conferences and has also published 3 book chapters.

His primary research activities lie in the areas of wireless communications and mobile networking. He is particularly interested in mobility management, IP multimedia over mobile networks, mobile network architectures and protocols, quality of service (QoS) for wireless networks, and radio modem design with digital signal processors (DSPs). Recently, he has been active in WLAN/3G interworking and with the evolution and standardization of GPRS and UMTS networks. Until 2002, he was an active participant and contributor in the Third-Generation Partnership Project (3GPP) standards development organization and was an editor of several 3GPP specifications.

Dr. Salkintzis is a member of the Institute of Electrical and Electronics Engineers (IEEE) as well as a member of the Technical Chamber of Greece.

Contributors

A.H. Aghvami
King's College London
London, England

Farooq Anjum
Telcordia Technologies, Inc.
Morristown, New Jersey

Jari Arkko
Ericsson Research Nomadiclab
Jorvas, Finland

Subir Das
Telcordia Technologies, Inc.
New Providence, New Jersey

Lila V. Dimopoulou
National Technical University
    of Athens
Athens, Greece

Frank H.P. Fitzek
Acticom GmbH
Berlin, Germany

Stefan Hendrata
Acticom GmbH
Berlin, Germany

Antonio Iera
University Mediterranea
    of Reggio Calabria
Italy

Christophe Janneteau
Motorola, Inc.
Gif-sur-Yvette, France

Hong-Yon Lach
Motorola, Inc.
Gif-sur-Yvette, France

Antonella Molinaro
University of Reggio Calabria
Italy

Bongkyo Moon
King's College London
London, England

Pekka Nikander
Ericsson Research Nomadiclab
Jorvas, Finland

Yoshihiro Ohba
Toshiba America Research, Inc.
Convent Station, New Jersey

Nikos Passas
University of Athens
Athens, Greece

Alexandru Petrescu
Motorola, Inc.
Gif-sur-Yvette, France

Martin Reisslein
Arizona State University
Tempe, Arizona

Imed Romdhani
Motorola, Inc.
Gif-sur-Yvette, France

Apostolis K. Salkintzis
Motorola, Inc.
Athens, Greece

Patrick Seeling
Arizona State University
Tempe, Arizona

Iakovos S. Venieris
National Technical University
    of Athens
Athens, Greece

Contents

1

The Evolution toward the Mobile Internet

Apostolis K. Salkintzis
Motorola, Inc.

1.1 Introduction

Simply stated, the mobile Internet can be thought of as the migration of standard Internet applications and services to a mobile environment. The introduction of mobility, however, raises a number of new considerations: What wireless technology is most appropriate for the provision of Internet services? Is this technology equally appropriate for applications with dissimilar requirements, such as e-mail and video broadcasting? How do we provide *ubiquitous* Internet services while users move across different locations, where the same wireless service may not be available? Will it be more appropriate to consider several wireless technologies, such as cellular data networks, wireless local area networks (WLANs), wireless personal area networks (WPANs), and so forth? If so, then how do we combine them into a seamless wireless service? How do we optimize the utilization of wireless resources to accommodate as many mobile Internet users as possible? How do we handle the security issues raised by the wireless transmission and possibly by the use of different wireless services provided by different operators?

These are just a few of the questions we must address in our attempt to make the mobile Internet a reality. One important clarification is in order here: The fact that today we can use our laptop or personal digital assistant (PDA) along with a wireless device (e.g., a cellular phone or WLAN adapter) to establish access to our Internet/intranet e-mail or other Internet Protocol (IP) services does not really mean that the mobile Internet is already available. In reality, what is defined as the *mobile Internet* is far more complex than that. By definition, the Internet is a network of millions of users who communicate by means of standard Internet protocols. Therefore, in the mobile Internet we also need to support millions of *always-connected* mobile/wireless users. The key phrases we need to bear in mind are "always connected" and "millions of users." To be "always connected" means that we do not have to make a connection before every transaction or after we change wireless service. Our assumption is that we always have connectivity to public Internet services and that we are always reachable through a public IP address, no

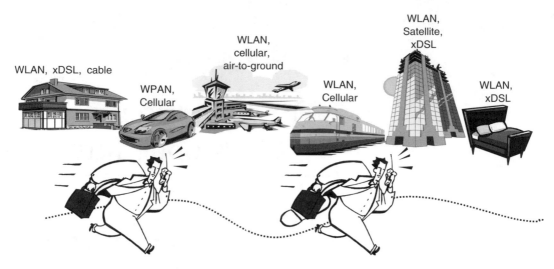

FIGURE 1.1   The concept of always being connected.

matter where we are. This calls for extensive mobility management and seamless handover across different wireless networks. And we do need several wireless networks (and several wireless technologies, as we will see later on) because no sole wireless network can provide ubiquitous wireless services, and therefore no sole wireless network can meet the always-connected requirement. A typical mobile Internet user is assumed to move *seamlessly* between different wireless networks (or even between fixed and wireless networks), which may or may not employ the same radio-access technology. This is schematically illustrated in Figure 1.1. The seamless fashion of movement suggests that our (virtual) connection to the public Internet is transferred from one access network to the other without any actions from the user's side. In effect, this creates a virtual wireless network from the user's perspective that provides ever-present Internet connectivity.

We need also to support millions of users wirelessly connected to the Internet in order to be compliant with the large scale of users supported by the Internet. This creates capacity concerns and better explains why no single wireless network is sufficient. The capacity concerns have direct implications in several design aspects. For instance, the wireless technology needs to be as spectrum efficient as possible; it has to implement a random access scheme that can accommodate a large number of users and degrade gracefully in overload conditions; the IP addressing scheme should be able to support the required user capacity; and the end-to-end transport schemes and applications should probably take mobility into account.

It is interesting to note that most research projects nowadays, as well as standardization activities, are moving around such goals. In this context, the correlation between the mobile Internet and the so-called beyond third-generation (3G) technologies is evident. In reality, most of the beyond-3G technologies are tailored to support the requirements for mobile Internet: increased capacity, quality of service (QoS), mobility, security, TCP/IP enhanced performance, and integration of diverse technologies into one ubiquitous virtual network.

Despite the fact that in this book we focus entirely on wireless networks as a potential means of access for the mobile Internet, it is important to note that fixed-access networks and their associated technologies (e.g., xDSL [x Digital Subscriber Line] and cable modems) do play a significant role. In fact, the vision of the mobile Internet entails both wireless and fixed-access technologies, as well as methods for seamlessly integrating them into an IP-based core network (see Figure 1.2). A typical usage scenario that shows the fundamental interworking requirements between fixed- and wireless-access networks is when a user downloads a file over the Internet by using, for instance, a cable modem, and in the middle of the file transfer takes his or her laptop to his or her car and drives to the airport (refer to Figure 1.1). In a mobile-Internet environment, the file download would be seamlessly switched from the cable connection to, say, a cellular data connection and would be carried on while the user is on the go.

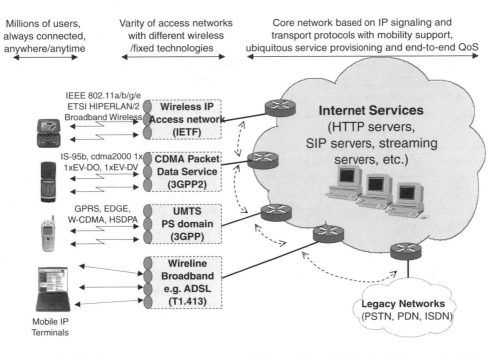

FIGURE 1.2 The architecture of the mobile Internet is tightly coupled with the beyond-3G network architecture.

In our mobile-Internet environment, as defined above, it is evident that enhanced mobility is a key requirement, and Chapter 5 discusses it in detail. Without a doubt, the technologies that can provide wireless access to the Internet are also of paramount importance. We will look at several of them, including WLANs in Chapter 2, the General Packet Radio Service (GPRS) in Chapter 3, and satellite access in Chapter 4. The issues of IP QoS and multicast in mobile environments are addressed in Chapters 6 and 7, respectively. The remaining chapters deal with several other equally important issues, such as content streaming, security, and integration between WLANs and 3G networks.

1.2 Key Aspects of the Evolution toward the Mobile Internet

Bearing in mind the above discussion, we list below the key aspects of the evolution toward the mobile Internet. In the next section, we look more closely at how these aspects are handled by several organizations worldwide, such as the Third-Generation Partnership Project (3GPP), the Third-Generation Partnership Project 2 (3GPP2), the Internet Engineering Task Force (IETF), and the Institute of Electrical and Electronics Engineers (IEEE), among others.

- Mobile networks will evolve to an architecture encompassing an **IP-based core network** and many wireless-access networks. The key feature in this architecture is that signaling with the core network is based on IP protocols (more correctly, on protocols developed by IETF), and it is independent of the access network (be it Universal Mobile Telecommunications System [UMTS], Code Division Multiple Access [CDMA], CDMA2000, WLAN, or whatever). Therefore, the same IP-based services could be accessed over any networked system. An IP-based core network uses IP-based protocols for *all* purposes, including data transport, networking, application-level signaling, and mobility. The first commercial approach toward this IP-based core network is the *IP Multimedia Core Network Subsystem (IMS)* standardized by 3GPP and 3GPP2. We further discuss the concept of IP-based network in Section 1.2.1. In addition, the IMS is discussed in Section 1.3.1.2.1.

- The long-term trend is toward **all-IP mobile networks**, where not only the core network but also the radio-access network is solely based on IP technology. In this approach, the base stations in a cellular system are IP access routers and mobility/session management is carried out with IP-based protocols (possibly replacing the cellular-specific mobility/session management protocols).
- Enhanced **IP multimedia applications** will be enabled by means of application-level signaling protocols standardized by IETF (e.g., Session Initiation Protocol [SIP] and Hypertext Transfer Protocol [HTTP]). This again is addressed by the 3GPP/3GPP2 IMS.
- **End-to-end QoS** provisioning will be important for supporting the demanding multimedia applications. In this context, extended interworking between, say, UMTS QoS and IP QoS schemes is needed, or more generally, interworking between Layer 2 QoS schemes and IP QoS is required for end-to-end QoS provisioning.
- **Voiceover IP (VoIP)** will be a key technology. As discussed in the next section, several standards organizations are specifying technology that will enable VoIP (e.g., the European Telecommunications Standards Institute [ETSI] Broadband Radio Access Networks [BRAN] and Telecommunications Internet Protocol Harmonization over Networks [TIPHON] projects and the IETF SIP working group).
- The mobile terminals will be based on **software-configurable radios** with capabilities for supporting many radio-access technologies across several frequency bands.
- The ability to move across hybrid access technologies will be an important requirement that calls for efficient and fast vertical handovers and **seamless session mobility**. As explained later, the IETF Seamoby and Mobile IP working groups are addressing some of the issues related to seamless mobility. Fast Mobile IP and micromobility schemes are chief technologies in this area.
- In the highly hybrid access environment of the mobile Internet, **security** will also play a critical role. IEEE 802.11 task group I (TGi) is standardizing new mechanisms for enhanced security in WLANs. IETF Seamoby also addresses the protocols that deal with security context transfer during handovers.
- For extended roaming between different administrative domains or different access technologies, **advanced Authentication, Authorization, and Accounting (AAA) protocols** and **AAA interworking mechanisms** will be required. AAA interworking between WLANs and 3GPP networks is currently being studied by 3GPP and 3GPP2, and it is also discussed in Chapter 12.
- Enhanced **networking application programming interfaces (APIs)** for QoS-, multicast-, and location-aware applications will be needed.
- **WPANs** will start spreading, initially based on Bluetooth technology (see www.bluetooth.com) and later on IEEE 802.15.3 high-speed WPAN technology, which satisfies the requirement of the digital consumer electronic market (e.g., wireless video communications between a personal computer [PC] and a video camera).
- Millions of users are envisioned to be always connected to the IP core infrastructure. Market forecasts suggest that about 1 billion addresses are needed by 2005 for integration of Internet-based systems into means of transportation (cars, aircraft, trains, ships, and freight transport) and associated infrastructures for mobile e-commerce. This indicates that there is a strong **need for IP version 6** (IPv6 is being adapted in 3GPP). The European Union in particular is pushing for the fast adoption of IPv6.
- Wireless communication technology will evolve further and will support higher bit rates. For instance, WLANs will soon support **bit rates of more than 100 Mbps**. This issue is being addressed by the IEEE Wireless Next Generation Standing Committee (see www.ieee802.org/11 and Section 1.3.3.1). Higher bit rates are typically accompanied by smaller coverage areas; therefore, efficient, fast, and secure horizontal handovers will be required.

These evolutionary aspects lead to a high-level network architecture like the one shown in Figure 1.3. Note that each administrative domain represents a network part that is typically operated by a single mobile network operator.

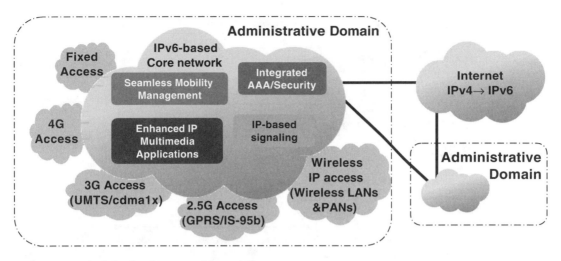

FIGURE 1.3 A high-level architecture of the mobile Internet.

1.2.1 Evolution to IP-Based Core Networks

Without a doubt, the most widely supported evolution is toward IP-based core networks, also referred to as *all-IP* core networks. The term "all-IP" emphasizes the fact that IP-based protocols are used for *all* purposes, including transport, application-level signaling, mobility, security, and QoS. Typically, several wireless and fixed-access networks are connected to an all-IP core network, as illustrated in Figure 1.2 and Figure 1.3. Users will be able to use multimedia applications over terminals with software-configurable radios capable of supporting a vast range of radio-access technologies, such as WLANs, WPANs, UMTS, and CDMA2000. In this environment, seamless mobility across the various access technologies is considered a key issue by many vendors and operators.

In the all-IP network architecture, the mobile terminals use the IP-based protocols defined by IETF to communicate with the core network and perform, for example, session/call control and traffic routing. All services in this architecture are provided on top of the IP protocol. As shown in the protocol architecture of Figure 1.4, the mobile networks, such as UMTS and CDMA2000, turn into access networks that provide only mobile-bearer services. The teleservices in these networks (e.g., cellular voice) are used to support only the legacy second-generation (2G) and 3G terminals, which do not support IP-based applications (e.g., IP telephony). For the provision of mobile bearer services, the access networks mainly implement micromobility management, radio resource management, and traffic management for provisioning of QoS. Micromobility management in 3GPP access networks is based on the GPRS Tunneling Protocol (GTP) and uses a hierarchical tunneling scheme for data forwarding. On the other hand, micromobility management in 3GPP2 access networks is typically based on IP micromobility protocols. Macromobility (inter-domain mobility) is typically based on Mobile IP, as per Request for Comment (RFC) 3220. All these mobility schemes are discussed in more detail in Chapter 5.

In the short term, the all-IP core network architecture would be based on the IMS architecture specified by 3GPP/3GPP2 (see Section 1.3.1.2.1), which in turn is based on the IP multimedia architecture and protocols specified by IETF. This IMS architecture would provide a new communications paradigm based on integrated voice, video, and data. You could call a user's IMS number, for instance, and be redirected to his or her home page, where you could have several options, such as writing a message, recording a voice message, or clicking on an alternative number to call if he or she is on vacation. You could place an SIP call to a server and update your communication preferences — for example, "Only my manager can call me, all others are redirected to my home page" (or vice versa!). At the same time, you could be on a conference call.

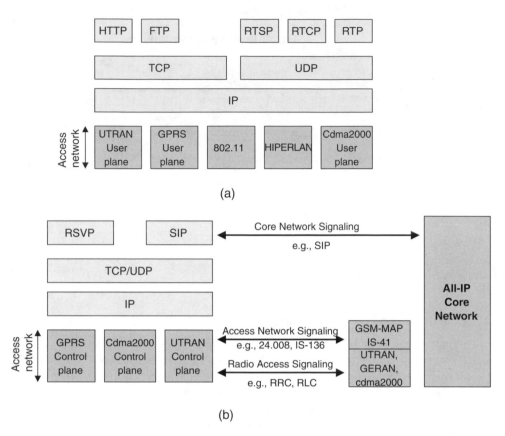

FIGURE 1.4   A simplified protocol architecture in an all-IP network architecture: (a) user plane and (b) control plane.

1.2.2   IP-Based Core Networks in the Enterprise

Figure 1.5 shows how an enterprise could take advantage of an all-IP core network to minimize its communication costs and increase communications efficiency. The typical IP network of the enterprise could be evolved to an IP multimedia network, which would support the following:

- IP signaling with the end terminals for establishing and controlling multimedia sessions
- Provisioning of QoS
- Policy-based admission control
- Authentication, authorization, and possibly accounting

The all-IP network provides an integrated infrastructure for efficiently supporting a vast range of applications with diverse QoS requirements, and, in addition, provides robust security mechanisms. The architecture of this all-IP network could be based on the IMS architecture specified by 3GPP/3GPP2 (see specification 3GPP TS 23.228 at www.3gpp.org/ftp/specs/2003-09/Rel-5/23_series/23228-5a0.zip for a detailed description).

In the example shown in Figure 1.5, an employee in the European office requests a voice call to another employee, for example, in the U.S. office. This request is then routed to the default Proxy Call State Control Function (P-CSCF) that serves the European office. This P-CSCF relays the request to the Serving CSCF (S-CSCF) of the calling employee — that is, to the CSCF with which this employee has previously registered. This S-CSCF holds the subscription information of the calling employee and can verify whether that individual is allowed to place the requested call. In turn, the S-CSCF finds another S-CSCF, with which the called subscriber has registered, and relays the request to this S-CSCF. Note that if the calling employee in the European office were calling a normal Public Switched Telephone Network (PSTN)

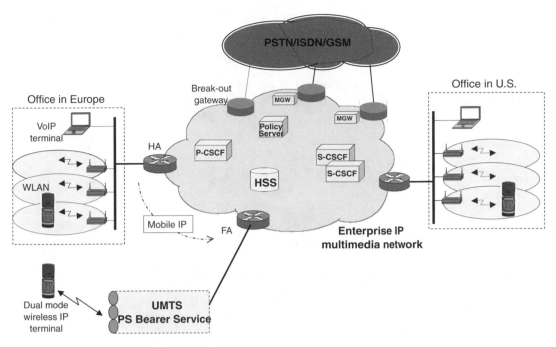

FIGURE 1.5 Deployment of all-IP networks in the enterprise.

number in the U.S., the call would be routed through the IP network to a break-out gateway that is closest to the called PSTN number. This way, the long-distance charges are saved.

The S-CSCF of the called U.S. employee holds information on the employee's whereabouts and can route the call to the correct location. In case that employee happens to be roaming in Europe, the call would be routed to the appropriate P-CSCF, which currently serves this employee. It is important to note that, although signaling can travel a long path in such a case (say, from Europe to the U.S. and then back to Europe), the user-plane path would be the shortest possible.

The support of roaming is another important advantage of the above architecture. For instance, a European employee could take his dual-mode mobile phone to the U.S. office. After powering on his mobile and registering his current IP address (with his S-CSCF), he would be able to receive calls at his standard number.

Even when the employee is away from his office and cannot directly attach to the enterprise network (e.g., he is driving on the highway), it is still possible to reach him on his standard number. In such a case, the employee uses, for example, the UMTS network to establish a mobile signaling channel to his all-IP enterprise network, which assigns him an IPv6 address. The employee registers this IP address with an S-CSCF (via the appropriate P-CSCF), and thereafter he can receive calls at his standard number. The signaling mobile channel remains active for as long as the employee uses UMTS to access the enterprise network. In such a scenario, the UMTS network is used only to provide access to the enterprise network and to support the mobile bearers required for IP multimedia services. To establish IP multimedia calls, the employee would need to request the appropriate UMTS bearers, each one with the appropriate QoS properties. For instance, to receive an audio-video call, two additional UMTS bearers would be required, one for each media component. The mapping between the application-level QoS and the UMTS QoS, as well as the procedures required to establish the appropriate UMTS bearers, are identified in 3GPP Rel-5 specifications, in particular 3GPP TS 24.008 and 3GPP TS 24.229 (available at www.3gpp.org/ftp/specs/2003-09/Rel-5).

If the enterprise network supports a macromobility protocol, such as Mobile IP, providing session mobility across the enterprise WLAN and the UMTS network becomes possible. In this case, when the employee moves from the WLAN to the UMTS, he uses Mobile IP to register his new IPv6 address with

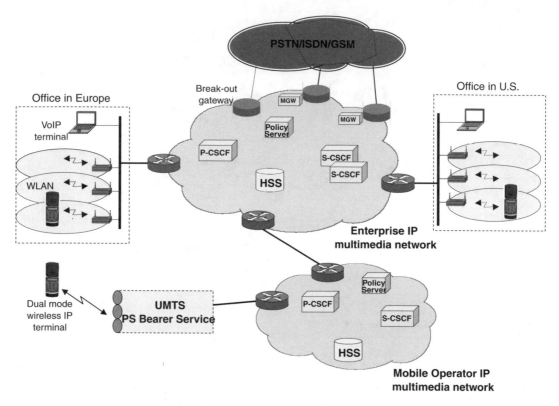

FIGURE 1.6 Deployment of all-IP networks in the enterprise (in which the mobile operator uses IMS).

his home agent. After that, any subsequent terminating traffic is tunneled from the employee's home agent to the foreign agent that serves the employee over the UMTS network.

Another roaming scenario is illustrated in Figure 1.6, which involves interworking between the enterprise all-IP network and the mobile operator's all-IP network. In this scenario, the key aspect is that the employee uses a P-CSCF in the mobile operator's domain — that is, it uses the operator's IMS system. Note that this scenario corresponds to the roaming scenario considered in 3GPP Rel-5 specifications and therefore might be the most frequently used in practice.

1.3   Key Players in the Evolution toward the Mobile Internet

The goal of this section is to provide an overview of the most important players in the evolution toward the mobile Internet. We focus primarily on the roles and activities of such organizations as the 3GPP, 3GPP2, IEEE, the European Telecommunications Standards Institute (ETSI), and IETF and explain how these organizations are involved with the aforementioned aspects of the evolution toward the mobile Internet. Throughout this discussion, we provide a general overview that describes who is driving the technology and in what directions. More important, we supply numerous references, which the interested reader could use to find the most up-to-date information.

1.3.1   Third-Generation Partnership Project (3GPP)

1.3.1.1   Overview

The Third-Generation Partnership Project (3GPP, at www.3gpp.org) is a collaboration agreement that was established in December 1998. The collaboration agreement brings together a number of telecommunications standards bodies, which are known as "Organizational Partners." The 3GPP Organizational Partners are the following:

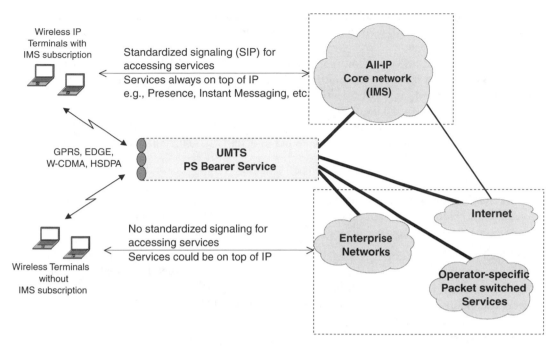

FIGURE 1.7 The all-IP core network provides a consistent signaling for accessing IP-based services.

- ARIB, the Association of Radio Industries and Businesses (Japan — www.arib.or.jp)
- CWTS, the China Wireless Telecommunication Standards Group (China — www.cwts.org/cwts/index_eng.html)
- TTA, the Telecommunications Technology Association (Korea — www.tta.or.kr)
- TTC, the Telecommunications Technology Committee (Japan — www.ttc.or.jp)
- T1 (North America)
- ETSI, the European Telecommunication Standards Institute (Europe — www.etsi.org)

The original scope of 3GPP was to produce globally applicable technical specifications for a 3G mobile network, called the UMTS, based on an evolved Global System for Mobile (GSM) core network and the Universal Terrestrial Radio Access (UTRA) network. The scope was subsequently amended to include the maintenance and development of the GSM network, including its related radio-access technologies, such as General Packet Radio Service (GPRS) and Enhanced Data for GSM Evolution (EDGE).

1.3.1.2   Key Standardization Activities in 3GPP

This section briefly discusses some key standardization activities undertaken by 3GPP regarding the mobile Internet. You can find more up-to-date information at www.3gpp.org.

1.3.1.2.1   IP Multimedia Services

The purpose of this activity is to specify an IP-based core network, formally called *IP Multimedia Core Network Subsystem (IMS)*, that provides a standard IP-based interface to wireless IP terminals for accessing a vast range of services independently from the access technology used by the terminals (see Figure 1.7). This interface uses the *Session Initiation Protocol (SIP)* specified by IETF for multimedia session control (see RFC 3261). In addition, SIP is used as an interface between the IMS session control entities and the service platforms. The goal of IMS is to enable the mobile operators to offer to their subscribers multimedia services based on, and built on, Internet applications, services, and protocols. Note that the IMS architecture is identical between 3GPP and 3GPP2, and it is based on IETF specifications. Thus, IMS forms a single core-network architecture that is *globally* available and that can be accessed through a variety of technologies, such as mobile data networks, wireless LANs, and fixed broadband (e.g., xDSL). No matter what technology is used to access IMS, the user always uses the same signaling protocols and accesses the same services.

In a way, the goal of IMS is to allow mobile operators to offer popular Internet-alike services, such as instant messaging and Internet telephony. IMS offers versatility that lets us quickly develop new applications. In addition, IMS is global (identical across 3GPP and 3GPP2), and it is the first convergence between the mobile world and the IETF world.

A general description of IMS architecture can be found in 3GPP Technical Standard (TS) 23.228, and the definition of IMS functional elements can be found in 3GPP TS 23.002. All 3GPP specs are available at www.3gpp.org/ftp/specs/.

1.3.1.2.2 IMS Messaging

This standardization work deals with the support of messaging in the IMS as part of the Release 6 timeframe. By combining the support of messaging with other IMS service capabilities, such as Presence (see Section 1.3.1.2.3), we can create new rich and enhanced messaging services for end users.

> The goal of 3GPP is to extend messaging to the IP Multimedia Core Network Subsystem, which is about to be deployed whilst also interoperating with the existing SMS, EMS, and MMS wireless messaging solutions as well as SIP-based Internet messaging services. The SIP-based messaging service should support interoperability with the existing 3GPP messaging services SMS, EMS and MMS as well as enable development of new messaging services, such as Instant Messaging, Chat, etc.
>
> It should be possible in a standardized way to create message groups (chat rooms) and address messages to the group of recipients as a whole, as well as individual recipients. Additional standardized mechanisms are expected to be needed to create, and delete message groups, enable and authorize members to join and leave the group and also to issue mass invitations to members of the group.
>
> SIP based messaging should also integrate well with the Presence service also being developed by 3GPP.

Support of messaging in IMS will provide various messaging services, such as instant messaging, chat, and store and forward messaging, with rich multimedia components.

1.3.1.2.3 Presence

This work addresses the concept of presence (see Figure 1.8), whereby users make themselves "visible" or "invisible" to other parties of their choice, allowing services such as group and private chats to take place. Presence is an attribute related to mobility information and provides a different capability that can be exploited by other services. The concept of the Presence service enables other multimedia services; Presence serves mainly as an enabling technology that supports other multimedia services and communications. Among the multimedia services that could potentially exploit the Presence capability are chats, e-mail, multimedia messaging, and instant messaging.

The architectural and functional specification of Presence is included in 3GPP TR 23.841 (www.3gpp.org/ftp/specs/2003-09/Rel-6/23_series/23841-600.zip).

1.3.1.2.4 Packet-Switched Streaming Service

The aim of this activity is to provide the ability to *stream* sound and movies to mobile devices. Streaming is a mechanism whereby media content can be reproduced in the client while it is being transferred over the data network. Therefore, streaming services enable instant access to media content in contrast to other multimedia services (such as Mobile Multimedia Service [MMS]), where the media content is presented to the user after being downloaded within a "message."

Streaming services enable the delivery of information *on-demand*. For instance, users are able to request music, videos, and news on-demand. In addition, users can request live media content, such as the delivery of radio and television programs. Such services are already supported in the Internet, but they need special handling to be available in mobile terminals.

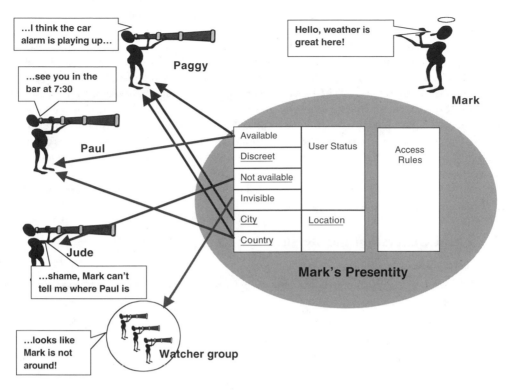

FIGURE 1.8 The concept of Presence. (Taken from 3GPP publicly available documents.)

In addition to providing the streaming mechanism, packet-switched streaming encompasses the composition of media objects, thereby allowing compelling multimedia services to be provided. For instance, a mobile cinema ticketing application would allow the user to view the film trailers.

Relevant specifications include 3GPP TS 22.233, "Transparent End-to-End Packet-Switched Streaming Service: Stage 1" and 3GPP TS 26.233, "Transparent End-to-End Packet-Switched Streaming Service: General Description."

1.3.1.2.5 Interworking with Wireless LANs

The purpose of this activity is to identify and analyze interworking architectures between 3GPP cellular systems and WLANs, such as the IEEE 802.11 family, HiperLAN/2, and MMAC HISWAN. The motivation behind this work is to make it possible for mobile operators to provide WLAN subscriptions to new subscribers (without Universal Subscriber Identity Module [USIM] cards in their WLAN terminals), and to enable the existing mobile subscribers to use their USIM cards for accessing WLAN services. This way, the mobile operators aim at getting into the WLAN business and exploit the WLAN market, which is projected to grow exponentially. Recent market analysis forecasts 20 million users of public WLAN in Europe by 2006 and 21 million Americans by 2007; this is compared with 600,000 WLANs users in 2002.

More details on interworking between WLAN and 3G networks are provided in Chapter 12, which is entirely devoted to this subject. The interested reader could also refer to the relevant 3GPP specifications, mainly 3GPP TS 23.234.

1.3.1.2.6 Multimedia Broadcast and Multicast Services

Multimedia Broadcast and Multicast Services (MBMS) deals with situations in which multiple users need to receive the same data at the same time. The benefit of MBMS in the network is that the data is sent once on each link. For example, the core network sends a single multicast packet to the radio-access network regardless of the number of users who wish to receive it. The benefit of MBMS on the air interface is that many users can receive the same data on a common channel; this strategy avoids congesting the air interface with multiple transmissions of the same data.

MBMS is considered a key enabler of mobile-Internet applications and a congestion-relief mechanism if certain types of multimedia services become popular and start to flood the mobile-access networks. Typical examples of such services could be video clips with sports highlights. MBMS contains two elements: broadcast and multicast. The broadcast element is the traditional type of broadcast in which the network "broadcasts" the information in a given geographical area, irrespective of the presence of users in that area. On the other hand, for the multicast element the network has knowledge about the users who subscribe to the service and their geographical location. With this information, the network can then choose to transmit the information only on those cells where there actually are mobiles expecting to receive it.

The functional requirements of MBMS are included in 3GPP TS 22.146 and the respective architectural description is included in report 3GPP TS 23.846.

1.3.2   Third-Generation Partnership Project 2 (3GPP2)

1.3.2.1   Overview

The Third-Generation Partnership Project 2 (www.3gpp2.org) is a collaborative standards-setting project, which does the following:

- Specifies the evolution of ANSI/TIA/EIA-41 "Cellular Radio-Telecommunication Intersystem Operations" network toward a 3G telecommunications system
- Develops global specifications for the radio transmission technologies supported by ANSI/TIA/EIA-41

3GPP2 was born out of the International Telecommunication Union's International Mobile Telecommunications IMT-2000 initiative (www.itu.int/imt/). 3GPP2 is a collaborative effort between the following five organizations:

- ARIB, the Association of Radio Industries and Businesses (Japan — www.arib.or.jp)
- CWTS, the China Wireless Telecommunication Standards Group (China — www.cwts.org/cwts/index_eng.html)
- TIA, the Telecommunications Industry Association (North America — www.tiaonline.org)
- TTA, the Telecommunications Technology Association (Korea — www.tta.or.kr)
- TTC, the Telecommunications Technology Committee (Japan — www.ttc.or.jp)

3GPP2 specifications (www.3gpp2.org/Public_html/Specs/index.cfm) are developed by five Technical Specification Groups (TSGs) consisting of representatives from the project's individual member companies. The 3GPP2 TSGs and their respective areas of work are the following:

- TSG-A (Access Network Interfaces; www.3gpp2.org/Public_html/A/index.cfm)
  - Physical links, transports, and signaling
  - Support for access network mobility
  - 3G capability (e.g., high-speed data support)
  - The Abis interface
  - Interoperability specification
  - Support for 3GPP2 radio-access technologies
- TSG-C (CDMA2000; www.3gpp2.org/Public_html/C/index.cfm)
  - Radio Layer 1 specifications
  - Radio Layer 2 specifications
  - Radio Layer 3 specifications
  - Mobile station/base station radio-performance specifications

- • Radio-link protocol
  - • Support for enhanced privacy, authentication, and encryption
  - • Digital speech codecs and related minimum performance specifications
  - • Video codec selection and specification of related video services
  - • Data and other ancillary services support
  - • Conformance test plans
  - • Removable User Identity Module (R-UIM)
  - • Location-based services support
- • TSG-N (Intersystem Operations; www.3gpp2.org/Public_html/N/index.cfm)
  - • Evolution of core network to support interoperability and intersystem operations
  - • User Identity Module (UIM) support (detachable and integrated)
  - • Support for enhanced privacy, authentication, encryption, and other security aspects
  - • VHE (Virtual Home Environment)
  - • Support for new supplemental services (including Integrated Services Digital Network [ISDN] interworking)
  - • Optimal interoperability specification for international roaming (e.g., selection of required parameters options)
  - • New features for internal roaming (global emergency number, optimal routing)
  - • IMT-2000 issues, as necessary, to ensure support of the ANSI-41 family member
- • TSG-P (Wireless Packet Data Networking; www.3gpp2.org/Public_html/P/index.cfm)
  - • Wireless IP services (including IP mobility management)
  - • Voiceover IP
  - • AAA and security
  - • Private network access
  - • Internet/intranet access
  - • Multimedia support
  - • QoS support
- • TSG-S (Services and Systems Aspects; www.3gpp2.org/Public_html/S/index.cfm)
  - • Development and maintenance of 3GPP2 System Capabilities Guide
  - • Development, management, and maintenance of 3GPP2 Work Plan
  - • Stage 1 services and features requirements definition
  - • High-level functionality description development for system services and capabilities
  - • Management and technical coordination as well as architectural and requirements development associated with all end-to-end features, services, and system capabilities including, but not limited to, security and QoS
  - • Requirements for international roaming
  - • Development of 3GPP2 Operations, Administration, Maintenance & Provisioning (OAM&P) across all TSGs including (1) Stage 1 high-level requirements and (2) Stage 2 and Stage 3 for the interface between network management system and element management functions
  - • High-level coordination of the work performed in other TSGs and monitoring of progress
  - • Coordination to resolve technical discrepancies between the works undertaken by other TSGs

Each TSG meets about 10 times a year to produce technical specifications and reports. Because 3GPP2 has no legal status, ownership and copyright of these output documents is shared between the Organizational Partners. The documents cover all areas of the project's charter, including CDMA2000 and its enhancements.

1.3.2.2 Standardization Activities in 3GPP2

1.3.2.2.1 *CDMA2000 Air Interface*

CDMA2000 is the radio-transmission technology standardized by 3GPP2 (TSG-C) to meet the IMT-2000 requirements for 3G mobile systems. CDMA2000 comes in two versions: CDMA2000 1x and

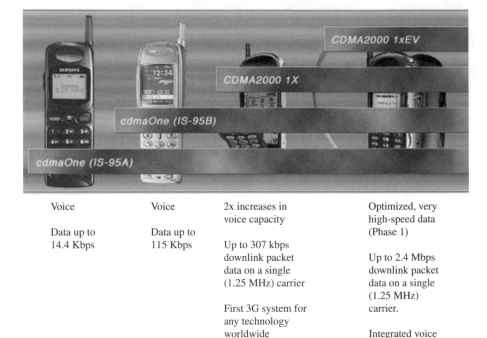

Voice   Voice   2x increases in   Optimized, very
                                                  voice capacity   high-speed data
Data up to   Data up to   (Phase 1)
14.4 Kbps   115 Kbps   Up to 307 kbps
                                                  downlink packet   Up to 2.4 Mbps
                                                  data on a single   downlink packet
                                                  (1.25 MHz) carrier   data on a single
                                                                           (1.25 MHz)
                                                  First 3G system for   carrier.
                                                  any technology
                                                  worldwide   Integrated voice
                                                                           and data (Phase 2);
                                                                           up to 3.09 Mbps

FIGURE 1.9   Evolution of CDMA technology in mobile data networks (source: www.cdg.org).

CDMA2000 3x (see Figure 1.9). 3GPP2 has completed the specifications of CDMA2000 1x, which operates on 1.25-MHz carriers. CDMA2000 3x is a multicarrier CDMA (MC-CDMA) technology that operates on three carriers of 1.25 MHz each ($3 \times 1.25 = 3.75$ MHz).

Basically, CDMA2000 1x is an enhancement of CDMAOne, specified in TIA/EIA-95-B, "Mobile Station — Base Station Compatibility Standard for Dual-Mode Wideband Spread Spectrum System." CDMAOne also uses carriers of 1.25 MHz. The relationship between CDMAOne and CDMA2000 1x is quite similar to the relationship between the conventional GSM radio technology and the EDGE radio technology. CDMA2000 1x can support up to 307 kbps on the downlink and up to 140 kbps on the uplink. Although CDMA2000 1x and EDGE seem to have similar capabilities in terms of bit rate, their commercial deployment has not found similar acceptance. In particular, the deployment of EDGE radio networks today is very limited, but it is progressively evolving.

The first 3G networks commercially deployed were launched in Korea in October 2000 with CDMA2000 1x technology. CDMA2000 1x radio-access networks are widely deployed, mainly in Asia and America.

CDMA2000 1x has further evolved to CDMA2000 1xEV-DO (1x Evolution — Data Only) and CDMA2000 1xEV-DV (1x Evolution — Data and Voice), both of them still using 1.25-MHz carriers. CDMA2000 1xEV-DO, officially known as IS-856, is an enhancement of CDMA2000 1x that puts voice and data on separate channels in order to provide data delivery at 2.4 Mbps on the downlink. 1xEV-DO is also known as CDMA2000 High-Rate Packet Data (HRPD) Air Interface, and it is specified in 3GPP2 C.S0024 specification (www.3gpp2.org/Public_html/Specs/C.S0024-0_v3.0.pdf). Conceptually, it is similar to the High-Speed Downlink Packet Access (HSDPA) specified by 3GPP.

CDMA 1xEV-DV is a further evolution that integrates voice and data on the same 1.25-MHz carrier and offers data speeds of up to 4.8 Mbps. On May 2002, the 1xEV-DV specification in IS-2000 Release C was approved by TIA TR45.5 as a complete standard.

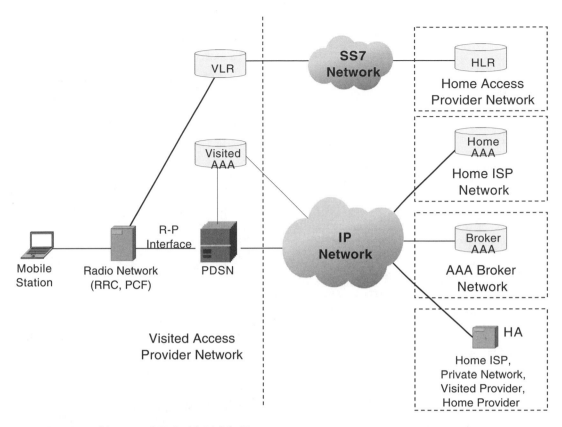

FIGURE 1.10 Architecture of PDS with Mobile IP.

1.3.2.2.2   *Packet Data Services in CDMA2000*

The goal of the packet data services (PDS) in CDMA2000 is to offer services similar to those provided by the GPRS network — that is, packet switched services. However, the architecture model in CDMA2000 is *far simpler* than the equivalent of GPRS, and it provides only wireless IP services (as opposed to *general* packet radio services). Furthermore, PDS is primarily based on IETF protocols for authentication, authorization, accounting, and mobility/session management, whereas GPRS is based on 3GPP-specific protocols (such as GTP, GMM, and SM).

The architecture of PDS in CDMA2000 is illustrated in Figure 1.10 for the case where Mobile IP is used for mobility management (for more details, see 3GPP2 P.R0001, available at www.3gpp2.org/ Public_html/Specs/P.R0001-0_v1.0.pdf). In this architecture, the user establishes a Point-to-Point Protocol (PPP) link to the Packet Data Serving Node (PDSN), and the operation resembles the access over dial-up lines. Figure 1.11 shows the protocol architecture for the case where Mobile IP and Internet Key Exchange (IKE) are used.

1.3.2.2.3   *All-IP Network*

In December 2000, 3GPP2 formed an all-IP ad hoc group (www.3gpp2.org/Public_html/AllIP/index.cfm) under TSG-S for defining the high-level requirements and network architectures required to support the future wireless Internet.

The All-IP architecture specified by the all-IP ad hoc group is included in document 3GPP2 S.R0037 (ftp.3gpp2.org/TSGS/Working/All_IP_Source_Documents/TSG-S_AllIP_NAM/NAM_Version_1.2.1/ 3GPP2 TSG-S S.R0037-NAM_rev-1-2-1.pdf) and supports both the *Legacy MS Domain (LMSD)* and the *IP Multimedia Domain (MMD)*. LMSD supports TIA/EIA-41 signaling with the mobiles (for call control and session control) and is designed to support the existing (legacy) 2G and 3G mobiles. On the other hand, MMD supports only IP-based signaling. SIP is typically used between the mobiles, and MMD is

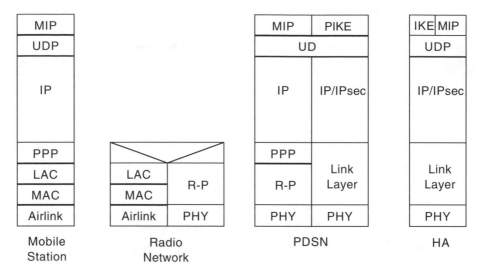

FIGURE 1.11 A protocol model of PDS with Mobile IP and IKE.

used for multimedia session control. As discussed before, 3GPP and 3GPP2 have agreed to align their terminology in the context of an all-IP core network. Therefore, MMD is now referred to as IMS (see Section 1.3.1.2.1).

1.3.3 Institute of Electrical and Electronics Engineers (IEEE)

1.3.3.1 Standardization Activities in IEEE

In the context of the IEEE 802 project, there are mainly three working groups (WGs) delivering standards for WLANs, WPANs, and fixed broadband wireless access (BWA) networks. These working groups, namely, 802.11, 802.15, and 802.16, are briefly discussed next.

1.3.3.1.1 IEEE WG 802.11 (Wireless LANs)

The IEEE 802.11 WG (www.ieee802.org/11/) develops WLAN consensus standards for short-range wireless networks. At the time of this writing, the IEEE 802.11 WG has published the following standards:

- IEEE 802.11-1999, "Wireless LAN Medium Access Control (MAC) and Physical Layer (PHY) Specifications." This standard specifies the 802.11 MAC protocol as well as three physical layers (Frequency Hopping Spread Spectrum, Direct Sequence Spread Spectrum, and Infrared) operating at speeds of 1 Mbps and 2 Mbps in the 2.4-GHz frequency range.
- IEEE 802.11a-1999, "High-Speed Physical Layer in the 5GHz Band." This standard provides changes and additions to IEEE 802.11-1999 to support a physical layer (based on Orthogonal Frequency Division Multiplexing [OFDM]) operating at speeds up to 54 Mbps in the 5-GHz frequency band.
- IEEE 802.11b-1999, "Higher-Speed Physical Layer Extension in the 2.4GHz Band." This standard provides changes and additions to IEEE 802.11-1999 to support a physical layer (based on Complementary Code Keying) operating at speeds up to 11 Mbps in the 2.4-GHz frequency band.
- IEEE 802.11d-2001, "Operation in Additional Regulatory Domains." This amendment specifies the extensions to IEEE 802.11 for WLANs providing specifications for conformant operation beyond the original six regulatory domains of that standard. These extensions provide a mechanism for an IEEE 802.11 access point to deliver the required radio-transmitter parameters to an IEEE 802.11 mobile station, which allows that station to configure its radio to operate within the

applicable regulations of a geographic or political subdivision. This mechanism is applicable to all IEEE 802.11 PHY types. A secondary benefit of the mechanism described in this amendment is the ability for an IEEE 802.11 mobile station to roam between regulatory domains.

The IEEE 802.11 WG continues its work for enhancing the published 802.11 specifications. The work is carried out in several task groups (TGs), which are tasked to deliver additional 802.11 standards. Let us look at some of the most important TGs:

IEEE 802.11 TGe — MAC Enhancements for Quality of Service — The purpose of TGe is to enhance the current 802.11 MAC in order to support applications with QoS requirements and to expand the capabilities and efficiency of the protocol. TGe is responsible for the IEEE 802.11e standard.

IEEE 802.11 TGf — Inter-Access Point Protocol — The purpose of TGf is to describe recommended practices for implementation of an Inter-Access Point Protocol (IAPP) on a distribution system (DS) supporting IEEE 802.11, WLANs. The recommended DS utilizes an IAPP that provides the necessary capabilities to achieve multivendor access point (AP) interoperability within the DS. This IAPP is described for a DS consisting of IEEE 802 LAN components utilizing an IETF IP environment. TGf is responsible for the IEEE 802.11f standard.

IEEE 802.11 TGg — Further Higher-Speed Physical Layer Extension in the 2.4-GHz Band — The purpose of TGf is to specify a new physical layer (based on Orthogonal Frequency Division Multiplexing) operating at up to 54 Mbps in the 2.4-GHz frequency band. TGg is responsible for the IEEE 802.11g standard.

IEEE 802.11 TGi — Enhanced Security — The purpose of TGi is to enhance the IEEE 802.11 standard in order to enable advanced security features. TGi has defined the concept of the robust security network (RSN), which provides a number of additional security features not present in the basic IEEE 802.11 architecture. TGi is responsible for the IEEE 802.11i standard.

IEEE 802.11 — Next Generation WLANs — In the May 2002 meeting of the IEEE 802.11 WG in Sydney, Australia, the Wireless Next Generation Standing Committee (WNG) moved to form two new study groups: the Radio Resources Measurements Study Group (RMSG) and the High Throughput Study Group (HTSG). The WNG and its study groups are investigating the technology for next-generation WLANs (with bit rates greater than 100 Mbps), including interworking schemes with other access technologies, such as HiperLAN/2. In this context, WNG collaborates with ETSI BRAN (see Section 1.3.4.1).

1.3.3.1.2 IEEE WG 802.15 (Wireless PANs)

The IEEE 802.15 WG (www.ieee802.org/15/) develops WPAN consensus standards for short-distance wireless networks. These WPANs address wireless networking of portable and mobile computing devices, such as PCs, PDAs, peripherals, cell phones, pagers, and consumer electronics and allows these devices to communicate and interoperate with one another. The goal is to publish standards, recommended practices, or guides that have broad market applicability and that deal effectively with the issues of coexistence and interoperability with other wired and wireless networking solutions.

The IEEE 802.15 WG is divided into the following four TGs:

IEEE 802.15 TG1 — Bluetooth — The IEEE 802.15 TG1 (TG1) has delivered a WPAN standard (802.15.1) based on the Bluetooth v1.1 specifications (see www.bluetooth.com). In particular, IEEE has licensed wireless technology from the Bluetooth Special Interest Group (SIG) to adapt and copy a portion of the Bluetooth specification as base material for IEEE Standard 802.15.1. This standard, which is fully compatible with the Bluetooth v1.1 specification, was conditionally approved on March 21, 2002.

IEEE 802.15 TG2 — Coexistence — The IEEE 802.15 TG2 (TG2) is developing Recommended Practices to facilitate coexistence of 802.15 WPANs and 802.11 WLANs. The TG is developing a coexistence model to quantify the mutual interference of a WLAN and a WPAN. TG2 is also developing a set of coexistence mechanisms to facilitate coexistence of WLAN and WPAN devices.

IEEE 802.15 TG3 — High-Rate WPAN — The IEEE 802.15 TG3 (TG3) is tasked to provide a new standard for high-rate (20 Mbit/sec or greater) WPANs. Besides a high data rate, the new standard will provide for low-power, low-cost solutions that address the needs of portable consumer digital imaging and multimedia applications. TG3 has adopted a physical layer (PHY) proposal based on a 2.4-GHz Orthogonal Quadrature Phase Shift Keying (OQPSK) radio design. The IEEE 802.15.3 specification features high data rates (11, 22, 33, 44, and 55 Mbps), a quality of service isochronous protocol, security mechanisms, low-power consumption, and low cost.

IEEE 802.15 TG4 — Low-Rate WPAN — The IEEE 802.15 TG4 (TG4) is tasked to provide a standard for a low data rate (from 20 to 250 kbps) WPAN solution with multimonth to multiyear battery life and very low complexity. It is intended to operate in an unlicensed, international frequency band (mainly in the 2.4-GHz band). Potential applications are sensors, interactive toys, smart badges, remote controls, and home automation.

1.3.3.1.3 IEEE WG 802.16 (Fixed BWA)

Since July 1999, the IEEE 802.16 WG (www.ieee802.org/16/) on BWA has been developing standards for wireless metropolitan area networks with global applicability. IEEE 802.16 provides standardized solutions for reliable, high-speed network access in the so-called last mile by homes and enterprises, which could be more economical than wireline alternatives.

The IEEE 802.16 WG has completed two IEEE standards:

1. The IEEE 802.16 WirelessMAN Standard (Air Interface for Fixed Broadband Wireless Access Systems), which addresses wireless metropolitan area networks. The initial standard, covering systems between 10 and 66 GHz, was approved in December 2001. After that, the work has been expanded to cover licensed and license-exempt bands as well as in the range from 2 to 11 GHz. Note that a fixed BWA system in this frequency range is also being developed by the ETSI BRAN project (see Section 1.3.4.1.3).
2. The IEEE Standard 802.16.2 is a Recommended Practice (Coexistence of Fixed Broadband Wireless Access Systems) covering between 10 and 66 GHz. The IEEE Standard 802.16.2 was published on September 10, 2001, and is available for download without charge at http://standards.ieee.org/getieee802/download/802.16.2-2001.pdf.

The WirelessMAN Medium Access Control (MAC) provides mechanisms for differentiated QoS support in order to address the needs of various applications. For instance, voice and video require low latency but tolerate some error rate. In contrast, generic data applications cannot tolerate error, but latency is not critical. The standard accommodates voice, video, and other data transmissions by using appropriate features in the MAC layer.

The WirelessMAN standard supports both frequency and Time-Division Duplexing (TDD). Frequency-Division Duplexing (FDD) requires two channel pairs, one for transmission and one for reception, with some frequency separation between them to mitigate self-interference. On the contrary, TDD provides a highly flexible duplexing scheme where a single channel is used for both upstream and downstream transmissions. A TDD system can dynamically allocate upstream and downstream bandwidth, depending on traffic requirements.

1.3.4 European Telecommunications Standards Institute (ETSI)

ETSI is a nonprofit organization whose mission is to produce the telecommunications standards that will be used throughout Europe and beyond. ETSI is composed of numerous technical committees, each one working on a particular technical area and which produces its own set of standards. ETSI has also established a number of projects; the most important in the context of Mobile Internet include the following:

- BRAN (Broadband Radio Access Networks)
- TIPHON (Telecommunications Internet Protocol Harmonization Over Networks)

Let us briefly discuss each project.

1.3.4.1 ETSI Broadband Radio Access Networks (BRAN)

The BRAN project prepares standards for equipment providing broadband (25 Mbit/sec or more) wireless access to wireline-based networks in both private and public environments, operating in either licensed or license-free spectrum. These systems address both business and residential applications. Fixed wireless access systems are intended to be high-performance, easily set up, competitive alternatives to wireline-based access systems.

The BRAN specifications address the PHY as well as the data link control (DLC) layer. Interworking specifications that allow broadband radio systems to interface to existing wired networks, such as ATM, TCP/IP, and UMTS, are also considered.

ETSI BRAN produces specifications for three major standard areas:

- HiperLAN/2, a mobile broadband short-range access network
- HiperAccess, a fixed wireless broadband access network
- HiperMAN, a fixed wireless access network that operates below 11 GHz

1.3.4.1.1 HiperLAN/2

HiperLAN/2 is intended for private WLANs as well as serving as a complementary access mechanism in hotspot areas for public mobile network systems. It gives consumers in corporate, public, and home environments wireless access to the Internet and future multimedia, as well as real-time video services at speeds of up to 54 Mbit/sec in the 5-GHz frequency band. HiperLAN/2 is quick and easy to install, and it provides interworking with several core networks, including Ethernet, IEEE 1394, and Asynchronous Transfer Mode (ATM).

ETSI BRAN has completed a technical report (see ETSI TR 101 957) on the requirements and the architecture for interworking between HiperLAN/2 and UMTS or other 3G networks. By cooperating closely with the Multimedia Mobile Access Communications Promotion Council (MMAC) in Japan and the IEEE in the U.S., the BRAN project aims to produce "generic" WLAN-3G interworking solutions independent of the access techniques used in the WLAN standard.

More information about HiperLAN/2 can be found at http://portal.etsi.org/bran/kta/Hiperlan/hiperlan2.asp.

1.3.4.1.2 HiperAccess

The HiperAccess project produces standards for broadband multimedia fixed wireless access. The HiperAccess specifications allow for a flexible and competitive alternative to wired access networks. It is an interoperable standard in order to promote mass market, and thereby low-cost, products.

HiperAccess is targeting high-frequency bands, especially for the 40.5- to 43.5-GHz band. For these frequency bands, Time Division Multiple Access (TDMA) is used as multiple access scheme and a single carrier modulation scheme is used.

Note that ETSI BRAN is closely cooperating with IEEE WG 802.16 (see Section 1.3.3.1.3) to harmonize the interoperability standards for broadband multimedia fixed wireless access networks.

1.3.4.1.3 HiperMAN

HiperMAN is aiming principally for the same usage as HiperAccess but is targeted at different market segments and uses a different part of the spectrum. In particular, HiperMAN standardization focuses on solutions optimized for frequency bands below 11 GHz. HiperMAN is an interoperable broadband fixed wireless access system operating at radio frequencies between 2 GHz and 11 GHz.

The HiperMAN standards specify the PHY and DLC layers, which are core-network independent and the core-network–specific convergence sublayers. It should be noted that to specify a complete system, other specifications, such as the Network and higher layers, are required. These specifications are assumed to be available or to be developed by other bodies. The implementation includes at least one subscriber unit that communicates with a base station via an interoperable radio air interface, the interfaces to external networks, and services transported by the DLC and PHY protocol layers.

In support of a single worldwide standard for fixed wireless access systems operating below 11 GHz, BRAN decided to use the IEEE 802.16a Orthogonal Frequency Division Multiplexing (OFDM) PHY and MAC as a baseline for PHY and DLC, respectively.

1.3.4.2 ETSI TIPHON (Standardization of Voice over the Internet)

The objective of the TIPHON project is to ensure that users connected to IP-based networks can establish voice and data communications with users in legacy circuit-switched networks (such as PSTN, ISDN, and GSM), and vice versa. Apart from delivering the appropriate ETSI technical specifications and reports, the activity also includes validation and demonstrations in order to confirm the appropriateness of the proposed solutions.

At the end of 2001, the TIPHON project completed the third release of standardization, providing for interoperability of telecommunication services in mixed technology and administrative environments. This release completes the specification of the basic call service across multiple transport networks made up of different technologies (packet, circuit-switched, or wireless) and under different administrative controls and policies.

The TIPHON approach uses a new functional architecture, which separates the roles of application service provider (ASP) and transport network operator. By using this approach, the ASP can offer unique application services consisting of a number of standardized building blocks or service capabilities. A further innovation is the introduction of a technology-independent protocol framework, known as the TIPHON meta-protocol. This protocol is used to generate profiles for widely deployed industry protocols, such as H.323, SIP, and H.248, thus enabling service interoperability in mixed-protocol environments. In addition, specifications and technical reports have been produced, including test suites and requirements for naming and addressing, end-to-end QoS, security, and lawful interception.

1.3.5 Internet Engineering Task Force (IETF)

In this section we briefly review the most important activities in IETF related to the mobile Internet.

IETF is separated into numerous WGs (www.ietf.org/html.charters/wg-dir.html), each one addressing a specific technical area. In general, IETF addresses the problems associated with the specification of all-IP mobile networks and applications. In this context, the IP protocol suite is enhanced in order to provide mobility management, paging, and other features required in mobile IP networks. In addition, several Application layer protocols are being specified to aid in the provision of IP multimedia services.

1.3.5.1 IETF MOBILE IP Working Group (Mobility)

The Mobile IP WG has developed routing support to permit IP nodes (hosts and routers) using either IPv4 or IPv6 to seamlessly "roam" among IP subnetworks and media types. The Mobile IP method (see RFC 3220) supports transparency above the IP layer, including the maintenance of active TCP connections and User Datagram Protocol (UDP) port bindings. Where this level of transparency is not required, solutions such as Dynamic Host Configuration Protocol (DHCP) and dynamic Domain Name System (DNS) updates may be adequate, and techniques such as Mobile IP may not be needed.

The Mobile IP WG focuses on deployment issues in Mobile IP and provides appropriate protocol solutions to address known deficiencies and shortcomings. For example, the wireless/cellular industry is considering using Mobile IP as one technique for IP mobility for wireless data. The working group is attempting to gain an understanding of data services in cellular systems, such as GPRS, UMTS, and CDMA2000, and interact with other standards bodies that are trying to adopt and deploy Mobile IP protocols.

The Mobile IP WG addresses the following:

- Using network address identifiers to identify mobile users/nodes
- Specifying how Mobile IP should use AAA functionality to support interdomain and intradomain mobility
- Developing solutions for IPv4 private address spaces for the scenarios needed for deployment
- Documenting any requirements specific to cellular/wireless networks
- Guaranteeing QoS in the mobile IP environment using Differentiated Services (DiffServ) and/or Integrated Services (IntServ)/Resource Reservation Protocol (RSVP)
- Ensuring location privacy

1.3.5.2 IETF Seamoby Working Group

During the fast handoff discussions within the Mobile IP WG, a need for a new protocol was identified that would allow state information to be transferred between edge mobility devices. Examples of state information that could be useful to transfer are AAA information, security context, QoS properties assigned to the user, and robust header compression information.

Several standards-defining organizations, such as the ones we have already discussed (3GPP, 3GPP2, IEEE, ETSI), among others, rely on IETF to develop a set of protocols that will enable them to provide real-time services over an IP infrastructure. Seamoby is expected to provide such protocols. Furthermore, these protocols must allow for real-time services to work with minimal disruption across heterogeneous wireless and wired technologies.

In addition to *context transfer*, the SEA Seamoby MOBY WG has identified two more technologies that are important for use as tools for providing real-time services over IP wireless infrastructure: *Handoff Candidate Discovery* and *Dormant Mode Host Alerting* (also known as *IP paging*). Another technology, the so-called micromobility, in which routing occurs without the Mobile IP address change, was determined by Seamoby to require additional research. However, micromobility has been addressed by the IRTF Routing Research Group.

1.3.5.3 IETF SIP Working Group (SIP)

The SIP WG (www.softarmor.com/sipwg/) is tasked with continuing the development of the SIP protocol, specified in RFC 3261 (www.ietf.org/rfc/rfc3261.txt). SIP is a text-based protocol, similar to HTTP and SMTP, for initiating interactive communication sessions among users; examples include voice, video, chat, interactive games, and virtual reality. SIP was first developed within the Multiparty Multimedia Session Control (MMUSIC) WG. In addition, the main Multipurpose Internet Mail Extension (MIME) type carried in SIP messages — that is, the Session Description Protocol (SDP), specified in RFC 2327 — was developed by MMUSIC.

The main work of SIP WG involves bringing SIP from a proposed to a draft standard, as well as specifying and developing proposed extensions that arise from strong requirements. The SIP WG concentrates on the specification of SIP and its extensions and does not explore the use of SIP for specific environments or applications.

IETF SIP maintains numerous Internet drafts, which you can find at www.ietf.org/ids.by.wg/sip.html.

1.3.5.4 IETF IMPP Working Group

The Instant Messaging and Presence Protocol (IMPP) WG (http://www.imppwg.org/) defines protocols and data formats necessary to build an Internet-scale end-user presence awareness, notification, and instant messaging system. Instant messaging differs from e-mail primarily in that its main focus is on immediate end-user delivery. IMPP provides an architecture for simple instant messaging and presence awareness/notification. It specifies how authentication, message integrity, encryption, and access control are integrated.

The Instant Messaging and Presence architecture of IMPP has been adopted by the 3GPP and 3GPP2 standardization bodies in order to offer instant messaging and presence services over their IP Multimedia Subsystem (see Section 1.3.1.2.3).

1.3.5.5 IETF AAA Working Group (AAA)

The AAA WG focuses on the development of requirements for authentication, authorization, and accounting as applied to network access. Requirements were gathered from the IETF Network Access Server Requirements (NASreq), Mobile IP, and Roaming Operations (ROAMOPS) WGs as well as Telecommunications Industry Association (TIA) 45.6.

1.3.5.6 IETF IPv6 Working Group

IPv6 (also formerly known as IP Next Generation or IPng) is intended to support the continued growth of the Internet, both in size and capabilities, by offering a greatly increased IP address space and other enhancements over IPv4. The IP Next Generation (IPng) WG was originally chartered by the Internet Engineering Steering Group (IESG) to implement the recommendations of the IPng Area Directors as outlined at the July 1994 IETF meeting and in "The Recommendation for the IP Next Generation Protocol," or RFC1752. Most of the tasks in that original charter have been completed, and the core IPv6 protocol specifications are now on the IETF standards track.

The Internet drafts and RFCs provided by IPv6 WG can be found at www.ietf.org/html.charters/ipv6-charter.html.

1.3.5.7 IETF IPSec Working Group (Security)

The Secure Internet Protocol (IPSec) WG focuses on the following short-term goals to improve the existing key management protocol (IKE) and IPSec encapsulation protocols:

- Changes to IKE to support network address translation (NAT)/firewall traversal
- Changes to IKE to support Stream Control Transmission Protocol (SCTP)
- New cipher documents to support AES-CBC, AES-MAC, SHA-2, and a fast AES mode suitable for use in hardware encryptors
- IKE Management Information Base (MIB) documents
- Sequence number extensions to Encapsulation Security Payload (ESP) to support an expanded sequence number space
- Clarification and standardization of rekeying procedures in IKE

The working group also updates IKE to clarify the specification and to reflect implementation experience, new requirements, and protocol analysis of the existing protocol. The requirements for IKE v2 were revised and updated as the first step in this process.

The Internet drafts and RFCs provided by IPv6 can be found at http://www.ietf.org/html.charters/ipsec-charter.html.

1.3.5.8 IETF DiffServ Working Group (QoS)

The DiffServ WG is dealing with the provision of differentiated classes of service for Internet traffic to support various types of application and specific business requirements. This WG has specified the DiffServ architecture for differentiated services in IP networks; it has standardized a small number of specific per-hop behaviors (PHBs) and recommended a particular bit pattern, or "code-point," of the DS field for each one, in RFC 2474, RFC 2597, and RFC 2598.

The Internet drafts and RFCs provided by IPv6 can be found at http://www.ietf.org/html.charters/diffserv-charter.html.

1.4 Concluding Remarks

In this introductory chapter we discussed the main aspects of the evolution toward the mobile Internet, including the key players in this evolution. In this context, we briefly discussed the activities of 3GPP, 3GPP2, IEEE, ETSI, and IETF, which relate to the technologies that enable the mobile Internet. We provided numerous references to enable the interested reader to find the most up-to-date information. We pointed out that the mobile-Internet technologies are tightly related to the so-called beyond-3G technologies. In addition, we described the main characteristics of the mobile Internet and clarified the wireless IP services that are already available.

In short, we concluded that the evolution toward the mobile Internet is characterized by several issues:

- An evolution toward all-IP-based networks — that is, networks that support IP-based application signaling, mobility management, and QoS, as well as IP-based transport in the core and network access.
- Advanced IP multimedia applications, which are tightly integrated with the equivalent applications in the Internet. Voiceover IP, video/audio streaming, instant messaging, presence, press-to-talk, and other services will be enabled through the use of IP-based signaling protocols, such as SIP.

- Provisioning of end-to-end QoS, in order to support the demanding multimedia applications over diverse access media.
- An evolution of mobile terminals toward software-configurable radios with the capability to be reprogrammed over the air and support many radio-access technologies across many frequency bands.
- An evolution toward highly heterogeneous networks with several access technologies (both wired and wireless) with enhanced mobility management that supports fast vertical handovers and seamless session mobility.
- Robust and highly sophisticated security and AAA mechanisms and protocols.
- Adaptation of IPv6.
- Integration with wireless personal area network technologies.

In the chapters that follow, we will look more closely at various technologies that serve as key mobile-Internet enablers.

2

Internet Access over Wireless LANs

Nikos Passas
University of Athens

2.1   Introduction

In the past few years, the Internet Protocol (IP) has enjoyed a tremendous attention, and it is considered today the major candidate for the next-generation networks. The IP protocol, up to its current version 4, was designed for fixed networks and "best effort" applications with low network requirements, such as e-mail and file transfer, and accordingly, it offers an unreliable service that is subject to packet loss, reordering, packet duplication, and unbounded delays. This service is completely inappropriate for such demanding services as videoconferencing and Voiceover IP (VoIP). Additionally, no mobility support is provided, making it difficult for pure IP to be used for mobile communications. The expected new version 6 of IP provides some means for quality of service (QoS) and mobility support, but it still needs supporting mechanisms to fulfill the increasing user requirements. On the other hand, the growing demand of users for mobility support has led to the considerable development of wireless communications. In particular, Wireless Local Area Networks (WLANs), offering indoor and limited outdoor communications, are becoming more popular and tend to replace the traditional wired LANs. Toward this direction, many standards from various organizations and forums have been created to extend the capabilities of wireless networks and offer more advanced services. As a result, the transmission speed has been increased to tens of megabits per second (Mbps), making it possible to support a wide range of applications.

The efficient support of IP communications over WLANs is considered a key issue for next-generation networks. WLANs can offer high-speed communications in indoor and limited outdoor environments, providing efficient solutions for advanced applications. They can either act as alternative in-building network infrastructures or complement wide-area mobile networks as alternative access systems in hotspots, where a large density of users is expected (e.g., metro stations, malls, and airports). Although the main problems to be solved are similar to those in other wireless environments (security, QoS, and mobility), the particularities of WLANs demand specialized solutions. These particularities include unlicensed operation bands, fast transmission speed, and increased interference and fading, combined with node movement. In particular, the lack of reliable security mechanisms and interworking solutions with existing and future Public Land Mobile Networks (PLMNs) prevent the adoption of WLANs in commercial applications.

For QoS support in WLANs, existing proposals include performance-enhancing proxies (PEPs), whose goal is to provide a wired equivalent behavior to the upper layers. These PEPs can include enhanced error-detection and -correction mechanisms, header compression techniques, and prioritization capabilities. Additionally, designated exclusively from access networks, WLANs have to incorporate the Integrated Services (IntServ) framework, designed by the Internet Engineering Task Force (IETF) for that purpose. IntServ uses the Resource Reservation Protocol (RSVP) for requesting QoS per flow and setting up reservations end-to-end upon admission. This approach is problematic in wireless links, due to the variable available bandwidth they provide, which results in a need for specific enhancements. Concerning mobility support, well-known macromobility solutions, such as Mobile IP, do not perform well in WLANs because of the long delays and increased packet losses they introduce in handover. The problem is amplified in real-time applications, where the tolerances in delay and packet losses are very strict. Accordingly, a number of proposals can be found in the literature, referred to as *micromobility solutions*, targeted at WLANs environments. We describe major representatives of these solutions later in this chapter. As for security support, WLANs suffer, like every radio-broadcasting technique, from potential interceptions and unauthorized use of the system. This chapter focuses mainly on security solutions provided by the 802.11 family, which is considered the most promising effort for a complete set of standards for WLANs.

2.2   WLANs Basics

A WLAN is a flexible data communication system, implemented as an extension to a wired LAN within a limited area of coverage. Using high-frequency electromagnetic waves, WLANs transmit and receive data over the air, minimizing the need for wired connections. They combine data connectivity with user mobility and, through simplified configuration, enable communication of movable terminals. Over the past decade, WLANs have gained strong popularity in a number of environments, including health care, retail, manufacturing, warehousing, and academia. Today, they are becoming more widely recognized as a general-purpose connectivity alternative for a wide range of applications.

End users access a WLAN via proper adapters, which can be implemented as external modules in notebooks and desktop computers or integrated modules within personal digital assistants (PDAs) and handheld computers. We usually refer to these terminals as *Mobile Nodes (MNs)*. The access to the wired LAN infrastructure is accomplished via a transceiver device, referred to as an *access point (AP)*, which includes a WLAN adapter and a connection to the wired LAN using standard Ethernet cable. The coverage area of an AP is usually referred to as a *cell*. The AP receives, buffers, and transmits data between MNs of its cell and the wired network infrastructure. Furthermore, the AP serves as a distribution node in case a WLAN user wishes to communicate with another WLAN user associated with the same or a different AP. In such a case, the data is transferred between the relevant APs to the intended receiver. A single AP can support a small number of users and can function within a comparatively limited area of a few meters. Multiple APs can provide wireless coverage for an entire building or campus. Alternatively, many WLAN technologies support direct communication of end users without the intervention of an AP. This kind of operation, known as *ad-hoc networking*, provides freedom of communication among compatible MNs that are placed in a limited range within each other, in regions where a supporting structure of APs is not available. On the other hand, the lack of a coordinating entity usually results in poorer or unpredictable performance, especially in heavy-load conditions. Ad-hoc networking is used mainly for special-purpose applications where an infrastructure is not available (e.g., military). In the rest of the chapter, we focus mainly on cell-based WLANs.

2.2.1   802.11 Brief Description

The Institute of Electrical and Electronics Engineers (IEEE) is developing an international WLAN standard identified as IEEE 802.11[1] whose scope is "to develop a Medium Access Control (MAC) and Physical Layer (PHY) specification for wireless connectivity for fixed, portable and moving stations within a local area." More specifically, the purpose of the standard is twofold:

- "To provide wireless connectivity to automatic machinery, equipment, or stations that require rapid deployment, which may be portable, or hand-held or which may be mounted on moving vehicles within a local area."
- "To offer a standard for use by regulatory bodies to standardize access to one or more frequency bands for the purpose of local area communication."

The IEEE 802.11 standard describes two transmission rates at 1 Mbps (mandatory) and 2 Mbps (optional). Mandatory support for asynchronous data transfer is specified, as well as optional support for Distributed Time-Bounded Services (DTBS). Asynchronous data transfer refers to traffic that is relatively insensitive to time delay. Time-bounded traffic, on the other hand, is bounded by specified time delays to achieve an acceptable QoS (e.g., packetized voice and video). The Basic Service Set (BSS) is the fundamental building block of the IEEE 802.11 architecture. A BSS is defined as a group of stations that are under the direct control of a single coordination function (i.e., a Distributed Coordination Function [DCF] or Point Coordination Function [PCF]). For example, the MNs and the AP of a cell are considered to be one BSS.

The standard also specifies three different physical-layer implementations: Frequency Hopping Spread Spectrum (FHSS), Direct Sequence Spread Spectrum (DSSS), and Infrared (IR). The FHSS utilizes the 2.4-GHz Industrial, Scientific, and Medical (ISM) band (i.e., 2.4000–2.4835 GHz), where a maximum of 79 channels are specified in the hopping set. The first channel has a center frequency of 2.402 GHz, and all subsequent channels are spaced 1 MHz apart. The channel separation corresponds to 1 Mbps of instantaneous bandwidth. The DSSS also uses the 2.4-GHz ISM frequency band, where the 1-Mbps basic rate is encoded using Differential Binary Phase Shift Keying (DBPSK), whereas the 2-Mbps enhanced rate uses Differential Quadrature Phase Shift Keying (DQPSK). The spreading is done by dividing the available bandwidth into 11 subchannels, each 11-MHz wide, and using an 11-chip Barker sequence to spread each data symbol. The maximum channel capacity is therefore (11 chips/symbol)/(11 MHz) = 1 Mbps if DBPSK is used. Finally, the IR specification identifies a wavelength range from 850 to 950 nm. The IR band is designed for indoor use only and operates with nondirected transmissions. The IR specification was designed to enable MNs to receive line-of-site and reflected transmissions. Encoding of the basic access rate of 1 Mbps is performed using 16-Pulse Position Modulation (PPM), where 4 data bits are mapped to 16 coded bits for transmission. The enhanced access rate (2 Mbps) is performed using 4-PPM modulation, where 2 data bits are mapped to 4 coded bits for transmission.

IEEE 802.11 defines two access modes for the MAC layer, namely the DCF and the PCF. DCF uses Carrier Sense Multiple Access/Collision Avoidance (CSMA/CA), requiring each MN to listen for other users. If the channel is idle, the MN may transmit. However, if the channel is busy, the MN waits until transmission stops and then enters into a random backoff procedure. This prevents multiple MNs from seizing the medium immediately after completion of the preceding transmission. An MN performs virtual carrier sensing by sending MAC Protocol Data Unit (MPDU) duration information in the header of Request to Send (RTS), Clear to Send (CTS), and data frames. An MPDU is a complete data unit that is passed from the MAC sublayer to the physical layer. The MPDU contains header information, payload, and a 32-bit Cyclic Redundancy Check (CRC). The duration field indicates the amount of time (in microseconds) after the end of the present frame that the channel will be utilized to complete the successful transmission of the data or management frame. Other MNs can use the information in the duration field to adjust their Network Allocation Vector (NAV), which indicates the amount of time that must elapse until the current transmission session is complete and the channel can be sampled again for idle status. The channel is marked busy if either the physical or virtual carrier sensing mechanisms indicate that the channel is busy.

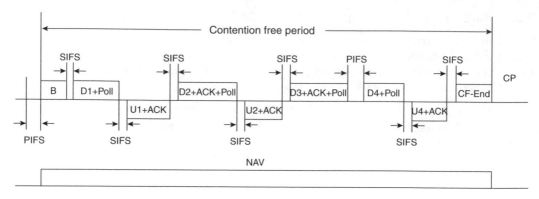

FIGURE 2.1 Transmission during the CFP.

As an optional access method, the 802.11 standard defines the PCF, which enables the transmission of time-sensitive information. With PCF, a Point Coordinator (PC) within the AP controls which MNs can transmit during any given period of time. Within a time period called the Contention Free Period (CFP), the PC steps through all MNs operating in PCF mode and polls them one at a time. For example, the PC may first poll MN A, allowing it to transmit data frames buffered in its output queues, without interference from other MNs. The PC will then poll the next MN and continue down the polling list, in order to give all MNs a chance to send their data. In brief, PCF is a contention-free protocol that enables MNs to transmit data frames synchronously, thereby controlling time delays between transmissions. This makes it possible for more effective support of information flows generated from applications with strict delay requirements, such as video and voice.

More specifically, the operation in PCF mode is as follows: At the nominal start of the CFP, the PC senses the medium. If the medium is idle, the PC transmits a beacon frame to initiate the CFP. The PC starts contention-free transmission by sending a CF-Poll (no data), Data, or Data + CF-Poll frame. The PC can immediately terminate the CFP by transmitting a CF-End frame, which is common if the network is lightly loaded and the PC has no traffic buffered. If a CF-aware MN receives a CF-Poll (no data) frame from the PC, the MN can respond to the PC with a CF-ACK (no data) or a Data + CF-ACK frame. If the PC receives a Data + CF-ACK frame from an MN, the PC can send a Data + CF-ACK+ CF-Poll frame to a different MN, where the CF-ACK portion of the frame is used to acknowledge receipt of the previous data frame. The ability to combine polling and acknowledgment frames with data frames, transmitted between MNs and the PC, was designed to improve efficiency. If the PC transmits a CF-Poll (no data) frame and the destination MN does not have a data frame to transmit, the MN sends a Null Function (no data) frame back to the PC. Figure 2.1 illustrates the transmission of frames between the PC and a MN, and vice versa. If the PC fails to receive an ACK for a transmitted data frame, the PC continues transmitting to the next station in the polling list. After receiving the poll from the PC, as described above, the MN may choose to transmit a frame to another MN in the BSS. When the destination MN receives the frame, a DCF-ACK is returned to the source MN, and after that the PC starts transmitting any additional frames.[2]

Two physical-layer extensions of the 802.11 have been standardized, first the 802.11b followed by 802.11a. The IEEE 802.11b physical layer can support higher data rates of 5.5 and 11 Mbps by using Complementary Code Keying (CCK) with Quadrature Phase Shift Keying (QPSK) modulation and DSSS technology at the 2.4-GHz band. In addition, IEEE 802.11b defines dynamic rate shifting, which allows data rates to be automatically adjusted to lower speed for noisy conditions. Most commercial products today are based on the 802.11b technology. 802.11a uses the 5-GHz unlicensed band and Orthogonal Frequency Division Multiplexing (OFDM) modulation to deliver data up to 54 Mbps at a shorter range of 50 to 70 meters. OFDM works by splitting the radio signal into multiple smaller subsignals that are then transmitted simultaneously at different frequencies to the receiver. 802.11a is expected to be the next logical step forward from 802.11b, as the need for increased data transfer rate accelerates.

Other standards in the 802.11 alphabet have been developed, including the following:

802.11g (20+ Mbps at 2.4 GHz): 802.11g uses the same OFDM scheme as 802.11a and will potentially deliver speeds comparable with 802.11a. Its main characteristic is that it operates in the 2.4-GHz band that 802.11b equipment occupies and for this reason should be compatible with existing WLAN infrastructures. Like 802.11b, 802.11g is limited to three nonoverlapping channels.

802.11f (Inter-Access Point Protocol): The existing 802.11 standard does not specify the communications among APs for supporting users during roaming from one AP to another. Vendor-specific solutions could work only for APs of the same vendor, limiting the implementation choices for present and future networks. The 802.11f task group is currently working on specifying the Inter-Access Point Protocol (IAPP) that provides the necessary mechanism for information exchange among APs needed to support the 802.11 distribution system functions (e.g., roaming). In the absence of 802.11f, APs of the same vendor should be utilized to ensure interoperability for roaming users. In some cases, a mix of AP vendors will still work, especially if the APs are Wireless-Fidelity (Wi-Fi) certified, but only the inclusion of 802.11f in AP design will eventually open up the options and add interoperability assurance when selecting AP vendors.

802.11e (MAC enhancements for QoS): Without strong QoS, the existing version of the 802.11 standard cannot support the operation of real-time traffic, such as voice and video. Even the PCF model cannot guarantee absolute values for packet delays. For this reason, the 802.11e task group is currently adding extra functionality to the 802.11 MAC layer to improve QoS for better support of a larger set of applications. The 802.11e standard falls within the MAC layer, and it will be common to all 802.11 PHY layers, while staying backward compatible with existing 802.11 WLANs. The main enhancements introduced with 802.11e are two new transmission modes: the Enhanced Distributed Coordination Function (EDCF) and the Hybrid Coordination Function (HCF), described later in this chapter.

802.11i (MAC enhancements for security in 802.11): The existing 802.11 standard specifies the use of relatively weak, static encryption keys without any form of key distribution management. This makes it possible for outsiders to access and decipher Wired Equivalent Privacy (WEP)-encrypted data on a WLAN. 802.11i will incorporate 802.1x and stronger encryption techniques, such as AES (Advanced Encryption Standard), to enhance the security of 802.11 in order to be suitable for confidential information exchange.

2.2.2 HiperLAN/2 Brief Overview

HiperLAN/2 is the result of the Broadband Radio Access Networks (BRAN) project of the European Telecommunications Standards Institute (ETSI), and its goal is to provide a high-speed radio-access standard for local area communications. Two basic modes of operation are supported: centralized mode (CM) and direct mode (DM). The CM of operation applies to the cellular networking topology where an AP controls each radio cell, which covers a certain geographical area. In this mode, MNs communicate with one another or with the core network through the AP. The CM of operation is mainly used in indoor and outdoor environments where the area to be covered is larger than a radio cell and the infrastructure allows the development of a distribution system. The DM of operation applies to ad-hoc networks, where the entire serving area is covered by one radio cell. In this mode, MNs can communicate directly with one another. In this mode, one MN has the role of the AP in order to simply assign radio resources to the rest of the MNs.

The HiperLAN/2 system operates at the 5-GHz unlicensed band (the same as IEEE 802.11a) and includes three major layers:

- A flexible PHY layer able to support multiple modes of transmission
- A Data-Link Control (DLC) layer, consisting of the MAC sublayer and error control capabilities
- A set of Convergence Layers (CLs), facilitating access to various core networks (Asynchronous Transfer Mode [ATM], IP, third-generation networks, and so forth)

TABLE 2.1   PHY-Layer Modes Defined for HiperLAN/2

Mode	Modulation	Code Rate	PHY Bit Rate	PHY	Bytes/OFDM Symbol
1	BPSK	—	6 Mbps	Mbps	3.0
2	BPSK	—	9 Mbps	Mbps	4.5
3	QPSK	—	12 Mbps	Mbps	6.0
4	QPSK	—	18 Mbps	Mbps	9.0
5	16QAM	9/16	27 Mbps	Mbps	13.5
6	16QAM	—	36 Mbps	Mbps	18.0
7	64QAM	—	54 Mbps	Mbps	27.0

For the PHY layer, the same OFDM scheme used in IEEE 802.11a has been selected as well for HiperLAN/2, due to its good performance on highly dispersive channels. A 20-MHz channel raster provides a reasonable number of channels in a 100-MHz bandwidth. To avoid unwanted mixed frequencies in implementations, the sampling frequency is also 20 MHz (at the output of a 64-point inverse fast Fourier transform[IFFT] in the modulator). The obtained subcarrier spacing is 312.5 kHz. To facilitate the implementation of filters and to achieve sufficient adjacent channel suppression, 52 subcarriers are used per channel; 48 subcarriers carry data, and 4 are pilots that facilitate coherent demodulation.[3] A key feature of the PHY layer is to provide several modulation and coding alternatives. The intent is both to adapt to current radio-link quality and to meet the requirements for different PHY-layer properties as defined for the transport channels within the DLC layer. BPSK, QPSK, and 16QAM are the supported subcarrier modulation schemes (64QAM is optional). Forward error control is performed by a convolutional code with a rate of 1/2 and a constraint length of 7. The three code rates — 1/2, 9/16, and 3/4 — are obtained by puncturing. Seven PHY-layer modes are specified in Table 2.1.

The DLC layer represents the logical link between an AP and its associated MTs. It includes functions for MAC and data transmission (in the user plane) as well as functions for terminal/user and connection control (in the control plane). Thus, the DLC layer includes the following functions (which can also be considered as sublayers):

- Medium Access Control (MAC)
- Error Control (EC)
- Radio Link Control (RLC), handling association, radio resource management, and connection setup/release

The DLC layer implements a service policy that takes into account such factors as QoS characteristics of each connection, channel quality, the number of terminal devices, and medium sharing with other access networks operating in the same area. DLC operates on a per-connection basis, and its main objective is to maintain QoS on a virtual circuit basis. Depending on the type of required service and the channel quality, capacity, and utilization, the DLC layer can implement a variety of means, such as Forward Error Correction (FEC), Automatic Repeat Request (ARQ), and flow pacing, to optimize the service provided and maintain QoS.

Two major concepts of the DLC layer are the *logical channels* and the *transport channels*. A logical channel is a generic term for any distinct data path, identified by the type of information it conveys and the interpretation of the values in the corresponding messages. Logical channels can be viewed as logical connections between logical entities and are mostly used when referring to the meaning of message contents. The names of the logical channels consist of four letters. The HiperLAN/2 DLC layer defines the following logical channels:

- **Broadcast Control Channel (BCCH):** Conveys downlink broadcast control data for the whole AP coverage area.
- **Frame Control Channel (FCCH):** Conveys the structure of the MAC frame on the downlink.

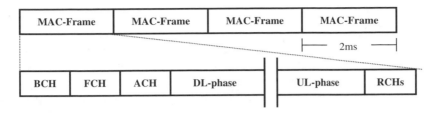

FIGURE 2.2 HiperLAN/2 MAC frame structure.

- **Random Access Feedback Channel (RFCH):** Used to inform the MTs about the result of their access to RCH.
- **RLC Broadcast Channel (RBCH):** Conveys broadcast control information on the downlink.
- **Dedicated Control Channel (DCCH):** Carries RLC messages and is established implicitly during MT association.
- **User Broadcast Channel (UBCH):** Used to transmit user broadcast data from the CL.
- **User Multicast Channel (UMCH):** Used to transmit user multicast data from the CL.
- **User Data Channel (UDCH):** Used to transmit user data on both directions.
- **Link Control Channel (LCCH):** Used to transmit ARQ feedback and discard messages in both directions.
- **Association Control Channel (ASCH):** Conveys new association and handover requests on behalf of the RLC on the uplink.

The logical channels are mapped onto different transport channels, which provide the basic elements for constructing Protocol Data Units (PDUs) and describe the format of the various messages (length, value representation, etc.). The message contents and their interpretation, however, are subject to the logical channels. The transport channels are named and referred to using three-letter abbreviations. The following transport channels are defined in the DLC layer:

- **Broadcast Channel (BCH):** Carries the BCCH on the downlink.
- **Frame Channel (FCH):** Carries the FCCH on the downlink.
- **Access Feedback Channel (ACH):** Used for transmitting the RFCH on the downlink.
- **Long Transport Channel (LCH):** Transports user data for the connections related to the UDCHs, UBCHs, and UMCHs, as well as control information for the connections related to DCCH and RBCH.
- **Short Transport Channel (SCH):** Transports short control information for DCCH, LCCH, and RBCH.
- **Random Channel (RCH):** It is used for sending control information when no granted SCH is available. It carries Resource Requests (RRs), ASCH, and DCCH data.

The MAC protocol of HiperLAN/2 is based on the Time-Division Duplex (TDD) and Time Division Multiple Access (TDMA) schemes. Time is divided in MAC frames, which are further subdivided in time slots. Time slots are allocated to the connections dynamically and adaptively, depending on the current needs of each connection. The duration of each MAC frame is fixed to 2 ms. Each frame consists of transport channels for broadcast control, frame control, access control, downlink (DL) and uplink (UL) data transmission, and random access. All data between the AP and the MTs is transmitted in the dedicated time slots, except for the random access channel, where contention for the same time slot is allowed. The duration of the broadcast control field is fixed, whereas the duration of the other field may be variable according to the current traffic needs. Slot allocation is performed by a MAC scheduler, located at the AP, taking into account QoS requirements and traffic characteristics of each connection. A MAC scheduling algorithm is not specified by the HiperLAN/2 standards, and it is left to the implementers to develop an efficient scheduling algorithm that will be able to incorporate all diverse characteristics of QoS provision. The basic structure of the HiperLAN/2 MAC frame is shown in Figure 2.2.

The EC entity supports three modes of operation:

- **Acknowledged mode** provides reliable transmissions using retransmissions to compensate for the poor link quality. The retransmissions are based on acknowledgments from the receiver. In order to support QoS for delay-critical applications (e.g., voice or real-time video), the EC may also utilize a discard mechanism for eliminating LCHs that have exceeded their lifetime.
- **Repetition mode** provides a reliable transmission by arbitrarily repeating LCHs without any feedback from the receiver. Repetition mode is used for the transmission of UBCH.
- **Unacknowledged mode** provides an unreliable, low-latency transmission, where no retransmission control or discard of messages is supported.

2.3 IP Quality of Service over WLANs

2.3.1 Problem Statement

The IP protocol and its main transport layer companions (TCP and UDP) were designed for fixed networks, under the assumption that the network consists of point-to-point physical links with stable available capacity. When a WLAN is used as an access system, it introduces at least one multiple-access wireless link with variable available capacity, resulting in low protocol performance. Here are the main weaknesses of the IP-over-wireless links:

- **No error-detection/correction mechanisms are available.** Assuming reliable links, the only logical explanation for undelivered IP packets is congestion at some intermediate nodes, which should be treated in higher layers with appropriate end-to-end congestion-control mechanisms. UDP, targeted mainly for real-time traffic, does not include any congestion control, because this would introduce unacceptable delays. Instead, it simply provides direct access to IP, leaving applications to deal with the limitations of IP's best-effort delivery service. TCP, on the other hand, dynamically tracks the round-trip delay on the end-to-end path and times out when acknowledgments are not received in time, thus retransmitting unacknowledged data. Additionally, it reduces the sending rate to the minimum and then increases it gradually in order to probe the network's capacity. In WLANs, where errors can occur due to temporary channel quality degradation, both these actions (TCP retransmissions and rate reduction) can lead to increased delays and low utilization of the scarce available bandwidth.
- **A lack of traffic prioritization.** Designed as a best-effort protocol, IP does not differentiate treatment depending on the kinds of traffic. For example, delay-sensitive real-time traffic, such as VoIP, will be treated in the same way as FTP or e-mail traffic, leading to unreliable service. In fixed networks, this problem can be relaxed with overprovisioning of bandwidth wherever possible (e.g., by introducing high-capacity fiber optics). In WLANs, this is not possible because the available bandwidth can be as high as a few tens of Mbps. But even if bandwidth were sufficient, multiple access could still cause unpredictable delays for real-time traffic. For these reasons, the introduction of scheduling mechanisms is required for IP over WLANs in order to ensure reliable service under all kinds of conditions.

Current approaches for supporting IP QoS over WLANs fall into the following categories:[4]

1. **Pure end-to-end:** This category focuses on the end-to-end TCP operation and the relevant congestion-avoidance algorithms that must be implemented on end hosts so as to ensure transport stability. Furthermore, enhancements for fast recovery, such as TCP's selective acknowledgment (SACK) option and NewReno, are also recommended.
2. **Explicit notification–based:** This category considers explicit notification from the network to determine when a loss is due to congestion[5, section 4.7] but, as expected, requires changes in the standard Internet protocols.

3. **Proxy-based:** Split-connection TCP and Snoop are proxy-based approaches, applying the TCP error-control schemes in only the last host of a connection. For this reason, they require the AP to act as a proxy for retransmissions.

4. **Pure link layer:** Pure link layer schemes are based on either retransmissions or coding overhead protection at the link layer (i.e., automatic repeat request [ARQ] and forward error correction [FEC], respectively), so as to make errors invisible at the IP layer. The error-control scheme applied is common to every IP flow irrespective of its QoS requirements.

5. **Adaptive link layer:** Finally, adaptive link layer architectures can adjust local error-recovery mechanisms according to the applications requirements (e.g., reliable flows versus delay-sensitive) and channel conditions.

Next, we discuss specific solutions aimed at improving the QoS offered to IP traffic over WLANs.

2.3.2 The Wireless Adaptation Layer (WAL)

To give a more detailed view of IP performance improvements over WLANs, we look at the Wireless Adaptation Layer (WAL), a proposal that combines a number of techniques in order to handle requirements imposed by various IP applications.[4] The WAL, designed and developed by the European project Wireless Internet Network (WINE), allows for optimized and QoS-aware Internet communications over WLANs to support multimedia services. The architecture is also flexible in that it accommodates a range of popular wireless access networks and can accommodate new technologies in the future.

From a functional point of view, the WAL could be described as a PEP,[6] implemented as an intermediate layer between the IP and lower layers of APs and MNs, that improves the performance of Internet protocols operating over wireless shared-access LANs. Two of the main requirements to be undertaken by the WAL are as follows:

1. **Adaptation to the observed link conditions:** Because the quality of a radio link varies over time, the WAL applies an adaptation scheme that (1) invokes the appropriate link layer service modules to improve channel reliability, and (2) sets parameters for the corresponding algorithms dynamically (e.g., new weights to the packet scheduler or new code ratio for the FEC). The WAL adapts itself to the channel conditions by exploiting feedback from the underlying WLAN technologies. This feedback is obtained by measurements requested by the WAL itself. These measurements are both wireless-interface dependent (such as bit-error ratio [BER] or signal strength) and wireless-interface independent (such as effective throughput for a given link layer module).

2. **IP QoS awareness:** End-to-end IP QoS architectures like IntServ and DiffServ can be applied both to Internet routers and to hosts. The WAL is a local boosting element, and in that sense, it complements these architectures by mapping the application requirements to the appropriate link layer service modules. That is, it operates a differentiated management of IP flows/streams at link layer level, mainly in terms of error control when coupled with queue management and packet scheduling.

A main feature of the WAL is an abstraction used for service provisioning at the link layer. Each IP datagram is classified into *classes* and *associations*. Service provision in the WAL is based on these two concepts. A WAL *class* defines the service offered to a particular type of application traffic (e.g., audio/video streaming, bulk transfer, interactive transfer, Web) and the sequence of link-layer modules (protocol components) that provide such a service. The module list for every class is completely defined so that every WAL MN uses the same module order within the same class. This approach allows the WAL packet classification to be mapped onto existing Internet QoS classes. Specifically, DiffServ is considered here, and the WAL packet class is determined by the DiffServ field, the Type of Service (ToS) and Traffic Class octet fields in IPv4 and IPv6, respectively) or the protocol type field (Protocol and Next Header octet fields in IPv4 and IPv6, respectively) of the IP packet header.

An *association* identifies a stream of IP datagrams belonging to the same class and destined to a specific MN — that is, *WAL_Association = <WAL_Class, MN_Id>*. A fair allocation of bandwidth can be easily

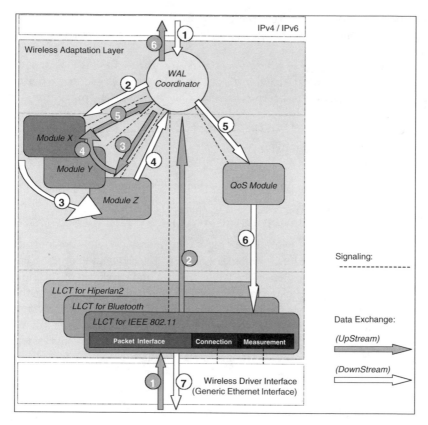

FIGURE 2.3 WINE internal architecture.

achieved if it is based on a per-user operation. In addition, services for particular users can be customized to meet their QoS requirements and to implement a differentiated-charging policy. Another advantage of distinguishing IP streams with respect to their destination is that channel-state conditions can be taken into account. In fact, as the condition of each wireless channel varies independently, the parameters of the modules defined for a class will be adjusted dynamically to adapt them to changes that take place in a channel.

The WAL coordinator shown in Figure 2.3 may be viewed as the central "intelligence" of the WAL. Both downstream and upstream traffic passes through the coordinator before being processed by other modules. In the downstream flow, the coordinator intercepts IP datagrams and decides on the sequence of modules that these datagrams should pass through, as well as the parameters of these modules. The sequence of modules for each IP flow is chosen on the basis of specific fields in IP datagrams' headers, identifying the "class" to which the datagrams belong. In the upstream flow, the WAL coordinator accepts WAL frames (encapsulated IP datagrams) and passes them through the sequence of modules associated with the class in the reverse order. Information about the modules' sequence as well as the required module parameters is contained in the WAL header (described later in this section). To determine the optimum module parameters, the WAL coordinator monitors the channel conditions through continuous measurements. The WAL configuration parameters can be set up remotely via Simple Network Management Protocol (SNMP) in the local "wireless" management information base (MIB).

The WAL coordinator maps the Internet DiffServ QoS classes onto WAL classes in order to provide flow isolation and fairness guarantees through traffic shaping and scheduling. The module performing flow regulation and scheduling is referred to as the *QoS module*. A packet scheduler is the core of the QoS module; it allows wireless resource sharing among traffic classes, according to their association. It is divided into two levels. The first level of the scheduler is implemented in both the AP and the MN and consists of a class-based queuing mechanism preceded by a traffic shaper for each traffic class. The second-level or main-level, scheduler is implemented in only the AP and is responsible for allocating the wireless network (or MAC level) bandwidth to each MN. The objective of the main scheduler is twofold:

- **Throughput maximization,** by taking into account the channel state for each MN and reducing the bandwidth allocated to it when the channel condition degrades.
- **Fairness in bandwidth allocation,** which requires an estimate of the effective gross throughput of each MN under the wireless link environment, and adjustment of the bandwidth offered to any nonconforming MN, in order to provide short- and long-term fairness.

The main scheduler has also to consider the state of the wireless link to every MN in order to avoid increased bit-error ratio, which reduces throughput and QoS. If a backlogged association, which should be serviced at time *t*, perceives a bad condition, the main scheduler will stop the service to this association and thus increase the *lag* of this association. This represents the amount of service that this association has lost, which must be made up at a future time. The scheduler then selects a packet for transmission among those backlogged association queues that currently enjoy a good wireless channel and increases the *lead* of the selected association. This reflects the number of the services that must be relinquished by that association at some time in the future.

Modules X/Y/Z comprise a pool of modules, aiming to improve performance in several ways. This pool includes error-control modules such as FEC, a Snoop module for TCP performance improvement, the Header Compression module, an ARQ module, and a fragmentation module that reduces packet-loss probability. Other modules may be included in later versions of the WAL to further improve the overall performance.

Finally, in order to interface with a number of wireless drivers of various platforms, a wireless tech-nology–specific Logical Link Control Translator (LLCT) module for each different platform has been introduced. The main functions of this module are (1) to manage the connection status with the wireless driver, (2) to ensure the stream conversions toward the wireless driver, (3) to perform channel measure-ments via the driver, and (4) to control MN registration and termination processes.

2.3.3 The IP2W Interface

The IP-to-Wireless Interface (IP2W), defined in the European project BRAIN, is another proposal for a PEP whose goal is to provide a way for the IP layer to interface with a number of WLAN technologies. Motivation for the interface stems from the characteristics of wireless links, which pose special require-ments on the interworking between the network layer and the wireless link layer. IP2W defines interfaces for controlling specific link layer features and identifies a set of functions and requirements, as well as a set of recommendations on how to design wireless link layers in a way that facilitates IP mobility, IP QoS, and efficient transmission of IP traffic in general.[7]

The interface is separated into a Data Interface and a Control Interface (i.e., a separation between the control plane and the user plane is identified), as shown in Figure 2.4. Each interface offers access to a set of functionality on the link layer. Several distinct functions have been identified under the interfaces, represented by the small ovals. The ovals surrounded by broken lines represent optional functions, which may or may not be supported by the wireless link. If supported, however, the requirements and recom-mendations given in the IP2W specification should be applied. The functional blocks below the IP2W Control Interface indicate functions that can be configured and controlled through the interface. These include *Configuration Management*, an interface for querying the capabilities of the link layer; *Address Management*; *Quality-of-Service (QoS) Control*; *Handover Control*; *Idle Mode Support*; and *Security Man-agement*. The last of these refers to controlling link-layer encryption and security facilities, which is still an open issue at the moment. The Data Interface consists of *Error Control*, mechanisms used to detect and correct errors on the link layer. *BufferManagement* refers to how buffers are managed on the link layer with regard to congestion, flow control, and other issues. *QoS Support* schedules packets to radio-link channels. *Segmentation* and *Reassembly* refer to supporting links, which do not support minimum

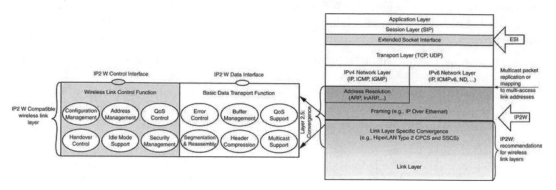

FIGURE 2.4 The IP2W interface.

MDPU-sized packets. *Header Compression* can optionally be performed on the link layer, and *Multicast Support* refers to supporting native multicast transmission.

The IP2W interface aims to be generic enough to be applicable to different wireless link layer technologies, yet detailed enough to preclude the need at upper layers to utilize any functionality or information that is specific to a particular wireless technology. This flexibility is achieved by dividing the IP2W functional blocks into specific capabilities, some of which are considered optional. An IP2W-compliant link layer advertises the capabilities it supports through a configuration function, allowing the higher layers to adjust to the characteristics and capabilities of the link layer. Finally, it is worth noting that IP2W is always coupled with an implementation of a "convergence layer" for a specific link layer technology.

2.3.4 RSVP over WLANs

As part of a possibly broad IP network, WLANs have to incorporate QoS frameworks that have been introduced for fixed IP networks. This should be done regardless of the specific techniques used for ensuring the anticipated QoS level. To treat the problem of QoS in IP networks in general, the IETF has introduced two main frameworks, namely the Integrated Services (IntServ)[8] and the Differentiated Services (DiffServ).[9] DiffServ classifies and possibly conditions the traffic in order to ensure similar behavior throughout the network. It performs well in core networks because of its scalability in supporting large numbers of flows. IntServ, on the other hand, is targeted mainly at the access systems and provides a means to request and obtain end-to-end QoS per flow. Designated exclusively for access systems, WLANs have to incorporate IntServ to be compatible with the QoS schemes followed in fixed-access systems.

The RSVP[10] is considered the major signaling protocol for the IntServ framework, and it aims to establish virtual circuits that provide per-flow QoS in an IP network. More specifically, it defines a communication *session* as a data flow with a particular destination and transport-layer protocol, identified by the triplet (destination address, transport-layer protocol type, destination port number). Its operation applies only to packets of a particular session, and therefore every RSVP message must include details of the session to which it applies. In the rest of this chapter, the term "flow" stands for "RSVP session." The two most important messages of the RSVP protocol are the PATH and the RESV messages. The PATH message is initiated by the sender and travels toward the receiver to provide characteristics of the anticipated traffic, as well as measurements for the end-to-end path properties. The RESV message is initiated by the receiver upon receiving the PATH message and carries reservation requests to the routers along the communication path between the receiver and the sender. After the path establishment, PATH and RESV messages should be issued periodically to maintain the so-called *soft states* that describe the reservations along the path.

In contrast to the stable links used in fixed networks, bandwidth of wireless links is subject to variations, resulting in low performance of static resource reservation-based schemes such as RSVP. For example, if available resources are reduced after admission control, the network may not be able to meet commitments for flows that have been successfully admitted. To address this problem, RSVP extensions and modifications have been proposed in the literature, and one of the most promising ones is the so-called dynamic RSVP (dRSVP);[11] dRSVP modifies the existing RSVP standard in order to request ranges of QoS instead of specific values. The network guarantees that the provided QoS will be within the agreed range but will not always have the maximum value. In case the channel quality falls to a level that not even the lowest values can be guaranteed for the admitted flows, the network has to reject one or more of them to maintain QoS to the rest. In both cases of RSVP and dRSVP over wireless, a flow rejection algorithm is necessary to handle unpredictable degradations of the link quality.

The main extensions/modifications of dRSVP, compared with the standard RSVP, are listed here:

1. An additional flow specification object (FLOWSPEC) in RESV messages and an additional traffic specification object (SENDER TSPEC) in PATH messages have been introduced to describe ranges of traffic flows.
2. A measurement specification object (MSPEC) has been added in the RESV messages, used to allow nodes to learn about downstream resource bottlenecks.
3. A measurement specification object (SENDER MSPEC) has been introduced in the PATH messages, used to allow nodes to learn about upstream resource bottlenecks.
4. An admission-control process has been added, able to handle ranges of required bandwidth.
5. A bandwidth-allocation algorithm has been introduced that divides up available bandwidth among admitted flows, taking into account the desired range for each flow as well as any upstream or downstream bottlenecks.
6. Finally, a flow-rejection algorithm has been added for determining the flows that have to be rejected when the available bandwidth is insufficient to fulfill all requirements.

Flow rejection can be performed through standard RSVP signaling (RESV Tear, PATH Tear messages). A simple algorithm that rejects flows randomly until the total requested minimum bandwidth is reduced below the available capacity can lead to low efficiency, in terms of flow-dropping probability and bandwidth utilization, as it might tear down the following:

1. More flows than necessary (this can happen when a large number of flows with low-bandwidth requirements is randomly selected for rejection, instead of a smaller number of high-bandwidth flows)
2. High-priority flows (if the algorithm does not differentiate the flows, based on their importance, it can reject high-priority instead of low-priority flows)
3. Flows that utilize a large portion of the available bandwidth (this could lead to low-bandwidth utilization)

Accordingly, a more efficient algorithm is needed that will avoid these situations. Three of the main criteria that could be considered by this algorithm are as follows:

1. **Reject the lower priority flows first.** To achieve this, a classification scheme is needed. In its simplest form, this scheme can use the transport-layer protocol type (included in every session identification triple) to classify flows. For example, UDP flows can be considered high priority because they usually carry real-time data, whereas TCP flows can be considered low priority. Alternatively, an extra specification parameter can be introduced in every PATH and RESV message, referred to as CLASS SPEC, containing the flow priority. Considering M priority classes, CLASS SPEC can take values in the range [1,...M]. This parameter could be set by the sender when a communication session is initiated.
2. **Minimize the number of rejected flows.** If there are more than one set of flows that can be rejected, then the set with the fewer members (i.e., flows) should be selected for rejection. This criterion prevents rejection of a large number of low-bandwidth flows.
3. **Prevent underutilization.** This could happen if the algorithm chose to reject one or more flows with large bandwidth requirements, leading to a total minimum required bandwidth that is much lower than the available bandwidth. Accordingly, the algorithm should reject a set of flows that

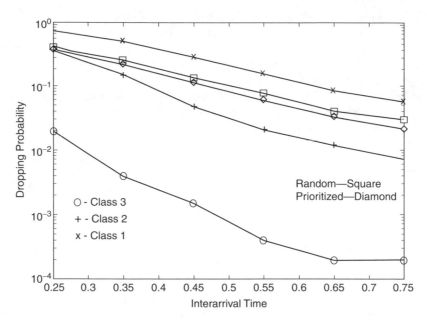

FIGURE 2.5 Dropping probability versus mean interarrival time.

leave a total minimum required bandwidth lower but as close to the available bandwidth as possible.

It is clear that these three criteria could conflict with one another. For example, there could be a case where the smaller set of flows that could be rejected belongs to a high-priority class or results in poor utilization of the available bandwidth. For this reason, possible solutions should consider some sort of ordering of the criteria.

A possible algorithm for flow rejection based on the above criteria could work as follows.[12] The algorithm takes effect when the available bandwidth of the wireless link falls below the total required bandwidth of all flows. Assuming an ascending list of flows $\{f_{i,1}, f_{i,2}, \cdots, f_{i,n_i}\}$ in each priority class C_i, according to their bandwidth requirements, the algorithm starts with the first flow $f_{1,1}$ of the lowest priority list C_1 and checks if, by rejecting it, the total required bandwidth falls below the available. It continues traversing the list until either:

(i) Such a flow is found.
(ii) The end of the list is reached.

In the first case (i), it rejects the flow and stops because the total required bandwidth is below the available. In the other case (ii), it rejects the last flow of the list f_{1,n_1} (the one with the maximum bandwidth reservations in the list) and starts over from the beginning of the list. If all the flows in the list are rejected (i.e., all the flows of the particular priority class), it continues with the flows of the next higher-priority class C_2, and so on, until the total required bandwidth falls below the available, or until all flows have been rejected. At the end, the difference between the total minimum requested bandwidth of the remaining flows and the available bandwidth is proportionally shared among flows, as long as none of the flows gets more than the maximum bandwidth requested. The same flow-rejection algorithm can also be used in the case of standard RSVP over wireless by simply assuming that the range of bandwidth requirements degenerates to a single value (i.e., the minimum and maximum requested bandwidth have the same value).

Performance evaluation results have shown that the total flow-dropping probability can be reduced up to 50% with the use of the algorithm, whereas the improvement for high-priority flows can be even more impressive.[12] For example, Figure 2.5 presents the simulation results for a channel with a capacity of 60 and 40 and a mean dwell time of 8 and 1 in the good and bad state, respectively. The model considers flows belonging to three different priority classes, $C1$ = Low, $C2$ = Medium, and $C3$ = High, although the algorithm can work with any number. A range of comparatively large mean interarrival times resulted in a relatively small number of active flows at any instance of time. The mean flow duration was equal to 4 for all flows, whereas six experiments were performed with different interarrival times per class in the range [0.25, 0.75]. The algorithm described earlier was compared against an algorithm that randomly discards a number of flows to reduce the total bandwidth requirements below the offered limit. As you can see, the attained overall dropping probability is much lower in all experiments, whereas for high-priority classes C2 and C3 the performance is significantly improved. The increased dropping probability for C1 is considered acceptable because this is the low-priority class, including mostly best-effort flows.

2.3.5   802.11e

In 802.11, none of the two standard transmission modes (DCF and PCF) can provide strict QoS.[13] In DCF, all the MNs in one BSS and all the flows in each MN compete for the resources and channel with the same priorities. There is no differentiation mechanism to guarantee bandwidth, packet delay, and delay jitter for high-priority MNs or multimedia flows. Throughput degradation and high delay are also caused by the increasing time used for channel-access contention. PCF, on the other hand, experiences three main problems that lead to poor QoS performance. The first problem involves the complex centralized polling scheme, which deteriorates performance in heavy-traffic load conditions, as MNs are polled sequentially without any knowledge of their current transmission requirements. The second problem is a result of incompatible cooperation between Contention Period (CP) and Contention Free Period (CFP) modes, which lead to unpredictable beacon delays. At Target Beacon Transition Time (TBTT), the Point Coordinator (PC) schedules the beacon as the next frame to be transmitted, but the beacon can be transmitted only when the medium has been found idle for greater than a predefined interval, referred to as the PCF Interframe Space (PIFS) interval. Depending on whether the wireless medium is idle or busy around the TBTT, the beacon frame may be delayed significantly. This may severely reduce the QoS performance by introducing unpredictable time delays in each CFP. Finally, the third problem is that the transmission time of the polled MNs is unknown. A MN that has been polled by the PC is allowed to send a MAC Service Data Unit (MSDU) that may be fragmented into a different number of smaller fragments. Furthermore, various modulation and coding schemes are specified in 802.11a, so the transmission time of the MSDU can change and is not under the control of the PC. This prevents the PC from providing QoS guarantees to other MNs that are polled during the remaining CFP.

To address these problems, and to enhance the operation of 802.11 to handle QoS-demanding applications, the IEEE established the task group 802.11e in September 1999. This group's goal is to enhance the ability of all the physical layers of 802.11 (802.11b, 802.11a, 802.11g) to deliver time-critical multimedia data, together with traditional data packets. The 802.11e standard defines two additional transmission modes: the Enhanced Distributed Coordination Function (EDCF) and the Hybrid Coordination Function (HCF), which we briefly describe below. The centralized controller, usually implemented within the AP, required to coordinate transmission is referred to as the Hybrid Coordinator (HC). An MN that supports these transmission modes is referred to as QSTA (QoS Station) in 802.11e, whereas a BSS that supports the new priority schemes of the 802.11e is referred to as QoS supporting BSS (QBSS). A superframe is composed of a CFP and a CP, controlled by the HC. The EDCF is used in the CP mode only, whereas the HCF is used in both modes. One crucial feature of 802.11e MAC is the Transmission Opportunity (TXOP). A TXOP is defined as an interval of time when an MN has the right to initiate transmissions, defined by a starting time and a maximum duration. TXOPs are allocated via contention during EDCF (EDCF-TXOP) or granted through HCF (polled-TXOP). The duration of an EDCF-TXOP is limited by a QBSS-wide TXOP limit distributed in beacon frames, whereas the duration of a polled TXOP is specified by the duration field inside the poll frame.

The QoS support is realized in EDCF through the introduction of up to eight Traffic Categories (TCs) per MN (see Figure 2.6). In the CP, each TC within the MNs independently starts a backoff after detecting

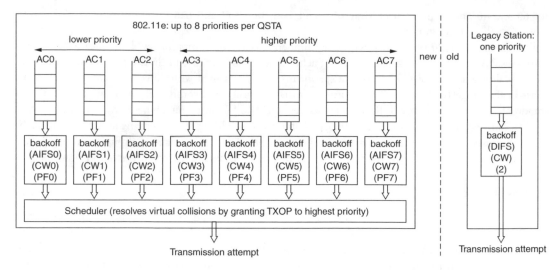

FIGURE 2.6 EDCF versus legacy DCF.

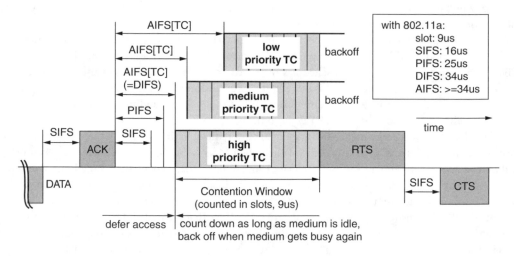

FIGURE 2.7 Contention window and AIFS for different priorities.

the channel being idle for an Arbitration Interframe Space (AIFS); the AIFS of each TC is at least equal to the DCF Interframe Space (DIFS) and can be defined individually for each TC. DIFS is the time interval that the medium has to stay idle before MNs operating in DCF can have direct access to it. Higher-priority TCs usually have shorter AIFS because this allows them to start the backoff earlier than others. After waiting for AIFS, each backoff sets a counter to a random number drawn from the range of its contention window (CW) (*1, CW* + 1). The minimum size (*CWmin[TC]*) of the first CW is another parameter dependent on the TC. Priority over legacy MNs can provided by setting *CWmin[TC]* < *15* (in case of 802.11a PHY) and *AIFS = DIFS*. As in legacy DCF, when the medium is determined to be busy before the counter reaches 0, the backoff has to wait for the medium to become idle for again before continuing to count down the counter. EDCF uses the contention window to change the priority of each TC. Assigning a large contention window to a high-priority class provides this class with a greater probability of successful transmission (i.e., getting a TXOP); see Figure 2.7.[14]

A big difference from the legacy DCF is that when the medium is determined to have been idle for the AIFS period, the backoff counter is reduced by 1 beginning with the last slot interval of the AIFS period. Note that with the legacy DCF, the backoff counter is reduced by 1 beginning with the first slot

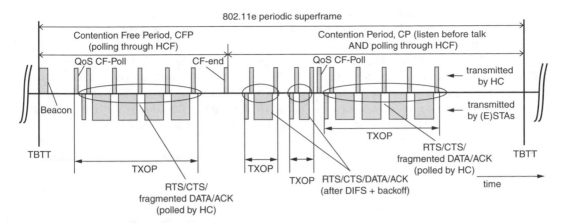

FIGURE 2.8 A typical 802.11e superframe.

interval after the DIFS period. After a collision, a new CW is calculated with the help of the persistence factor *PF[TC]*, and another uniformly distributed backoff counter from this new and enlarged CW is drawn to reduce the probability of a new collision. Whereas in legacy 802.11 CW is always doubled after any unsuccessful transmission (equivalent to $PF = 2$), 802.11e uses the PF to increase the CW different for each TC:

$$newCW\,[TC] >= ((oldCW[TC] + 1) * PF) - 1$$

The CW never exceeds the parameter *CWmax[TC]*, which is the maximum possible value for CW. "Virtual" collisions that occur when the counters of two or more TCs of the same MN reach 0 at the same time are solved internally by the MN, which allows the highest TC to transmit. After obtaining the channel, an MN is allowed to transmit as many packets as it wishes, provided that the total access time does not exceed a maximum value. It can also transmit packets from TCs other than the one that obtained the channel.

The HCF extends the EDCF by providing the HC with the power to allocate TXOPs to MNs during both the CP and the CFP of a superframe. During the CP, a special CF-Poll frame can be sent by the HC after a PIFS idle period without any backoff. With this frame the HC can explicitly allocate a TXOP to a specific MN for a set duration of time. The PIFS is shorter than any AIFS and DIFS in order to give priority to the HC rather than the competing MNs. During the CFP, the starting time and maximum duration of each TXOP is specified by the HC, again using the QoS CF-Poll frames. MNs will not attempt to get medium access on their own during the CFP, so only the HC can grant TXOPs by sending QoS CF-Poll frames. The CFP ends after the time announced in the beacon frame or by a CF-End frame from the HC (see Figure 2.8).

Additionally, extra schemes are included in 802.11e specification aimed at further improving efficiency of the MAC protocol. For instance, the optional block acknowledgments allow a backoff entity to deliver a number of MSDUs consecutively during one TXOP and transmitted without individual ACK frames. The MPDUs that are transmitted during the TXOP are referred to as a *block* of MPDUs. At the end of the block, or in a later TXOP, all MPDUs are acknowledged by a bit pattern transmitted in the block acknowledgment frame, thus reducing the overhead of control-exchange sequences to a minimum of one acknowledgment frame per number of MPDUs delivered in a block. Moreover, any backoff entity can directly communicate with any other backoff entity in a QBSS without communicating via the AP. In the legacy 802.11 protocol, within an infrastructure-based BSS all data frames are sent and received through the AP. This, however, consumes at least twice the channel capacity compared with direct communication. Only in an independent BSS (IBSS) is station-to-station communication in the legacy protocol, due to the absence of the AP. On the other hand, direct communication is allowed in an 802.11e QBSS even if an AP is present and is referred to as *Direct Link (DiL)*. The setup of the communication is performed through a special-purpose protocol referred to as the *Direct Link Protocol (DLP)*. Direct communications are especially important for in-house applications, which are expected to be extremely important in the near future.

It is clear that the HCF provides the HC with much more control of the medium than the EDCF. This explains why this transmission mode is considered efficient for IP IntServ with strict delay and bandwidth requirements. The HC can allocate TXOPs to the respective Traffic Categories (TCs) in a way that fulfills these requirements. An important component of the HC toward this direction is the traffic scheduler, an entity that implements the algorithm that decides how TXOPs will be given. Although the standard does not describe a specific TS, manufacturers should carefully design and implement such an entity in their products because its operation can significantly influence the overall protocol performance. To provide the TS with real-time information about the TC's requirements, the HCF includes the Control Contention (CC). The CC, which was included in draft 2.0 of 802.11e, is a way for the HC to learn which MN needs to be polled, at which time, and for which duration. The CC mechanism allows MNs to request the allocation of polled TXOPs by sending a Reservation Request (RR) frame, without contending with other EDCF or DCF traffic. The Controlled Contention Interval (CCI) is started when the HC sends a specific control frame. This frame forces legacy MNs to silence and defines a number of controlled contention opportunities (CCOP, i.e., short intervals separated by Short Interframe Space [SIFS] intervals) and a filtering mask containing the TCs for which RRs may be placed. Each MN with queued traffic for a TC matching the filtering mask chooses one CCOP interval and transmits a RR frame containing its requested TC and TXOP duration or the queue size of the requested TC. For fast collision resolution, the HC acknowledges the RR frame by generating a control frame with a feedback field so that the requesting MNs can detect collisions during the CCI.

2.4 Micromobility in WLANs

Designed for fixed networks, the IP does not support mobility. End nodes are characterized by static IP addresses, used for packet routing throughout the network. Accordingly, when a node changes its point of attachment, the two straightforward solutions are to either obtain a new IP address or keep the same address and inform the routers throughout the network of its new position. Unfortunately, neither of these solutions is feasible. The former requires updating all entries containing the old IP address, starting with the domain name servers that keep the mappings between IP names and IP addresses. Because many corresponding nodes might store the IP address of an MN, this solution cannot work for all cases. The latter solution implies a huge amount of signaling and long delays for updating the routing tables of probably a large number of routers. Additionally, extra entries have to be added to the routing tables for moving nodes, resulting in large storage requirements and long packet processing delays. Accordingly, more sophisticated solutions have to be implemented in order to support IP mobility.

The large number of proposed solutions for IP mobility support are divided into two main categories, based on the range of mobility they provide. *Macromobility* solutions can support movement in large areas, where users have to connect to another access network. Although useful for roaming purposes, these solutions cannot provide seamless handover; they experience large connection reestablishment delays. The main representative of this category is Mobile IP,[15] which is considered the *de facto* standard and is included in the next version of IP (IPv6). Mobile IP is optimized for macro-level mobility and relatively slow-moving hosts because it requires that after each migration a location update message be sent to a possibly distant home agent, potentially increasing handoff latency and load on the network. *Micromobility* solutions, on the other hand, can cover mobility in only small areas, where a mobile host moves inside the same access network, but they provide fast connection reestablishment suitable for seamless handover. It is clear today that only one solution is incapable of covering all mobility scenarios and that a combination of macro- and micromobility techniques is necessary. Here, we focus exclusively on micromobility proposals, which provide for mobility inside a WLAN. We present three main representatives of this category: the Cellular IP, HAWAII, and the Hierarchical Mobile IP. For more details, see Chapter 5, as well as the respective references.

2.4.1   Cellular IP

According to Cellular IP (which is discussed in more detail in Section 5.3.2), none of the nodes in the access network knows the exact location of a mobile host. Packets are routed to the mobile host on a hop-by-hop basis, where each intermediate node needs to know only on which of its outgoing ports it should forward packets.[16] To minimize control messaging, regular data packets transmitted by mobile hosts on the uplink direction are used to establish host-location information. The path taken by these packets is cached in the intermediate nodes to locate the mobile node's current position. To route downlink packets addressed to a mobile host, the path used by recent uplink packets transmitted by the mobile host is reversed. When the mobile host has no data to transmit, it periodically sends a route update packet to the gateway to maintain its downlink routing state.

Following the principle of passive connectivity, idle mobile hosts allow their respective soft-state routing cache mappings to time out. These hosts transmit paging-update packets at regular intervals defined by a paging-update time. The paging-update packet is an empty IP packet addressed to the gateway and is distinguished from a route-update packet by its IP type parameter. Paging-update packets are sent to the base station that offers the best signal quality. Similar to data- and route-update packets, paging-update packets are routed on a hop-by-hop basis to the gateway. Intermediate nodes may option-ally maintain paging caches that have the same format and operation as a routing cache except for two differences. First, paging-cache mappings have a longer timeout period called *paging timeout*. Second, paging-cache mappings are updated by any packet sent by mobile hosts, including paging-update packets. Paging cache is used to avoid broadcast search procedures found in cellular systems. Intermediate nodes that have paging cache will forward the paging packet only if the destination has a valid paging-cache mapping and only to the mapped interface(s). Without any paging cache, the first packet addressed to an idle mobile host is broadcast in the access network. Although the packet does not experience extra delay, it does, however, load the access network. Using paging caches, the network operator can restrict the paging load in exchange for memory and processing cost.[17]

2.4.2   HAWAII

HAWAII (Handoff-Aware Wireless Access Internet Infrastructure) segregates the network into a hierarchy of domains, loosely modeled on the autonomous system hierarchy used in the Internet.[18] The gateway into each domain is called the *domain root router*. Each host is assumed to have an IP address and a home domain. While moving in its home domain, the mobile host retains its IP address. Packets destined to the mobile host reach the domain root router based on the subnet address of the domain and are then forwarded over special dynamically established paths to the mobile host. When the mobile host moves into a foreign domain, we revert to traditional Mobile IP mechanisms. If the foreign domain is also based on HAWAII, then the mobile host is assigned a co-located care-of address from its foreign domain. Packets are tunneled by the home agent to the care-of address, according to Mobile IP. When moving within the foreign domain, the mobile host retains its care-of address unchanged, and connectivity is maintained using dynamically established paths. The protocol contains three types of messages for path setup: power-up, update, and refresh.

A mobile host that first powers up and attaches to a domain sends a *path setup power-up message*. This has the effect of establishing host-specific routes for that mobile host in the domain root router and any intermediate routers on the path toward the mobile host. Thus, the connectivity from that domain root router to the mobile hosts connected through it forms a virtual tree overlay. Note that other routers in the domain have no specific knowledge of this mobile host's IP address. While the mobile host moves within a domain, maintaining end-to-end connectivity to the mobile host requires special techniques for managing user mobility. HAWAII uses *path setup update messages* to establish and update host-based routing entries for the mobile hosts in selective routers in the domain so that packets arriving at the domain root router can reach the mobile host with limited disruption. The choice of when, how, and which routers are updated constitutes a particular path setup scheme. The HAWAII path state maintained in the routers is characterized as "soft state." This increases the robustness of the protocol to router and link failures. The mobile host infrequently sends periodic *path refresh messages* to the base station to which it is attached to maintain the host-based entries, failing which they will be removed by the base station. The base station and the intermediate routers, in turn, sends periodic *aggregate hop-by-hop* refresh messages to the domain root router. Path setup messages are sent to only selected routers in the domain, resulting in very little overhead associated with maintaining soft state.

A more detailed discussion of HAWAII is included in Section 5.3.3.

2.4.3 Hierarchical Mobile IP (HMIP)

HMIP[19] is an extension of the traditional Mobile IP protocol used to cover micromobility scenarios. It introduces a new function, the Mobility Anchor Point (MAP), and minor extensions to the mobile-node operation. The correspondent node and home agent operation are not affected. A MAP is a router located in a network visited by the mobile host, and it is used as a local home agent. Just like Mobile IP, HMIP is independent of the underlying access technology, allowing mobility within or among different types of access networks.

Let us take a brief look at the operation of the protocol. A mobile host entering a foreign network will receive router advertisements containing information on one or more local MAPs. The mobile host can bind its current location with a temporary address on the foreign subnet. Acting as a local home agent, the MAP will receive all packets on behalf of the mobile node it is serving and will encapsulate and forward them directly to the mobile node's current address. If the mobile node changes its current address within the foreign network, it needs to register the new address only with the MAP. Hence, only the regional address needs to be registered with correspondent nodes and the home agent, which does not have to change as long as the MN moves within the same network. This makes the mobile node's mobility transparent to the correspondent nodes it is communicating with and to its home network. An HMIP-aware mobile host with an implementation of Mobile IP should choose to use the MAP when discovering such capability in a visited network. However, in some cases the mobile node may prefer to simply use the standard Mobile IP implementation. For instance, the mobile host may be located in a visited network within its home site. In this case, the home agent is located near the visited network and could be used instead of a MAP. In this scenario, the mobile host would have to update the home agent whenever it moves.

For a comprehensive discussion on hierarchical Mobile IPv6, see Section 5.3.5.

2.5 QoS during Handovers in WLANs

To improve the QoS provided during handovers, the techniques we just described should be combined with IP QoS mechanisms such as RSVP. When an MN changes location (e.g., in handover), first it has to be reauthenticated in the new AP, then it must reestablish reservations with all its Corresponding Nodes (CNs) along the new paths. If RSVP is used, path reestablishment has to be performed end to end, resulting in considerable delays depending on the distance between peers. To facilitate fast handovers, the IEEE 802.11f task group is currently standardizing the InterAccess Point Protocol (IAPP) in order to provide the necessary capabilities for transferring context from one AP to another,[20] which can be used both for reauthentication and for path reestablishment. Especially for reservations reestablishment, a number of RSVP extensions have been proposed in the literature. In this section, we elaborate on both RSVP extensions for mobility and the use of the IAPP for faster handovers.

2.5.1 RSVP Extensions for Mobility

Designed for fixed networks, RSVP assumes fixed endpoints, and for that reason its performance is problematic in mobile networks. When an active MN changes its point of attachment with the network (e.g., in handover), it has to reestablish reservations with all its CNs along the new paths. For an outgoing flow, the MN must issue a PATH message immediately after the routing change and wait for the corresponding RESV message before starting data transmission through the new attachment point. Depending on the hops between the sender and the receiver, this approach can cause considerable delays and thus result in temporary service disruption. The effects of handover are even more annoying in an incoming flow because the MN has no power for invoking immediately the path-reestablishment procedure. Instead, it has to wait for a new PATH message, issued by the sender, before responding with a RESV message in order to complete the path reestablishment. Simply decreasing the period of the soft-state timers is not an efficient solution, because this strategy could increase signaling overhead significantly.

A number of proposals can be found in the literature, extending RSVP for either inter-subnet or intra-subnet scenarios. For intra-subnet scenarios, proposals that combine RSVP with micromobility solutions, such as Cellular IP, can reduce the effects of handover on RSVP because only the last part of the virtual circuit has to be reestablished. For inter-subnet scenarios, the existing proposals include advance reservations, multicasting, and RSVP tunneling, among others. At this point, we will focus in intra-subnet solutions, which can better fit in WLANs.

Talukdar et al.[21] proposed Mobile RSVP (MRSVP), an extension of RSVP that allows the MN to preestablish paths to all the neighboring cells. All reservations to these cells are referred to as *passive* reservations, in contrast to the *active* reservations in the cell that the MN actually is. When the MN moves from an old cell to a new one, the reservations in the new cell become active, whereas the reservations in the old cell change to passive. Although this proposal reduces the handover delays for path reestablishment, it requires RSVP to be enhanced to support a possibly large number of passive reservations, and each AP has to maintain much state information regarding active and passive reservations. Additionally, new real-time flows have to wait for all the necessary (passive and active) reservations before starting transmission, resulting in a possibly high blocking rate. Tseng et al.[22] proposed the Hierarchical MRSVP (HMRSVP) in an attempt to reduce the number of required passive reservations. According to HMRSVP, passive reservations are performed only when an MN is moving in the overlapping area of two or more cells.

According to Kuo and Ko,[23] RSVP is extended with two additional processes — a resource clear and a resource re-reservation — in order not to release and reallocate reservations in the common routers of the old and the new path. This solution performs well in reducing the path reestablishment time but modifies the RSVP protocol significantly.

Chen and Huang[24] proposed an RSVP extension based on IP multicast to support MNs. RSVP messages and actual IP datagrams are delivered to an MN using IP multicast routing. The multicast tree, rooted at each MN, is modified dynamically every time an MN roams to a neighboring cell. Hence, the mobility of an MN is modeled as a transition in multicast group membership. In this way, when the MN moves to a neighboring cell that is covered by the multicast tree, the flow of data packets can be delivered to it immediately. This method minimizes service disruption by rerouting the data path during handovers, but it introduces extra overhead for the dynamic multicast tree management and requires multiple reservations in every multicast tree.

All these approaches, while attempting to improve the performance of RSVP in mobile networks, either result in low resource utilization (due to advance reservations) or require considerable modifications of protocols and network components operation. In micromobility environments, only a small part of the path is altered, whereas the remaining circuit can be reused. Accordingly, a scheme for partial-path reestablishment can be considered that handles discovery and setup of the new part between the crossover router and the MN. The number of hops required to set up the partial path during a handover depends on the position of the crossover router. Such a scheme can reduce resource reservation delays and provide better performance for real-time services, all without affecting the operation of RSVP in any significant way.

Paskalis et al.[25] have recently proposed a scheme that reduces the delay in data-path reestablishment without reserving extra resources, and it requires modifications only in the crossover router between the old and the new path. According to this scheme, an MN may acquire different "local" care-of addresses while moving inside an access network but is always reachable by a "global" care-of address through tunneling, address translation, host routing, or any other routing variety, as suggested in various hierarchical mobility management schemes. The crossover router, referred to as *RSVP Mobility Proxy*, handles resource reservations in the last part of the path and performs appropriate mappings of the global care-of address to the appropriate local care-of address.

A similar approach for partial-path reestablishment has also been proposed by Moon and Aghvami.[26] According to this scheme, if an MN is a sender, it sends an RSVP PATH message after the route update has completed. When the RSVP daemon on the crossover router (which is determined by the route update message) receives an RSVP PATH message after a mobility event, it immediately sends an RSVP RESV message to the MN without delivering it to the original receiver. If an MN is a receiver, the RSVP daemon on the crossover router can trigger an RSVP PATH message immediately after detecting any changes to the stored PATH state or receiving a notification from the underlying routing daemon. This PATH message can be generated based on the PATH state stored for the flow during previous RSVP message exchanges.

2.5.2   The InterAccess Point Protocol (IAPP)

The need for a standard way of communication between 802.11 APs through the Distribution System (DS) was a consequence of the fact that the 802.11 standard specifies only the PHY and MAC layers of a WLAN system. Although this arrangement provides the required flexibility in designing such systems, it results in implementation-specific solutions, which means APs of a specific vendor are required and devices of different vendors are unlikely to interoperate. As 802.11 systems have grown in popularity, this limitation has become an impediment to WLAN market growth. At the same time, it has become clear that a small number of DS environments comprise the bulk of the commercial and private WLAN system installations.

The IAPP is a recommended practice that describes a Service Access Point (SAP), service primitives, a set of functions, and a protocol that will allow conformant APs to interoperate on a common DS, using the UDP over IP (UDP/IP) protocol to carry IAPP packets between APs, as well as describing the use of the Remote Authentication Dial-In User Service (RADIUS) protocol, so that APs may obtain information about one another. The devices in a network that might use the IAPP are 802.11 APs. Other devices in a network that are affected by the operation of the IAPP are Layer 2 networking devices, such as bridges and switches. The two most significant messages of the IAPP are the MOVE_notify, issued by the new AP to inform the old AP that the MN has moved to its area, and the MOVE_response, issued by the old AP and containing the context block. To find the IP address of the old AP, the new AP contacts the RADIUS server, using the BSS Identifier (BSSID) of the old AP carried in the reassociate message from the MN. If communication between APs needs to be encrypted, security blocks are exchanged before actual communication. The basic IAPP operation is illustrated in Figure 2.9. Although initially intended to contain authentication information (to allow the new AP to accept the MN without reauthentication), the context block has a flexible structure, able to support any information exchange.

More specifically, the IAPP supports two protocol sequences, one for association and the other for reassociation of a MN. Upon receiving an association sequence, the AP should send an IAPP ADD-notify packet and a Layer 2 Update Frame. The IAPP ADD-notify packet is an IP packet with a destination-IP-address of the IAPP IP multicast address and the source IP and MAC address of the AP. The message body contains the MAC address of the MN and the sequence number from the association request sent by the MN. On receiving this message, an AP should check its association table and remove any stale association with the MN if it exists and is determined to be older than the association indicated by the ADD-notify packet. The Layer 2 Update Frame carries the source MAC address of the associating MN and is used by the receiving APs and other Layer 2 devices to update their learning tables in order to forward data for the MN through the correct AP.

On receiving a standard 802.11 reassociate request from an MN, the new AP should start IAPP signaling with the old AP in order to receive context information. To learn the IP address of the old AP, the new AP uses the Old-BSSID contained in the reassociate request message and contacts a RADIUS server containing the mappings between BSSIDs and IP addresses for all APs in a DS. After obtaining the IP address of the old AP, the new AP issues an IAPP MOVE-notify packet over TCP/IP directly to the old AP. TCP is preferred over UDP as a result of the need for reliable message transmission. The data field of the MOVE-notify packet carries the MAC address and sequence number from the STA that has reassociated with the AP sending the packet. The format of the data field for this packet is as follows:

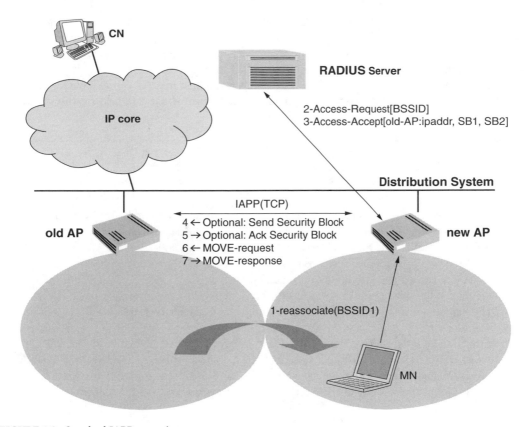

FIGURE 2.9 Standard IAPP operation.

Address Length	Reserved	MAC Address	Sequence Number	Length of Context Block	Context Block
Octets: 1	1	n = Address Length	2	2	m = Length of Context Block

The Address Length contains an 8-bit integer that indicates the number of octets in the MAC Address. The Reserved field in the current version should be set with a value of 0. The Reserved field should be ignored on reception. The MAC Address is the MAC address of the MN that has requested reassociation. The Sequence Number field contains the integer value of the sequence number of the reassociation request frame received by the AP from the MN that has requested reassociation. Allowable values for the Sequence Number field are between 0 and 4095. The Length of Context Block field contains a 16-bit integer that indicates the number of octets in the Context Block field. Context Block is a variable-length field that contains the context information being forwarded for the reassociated MN indicated by the MAC Address field. The content of the Context Block field should not be interpreted by the IAPP. Context Block contains information defined in other 802.11 standards that needs to be forwarded from one AP to another upon reassociation of an MN. This field contains a series of information elements. The format of the information element (IE) is as follows:

Element Identifier	Length	Information
Octets: 2	2	n = Length

The element identifiers and format of the IE content are defined by the standards that use the IAPP to transfer context from one AP to another. IEs are defined to have a common general format consisting of a 2-octet Element ID field, a 2-octet Length field, and a variable-length element-specific Information field. Each element is assigned a unique Element ID as defined in the standards that use the IAPP to transfer context between APs. The Length field specifies the number of octets in the Information field. Users of the IAPP service should ignore information elements whose element identifier they do not understand, rather than discarding the entire IAPP MOVE-notify packet.

The old AP responds to the MOVE-notify with a MOVE-response message carrying the MAC address of the reassociated MN and the context information. The format of the data field for this message is shown here:

Address Length	Status	MAC Address	Sequence Number	Length of Context Block	Context Block
Octets: 1	1	n = Address Length	2	2	m = Length of Context Block

The Address Length field contains an 8-bit integer that indicates the number of octets in the MAC Address field. The Status field is an 8-bit integer that indicates the status resulting from the receipt of the MOVE-notify packet. The MAC Address field specifies the MAC address of the MN that has reassociated. The Sequence Number field contains the integer value of the sequence number from the MOVE-notify packet that caused the generation of this packet. The Length of Context Block field contains a 16-bit integer that indicates the number of octets in the Context Block field. Context Block is a variable-length field that contains the context information being forwarded for the reassociated MN indicated by the MAC Address field.

Although initially intended to contain authentication information to allow the new AP to accept the MN without reauthentication and thus result in faster handover, the exchanged context block has a flexible structure able to support any information exchange, as described earlier. In the next section, we discuss a solution that uses this structure to transfer RSVP information in order to reduce reestablishment delays during handovers.

2.5.3   IAPP and RSVP

As you saw in the previous section, the Context Block field may include IEs with the required authentication information, in order to allow the new AP to accept the MN without reauthenticating it and thus reducing the handover delay. This flexible Context Block structure can also include IEs carrying RSVP information in order to accelerate the required path reestablishment process through the new AP and further reduce handover latency, which leads to better QoS. One, for example, is used per active outgoing or incoming flow and contains the details required for the new AP to create the corresponding PATH or RESV state, respectively. The aim is to allow the new AP to initiate RSVP messages to the network on behalf of the MN, before the completion of the reassociation process. The IE should contain the session identification triplet (destination address, transport-protocol type, destination port) and the traffic descriptor (SENDER_TEMPLATE and SENDER_TSPEC) objects of the RSVP. Similarly, for an incoming flow, the IE should contain the session identification, the flow descriptor (FLOWSPEC and FILTER_SPEC), and the reservation style (STYLE) objects.

More specifically, for an outgoing session, the IE contained in the MOVE-response message should contain all the information required by the new AP to create a PATH state for that session. This is part of the information that the new AP would get through the respective PATH message sent by the MN in the case after reassociation. For this reason, we refer to this IE as the *PATH IE*. This information is available in the PATH state of the old AP. In Braden et al.,[10] the structure of a PATH message is described as follows:

```
<Path Message> ::= <Common Header> [ <INTEGRITY>]
                                    <SESSION> <RSVP_HOP>
                                    <TIME_VALUES>
                                    [ <POLICY_DATA>...]
```

```
                              [ <sender descriptor>]
        <sender descriptor> ::= <SENDER_TEMPLATE> <SENDER_TSPEC>
                              [ <ADSPEC>]
```

We provide a justification on which objects should be included in the PATH IE and which should not. We demonstrate that the IE structure is flexible enough to easily include any number of objects. More detailed definitions of the objects can be found in Braden et al.[10]

COMMON HEADER: This object has the same structure in all RSVP messages:

```
          0               1               2               3
    +-------------+-------------+-------------+-------------+
    | Vers | Flags|  Msg Type   |       RSVP Checksum       |
    +-------------+-------------+-------------+-------------+
    |   Send_TTL  | (Reserved)  |       RSVP Length         |
    +-------------+-------------+-------------+-------------+
```

The RSVP version is considered known and can be omitted. The flags are not used yet in the current version. The message type is known from the IE identifier (PATH IE or RESV IE). The RSVP checksum is used for error control in the wireless medium, and there is no need to transmit it in the fixed distribution system. The Send_TTL is for identifying intermediate Non_RSVP nodes by comparing the TTL with which a message is *sent* to the Time to Live (TTL) with which it is *received*. Because no Non_RSVP nodes exist between the MN and the AP, it can be omitted in the PATH IE. Finally, the information for the RSVP Length field is not needed; it is included in the Length field of the IE. Consequently, none of the Common Header attributes has to be transmitted in the PATH IE.

INTEGRITY: This object is optional and contains cryptographic data that authenticates the originating node and verifies the contents of the message. Because the communication of the two APs through the IAPP is considered secure, especially when security blocks are exchanged (messages 4 and 5 in Figure 2.9), there is no need to include this object in the PATH IE.

SESSION: The SESSION object is required for identifying the session to which the RSVP information applies. It contains the destination address, the transport-layer protocol identifier, and an optional destination port.

RSVP_HOP: This object contains the previous hop address — that is, the IP address of the interface through which the PATH message was most recently sent. For MNs with only one interface in the wireless medium(which is the usual case in mobile networks), this information can be obtained by the SENDER_TEMPLATE object described below.

TIME_VALUES: This object specifies the time period used for refreshing the PATH state. For the IAPP-assisted handover, the new AP can set a maximum time period, within which it should receive the first PATH message from the MN after reassociation. If no PATH message is received within this period from the MN, the session is considered finished, and resources can be released.

POLICY_DATA: This object is optional and is used by the policy control in each node to determine whether a certain flow is permitted. Policy data may include credentials identifying users or user classes, account numbers, limits, quotas, and so forth, which are simply transferred by RSVP and delivered to policy control. Assuming a unique policy in the access network in which the MN moves, its active flows have already been accepted by the system, and therefore no policy control is required. Consequently, POLICY_DATA does not have to be included in the PATH IE.

SENDER_TEMPLATE: This object identifies the sender by specifying the sender IP address and optionally the UDP/TCP sender port, while it assumes the protocol identifier specified for the session in the SESSION object. Therefore, it is clearly needed in the PATH IE.

SENDER_TSPEC: This object contains the traffic parameters for the traffic flow that the corresponding sender will generate. This information is absolutely necessary for the new AP to create the PATH state and probably perform admission control.

ADSPEC: It is an optional object that carries information describing the properties of the data path, including the availability of specific QoS control services. The MN may construct an initial ADSPEC object that carries information about the QoS control capabilities and requirements of the sending application itself. Assuming that the MN never asks for a traffic flow that cannot be supported by the application, the ADSPEC object can be omitted in the PATH IE.

As in the case of outgoing sessions, the required information that should be contained for an incoming session in the corresponding IE is part of the information that the new AP would get through the respective RESV message sent by the MN. Consequently, we refer to this IE as the *RESV IE*. This information is available in the RESV state of the old AP. In Grossman et al.[9], the structure of a RESV message is described as follows:

```
<Resv Message> ::= <Common Header> [ <INTEGRITY>]
                              <SESSION> <RSVP_HOP>
                              <TIME_VALUES>
                              [ <RESV_CONFIRM>] [ <SCOPE>]
                              [ <POLICY_DATA>...]
                              <STYLE> <flow descriptor list>
           <flow descriptor list> ::= <empty> |
                        <flow descriptor list> <flow descriptor>
```

Next we justify which objects should be included in the RESV IE and which should not. Grossman et al.[9] provide additional information about the objects. For the objects of the Common Header, INTEGRITY, SESSION, RSVP_HOP, TIME_VALUES, and POLICY_DATA, the same justification as in the PATH case applies as well.

RESV_CONFIRM: When included in a RESV message, this object indicates a request for a reservation confirmation and carries the IP address of the upstream node where the confirmation should be sent. Here, the only upstream node is the MN, which does not actually send the RESV message. Consequently, this object can be omitted in the respective RESV IE.

STYLE: This object specifies the desired reservation style that determines the kind of reservation requested ("distinct" reservation for each upstream sender, or "shared" among all packets of selected senders) and the kind of senders ("explicit" list of all selected senders, or "wildcard" selection of all the senders of the session). Three styles are defined, according to the following table:

Sender Selection	Reservations	
	Distinct	Shared
Explicit	Fixed-Filter (FF) style	Shared-Explicit (SE) Style
Wildcard	(None defined) (WF) Style	Wildcard-Filter

Assuming that all the available reservation styles should be supported, this object should be included in the RESV IE because its value affects the processing of the RESV information by the new AP.

SCOPE: This object carries an explicit list of sender hosts to which the RESV message is to be forwarded. SCOPE objects are not necessary if the multicast routing uses shared trees or if the reservation style has explicit sender selection (meaning that it applies only in the case of Wildcard-Filter [WF] style and multiple senders). Attaching a SCOPE object to a reservation should be deferred to a node with more than one previous hop for the reservation state.[4] Considering that the MN has only one previous hop (the new AP), a SCOPE object is not needed in the RESV message originated by the MN and, consequently, in the RESV IE.

Flow descriptor list: The specific structure of this object depends on the reservation style, but in general, it consists of FLOWSPEC and FILTER_SPEC objects. FLOWSPEC defines the QoS to be provided for a flow, whereas FILTER_SPEC specifies the set of data packets of the sender that should receive this QoS. Data packets that are addressed to a particular session but that do not match any of the filter specs for that session are handled as best-effort traffic. It is clear that both these parameters are required to establish the RESV state, and therefore, the whole flow descriptor list should be included in the RESV IE.

From the previous discussion, the RSVP objects that should be included in both the PATH and the RESV IEs of the IAPP MOVE-response message are derived. In brief, the PATH IE should contain the SESSION, SENDER_TEMPLATE, and SENDER_TSPEC objects, whereas the RESV IE should contain the SESSION, STYLE, and flow descriptor list (consisting of FLOWSPEC and FILTER_SPEC objects). According to Braden et al.[10], each object consists of one or more 32-bit words with a one-word header, with the following format:

```
          0              1              2              3
    +-------------+-------------+-------------+-------------+
    |       Length (bytes)      |  Class-Num  |   C-Type    |
    +-------------+-------------+-------------+-------------+
    |                                                      |
    //                  (Object contents)                 //
    |                                                      |
    +-------------+-------------+-------------+-------------+
```

Length is a 16-bit field containing the total object length in bytes, *Class-Num* identifies the object class, whereas *C-Type* is unique within Class-Num and at the moment can take two values indicating IPv4 or IPv6 addressing.

Figure 2.10 and Figure 2.11 show examples for the structure of the PATH and RESV IEs, respectively. We assume IPv4 addressing, Fixed-Filter style, and Guaranteed Service. For IPv6, the only difference is that source and destination addresses occupy 16 bytes instead of 4. Guaranteed Service is considered for the FLOWSPEC in order to offer bounded end-to-end delays and bandwidth. For Controlled-Load Service, the FLOWSPEC object should not include the *Rate* and *Slack Term* parameters. Details for the specific fields can be found in Braden et al.[10]

It is clear that many fields inside the objects could be omitted or forced to consume less space. Nevertheless, the objects are included as defined for RSVP in order to be compatible with future versions of the RSVP protocol that will use more fields. Additionally, the above IE structure is flexible enough to include more objects if necessary. The only thing that is needed in order to include a new object is to add the object to the end of the respective IE, as defined in RSVP together with its object header, and update the IE's Length field accordingly. In the example presented earlier, the length of the PATH IE is 64 bytes, whereas the length of the RESV IE is 84 bytes. Note that using the IAPP this information is transmitted through a high-speed distribution system connecting the two APs. In a different case, the same information would have to be transmitted using standard RSVP signaling through the wireless medium, after reassociation of the MN with the new AP, which would result in considerable delays and signaling overhead.

More specifically, because the new AP might be more crowded than the old one, or can provide lower available bandwidth, some of the flows might not be supported. Consequently, an admission-control algorithm should be applied in the new AP to decide which flows can be accepted, based on the RSVP information received through the MOVE-response message. For an accepted outgoing flow, the new AP immediately creates the respective PATH state and issues a PATH message to the network without waiting to receive this message from the MN after the reassociation process is finished. For denied outgoing flows, the new AP can either omit the PATH messages and let reservations expire or send PATH_Tear messages

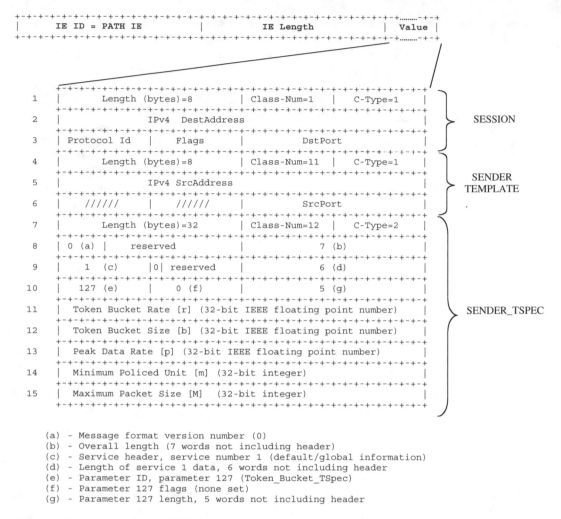

```
+-+-+-+-+-+-+-+-+-+-+-+-+-+-+-+-+-+-+-+-+-+-+-+-+-+-+-+-+-+-+........+-+-+
|        IE ID = PATH IE        |           IE Length         | Value |
+-+-+-+-+-+-+-+-+-+-+-+-+-+-+-+-+-+-+-+-+-+-+-+-+-+-+-+-+-+-+........+-+-+
```

```
      +-+-+-+-+-+-+-+-+-+-+-+-+-+-+-+-+-+-+-+-+-+-+-+-+-+-+-+-+-+-+-+
  1   |        Length (bytes)=8       | Class-Num=1   |  C-Type=1   |
      +-+-+-+-+-+-+-+-+-+-+-+-+-+-+-+-+-+-+-+-+-+-+-+-+-+-+-+-+-+-+-+
  2   |              IPv4   DestAddress                             |        }  SESSION
      +-+-+-+-+-+-+-+-+-+-+-+-+-+-+-+-+-+-+-+-+-+-+-+-+-+-+-+-+-+-+-+
  3   | Protocol Id   |    Flags      |         DstPort             |
      +-+-+-+-+-+-+-+-+-+-+-+-+-+-+-+-+-+-+-+-+-+-+-+-+-+-+-+-+-+-+-+
  4   |        Length (bytes)=8       | Class-Num=11  |  C-Type=1   |
      +-+-+-+-+-+-+-+-+-+-+-+-+-+-+-+-+-+-+-+-+-+-+-+-+-+-+-+-+-+-+-+        SENDER
  5   |              IPv4 SrcAddress                                |        TEMPLATE
      +-+-+-+-+-+-+-+-+-+-+-+-+-+-+-+-+-+-+-+-+-+-+-+-+-+-+-+-+-+-+-+
  6   |  //////       |   //////      |         SrcPort             |
      +-+-+-+-+-+-+-+-+-+-+-+-+-+-+-+-+-+-+-+-+-+-+-+-+-+-+-+-+-+-+-+
  7   |        Length (bytes)=32      | Class-Num=12  |  C-Type=2   |
      +-+-+-+-+-+-+-+-+-+-+-+-+-+-+-+-+-+-+-+-+-+-+-+-+-+-+-+-+-+-+-+
  8   | 0 (a) |     reserved          |             7 (b)           |
      +-+-+-+-+-+-+-+-+-+-+-+-+-+-+-+-+-+-+-+-+-+-+-+-+-+-+-+-+-+-+-+
  9   |   1  (c)    |0| reserved      |             6 (d)           |
      +-+-+-+-+-+-+-+-+-+-+-+-+-+-+-+-+-+-+-+-+-+-+-+-+-+-+-+-+-+-+-+
 10   |   127 (e)     |    0 (f)      |             5 (g)           |
      +-+-+-+-+-+-+-+-+-+-+-+-+-+-+-+-+-+-+-+-+-+-+-+-+-+-+-+-+-+-+-+
 11   |  Token Bucket Rate [r] (32-bit IEEE floating point number)  |        SENDER_TSPEC
      +-+-+-+-+-+-+-+-+-+-+-+-+-+-+-+-+-+-+-+-+-+-+-+-+-+-+-+-+-+-+-+
 12   |  Token Bucket Size [b] (32-bit IEEE floating point number)  |
      +-+-+-+-+-+-+-+-+-+-+-+-+-+-+-+-+-+-+-+-+-+-+-+-+-+-+-+-+-+-+-+
 13   |  Peak Data Rate [p] (32-bit IEEE floating point number)     |
      +-+-+-+-+-+-+-+-+-+-+-+-+-+-+-+-+-+-+-+-+-+-+-+-+-+-+-+-+-+-+-+
 14   |  Minimum Policed Unit [m] (32-bit integer)                  |
      +-+-+-+-+-+-+-+-+-+-+-+-+-+-+-+-+-+-+-+-+-+-+-+-+-+-+-+-+-+-+-+
 15   |  Maximum Packet Size [M]  (32-bit integer)                  |
      +-+-+-+-+-+-+-+-+-+-+-+-+-+-+-+-+-+-+-+-+-+-+-+-+-+-+-+-+-+-+-+
```

```
(a) - Message format version number (0)
(b) - Overall length (7 words not including header)
(c) - Service header, service number 1 (default/global information)
(d) - Length of service 1 data, 6 words not including header
(e) - Parameter ID, parameter 127 (Token_Bucket_TSpec)
(f) - Parameter 127 flags (none set)
(g) - Parameter 127 length, 5 words not including header
```

FIGURE 2.10 An example of a PATH IE structure.

to release resources along the paths. Concerning an active incoming flow, as soon as the new AP receives the PATH message from the network, it immediately creates a RESV state and responds with the respective RESV message without waiting for the MN to issue it, provided that the specific flow was accepted during admission control. In a different case, either the PATH message is ignored or a RESV_Tear message is issued to release the resources along the path.

According to this procedure, the new AP does not have to wait for the MN to send the required RSVP messages after reassociation. Additionally, assuming the MN is aware of the IAPP usage to transfer the RSVP context to the new AP, it can omit sending the PATH and RESV messages (messages 10 and 11, respectively) for the active flows and just wait for reestablishment of the paths. In this case, for an outgoing flow the MN can start transmitting data as soon as it receives the RESV message from the new AP, whereas for an incoming flow, the MN can simply wait to receive data packets. In this way, signaling overhead and reestablishment delays are reduced during handover. Figure 2.12 and Figure 2.13 present examples of a standard and proposed signaling sequence for an outgoing session. In Figure 2.12, the IAPP is used to transfer only authentication information; in Figure 2.13 it transfers RSVP information as well. The proposed scheme can also work together with a partial-path reestablishment solution like Paskalis et al.[25] and Moon and Aghvami.[26] In these cases, the new AP communicates with the crossover router instead of the CN. A side effect of the proposed scheme is that it transmits the full set of RSVP traffic and QoS information for active flows to the new AP inside a single message (MOVE-response). This allows for

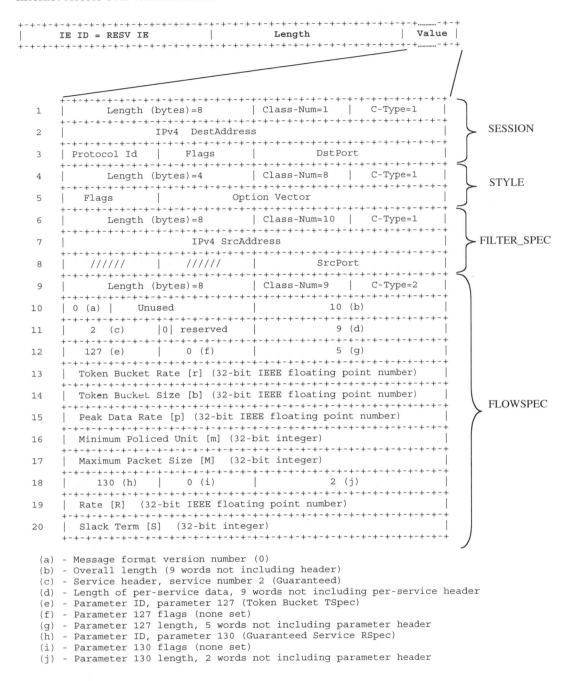

```
+-+-+-+-+-+-+-+-+-+-+-+-+-+-+-+-+-+-+-+-+-+-+-+-+-+-+-+-+-+........-+-+
|        IE ID = RESV IE        |            Length           | Value |
+-+-+-+-+-+-+-+-+-+-+-+-+-+-+-+-+-+-+-+-+-+-+-+-+-+-+-+-+-+........-+-+
```

```
     +-+-+-+-+-+-+-+-+-+-+-+-+-+-+-+-+-+-+-+-+-+-+-+-+-+-+-+-+-+-+-+-+
 1   |        Length (bytes)=8       |  Class-Num=1  |   C-Type=1    |   ⎫
     +-+-+-+-+-+-+-+-+-+-+-+-+-+-+-+-+-+-+-+-+-+-+-+-+-+-+-+-+-+-+-+-+   |
 2   |                    IPv4  DestAddress                          |   ⎬ SESSION
     +-+-+-+-+-+-+-+-+-+-+-+-+-+-+-+-+-+-+-+-+-+-+-+-+-+-+-+-+-+-+-+-+   |
 3   | Protocol Id |     Flags       |            DstPort            |   ⎭
     +-+-+-+-+-+-+-+-+-+-+-+-+-+-+-+-+-+-+-+-+-+-+-+-+-+-+-+-+-+-+-+-+
 4   |        Length (bytes)=4       |  Class-Num=8  |   C-Type=1    |   ⎫
     +-+-+-+-+-+-+-+-+-+-+-+-+-+-+-+-+-+-+-+-+-+-+-+-+-+-+-+-+-+-+-+-+   ⎬ STYLE
 5   |    Flags    |             Option Vector                       |   ⎭
     +-+-+-+-+-+-+-+-+-+-+-+-+-+-+-+-+-+-+-+-+-+-+-+-+-+-+-+-+-+-+-+-+
 6   |        Length (bytes)=8       |  Class-Num=10 |   C-Type=1    |   ⎫
     +-+-+-+-+-+-+-+-+-+-+-+-+-+-+-+-+-+-+-+-+-+-+-+-+-+-+-+-+-+-+-+-+   |
 7   |                    IPv4  SrcAddress                           |   ⎬ FILTER_SPEC
     +-+-+-+-+-+-+-+-+-+-+-+-+-+-+-+-+-+-+-+-+-+-+-+-+-+-+-+-+-+-+-+-+   |
 8   |   //////    |   //////    |            SrcPort                |   ⎭
     +-+-+-+-+-+-+-+-+-+-+-+-+-+-+-+-+-+-+-+-+-+-+-+-+-+-+-+-+-+-+-+-+
 9   |        Length (bytes)=8       |  Class-Num=9  |   C-Type=2    |   ⎫
     +-+-+-+-+-+-+-+-+-+-+-+-+-+-+-+-+-+-+-+-+-+-+-+-+-+-+-+-+-+-+-+-+   |
10   | 0 (a) |    Unused           |            10 (b)             |   |
     +-+-+-+-+-+-+-+-+-+-+-+-+-+-+-+-+-+-+-+-+-+-+-+-+-+-+-+-+-+-+-+-+   |
11   |   2 (c)   |0| reserved      |             9 (d)             |   |
     +-+-+-+-+-+-+-+-+-+-+-+-+-+-+-+-+-+-+-+-+-+-+-+-+-+-+-+-+-+-+-+-+   |
12   |  127 (e)  |   0 (f)         |             5 (g)             |   |
     +-+-+-+-+-+-+-+-+-+-+-+-+-+-+-+-+-+-+-+-+-+-+-+-+-+-+-+-+-+-+-+-+   |
13   | Token Bucket Rate [r] (32-bit IEEE floating point number)    |   |
     +-+-+-+-+-+-+-+-+-+-+-+-+-+-+-+-+-+-+-+-+-+-+-+-+-+-+-+-+-+-+-+-+   |
14   | Token Bucket Size [b] (32-bit IEEE floating point number)    |   |
     +-+-+-+-+-+-+-+-+-+-+-+-+-+-+-+-+-+-+-+-+-+-+-+-+-+-+-+-+-+-+-+-+   |
15   | Peak Data Rate [p] (32-bit IEEE floating point number)       |   ⎬ FLOWSPEC
     +-+-+-+-+-+-+-+-+-+-+-+-+-+-+-+-+-+-+-+-+-+-+-+-+-+-+-+-+-+-+-+-+   |
16   | Minimum Policed Unit [m] (32-bit integer)                    |   |
     +-+-+-+-+-+-+-+-+-+-+-+-+-+-+-+-+-+-+-+-+-+-+-+-+-+-+-+-+-+-+-+-+   |
17   | Maximum Packet Size [M]  (32-bit integer)                    |   |
     +-+-+-+-+-+-+-+-+-+-+-+-+-+-+-+-+-+-+-+-+-+-+-+-+-+-+-+-+-+-+-+-+   |
18   |  130 (h)  |    0 (i)       |             2 (j)             |   |
     +-+-+-+-+-+-+-+-+-+-+-+-+-+-+-+-+-+-+-+-+-+-+-+-+-+-+-+-+-+-+-+-+   |
19   | Rate [R]  (32-bit IEEE floating point number)                |   |
     +-+-+-+-+-+-+-+-+-+-+-+-+-+-+-+-+-+-+-+-+-+-+-+-+-+-+-+-+-+-+-+-+   |
20   | Slack Term [S]  (32-bit integer)                             |   ⎭
     +-+-+-+-+-+-+-+-+-+-+-+-+-+-+-+-+-+-+-+-+-+-+-+-+-+-+-+-+-+-+-+-+
```

(a) - Message format version number (0)
(b) - Overall length (9 words not including header)
(c) - Service header, service number 2 (Guaranteed)
(d) - Length of per-service data, 9 words not including per-service header
(e) - Parameter ID, parameter 127 (Token Bucket TSpec)
(f) - Parameter 127 flags (none set)
(g) - Parameter 127 length, 5 words not including parameter header
(h) - Parameter ID, parameter 130 (Guaranteed Service RSpec)
(i) - Parameter 130 flags (none set)
(j) - Parameter 130 length, 2 words not including parameter header

FIGURE 2.11 An example of PATH IE structure.

advanced admission-control algorithms in the new AP in order to decide which flows can be accepted in case bandwidth is insufficient. This is not possible in the regular case, where path reestablishment requests arrive sporadically from the MN after reassociation.

2.6 IP Security over WLANs

The problem with security in WLANs comes as a direct result of their nature to transmit in the air. Traditional LANs share a single medium (copper cable, fiber optics, etc.) and passive hubs or routers.

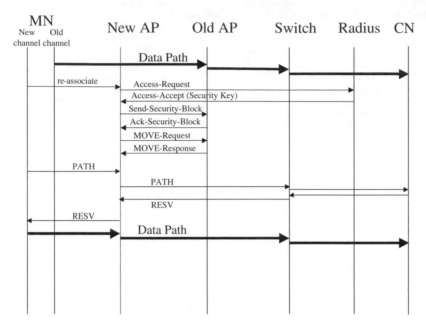

FIGURE 2.12 The standard message exchange for outgoing sessions.

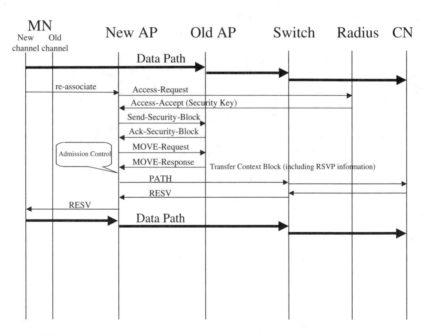

FIGURE 2.13 An IAPP-assisted RSVP message exchange for outgoing sessions.

Hub ports and cable taps are almost always located within a facility with some physical security that makes it difficult for unauthorized users to have access to them. On the other hand, WLANs share a broadcast medium in free space, which almost certainly covers areas outside the physical control of WLAN administrators, such as a company parking lot, other floors of the facility, or nearby high-rise buildings, making them vulnerable to interceptions.

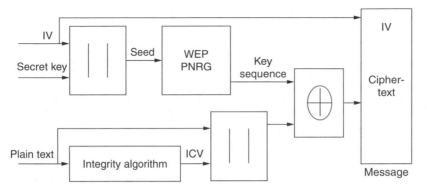

FIGURE 2.14　WEP operation.

Interference, on the other hand, is also easy in WLANs, especially when WLANs operate in public frequencies such as 2.4 or 5 GHz. A simple jamming transmitter within close distance can block all data communications. For example, consistently hammering an AP with access requests, whether successful or not, will eventually exhaust its available radio-frequency spectrum. Other wireless technologies in the same frequency range can also reduce the range and usable bandwidth of a WLAN. Bluetooth, for example, which is used to communicate between handsets and other information appliances, is one of many technologies today that use the same radio frequency as WLAN devices. These intentional or unintentional Denial-of-Service (DoS) attacks can make WLAN devices practically unusable.

2.6.1　Security in 802.11

The inability of 802.11 to provide a complete framework for secure transmission over the air was one of the basic reasons for delaying its widespread adoption in the past few years. The foundation of the security of 802.11 is based on a frame encryption protocol called Wired Equivalent Privacy (WEP). WEP is a framework that defines an operation mode for the RC4 algorithm, and it specifies how to use the Initialization Vector (IV), how to work out a key stream, and how to encapsulate encrypted information. Both encryption and authentication are based on WEP. Many administrators use WEP keys to configure rudimentary wireless encryption. These keys come in two sizes: 40-bit and 128-bit. In the next few paragraphs we discuss the problems with WEP in some detail. As described in Casole,[27] the IV, which has a length of 24 bits, is concatenated to the secret key (40 or 104 bits), thus resulting in a seed for the WEP Pseudo-Random Number Generator (PRNG) of 64 bits. The WEP PRNG is based on the RC4 algorithm.[28] The output of the WEP PRNG is a key sequence of the same length as the text to be encrypted, given by the actual plaintext and an Integrity Check Value (ICV) corresponding to a Cyclic Redundancy Check (CRC-32) of the plaintext. The key sequence and the text to be encrypted are then *ex-ored*; the result of this operation is sent over the medium, concatenated with the IV (Figure 2.14). The standard specification suggests changing the IV for every PDU to be sent over the air, but no clear directions are given. This means that every implementation can choose its own method, which often opens the implementation to many security threats.

As for authentication, IEEE 802.11 defines two methods:

- The *open system authentication* method is in fact a NULL authentication method, based on the transmission of the identity from the MN being authenticated to the MN performing authentication. The MN performing authentication returns the authentication result, which, if successful, means that the two MNs are mutually authenticated.
- The *preshared key authentication* mechanism is a method based on challenge and response. For example, MN A starts by sending an authentication request and its MN identifier to MN B, who replies with an authentication message containing a random text of 128 bits (the challenge). Then, MN A copies the challenge in a new message, encrypts it with the shared WEP key, and sends it

to MN B (response). MN B decrypts the text and replies with the outcome of the authentication procedure, depending on whether the received text is identical to the initial. It is clear that this method requires that the common shared key have been previously delivered to the participating MNs through a secure channel that is independent of IEEE 802.11. Note that the key used for authentication is the same used for encryption, making the security framework considerably vulnerable.

The standard makes it possible to use up to four different global WEP keys, shared by all the MNs within a BSS; the same WEP keys are used for encryption and authentication by all MNs in a BSS. Because WLANs assume the possibility for a MN to move between different radio cells, this has led to a widely used poor practice of setting up the same WEP key in all cells belonging to the same infrastructure. The loss of this single key thus compromises the security of the whole network (referred to as a single point of risk). The standard considers the possibility to have per-user WEP keys, or better per-MAC address WEP keys, but currently only a few products implement this part of the standard because of the administration overhead.

A number of weaknesses have been recently discovered in the RC4 algorithm, as well as in the WEP framework. Fluhrer et al.[29] identify a large class of weak RC4 keys, weaknesses in the key scheduling algorithm, and many flaws in the Pseudo-Random Generator Algorithm (PRGA); furthermore, the deployment of the RC4 PRGA within the WEP framework turned out to be completely insecure, made even worse by the common practice of reusing the same IV for multiple frames. Other descriptions of WEP weaknesses and how an attack can be launched have been described in Borisov et al.[30] and Stubblefield et al.,[31] where an attack was implemented and the WEP key was discovered after 5 million packets.

2.6.2 802.11i

The 802.11i task group is the official attempt of the IEEE 802.11 group to fill the gaps in the 802.11 security framework. Work in 802.11i is currently in progress, and so far two new algorithms have been proposed for enhancing WEP encryption: the Temporal Key Integrity Protocol (TKIP) and the Wireless Robust Authenticated Protocol (WRAP).

The TKIP is a new protocol that tries to fix the known problems with WEP. TKIP uses the same ciphering kernel as WEP (RC4) but adds a number of functions:[32]

- A 28-bit encryption key
- A 48-bit Initialization Vector
- The New Message Integrity Code (MIC)
- Initialization Vector (IV) sequencing rules
- A per-packet key mixing algorithm that provides a RC4 seed for each packet
- Active countermeasures

The purpose of TKIP is to provide a fix for WEP for existing 802.11b products. It is believed that essentially all existing 802.11b products can be software-upgraded with TKIP. The TKIP MIC was designed with the constraint that it must run on existing 802.11 hardware. It does not offer very strong protection but was considered the best that could be achieved with the majority of legacy hardware. It is based on an algorithm called Michael that is a 64-bit MIC with 20-bit design strength.

The IV sequence is implemented as a monotonically incrementing counter that is unique for each key. This ensures that each packet is encrypted with a unique (key, IV) pair; in other words, it ensures that an IV is not reused for the same key. The receiver will also use the sequence counter to detect replay attacks. Because frames may arrive out of order due to traffic-class priority values, a replay window (16 packets) has to be used. A number of "weak" RC4 keys have been identified for which knowledge of a few RC4 seed bits makes it possible to determine the initial RC4 output bits to a non-negligible probability. This makes it easier to crypt-analyze data encrypted under these keys. The per-packet mixing function is designed to defeat weak-key attacks. In WEP, the IV and the key are concatenated and then used as seed to RC4. In TKIP, the cryptographic per-packet mixing function combines the key and the IV into a seed for RC4.

Because the TKIP MIC is relatively weak, TKIP uses countermeasures to compensate for this. If the receiver detects a MIC failure, the current encryption and integrity protection keys will not be used again. To allow a follow-up by a system administrator, the event should be logged. The rate of MIC failure must also be kept below one per minute, which means that new keys should not be generated if the last key update due to a MIC failure occurred less than a minute ago. To minimize the risk of false alarms, the MIC should be verified after the CRC, IV, and other checks have been performed. Through these enhancements, TKIP tries to address all known vulnerabilities of WEP. Nevertheless, TKIP is an interim solution for supporting 802.11i on legacy hardware. It is not considered as secure as WRAP but much better than WEP.

WRAP, on the other hand, uses the Advanced Encryption Standard (AES)[33] and Offset Codebook (OCB),[34] and its operation is based on the Rijndael cipher. It is much stronger than WEP, uses different encryption (per-user) keys for received and transmitted data, and allows an efficient parallel implementation in both software and hardware. Furthermore, the WRAP encapsulation technique protects the integrity of the transmitted data.

More specifically, the WRAP privacy consists of three parts:

1. **The key derivation procedure.** Once an association is established and a temporal key for the association is configured with the use of 802.1X, the 802.11 MAC uses the key-derivation algorithm to derive a cryptographic key from the (Re)association Request and Response. This will produce the key used to protect the association data. Note that the above procedure requires 802.1X authentication and key management.
2. **The encapsulation procedure.** Once the key has been derived and its associated state initialized, the 802.11 MAC uses the WRAP encapsulation algorithm with the key and the state to protect all unicast MSDUs it sends to an associated station.
3. **The decapsulation procedure.** Similarly, once the key has been derived and its associated state initialized, the 802.11 MAC uses the WRAP decapsulation algorithm with the receive key and state to decapsulate all unicast MSDUs received from an associated station. Once the key is established, the MAC discards any MSDUs received over the association that are unprotected by the encapsulation algorithm.

IEEE 802.1X may also assign a broadcast/multicast key. The implementation uses this key as configured, without derivation. The MAC utilizes the broadcast/multicast key to protect all broadcast/multicast MSDUs it sends and discards any broadcast/multicast MSDUs received that are not protected by this key. Note that AES requires hardware support, meaning that the majority of legacy 802.11b products will not be able to run WRAP. You can learn more about security in WLANs in Chapter 9.

References

1. IEEE Standard for Wireless LAN Medium Access Control (MAC) and Physical Layer (PHY) Specifications, IEEE Stds. Dept. P802.11, Nov. 1997.
2. Crow, B.P., Widjaja, I., Kim, J.G., and Sakai, P.T., IEEE 802.11 wireless local area networks, *IEEE Commn. Mag.*, Sep. 1997.
3. Khun-Jush, J., Malmgren, G., Schramm, P., and Tornster, J., "HIPERLAN Type 2 for Broadband Wireless Communication," *Ericsson Rev.*, 2, 2000.
4. Mähönen, P., Saarinen, T., Passas, N., Orphanos, G., Muñoz, L., García, M., Marshall, A., Melpignano, D., Inzerilli, T., Lucas, F., and Vitiello, M., Platform-independent IP transmission over wireless networks: the WINE approach, *IEEE P. Commn. Mag.*, 8, 6, Dec. 2001.
5. Montenegro, G., Dawkins, S., Kojo, M., Magret, V., and Vaidya, N., Long thin networks, RFC 2757, Jan. 2000.
6. Border, J., Kojo, M., Griner, J., Montenegro, G., and Shelby, Z., Performance enhancing proxies, Internet Draft (work in progress), draft-ietf-pilc-pep-04.txt, Oct. 2000.
7. Laukkanen, A., IP to wireless convergence interface, BRAIN London workshop, Nov. 2000.

8. Braden, R. et al., Integrated services in the Internet architecture: an overview, RFC 1633, June 1994.
9. Grossman, D. et al., New terminology and clarifications for DiffServ, RFC 3260, Apr. 2002.
10. Braden, R. et al., Resource ReSerVation Protocol (RSVP) version 1 functional specification, RFC 2205, Sep. 1997.
11. Mirhakkak, M., Schult, N., and Thomson, D., Dynamic bandwidth management and adaptive applications for a variable bandwidth wireless environment, *IEEE J. Sel. Areas Commn.*, 19, 10, Oct. 2001.
12. Passas, N., Zervas, E., Hortopan, G., and Merakos, L., Providing for QoS maintenance in wireless IP environments, Global Communications Conference (GLOBECOM) 2003, San Francisco, Dec. 2003.
13. Ni, Q., Romdhani, L., Turletti, T., and Aad, I., QoS issues and enhancements for IEEE 802.11 wireless LAN, INRIA, TR4612, Nov. 2002.
14. Mangold, S., Choi, S., May, P., Klein, O., Hiertz, G., and Stibor, L., IEEE 802.11e Wireless LAN for Quality of Service (invited paper), in Proc. of the European Wireless, 1, Feb. 2002, pp. 32–39, Florence, Italy.
15. Perkins, C., Ed., IP mobility support, Internet RFC 2002, Oct. 1996.
16. Valko, A.G., Cellular IP — a new approach to Internet host mobility, *ACM Comput. Commn. Rev.*, Jan. 1999.
17. Campbell, A.T., Gomez, J., and Valko, A.G., An overview of Cellular IP, IEEE Wireless Communications and Networking Conference (WCNC'99), New Orleans, Sep. 1999.
18. Ramjee, R., Varadhan, K., Salgarelli, L., Thuel, S., Wang, S. Y., and La Porta, T., HAWAII: a domain-based approach for supporting mobility in wide-area wireless networks, IEEE/ACM Transactions on Networking, June 2002.
19. Soliman, H., Castelluccia, C., El-Malki, K., and Bellier, L., Hierarchical MIPv6 mobility management (HMIPv6), Internet Draft (work in progress), Oct. 2002. http://www.ietf.org/internet-drafts/draft-ietf-mobileip-hmipv6-07.txt.
20. Draft Recommended Practice for Multi-Vendor Access Point Interoperability via an Inter-Access Point Protocol Across Distribution Systems Supporting IEEE 802.11 Operation, IEEE Std. 802.11f/D3.1, Apr. 2002.
21. Talukdar, A. et al., MRSVP: a resource reservation protocol for an integrated services network with mobile hosts," *J. Wireless Networks*, 7, 1, 2001.
22. Tseng, C.-C. et al., HMRSVP: a hierarchical mobile RSVP protocol, in Proc. International Workshop on Wireless Networks and Mobile Computing (WNMC) 2001, Valencia, Spain, Apr. 2001.
23. Kuo, G.-S. and Ko, P.-C., Dynamic RSVP for mobile IPv6 in wireless networks, in Proc. IEEE Vehicular Technology Conference (VTC) 2000, Tokyo, Japan, May 2000.
24. Chen, W.-T. and Huang, L.-C., RSVP mobility support: a signalling protocol for integrated services internet with mobile hosts, in Proc. INFOCOM 2000, Tel Aviv, Israel, Mar. 2000.
25. Paskalis, S. et al., An efficient RSVP/Mobile IP interworking scheme, *ACM Mobile Networks J.*, 8, 3, June 2003.
26. Moon, B. and Aghvami, A.H., Reliable RSVP path reservation for multimedia communications under IP micromobility scenario, *IEEE Wireless Commn. Mag.*, Oct. 2002.
27. Casole, M., WLAN Security — Status, Problems and Perspective, in Proc. of the European Wireless, 2002, Florence, Italy, Feb. 2002.
28. Schneier, B., *Applied Cryptography: Protocols, Algorithms, and Source Code in C, 2nd ed.*, Wiley: New York, 1996.
29. Fluhrer, S., Mantin, I., and Shamir, A., Weaknesses in the key scheduling algorithm of RC4, 8th Annual Workshop on Selected Areas in Cryptography, Toronto, Aug.16–21, 2001.
30. Borisov, N., Goldberg, I., and Wagner, D., Intercepting mobile communications: the insecurity of 802.11, 7th Annual International Conference on Mobile Computing and Networking, July 16–17, 2001, ACM.

31. Stubblefield, A., Ioannidis, J., and Rubin, A.D., Using the Fluhrer, Mantin and Shamir attack to break WEP, AT&T Labs, 2001.

32. Ericsson, Introduction of IEEE 802.11 Security, 3GPP TSG SA WG3 Security Meeting, Helsinki, Finland, July 2002, available at http://www.3gpp.org/ftp/tsg_sa/WG3_Security/2002_meetings/ TSGS3_24_Helsinki/Docs/PDF/S3-020341.pdf.

33. Advanced Encryption Standard (AES), Federal Information Processing Standards Publication 197, Nov. 2001.

34. Rogaway, P., Bellare, M., Black, J., and Krovetz, T., OCB: a block-cipher mode of operation for e-client authenticated encryption, 8th ACM Conference on Computer and Communications Security (CCS-8), Philadelphia, Nov. 2001.

3

Internet Access over GPRS[1]

Apostolis K. Salkintzis
Motorola, Inc.

3.1 Introduction

In this chapter, we discuss how the *General Packet Radio Service (GPRS)* technology is used for providing mobile/wireless access to the Internet. We explain the fundamental GPRS concepts, protocols, and procedures and demonstrate the main functionality provided by the GPRS network. The key procedures examined are the registration procedure, the routing/tunneling procedure, and the mobility management procedure, all of which enable mobile/wireless IP sessions.

GPRS is a bearer service of the Global System for Mobile (GSM) communications, which offers packet data capabilities. The key characteristic of the data service provided by GPRS is that it operates in *end-to-end packet mode*. This means that no communication resources are exclusively reserved for supporting the communication needs of every individual mobile user. On the contrary, the communication resources are utilized on a demand basis and are statistically multiplexed between several mobile users. This characteristic renders GPRS ideal for applications with irregular traffic properties (such as Web browsing), because, with this type of traffic, the benefits of statistical multiplexing are exploited; that is, we obtain high utilization efficiency of the communication resources. A direct effect of this property is the drastically increased capacity of the system in the sense that we can support a large number of mobile users with only a limited amount of communication resources. The increased capacity offered by GPRS, combined with the end-to-end packet transfer capabilities, constitute the main factors that drive the use of GPRS in providing wide-area wireless Internet access.

In this chapter, we investigate the key operational and conceptual aspects of GPRS, and we demonstrate how it is used to provide wide-area wireless access to the Internet and other IP-based networks. We start our discussion with an introduction to GPRS technology and the necessary terminology. Further tutorial material that explains several GPRS aspects can be found in references 1 and 3 to 6. Another comprehensive resource is the 3GPP TS 22.060 specification[7] and the 3GPP TS 23.060,[8] available online at

---

[1]Parts of this chapter are used with permission from Dixit, S. and Prasad, R., *Wireless IP and Building the Mobile Internet*, ISBN 1-58053-354-X, Artech House, 2002.

www.3gpp.org/ftp/specs. In addition, reference 2 discusses some interesting issues related to the simultaneous provisioning of voice and packet data capabilities in a GSM/GPRS network.

3.2   GPRS Overview

In general, a GPRS network can be viewed as a special IP network, which offers IP connectivity to IP terminals on the go. To provide such a mobile connectivity service, the GPRS network must feature additional functionality compared with standard IP networks. However, from a high-level point of view, the GPRS network resembles a typical IP network in the sense that it provides typical IP routing and interfaces to the external world through one or more IP routers.

Figure 3.1 shows schematically this high-level conceptual view of a GPRS network. By using shared radio resources, mobile users gain access to remote Packet Data Networks (PDNs) through a remote access router, which in GPRS terminology is termed *Gateway GPRS Support Node (GGSN)*. You can think of access to a remote PDN as being similar to a typical dial-up connection. Indeed, as discussed in Section 3.3, a user establishes a virtual connection to the remote PDN. However, with GPRS a user may "dial up" to many remote PDNs simultaneously and can be charged by the volume of the transferred data, not by the duration of a connection.

GPRS can offer both *transparent* and *nontransparent* access to a PDN. With transparent access, the user is not authenticated by the remote PDN, and he or she is assigned an IP address (private or public) from the address space of the GPRS network. On the other hand, with nontransparent access the user's credentials are sent to the remote PDN and the user is permitted to access this PDN only if he or she is successfully authenticated. In this case, the user is typically assigned an IP address (private or public) from the address space of the PDN he or she is accessing. Note that, irrespective of the type of access to a PDN, a user is always authenticated by the GPRS network before being permitted access to GPRS services (this is further discussed in Section 3.2.3). The nontransparent access is particularly useful for accessing secure intranets (e.g., corporate networks) or Internet Service Providers (ISPs), whereas the transparent access is most appropriate for users who do not maintain subscriptions to third-party ISPs or intranets. As illustrated in Figure 3.1, the GPRS network forms an individual subnet, which (from an address-allocation point of view) contains all users who use transparent access to remote PDNs. External PDNs perceive this subnet as being a typical IP network.

Figure 3.2 illustrates some more detailed aspects of a GPRS network. On the left, it shows a Mobile Station (MS), and on the right we see the Gateway GPRS Serving Node (GGSN). Among other things, the GGSN offers IP routing functionality, and it is used for interfacing with external IP networks (also referred to as PDNs). From the MS point of view, the GGSN can be thought as a remote access router. Note that, in general, the GGSN may interface not only with IP PDNs but also with several other types of PDNs — for example, with X.25 networks.[7,8] However, in practice GPRS is commonly used to provide access primarily to IP PDNs, and consequently in this chapter we focus on IP only. Therefore, unless otherwise indicated, we assume that GPRS interfaces with IP PDNs only, such as the Internet.

3.2.1   GPRS Bearers

As illustrated in Figure 3.2, the GPRS network effectively provides a GPRS bearer — that is, it provides a communication channel with specific attributes between the MS (the terminal) and the GGSN (the router). Over the GPRS bearer, the MS may send IP packets to the GGSN, and it may receive IP packets from the GGSN. As explained below, the GPRS bearer is dynamically set up at the beginning of an IP session (when the user "dials" to a specific PDN), and it can be tailored to match the specific requirements of an application. In other words, it can be set up with specific Quality of Service (QoS) attributes, such as delay, throughput, precedence, and reliability.[8]

Figure 3.2 also illustrates the internal structure of a GPRS bearer, which includes the protocols and the GPRS nodes involved in the provisioning of this bearer. Here is a brief explanation: The MS communicates through the radio interface (the so-called *Um* reference point) with a *Base Transceiver Station*

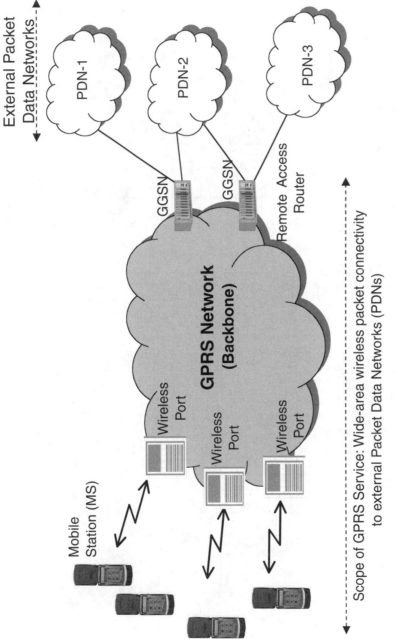

FIGURE 3.1 High-level conceptual view of a GPRS network.

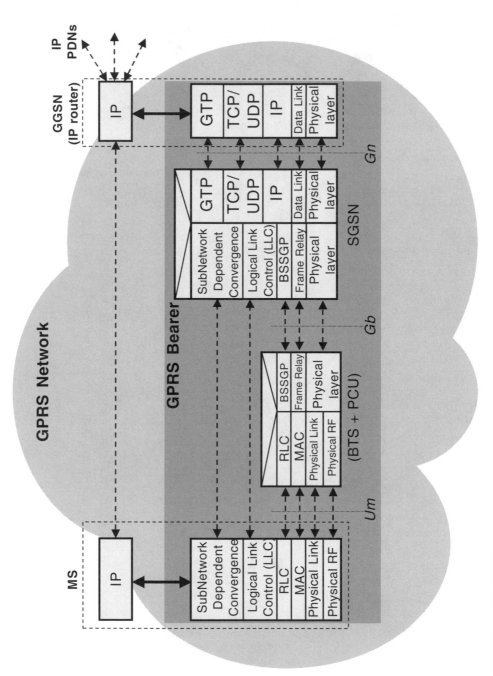

FIGURE 3.2   The GPRS bearer service.

(BTS), which provides mainly physical-layer functionality. In GPRS, the BTS handles the transmission and the reception of packet data on the GPRS physical channels. Data received by the BTS is processed (e.g., decoded and de-interleaved) and then relayed to the next hierarchical node in the GPRS architecture — that is, to the *Packet Control Unit (PCU)*. The PCU offers radio resource management and is responsible for allocating uplink and downlink resources to the various MSs on a demand basis. As we discuss later, the radio resource allocation is implemented with a packet-scheduling function that takes into account the QoS committed to each active MS.

The PCU communicates with the *Serving GPRS Support Node (SGSN)* over a frame relay interface (Gb). The SGSN provides mobility management functionality, session management, packet scheduling on the downlink, and packet routing/tunneling. The interface between the SGSN and the GGSN (Gn) is entirely based on IP, typically on IPv4. The GGSN provides mainly routing and optionally screening functionality and can be considered to be a remote access router interfacing with the external PDNs. The fact that we have two IP layers within the GGSN implies that some sort of IP-to-IP tunneling is applied across the Gn interface. This is discussed in more detail in Section 3.4.

Not all GPRS bearers feature the same attributes. The particular attributes of a GPRS bearer are specified mainly by the operational mode of each protocol and by the level of precedence applied in the scheduling procedures. For example (see Figure 3.2), in one GPRS bearer, the Logical Link Control (LLC) protocol may operate in acknowledged mode, whereas in another GPRS bearer it may operate in unacknowledged mode. By definition, the acknowledged mode of operation offers increased reliability compared with the unacknowledged mode of operation. Similar distinctions among different GPRS bearers may apply to the Radio Link Control (RLC) protocol and to the GPRS Tunneling Protocol (GTP). In addition, one GPRS bearer, which is given high precedence in the scheduling procedures, would typically feature lower delays compared with another GPRS bearer, which is given lower precedence in the scheduling procedures.

3.2.2   GPRS Protocols

Let us now examine the GPRS protocols shown in Figure 3.2. The *Subnetwork Dependent Convergence Protocol (SNDCP)* runs between the MS and the SGSN, and it is specified in reference 17. It is the first layer that receives the user IP datagrams for transmission. SNDCP basically provides (1) acknowledged and unacknowledged transport services, (2) compression of TCP/IP headers (conformant to RFC 1144; see reference 15), (3) compression of user data (conformant to either V.42bis [reference 16] or V.44), (4) datagram segmentation/reassembly, and (5) PDP context multiplexing (see Section 3.4). The segmentation/reassembly function ensures that the length of data units sent to LLC layer does not exceed a maximum prenegotiated value. For example, when this maximum value is 500 octets, then IP datagrams of 1500 octets will be segmented into three SNDCP data units. Each one will be transmitted separately and reassembled by the receiving SNDCP layer.

As we will learn in Section 3.3, a PDP context essentially represents a virtual connection between an MS and an external PDN. The PDP context multiplexing is a function that (1) routes each data unit received on a particular PDP context to the appropriate upper layer and (2) routes each data unit arriving from an upper layer to the appropriate PDP context. For example, let us assume a situation where the MS has set up two PDP contexts, both with type IP but with different IP addresses. One PDP context could be linked to a remote ISP, and the other could be linked to a remote corporate network. In this case, there are two different logical interfaces at the bottom of the IP layer, one for each PDP context. The SNDCP layer is the entity that multiplexes data to and from those two logical interfaces.

The *Logical Link Control (LLC)* protocol runs also between the MS and the SGSN, and it is specified in reference 12. LLC basically provides data-link services, as specified in the Open System Interconnection (OSI) seven-layer model.[18] In particular, LLC provides one or more separate logical links (LLs) between the MS and the SGSN, which can be broken down into *user*-LLs (used to carry user data) and *control*-LLs (used to carry signaling). There can be up to four simultaneous user-LLs, whereas there are basically three control-LLs: one for exchanging GPRS mobility management and session management signaling, another to support the Short Message Service (SMS),[19] and a third to support Location Services (LCS).[20] A user-LL is established dynamically, via the PDP context activation procedure (see Section 3.3), and its properties are negotiated between the MS and the SGSN during the establishment phase. Negotiated properties typically include (1) the data transfer mode (acknowledged versus unacknowledged), (2) the maximum length of transmission units, (3) timer values, (4) flow control parameters, and so forth. A part of these properties defines the QoS that will be provided by the user-LL. On the other hand, the control-LLs have predefined properties, and they are automatically set up right after the MS registers to the GPRS network (see Section 3.2.3). Note that one user-LL can carry user data pertaining to one or more PDP contexts, all sharing the same QoS.

Control-LLs operate only in unacknowledged mode, which basically provides an unreliable transport service. On the other hand, user-LLs operate either in unacknowledged mode or in acknowledged mode, depending on the reliability requirements. The latter mode provides reliable data transport by (1) detecting and retransmitting erroneous data units, (2) maintaining the sequential order of data units, and (3) providing flow control.

Another service provided by the LLC layer is ciphering. This service can be provided in both acknowledged and unacknowledged modes of operation and, therefore, all LLs can be secured and protected from eavesdropping.

The *Radio Link Control (RLC)* and *Medium Access Control (MAC)* protocols run between the MS and the PCU, and they are specified in reference 13. The RLC makes available the procedures for unacknowledged or acknowledged operation over the radio interface. It also provides segmentation and reassembly of LLC data units into *fixed-size* RLC/MAC blocks. In RLC acknowledged mode of operation, the RLC also provides error-correction procedures that enable the selective retransmission of unsuccessfully delivered RLC/MAC blocks. Additionally, in this mode of operation, the RLC layer preserves the order of higher-layer data units supplied to it. Note that whereas LLC provides transport services between the MS and the SGSN, the RLC offers similar transport services between the MS and the PCU.

The MAC layer implements the procedures that enable multiple MSs to share a common radio resource, which may consist of several physical channels. In particular, in the uplink direction (MS to network), the MAC layer provides the procedures, including contention resolution, for the arbitration among multiple MSs that simultaneously attempt to access the shared transmission medium. In the downlink direction (network to MS), the MAC layer supplies the procedures for queuing and scheduling of access attempts. More details are provided below.

The MAC function in the network maintains a list of active MSs, which are mobile stations with pending uplink transmissions (they have uplink data to transmit). These MSs have previously requested permission to content for uplink resources, and the network has responded positively to their request. Each active MS is associated with a set of committed QoS attributes, such as delay or throughput. These QoS attributes were negotiated when the MS requested uplink resources.

The main function of the MAC layer in the network is to implement a scheduling function (in the uplink direction), which successively assigns the common uplink resource to active MSs in a way that guarantees that each MS receives its committed QoS. A similar scheduling function is also implemented in the downlink direction.

From the previous discussion, it is obvious that every GPRS cell features a central authority (the PCU), which (1) arbitrates the access to common uplink resources by providing an uplink scheduling function, and (2) administers the transmission on the downlink resources by providing a downlink scheduling function. These scheduling functions are part of the functions required to guarantee the provisioning of QoS on the radio interface and are implementation dependent.

The *Base Station Subsystem GPRS Protocol (BSSGP)* runs across the Gb interface; it is specified in reference 21. BSSGP basically provides (1) unreliable transport of LLC data units between the PCU and the SGSN, and (2) flow control in the downlink direction. The flow control attempts to prevent the flooding of buffers in the PCU and to conform the transmission rate on Gb (from SGSN to PCU) to the transmission rate on the radio interface (from PCU to MS). Flow control in the uplink direction is not provided because it is assumed that uplink resources on the Gb interface are suitably dimensioned and are significantly greater than the corresponding uplink resources on the radio interface. BSSGP provides unreliable transport because the reliability of the underlying frame relay network is considered sufficient enough to meet the required reliability level on the Gb.

BSSGP also provides addressing services, which are used to identify a given MS in uplink and downlink directions, and a particular cell. In the downlink direction, each BSSGP data unit typically carries an LLC data unit, the identity of the target MS, a set of radio-related parameters (identifying the radio capabilities of the target MS), and a set of QoS attributes needed by the MAC downlink scheduling function. The identity of the target cell is specified by means of a BSSGP Virtual Channel Identifier (BVCI), which eventually maps to a frame relay virtual channel. In the uplink direction, each BSSGP data unit typically carries an LLC data unit, the identity of the source MS, the identity of the source cell, and a corresponding set of QoS attributes. The mobility management function in the SGSN uses the source cell identity to identify the cell where the source MS is located.

As shown in Figure 3.2, the *GPRS Tunneling Protocol (GTP)* runs between the SGSN and the GGSN. In general, however, GTP also runs between two SGSNs. GTP provides an unreliable data transport function (it usually runs on top of UDP) and a set of signaling functions primarily used for tunnel management and mobility management. The transport service of GTP is used to carry user-originated IP datagrams (or any other supported packet unit) into GTP tunnels. GTP tunnels are needed between the SGSN and the GGSN for routing purposes. (This is further explained in Section 3.4.) They are also necessary for correlating user-originated IP datagrams to PDP contexts. By means of this correlation, a GGSN decides how to treat an IP datagram received from an SGSN (e.g., to which external PDN to forward this datagram), and an SGSN decides how to treat an IP datagram received from a GGSN (what QoS mechanisms to apply to this datagram, to which cell to forward this datagram, and so forth).

Now that we have discussed the fundamental concepts and protocols of GPRS, let us investigate the typical procedures carried out to enable wireless IP sessions over GPRS. In particular, we describe the registration procedure, the routing/tunneling procedures, and the mobility management procedures.

3.2.3 The Attach Procedure

Before a mobile station can start a wireless IP session or any other packet data session over the GPRS network, it has to perform the registration procedure. In the GPRS specifications,[8] the registration procedure is formally referred to as an attach procedure. During this procedure, the mobile station is actually informing an SGSN that it wants to have access to the GPRS network, and at the same time it identifies its comprehensive set of capabilities. In response, the SGSN authenticates the mobile station, retrieves its subscription data, and checks whether it is authorized to have access to the GPRS network from its current routing area (i.e., one or more cells served by the same SGSN[8]). If none of the checks fails, the SGSN accepts the attach request of the mobile and it returns an accept message. After that, the SGSN becomes the *serving* SGSN of that particular mobile.

The entire attach procedure is schematically illustrated in Figure 3.3 in the form of a message sequence diagram. In step 1, the MS sends an *Attach Request* message to the SGSN (labeled as *new SGSN*), which serves the routing area where the mobile station is located. In the Attach Request message, the MS typically includes a temporary identifier, called the Packet Temporary Mobile Station Identity (P-TMSI). This P-TMSI has previously been allocated, possibly by another SGSN (e.g., the *old SGSN* shown in Figure 3.3) and, possibly, in another routing area. However, the P-TMSI is stored in the nonvolatile memory of the MS, and as long as it is valid, it is used as an MS identity. The use of a temporary identity instead of the permanent MS identity (i.e., the International Mobile Station Identity [IMSI]) provides user identity confidentiality.[9] As explained below, the GPRS network allocates a new P-TMSI value to the MS whenever appropriate. Along with the P-TMSI, the Attach Request includes the identity of the routing area where this P-TMSI was allocated, as well as information related to the MS capabilities — for example, supported frequency bands, multislot capabilities, and ciphering capabilities.

In step 2, the new SGSN tries to acquire the permanent MS identity — that is, its IMSI. If the P-TMSI included in the Attach Request message has previously been allocated by the new SGSN, then the new

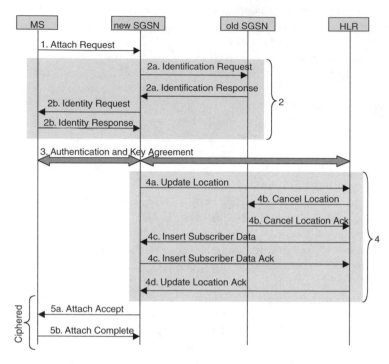

FIGURE 3.3 The GPRS attach procedure.

SGSN also knows the IMSI of the MS. However, in the example shown in Figure 3.3, we assume that the P-TMSI has previously been allocated by the old SGSN. Therefore, the new SGSN may try to contact the old SGSN and request the IMSI value that corresponds to the P-TMSI reported by the MS. This is accomplished in step 2a with the *Identification* messages exchanged between the two SGSNs. Note that the address of the old SGSN is derived by the new SGSN with the aid of the routing area identity (RAI) included in the Attach Request. The exact mapping between an RAI and an SGSN IP address is implementation specific and can typically be based on preconfigured mapping tables or DNS queries.

If the new SGSN cannot acquire the MS's IMSI value in step 2a (e.g., because the old SGSN has deleted the relevant information, or because the IP address of the old SGSN cannot be resolved), then the new SGSN requests that the MS send its permanent identity. This is accomplished in step 2b. The obvious drawbacks of this step are that it introduces additional signaling over the radio interface and that it compromises the user identity confidentiality (because the IMSI is transmitted unciphered on the radio interface).

In step 3, the authentication and key agreement procedure is executed. During this procedure, the new SGSN contacts a Home Location Register (HLR), which maintains the subscription data of the identified IMSI and requests from this HLR the authentication data required to authenticate the MS. The address of the appropriate HLR is derived by translating the routing information contained in IMSI value. Note that an HLR is typically accessible over the international SS7 network and either the Message Transfer Part (MTP) transport or IP transport can be used for SS7 signaling.[14] The authentication and key agreement procedure is identical to the one used in GSM, and more details about it can be found in the GSM technical specifications[9] as well as in other references, such as reference 10 or 11. Typically, after the authentication and key agreement procedure, ciphering is enabled on the radio interface, and therefore, further messages transmitted on this interface are enciphered. In GPRS, the ciphering function is performed at the LLC layer. Further details can be found in the LLC specification.[12]

In step 4 (see Figure 3.3), the new SGSN tries to update the HLR database with the new location of the MS. For this purpose, it sends an *Update Location* message to the HLR containing its own IP address, its own SS7 address, and also the IMSI value of the MS. Subsequently, the HLR informs the old SGSN that it can now release any information stored for this MS. This is done with the *Cancel Location* message. Typically, when the old SGSN receives this message, it will release the previously allocated P-TMSI for this MS (and make it available for reallocation), delete any other information possibly stored for this MS, and respond with a *Cancel Location Ack* message. In step 4c, the HLR sends to the new SGSN the GPRS subscription data of the MS. At this point, the new SGSN may perform several inspections; for example, it may check whether the MS is allowed to roam in its current routing area. If none of the checks fails, then the new SGSN builds up a GPRS Mobility Management (GMM) context for this MS and returns a positive acknowledgment to the HLR. On the other hand, if an inspection routine fails (e.g., due to roaming restrictions), the new SGSN sends a negative response to the HLR and subsequently it sends an *Attach Reject* message to the MS, including the specific reason for rejecting the attach request. The GMM context can be considered a database record that holds GPRS mobility management information pertaining to a specific MS. Such information includes the IMSI value of the MS, the routing area and the cell where the MS is currently located, the P-TMSI allocated to the MS, the ciphering algorithm used to encipher packets for this MS, the GPRS capabilities of the MS, and data that can be used to authenticate the MS in the future. In step 4d, the HLR acknowledges the location update that was requested earlier in step 4a.

In step 5a, the new SGSN sends an Attach Accept message to the MS to indicate that the MS has successfully been registered for GPRS services. Typically, with the Attach Accept message, the new SGSN assigns a new P-TMSI value to the MS. At the final step (5b), the MS responds with an *Attach Complete* message, which acknowledges the correct reception of the new P-TMSI value. Note that the messages transmitted in steps 5a and 5b are typically enciphered; therefore, the new P-TMSI value cannot be eavesdropped on.

3.3   Setting Up PDP Contexts

After a successful GPRS attach procedure, the mobile station is permitted to use the mobile GPRS services in a secure fashion (a security context is established between the mobile and the network). However, further actions are needed for accessing an external PDN. In particular, a virtual connection has to be set up with that PDN. This is accomplished with the (formally referred to) PDP context activation procedure. (The term *PDP [Packet Data Protocol] context*, used instead of *IP context*, emphasizes the fact that GPRS supports not only IP contexts but also other types of packet contexts, such as X.25 and PPP.) Roughly speaking, this procedure can be conceptually associated to the well-known dial-up procedure used over the Public Switched Telephone Network (PSTN) to establish connectivity (e.g., with ISPs). However, GPRS PDP contexts (virtual connections) operate in connectionless mode, as opposed to the connection-oriented mode of the PSTN dial-up connections. In this section, we discuss the concepts behind GPRS PDP contexts and the PDP context activation procedure.

As mentioned before, you can think of the GPRS network as an access network that offers connectivity between a number of mobile stations and a number of external PDNs. For this purpose, the GPRS network offers access ports where the mobile stations can be connected and access ports where the external PDNs can be connected. This concept is schematically illustrated in Figure 3.4. This figure also depicts two established PDP contexts, one for MS A and one for MS B.

Each connection between the GPRS network and an external PDN features a unique official name, similar to a domain name used in the Internet. This unique official name is formally called Access Point Name (APN), and it is represented as *PDN_name.PLMN_name.gprs*. *PDN_name* is a sequence of labels in the form *label1.label2. ...* and identifies an external PDN. *PLMN_name* identifies the Public Land Mobile Network (in this case, the GPRS network) that is used to provide access to the external PDN. The encoding of *PLMN_name* depends on the Mobile Country Code (MCC) and the Mobile Network Code (MNC) allocated to the given PLMN. For instance, for a PLMN with MCC = 10 and MNC = 202, *PLMN_name* is mnc202.mcc010. To simplify the illustration, however, *PLMN_name* in Figure 3.4 is shown either as PLMNA or PLMNB. The APN names shown in this figure are typical examples, used to explain the APN structure.

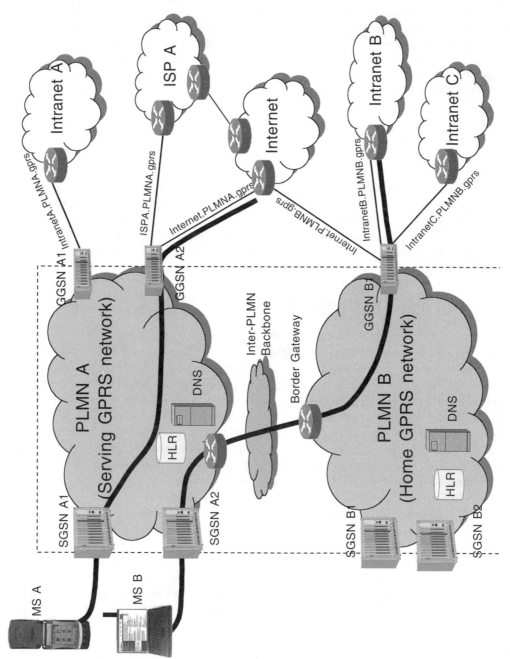

FIGURE 3.4   Establishment of PDP contexts for accessing external PDNs.

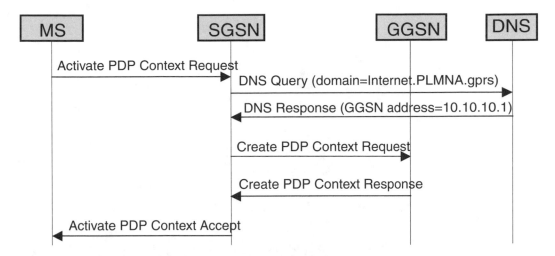

FIGURE 3.5 Establishing PDP context; a simplified message sequence diagram.

Each PDP context is characterized by:

- A specific PDP type, e.g., IPv4, IPv6, X.25, or PPP, which specifies the type of the payload transferred on the PDP context
- A specific APN, which represents an external PDN
- A specific GPRS bearer, i.e., by specific transmission properties

The GPRS bearer is a key characteristic of a PDP context because it specifies QoS properties such as the reliability, delay, throughput, and precedence of the packets transmitted on the PDP context.

It is important to note that a GPRS mobile may have one or more simultaneously active PDP contexts. This means that one GPRS mobile may simultaneously exchange data with one or more external PDNs — for instance, with one that provides Internet access and with another one that provides access to a corporate intranet. Of course, this is not possible with a single PSTN dial-up connection.

The message flow sequence for establishing a new PDP context is illustrated in Figure 3.5. When an MS wants to establish a new PDP context, it sends a specific signaling message (*Activate PDP Context Request*) to its serving SGSN. This message specifies all the previously mentioned characteristics of the requested PDP context. The SGSN checks the requested APN and identifies (e.g., by using the DNS system) the IP address of the GGSN that provides access to that APN. This GGSN may be located either in the serving GPRS network or in the home GPRS network (see Figure 3.4). If the MS specifies only the *PDN_name* (e.g., Internet) instead of the full APN name, the SGSN will first try to use a GGSN in the serving GPRS network. If that fails, it will then try to locate a GGSN in the home GPRS network. If, on the other hand, the MS specifies the full APN name in the *Activate PDP Context Request* (e.g., Internet.PLMNA.gprs), a GGSN in PLMN A may be used only to offer connectivity to the Internet. It is evident that, for establishing a PDP context like the one shown in Figure 3.4 for MS B, specific inter-PLMN connectivity means must exist and the operators of PLMN A and PLMN B must have established a roaming agreement.

After identifying a GGSN, the SGSN sends a GTP signaling message (*Create PDP Context Request*) to that GGSN to request the activation of the requested PDP context. Typically, the GGSN checks whether the MS is authorized to access the requested APN, and if so, it allocates a new IPv4 address to this PDP context (assuming that the requested PDP type is IPv4). It must be pointed out that the GGSN may request a new IPv4 address either from an internal Dynamic Host Configuration Protocol (DHCP) server or from an external DHCP server located in the requested PDN. In the first case, the MS is allocated an IPv4 address from the address space of the serving or home GPRS network, and the MS becomes a new IPv4 node within this network. In the latter case, however, the MS is allocated an IPv4 address from the

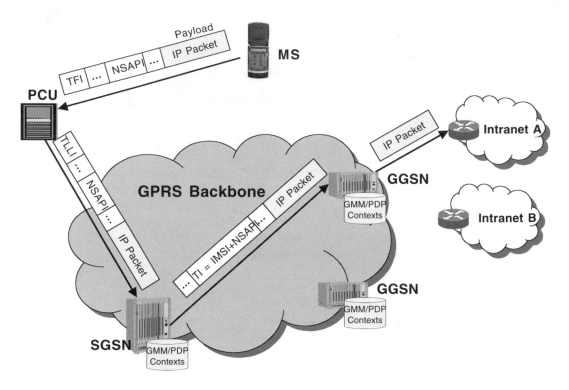

FIGURE 3.6   Tunneling IP packets from the MS to the GGSN.

address space of the external PDN, and effectively it becomes a new IPv4 node inside this PDN. This is equivalent to the case where access to the PDN is accomplished through a dial-up connection. It is typically used when the external PDN is an intranet, which may use private (rather than public) IP addresses.

Under normal conditions, the GGSN accepts the request to create the new PDP context and returns a positive GTP response (*Create PDP Context Response*) to the SGSN. Subsequently, the SGSN returns an accept message (*Activate PDP Context Accept*) to the MS, which includes the IPv4 address allocated to the new PDP context.

At this point, a new PDP context (i.e., a new virtual connection) has been established. Note that the establishment of a PDP context does not involve the reservation of dedicated communication resources in the GPRS network. This applies to both the radio interface and the wireline part of the GPRS network. The establishment of a PDP context involves only the storage of new information in the GPRS nodes (i.e., the creation of new PDP context records in the SGSN and the GGSN). This new information is subsequently used to route the packets correlated with that PDP context. We discuss this routing procedure in the next section.

3.4   Routing and Tunneling

After we have established a PDP context, we use a tunneling procedure to transfer PDP packets from the MS to the GGSN. Assume, for instance, that a PDP context of type IPv4 has been established between the MS and the GGSN shown in Figure 3.6.

In this case, each IP packet transmitted from the MS is put into an envelope that carries two important addressing identifiers: the Traffic Flow Identity (TFI; see reference 13) and the Network Service Access Point (NSAPI). The TFI effectively identifies an active GPRS MS in a certain cell, and the NSAPI identifies one of the PDP contexts that has been activated by that GPRS MS. The PCU that receives this packet translates the TFI into a Temporary Logical Link Identifier (TLLI) and forwards the packet to the SGSN. The TLLI is another MS identifier, which, as opposed to TFI, is decoupled from the cell where the MS is located. In particular, the TLLI is a unique identifier in the SGSN and is used to identify a specific MS served by that SGSN. It is essentially derived from P-TMSI, which is another identifier for the MS. The difference between TLLI and P-TMSI is their range of applicability: The first is applied as an identifier at the LLC protocol, whereas the second is applied as an identifier at the GMM protocol (this is a special signaling protocol that handles GPRS mobility management issues; see reference 8). Because TLLI is derived from P-TMSI, a unique TLLI is also assigned to every MS when it is registered with an SGSN.

The SGSN that receives the packet from the PCU tries to correlate this packet with a preestablished PDP context. For this purpose, the SGSN searches its PDP context database and identifies the PDP context that has stored TLLI and NSAPI values matching the TLLI and NSAPI values contained in the envelope of the received packet. From the information contained in the identified PDP context, the SGSN finds out the IP address of the GGSN associated with this PDP context. Subsequently, it makes up a new IP packet, addresses this packet to the identified GGSN, and encapsulates in it the original IP packet transmitted from the MS (this is called IP-IP encapsulation, or tunneling). Afterward, this new packet is transported through the GPRS IP backbone to the addressed GGSN. Note that, in general, the new IP packet may have encapsulated other types of payload, such as X.25, PPP, and IPv6. The type of the payload will match the type of the PDP context.

The envelope of the IP packet transmitted by the SGSN contains a Tunnel Identifier (TI), which is the concatenation of the MS's IMSI and the NSAPI. The TI is used by the receiving GGSN to correlate this packet with the correct PDP context. When the GGSN identifies the PDP context that has a stored TI that matches the TI in the envelope of the received packet, it discovers the APN associated with this packet and effectively knows the external PDN that the payload (i.e., the original IP packet transmitted by the MS) should be forwarded to.

In the downlink direction, the routing procedures carried out in the GGSN and the SGSN are similar. In this case, the GGSN identifies from the destination address of an inbound packet the PDP context associated with this packet. It then identifies the SGSN address associated with this PDP context and forwards this packet to that SGSN, after including in the header the correct TI value. In a sense, the packet is sent to the SGSN over a particular tunnel, identified by the TI value. The SGSN uses the TI to identify the associated PDP context record in its database. From the contents of the identified record, the SGSN finds out the TLLI of the target MS and finally forwards the packet to that MS through the correct PCU. The correct PCU is the one where the last uplink packet from that MS was received.

3.5 Mobility Handling

Throughout this section, we consider an MS, which is communicating with an Internet host; say, it is downloading a file from that host. Our main effort is to illustrate how the file transfer can be sustained when the MS is on the move and roams among different radio access points (i.e., base stations). In such situations, the GPRS network has to dynamically cope with the location changes of the MS and carry out procedures to modify the associated PDP contexts according to the identified location changes. All these procedures are typically termed *mobility management procedures* and are the basis for all mobile networks. In this section, we focus on only the mobility management procedures executed when an MS is in the packet transfer mode. The mobility management procedures executed when an MS is in the idle mode (i.e., involved with no packet transfer) are not discussed.

Figure 3.7 shows a network schematic diagram that will be used throughout the discussion. In this figure, the lines connecting the various network elements are used merely to illustrate the connectivity among these network elements. They do not imply, however, that the network elements are physically interconnected by point-to-point links. For efficiency and cost reasons, it is common in practice to deploy sophisticated transport means to interconnect the network elements. For instance, between an SGSN and a PCU there is typically a frame relay network. In this case, the line connecting an SGSN with a PCU corresponds to a permanent virtual circuit of the frame relay network.

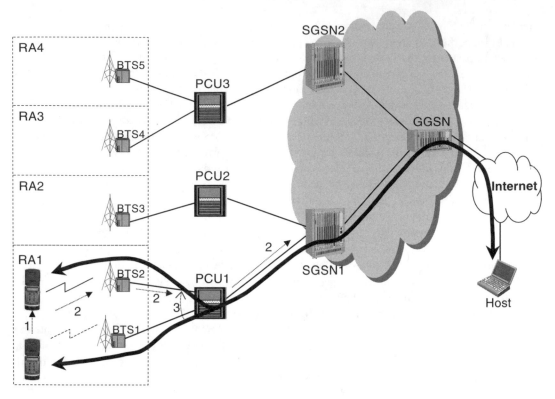

FIGURE 3.7   Cell change: a new cell in the same routing area.

At this point, we assume that the MS has established an appropriate PDP context and is currently within the coverage area of BTS1 receiving a series of downlink packets, each one belonging to the ongoing file transfer session. From Figure 3.7, we note that every downlink packet traverses a series of network nodes in order to be delivered to the MS — i.e., from the GGSN to SGSN1, to PCU1, and finally to BTS1. (The routing procedures used to transfer the packets among successive nodes were explained in Section 3.4.1.) The series of network nodes traversed by the packets belonging to the same PDP context define the *transmission path* of the PDP context. As you will see below, the transmission path of a PDP context changes dynamically (e.g., from one SGSN to another) in order to facilitate the location changes of the MS. However, the GGSN in the transmission path of a PDP context can never change and therefore serves as an anchor point. This anchor point effectively hides the mobility of the mobile stations and makes it possible for an external PDN to reach a specific MS through the same GGSN no matter where the MS is located.

3.5.1   Cell Change

Let us now assume that the MS starts moving toward the BTS2 (see arrow 1 in Figure 3.7). At some instant, the Radio Resource (RR) layer in the MS will recognize that BTS2 can provide better communications quality and will camp on a Radio Frequency (RF) channel controlled by BTS2. This will happen by suddenly switching RF channels and camping on a new one. This procedure is referred to as *mobile-originated handover* because the handover from one cell to another is decided by and performed by the MS alone. In GPRS, this procedure is also referred to as cell reselect. According to reference 13, "when the conditions are fulfilled to switch to the new cell, the mobile station shall immediately cease to decode the downlink, cease to transmit on the uplink, and stop all RLC/MAC timers except for the timers related to measurement reporting. The mobile station shall then switch to the new cell and shall obey the relevant RLC/MAC procedures on this new cell."

Note now that neither PCU1 nor SGSN1 will know that the MS has moved to another cell until the MS makes an uplink transmission in the new cell. Therefore, for some time, the connection with the MS is inevitably lost, and consequently, downlink packets that may be sent by PCU1 are not received by the MS. This means that, during a handover, the packets transmitted by SGSN1 in unacknowledged LLC mode will be lost. On the other hand, packets transmitted by SGSN1 in acknowledged LLC mode (remember that the LLC mode is specified during the establishment of a PDP context) will not be lost but will stay unacknowledged and will be retransmitted later, when the communication with the MS is made feasible. These recovery procedures are handled by the LLC layer, which copes with the occasional "blackouts" that may occur due to the MS mobility. Here we observe that, even when all single-hop links between the MS and the SGSN1 are perfectly reliable (which means they can transfer data with no errors), the link between the MS and the SGSN can still be unreliable. This observation explains the need for the LLC protocol, a reliable data-link protocol between the MS and the SGSN.

After the handover procedure, the RR layer monitors the *broadcast control channel* of the new cell. Over this channel, BTS2 transmits the cell ID of the new cell and a Routing Area ID (RAI), which identifies the Routing Area (RA) where this cell belongs. The RR layer will inform the GMM layer that the cell ID has changed but that the RAI is still the same (according to Figure 3.7, BTS1 and BTS2 belong to the same RA). In response, the GMM layer will command the LLC layer to transmit a *NULL* frame on the uplink (arrow 2). This is a special LLC frame, whose goal is to notify the network of the cell change. When this NULL frame is received by the LLC layer in SGSN1, the cell change is recorded and subsequent downlink packets are forwarded via BTS2 (arrow 3). This procedure is illustrated in Figure 3.7. Any downlink packets that were sent from SGSN1 to PCU1 during the blackout period are transmitted in the old cell and are never acknowledged by the RLC layer. Typically, these packets are discarded as soon as their "lifetime" expires.

3.5.2 Intra-SGSN Routing Area Change

Suppose now the MS moves further, and suddenly, the RR layer makes another handover, this time from BTS2 to BTS3 (see Figure 3.8). This is again a cell change, and what was mentioned in the previous section applies here as well. However, in this case, the RA changes, too, and the RR layer in the MS informs the GMM layer that the mobile station has entered into a new RA. In response, the GMM layer does not send a NULL frame but rather an *RA Update (RAU) Request* message (arrow 2). This is done because the MS does not know if the new RA is handled by the same SGSN (SGSN1), and therefore it has to include additional information in its uplink transmission. This additional information is included in the *RAU Request* message and can be used by a new SGSN to retrieve subscription and other mobility-related information about that MS. This is explained in more detail in the next sub-section. In the example shown in Figure 3.8, the RAU Request will reach SGSN1 and will be treated merely as a cell change. That is, it will simply notify the SGSN1 that the MS can now be reached in a new RA (i.e., through PCU2 and BTS3). The SGSN1 will confirm that the MS is eligible to roam to the new RA and will accept the RAU Request by replying with a *RAU Accept* message. Subsequently, it will change the transmission path of the PDP context from PCU1 to PCU2 (arrow 3). This means that further downlink packets related to that PDP context are sent via PCU2.

3.5.3 Inter-SGSN Routing Area Change

When the MS performs a handover from BTS3 to BTS4, as shown in Figure 3.9, it will again transmit a RAU Request message (arrow 2). This time, however, the new RA is controlled by another SGSN, namely SGSN2. At this point, SGSN2 has to acquire some information about the MS. For this purpose, it sends an *SGSN Context Request* message to SGSN1, asking for the needed information (arrow 3). The address of SGSN1 is effectively derived from the RAI parameter included in the RAU Request message. Now, however, SGSN1 recognizes that the MS has moved to another RA and stops sending downlink packets to the MS (point 3a).

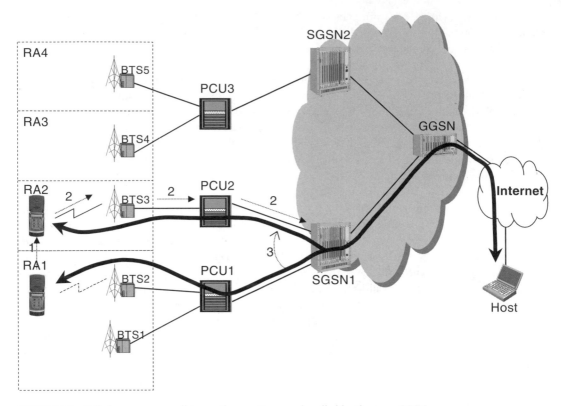

FIGURE 3.8 Cell change: a new cell in another routing area handled by the same SGSN.

Note that between the instant where the handover took place and the instant where SGSN1 received the SGSN Context Request message, another "blackout" period exists. During that period, SGSN1 could have been transmitting downlink packets to the MS in the context of the ongoing file transfer. These packets would be unacknowledged and would need to remain buffered at SGSN1. It is also important to note that the GGSN still believes that the MS is reachable through SGSN1 and could be transmitting new downlink packets to SGSN1. The latter would need to buffer these packets as well. Observe that if the transport protocol in the GPRS backbone is based on User Datagram Protocol (UDP) — which does not support any flow control — when SGSN1 is out of buffering resources, it has no means to signal that to the GGSN; therefore, downlink packets can actually be lost due to limited buffering capabilities in SGSN1. That observation justifies that the UDP transport is unreliable, even when no transmission errors occur in the backbone links. If during the PDP establishment no reliable transport in the GPRS backbone was requested, then the applications running at the MS and the host are responsible for correcting the packet drops that may take place in the GPRS backbone. In our example, the MS and the host could deal with such potential drops by carrying out the file transfer over a reliable transport protocol, such as the Transport Control Protocol (TCP). If the MS and the host could not provide their own reliable transport protocol and if they were running applications vulnerable to packet drops, they would have to establish a highly reliable PDP context, which would also support reliable transfer in the GPRS backbone (e.g., by using TCP instead of UDP to transport packets between the SGSN and the GGSN).

Again, any potential downlink packets that were transmitted to PCU2 before SGSN1 was informed about the RA change will be discarded after their lifetime expires.

An interesting situation arises when the RAU Request message sent by the MS (arrow 2) is lost due to bad radio conditions, for instance. (Typically, another RAU Request would be transmitted after 15 sec.) In this case, it would take quite a long while for SGSN1 to realize that the MS has changed routing

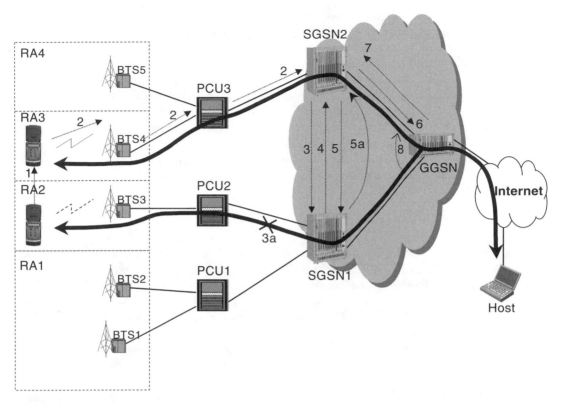

FIGURE 3.9 Cell change: a new cell in another routing area handled by another SGSN.

areas. During this period, SGSN1 would (1) keep buffering any new packets sent by the GGSN, and (2) periodically retransmit the buffered downlink packets that remain unacknowledged. To tackle such situations, a careful dimensioning of the buffering resources of the SGSN is required. In addition, if SGSN1 attempts the maximum number of LLC retransmissions before receiving the SGSN Context Request message, the LLC connection would be released and any activated PDP contexts would be deactivated. In this situation, the file transfer would effectively be dropped. When setting up the LLC parameters, we must take these situations into account.

Let us now continue with the normal message flow. As soon as SGSN1 receives the SGSN Context Request message (arrow 3), it will reply with an *SGSN Context Response* message (arrow 4) passing the requested information to SGSN2. The latter sends an *SGSN Context Ack* message (arrow 5), which verifies that it has received the requested information and is ready to receive any buffered packets for the MS that are still unacknowledged. At this point, SGSN1 forwards all the buffered packets for the MS to the SGSN2 (arrow 5a) within a new tunnel. At the same time, SGSN2 sends an *Update PDP Context Request* (arrow 6) to the GGSN to inform it that any further downlink packets for the MS should be forwarded to SGSN2. In response, the GGSN replies with an *Update PDP Context Response* (arrow 7). Now the transmission path of the PDP context changes (arrow 8) to accommodate the location change of the MS.

Note that in the case of inter-SGSN routing area change, all the LLC connections in the MS are released and new LLC connections are established with SGSN2. After the short interruption required for modifying the PDP context and for reestablishing the new LLC connections, the ongoing file transfer is resumed.

3.6 Summary

In this chapter, we discussed the key GPRS concepts and procedures, and we demonstrated how these procedures enable the provision of wireless packet data services, including wide-area wireless Internet access. In particular, we described the registration procedure, the activation of virtual connections (PDP contexts), the routing and the tunneling of the data in the GPRS backbone, and, finally, how wireless IP connectivity is sustained while a user roams between different areas.

At this point, it should be apparent that the GPRS system is a versatile and cost-effective solution for the provision of wireless Internet services, such as web browsing, e-mail retrieval, text messaging, and file downloading. It provides increased capacity and efficient utilization of the communication resources, and it can support various types of packet data protocols, such as X.25, IPv4, and IPv6. In addition, it supports access to both intranets and extranets, and it can offer roaming capabilities, which would ultimately provide for ubiquitous access to data facilities.

The GPRS technical specifications (available online at www.3gpp.org/ftp/specs) are continuously evolving in order to enable wireless access to more demanding Internet services, such as video streaming and voiceover IP.

References

1. Salkintzis, A.K., Mobile packet data technology, in *Encyclopedia of Telecommunications*,18, Article 346, pp. 65–103, Froehlich, F.E. and Kent, A., Eds., Marcel Dekker, New York, 1999.
2. Pecen, M. and Howell, A., Simultaneous voice and data operation for GPRS/EDGE: Class A dual transfer mode, *IEEE Pers. Commn. Mag.*, Apr. 2001, pp. 14–29.
3. Priggouris, G. et al., Supporting IP QoS in the General Packet Radio Service, *IEEE Network*, Sep./Oct. 2000, pp. 8–17.
4. Kalden, R., et al., Wireless Internet access based on GPRS, *IEEE Commn. Mag.*, Apr. 2000, pp. 8–18.
5. Cai, J. and Goodman, D.J., General Packet Radio Service in GSM, *IEEE Commn. Mag.*, Oct. 1997, pp. 122–131.
6. Brasche, G. and Walke, B., Concepts, services and protocols of the new GSM Phase 2+ General Packet Radio Service, *IEEE Commn. Mag.*, Aug. 1997, pp. 94–104.
7. 3GPP, General Packet Radio Service (GPRS); Service description; Stage 1, (Release 5), Technical specification 3GPP TS 22.060, ver. 5.3.0, June 2003.
8. 3GPP, General Packet Radio Service (GPRS); Service description; Stage 2, (Release 5), Technical specification 3GPP TS 23.060, ver. 5.6.0, June 2003.
9. 3GPP, 3G Security; Security architecture (Release 5), Technical specification 3GPP TS 33.102, ver. 5.2.0, June 2003.
10. Mehrotra, A., *GSM System Engineering*, Artech House Inc., Norwood, MA, 1997.
11. Salkintzis, A.K., *Broadband Wireless Mobile — 3Gwireless and Beyond*, esp. chap. 3, "Network Architecture," Willie, W. (ed.), John Wiley & Sons, Chichester, U.K., 2002.
12. 3GPP, General Packet Radio Service (GPRS); Mobile Station — Serving GPRS Support Node (MS-SGSN); Logical Link Control (LLC) layer specification (Release 5), Technical specification 3GPP TS 44.064, ver. 5.1.0, June 2002.
13. 3GPP, General Packet Radio Service (GPRS); Mobile Station (MS) — Base Station System (BSS) interface; Radio Link Control/Medium Access Control (RLC/MAC) protocol (Release 5), Technical specification 3GPP TS 44.060, ver. 5.6.0, Apr. 2003.
14. 3GPP, SS7 Signaling Transport in Core Network; Stage 3, Technical specification 3GPP TS 29.202, release 4, ver. 4.1.0, Sep. 2001.
15. Jacobson, V., Compressing TCP/IP headers for low-speed serial links, *IETF RFC 1144*, Feb. 1990.
16. ITU-T, Data compression procedures for data circuit- terminating equipment (DCE) using error correction procedures, Recommendation V.42bis, 1990.
17. 3GPP, General Packet Radio Service (GPRS); Mobile Station (MS) — Serving GPRS Support Node (SGSN); Subnetwork Dependent Convergence Protocol (SNDCP) (Release 5), Technical specification 3GPP TS 44.065, ver. 5.0.0, June 2002.
18. Tanenbaum S.A., *Computer Networks*, 3d ed., Prentice Hall, Englewood Cliffs, NJ, 1996.

19. 3GPP, Technical realization of the Short Message Service (SMS); Point-to-Point (PP), Technical specification 3GPP TS 23.040, ver. 3.9.0, June 2002.

20. 3GPP, Functional Stage 2 Description of Location Services (LCS) in GERAN (Release 5), Technical specification 3GPP TS 43.059, ver. 5.3.1, Apr. 2003.

21. 3GPP, General Packet Radio Service (GPRS); Base Station System (BSS) — Serving GPRS Support Node (SGSN); BSS GPRS Protocol (BSSGP) (Release 5), Technical specification 3GPP TS 48.018, ver. 5.6.1, Apr. 2003.

4

Internet Access over Satellites

Antonio Iera
*University Mediterranea
of Reggio Calabria*

Antonella Molinaro
*University Mediterranea
of Reggio Calabria*

The Internet is approaching full maturity, as witnessed by the significant rise in the number of users connected through this "network of networks." Currently, Internet hosts are distributed all over the world, scattered across heterogeneous network platforms and exploited by users to access heterogeneous services by relying on the common TCP/IP protocol as the *unifying network and transport solution*. The role of "core" in a global telecommunications infrastructure deploying pervasive and ubiquitous information superhighways will thus be "naturally" played by the Internet, which has recently emerged as the global solution for telecommunications.[1]

What the Internet will be required to provide is thus the differentiated handling of heterogeneous types of traffic, each one with its own Quality-of-Service (QoS) requirements, as well as flexibility in coping with various platforms explicitly conceived for high-speed transportation and large-scale distribution of multimedia traffic. In such a context, typical of a personal communications platform, we should not forget that ubiquity is perhaps the most relevant aspect that the research and standardization communities have to take into account. It is becoming clear that a modern Internet platform is not even conceivable without the appropriate features to match the emerging needs of a new population of nomadic mobile Internet users, accessing multimedia services regardless of the terminal used and the location. To achieve this further objective, the telecommunications community has recently converged toward the idea that an *Internet satellite* platform is a valid solution for accessing IP multimedia applications regardless of location and with a desired QoS.[2,3]

In the past few years, it has become apparent that the deployment of what some researchers[4] called a "growing armada" of satellites would have furnished "space access" to both traditional and novel (packet-based) telecommunications services. Furthermore, it has emerged that "the explosive expanding reach of the Internet inspired almost every prospective satellite network provider to include Internet access in its service offerings."[4] The "marriage" between satellite technology and the Internet in a typical scenario of personal communications is thus straightforward.

But you should not think that the interoperation between the deployed satellite IP networks and the preexisting terrestrial IP network is not lacking in difficulties and technical challenges. In fact, take into account that the two environments have experienced two completely independent evolutions, which resulted in the optimization of satellite platforms for very specific traditional telecommunications services (such as television broadcasting and connection-oriented transmission) and in the migration of the Internet toward the effective transport of highly heterogeneous packet-oriented traffic at differentiated QoS. Actually, as reported in reference 5, a point of convergence between the two technologies dates back to the years 1979–1985. Note that the communications satellite technology has been exploited by the Internet since its beginning; in those early years the Atlantic SAT-NET interconnecting the ARPANET (Advanced Research Projects Agency Network) with European networks was developed, and in subsequent years, other examples of satellite constellations have been used in the Internet network.

The difficulty in discussing an actual "integration" between the satellite and the Internet environments by referring to the cited experiences of the 1970s and 1980s is due to the realization that the satellite infrastructure was used just to provide backbone connections for regional computers. An actual integration has to pass through the delivery of Internet data directly to the end user; this is a particularly interesting challenge in large, isolated areas where it would be difficult and cost ineffective to deploy a terrestrial Internet infrastructure, or even in metropolitan areas when the user wants to access high-bandwidth multimedia services.[5] In such an evolutionary scenario, the satellite plays a starring role, thanks to its intrinsic broadcast transmission capability, its flexibility in handling on-demand bandwidth allocations, and, mainly, the possibility of supporting wide-range user mobility.

In the late '90s the first platforms for accessing multimedia applications via satellites began to appear, and this was just the beginning of a strong research activity aimed at achieving a solid and effective integration between the satellite and the Internet environments. Today a new generation of satellite systems is on its way to full deployment. These systems will be completely compatible with the Internet multimedia traffic, thanks to the following characteristics:

- Full connectivity among users connected to both fixed terrestrial and mobile stations, by means of terrestrial gateway stations[6]
- User-to-user direct satellite links without any intermediate connection to the gateway
- A wide range of applications, similar to those offered by terrestrial systems, such as access to the Internet, data distribution, video on demand, interactive television, videoconferences, telemedicine services, and remote education

Furthermore, to offer multimedia transmission capabilities, the majority of broadband satellite systems at geostationary (GEO), medium (MEO), and low (LEO) earth orbits are equipped with on-board signal switching and processing functionality (also known as On-Board Processing [OBP]).[7]

The advantages of satellite and terrestrial IP system integration are many. However, equally numerous are the issues that require further study in order to understand how to overcome the intrinsic limitations emerging in the satellite environment when it is called to interact with (or better, integrate into) the new-generation Internet global platform.

The main objectives of this chapter are to analyze the primary features that characterize the various satellite platforms currently available on the market, to highlight their points of strength and limitations when they are integrated with the Internet and thus called to handle packet-based Internet traffic with various QoS constraints and requirements, to describe how most of the emerging problems have been successfully faced by novel technological and protocol solutions, to make clear what projects are still under way and what needs further research efforts to come to a solution, and to examine the frontier of future research in the field of Internet via satellite.

Our aim is to give you a complete vision of the state of the art in this strategic field and to describe the instruments used.

4.1 The Integrated Satellite-Terrestrial Platform for Next-Generation Systems

4.1.1 Introduction to the New-Generation Satellite Systems

The telecommunication scenario emerging in the past few years can be described through its main attributes: extremely *complex*, extremely *dynamic*, *heterogeneous*, *multimedia* traffic oriented, and relying on many *interconnected* networks and systems, with the IP protocol as a common glue.

Furthermore, the "magic word" characterizing today's telecommunications scenario and clearly distinguishing it from past scenarios is undoubtedly "ubiquity." This is the concept of a system in which users are free to roam across different network segments, accessing the network from a generic terminal by means of a generic access modality (the one available at the location from where the user is currently connecting), a generic IP-based multimedia application in a personalized fashion. While approaching the age of global communications, we are thus witnessing a widespread diffusion of novel IP multimedia applications demanding a large amount of network bandwidth, due to the mix of component media of which they consist. At the same time, we are observing a remarkable and unexpected growth of wireless and mobile communication systems and technologies. These wireless access solutions are in fact the most valuable enabling factors for the full deployment of *personal communications platforms supporting the idea of ubiquitous IP.*

Thus, in such a scenario, we would not be surprised by the growing attention to satellite systems as an effective means to extend the potential of terrestrial networks. In this context, satellite systems play a very important role, thanks to their capability to extend terrestrial cellular network coverage[8] and provide both high-bit-rate access to information services and the ability to reach worldwide users. With a view to guaranteeing personal communications facility, wireless (terrestrial cellular, cordless, and satellite) networks are expected to be interfaced with cabled networks transparently to the user. The role envisioned for satellites within a Personal Communication System (PCS) infrastructure is therefore extremely significant and far from the one they have played up to now. But, to better understand both virtues and limitations of the current effort in merging satellite and Internet technologies within a PCS system, let us briefly revise the main characteristics of past and present satellite platforms. This will enable us to realize that within a platform for the Internet as a global telecommunications solution, the currently emerging idea is to consider satellites no longer as the simple objects we have been used to — that is, as "bent pipes" that receive information on the uplink and retransmit it on the downlink to the terminals within their spot-beam coverage.[9] The trend is toward the development of sophisticated broadband satellite systems exploiting on-board processing and switching techniques, medium access control algorithms, intersatellite link routing, and so forth in order to effectively interconnect to terrestrial (preferably based on the TCP/IP protocol) backbones.[10] In the U.S., Europe, and Japan, great efforts are being made to develop prototypes of such satellite platforms, ready for future telecommunication scenarios.

4.1.2 A Brief Classification of Satellites in Terms of Their Orbit Height

The most common approach followed to classify the satellite systems is one based on their orbit altitude above the earth's surface. The scheme uses the satellite classifications GEO, medium- MEO, and LEO. Furthermore, High-Altitude Platforms (HAPs) are currently being designed to supply broadband data services direct to the home along with third-generation mobile communications.[11] GEO satellites orbits are characterized by a revolution around the earth, which is synchronized with the earth's rotation itself and makes the GEO satellites appear fixed to an observer on Earth. The synchronization is achieved through the choice of an operational height of 35,786 km above the equator. In a different way, MEO and LEO satellites are closer to the earth's surface (from 3000 km up to the GEO orbit, and between 200 and 3000 km, respectively).[3] HAPs are intended to be located in the stratosphere 20 to 22 km above the earth and remain effectively in one fixed position during normal operation. Two HAP schemes can be considered: Airships (SkyStation[12] is an interesting example of advanced airship-based HAP systems) and Circling Aircraft (the High Altitude Long Operation [HALO] scheme[13] is an example).

The height of the satellites defines their behavior and their transmission characteristics. The GEO height allows each satellite to cover a wide surface; however, large antennas and transmission power are required. For MEO and LEO, the lower the orbit altitude, the greater the number of satellites required; furthermore, users may need to be handed off from satellite to satellite as they travel at high speeds relative to the earth's surface. GEO are characterized by a large propagation delay, whereas LEO and MEO have the advantage of offering short propagation delays (110–130 ms and 20–25 ms, respectively, which is comparable with that of a terrestrial link). HAPS systems can be deployed quickly with little ground-based infrastructure required, and their propagation delay will be small, as with LEO satellite systems, even if the HAPs will not suffer from the problems of handoff and Doppler shifts caused by the rapid movement of satellites traveling overhead.[11]

4.1.3   The Evolution of Satellites from the "Bent Pipe" to the "Switch" Concept

Since the first satellite has been exploited for telecommunications purposes, three generations of satellite communication have followed each other.[6] Each generation has its particular features that make it suited to the type of telecommunications services for which it has been developed. Each of them also plays a different role in supporting the Internet technology.

In this section, we briefly highlight the main features differentiating a satellite belonging to each generation and point out which relationship with the Internet each generation of satellites can have.

4.1.3.1   The First Generation: Fixed and Narrowband

The first generation is characterized by what we can call *fixed satellite communications systems*. By this, we mainly mean narrowband GEO satellites providing communication links among fixed terrestrial stations; the satellite, therefore, has the role of a mere repeater in the sky. These systems were a formidable means of extending the scope of the fixed Public Switched Telephone Networks (PSTNs) and of supporting television broadcasting services. An example of such satellites is the INTELSAT satellites dating back to the 1960s.

The role of these fixed narrowband platforms within a global Internet scenario has been the mere provision of backbone connectivity among remote fixed Internet systems segments on a wide geographical scale.

4.1.3.2   The Second Generation: Narrowband for Mobile Users

More interesting is what we can call the second generation of satellites. It includes the *narrowband satellite communications systems* exploited for mobile purposes. They can be GEO as well but must be able to supply communications between mobile stations and a fixed terrestrial gateway through satellite links. The International Marine/Maritime Satellite Organization (INMARSAT)[14] is an example of such a platform.

New families of satellites characterized by a LEO rotation orbit were added to this generation starting in the early 1990s. Most of these systems are designed to provide narrowband services (fax, paging, low-bit rate data, and voice communications) ubiquitously.[15] Iridium[16] and Globalstar[17,18] systems are examples of this category of satellites.

This generation of satellites supports different modalities for the Internet infrastructure. We report on two typical examples of access to distributed Internet information by relying on typical narrowband GEO/LEO satellite systems of this second generation. The one in Figure 4.1 refers to the so-called Internet scenario with bent-pipe satellites. It is feasible, for example, by means of a GEO (INMARSAT-like) or LEO (Globalstar-like) satellite, which supports access to Internet traffic for users who cannot connect to any Internet infrastructure. The example in Figure 4.2 refers to the architecture that in literature is usually named "network in the sky" (see reference 3) or "network in the space" (see reference 15) due to the exploitation of intersatellite links (ISL), such as those available in Iridium-like systems.

This second configuration is also typical of several Internet platforms based on LEO satellites of the next (third) generation (such as, for example, Teledesic[19]), which usually exploit ISLs. The main differences lie in

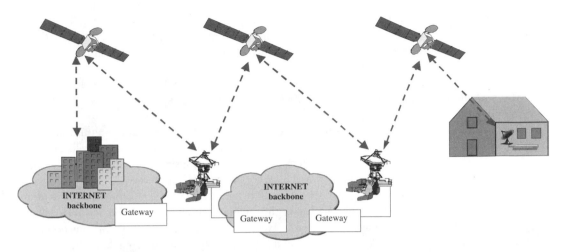

FIGURE 4.1 Internet scenario with bent-pipe satellites.

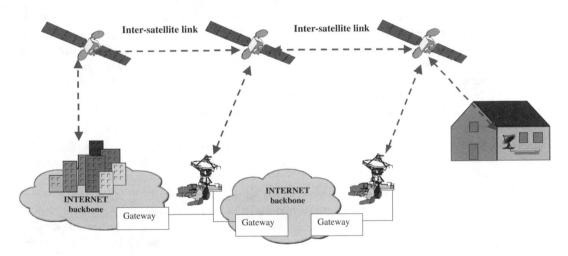

FIGURE 4.2 A network of satellites in the sky.

the narrowband capability of what we called the second generation of satellites, which is surely enough to support traditional IP traffic with non-real-time and low bit-rate requisites, but is not sufficient at all to support the "future" Internet (characterized by such features as Mobile IP capability, IP multimedia service support, and differentiated QoS guarantee through DiffServ and IntServ paradigms). For the latter, a new generation of satellite with *broadband* capability is required.

4.1.3.3 The Third Generation: Broadband for Personal Communications

Broadband satellite systems for personal communications represent the last generation of satellites, operative since the beginning of the twenty-first century. They are usually characterized by the possibility of providing the user with direct connections without an intermediate link to a gateway. These systems can be either GEO or LEO satellites and, in any case, are equipped with a sophisticated on-board switching and signal processing functionality. OBP includes demodulation/remodulation, decoding/recoding, transponder/beam switching, and routing to provide more efficient channel utilization. OBP can support high-capacity ISLs connecting two satellites within line of sight.[3] ISLs is the feature that, as we have already addressed, might allow connectivity in space without any terrestrial network support.

This type of satellite system is specifically thought to support broadband traffic coming from a wide range of multimedia applications. Most of these systems use very high frequency bands, such as the Ku band (10–18 GHz) and the Ka band (18–31 GHz), which allow the use of smaller antennas and enable more available bandwidth to be obtained.

Examples of platforms based on the use of GEO satellites are Astrolink, Cyberstar, Spaceway (using also MEO satellites), iSky, SkyPlexNet, and EuroSkyWay. Teledesic and Skybridge are systems operating by means of LEO constellations of satellites.

For an interesting paper that gives an exhaustive description of future broadband multimedia satellite systems, see reference 20. It classifies the most interesting examples of third-generation satellite systems, highlights the main features that distinguish them from the preceding systems, and traces some evolutionary trends. You will find equally interesting the survey presented in reference 15.

The way these systems interact with the terrestrial IP network is manifold. A straightforward idea is to push the IP technology to the satellite and make the satellite constellation become not only a system supporting mere pipes through which the IP traffic exchanged among terrestrial IP domains flows. The new generation of satellite constellation becomes an integral part of the network, thanks to the possibility of exploiting its OBP and ISL handling capability, as well as a higher computational power. Satellites can therefore carry IP switches that forward packets independently. These IP switches are connected to one another as well as to ground stations.[21] Interesting examples of such an approach have been studied, and we examine some of their advantages and limitations in the remaining part of this chapter.

A wider vision of an IP satellite system considers the satellite IP layer as a complement to both the fixed and the mobile terrestrial IP networks in a beyond-UMTS (Universal Mobile Telecommunications System) scenario. The so-called next-generation IP system has, in fact, the main objectives of overcoming the limitations of today's available Internet in terms of reachability, QoS differentiation, and personalization of offered services and access. This is favored by an approach based on the integration of different mobile and fixed networks through the exploitation of the common IP protocol at the network layer. Many projects (such as SUITED, for example) currently move in this direction because they aim to contribute to the design and deployment of the so-called *Global Mobile Broadband System (GMBS)*, a unique satellite/terrestrial infrastructure ensuring nomadic users access to Internet services with a negotiated QoS. The intended result is a multisegment access network with a federated Internet Service Provider (ISP).[2]

One of the most interesting aspects that marks a clear separation between the IP integration into first-generation satellite networks and the IP-satellite systems at the age of multimedia-broadband satellites is undoubtedly the possibility to create a sort of gateway to the Internet in space. This means that we are approaching broadband systems that, in addition to their traditional trunking services, will now enable the proliferation of personal communications services based on IP directly accessible by the Internet subscribers. Examples of enhanced end-to-end links between Internet users and ISPs that include satellites in the access network have been widely described in literature and are currently being deployed and made available on the market. For illustration purposes, we report here on an example taken from reference 4.

The more interesting aspect in the IP terrestrial-satellite integrated architecture envisaged by reference 4 is surely the possibility of the extension of the IP satellite link to the end user. This means that now the satellite enables efficient end-to-end user-ISP links. The example of a typical network architecture for satellite ISPs includes both the satellite network in the sky and the ground stations acting as a gateway to the terrestrial infrastructure through which the Internet is accessed.

Figure 4.3 illustrates how the whole system can be conceived. Three access modalities are designed to access a service offered by the ISP: directly through the satellite network provider itself, through a specific public network provider's embedded ISP service, or through a specific terrestrial ISP. The IP-satellite scenario reported in our earlier example can be defined as an "asymmetric" IP-satellite scenario. In fact, the return link through which the IP user communicates with the ISP exploits the terrestrial IP network. This is a typical scenario that meets the requirements of the typical Internet asymmetric applications (such as Web browsing) that need a larger amount of resources from the server to the end user than in the reverse direction. These applications support the trend in exploiting satellites with asymmetric

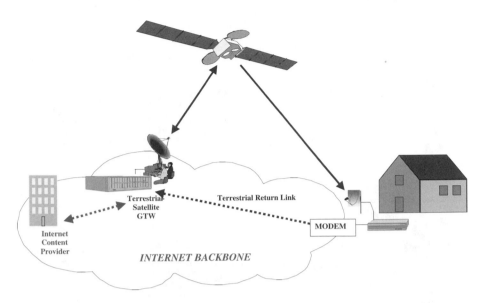

FIGURE 4.3 An asymmetric satellite communication system.

capability, such as Direct Broadcast Satellites (DBSs) typically used for television broadcasting. In this case, the user terminal is equipped with a receive-only satellite antenna, and a terrestrial link provides the reverse path to a specific server able to communicate with the satellite on the uplink direction (refer to Figure 4.3).

Nevertheless, we have to consider that the Internet is evolving into a more powerful network, able to support differentiation of services and thus also prepared to support such services as teleconferencing and other multimedia applications with a certain degree of interactivity. Contemporary satellites, as we have highlighted above, evolved into more powerful broadband platforms. Thus, the symmetrical IP satellite network approach is destined to evolve rapidly into platforms in which user terminals can be assumed to be *interactive*, which means they can directly transmit data up to and receive data from the satellite. This also means that the following assumptions hold: the network is *symmetric* with a two-way balanced load and identical link characteristics (refer to Figure 4.4).

4.2 Interconnecting Terrestrial and Satellite IP Segments: Enabling Technologies

At this point, it should be clear that the interconnection of terrestrial and satellite-based IP networks brings with it many research and technological issues that still need a thorough investigation. In this section, we examine the most relevant issues and highlight the intrinsic weaknesses related to the presence of satellite links (in any of the configurations and the platforms addressed in the previous sections) as well as the proposed solutions and the issues still open and thus deserving further study in the next few years. The implementation of an effective transport protocol for the Internet, which works efficiently across the satellite links as well, undoubtedly represents a potential weakness in any integrated platform.

At the moment, and almost certainly for a long time to come, the protocols used by the Internet network at the transport layer will be TCP and UDP. Due to their large diffusion in almost the totality of the segments of the terrestrial Internet, these protocols are expected to be exploited for conveying different kinds of traffic (with real-time and non-real-time, interactive and noninteractive nature) over IP satellite links. In literature it has been widely shown that the performance of the algorithms exploited at the transport layer by the Internet can be adversely affected by the long latency (for GEO satellites) and the error-prone nature of the satellite links. In particular, the first feature of the satellite channel has a negative impact on the TCP protocol. As you have learned in previous chapters, most of the features of TCP (such as, for example, the slow-start algorithm, bandwidth-delay products, congestion control,

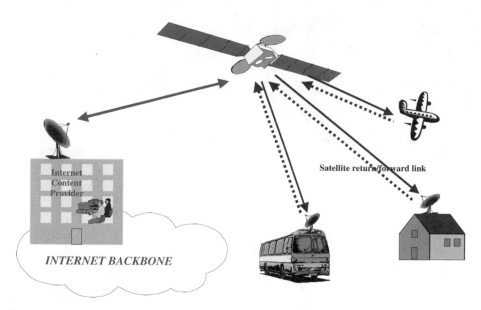

FIGURE 4.4 A symmetric satellite communication system.

acknowledgments, and error-recovery mechanisms) have become extremely inefficient in transmitting data over satellite links and are characterized by short transmission durations compared with the delay-bandwidth product, or high transmission rates compared with the round-trip delay.

A task group of the Internet Engineering Task Force (IETF) is currently focusing on different aspects of the problem and is developing an effective TCP protocol adaptation to the satellite environment.[22,23] Some of the results achieved by the IETF group and by the researchers involved in this popular topic are presented in the remaining part of this section.

One can think that the problem cited above can be easily solved by simply adopting LEO constellations instead of GEO constellation to profit from the shorter round-trip delay offered by such platforms. Nevertheless, keep in mind that LEO satellites are mobile; this introduces other problems, different from those affecting the GEO platforms, but not less disadvantageous to the Internet protocols (both at the TCP and at the IP layer).

TCP, in fact, is not the only weak link in the chain; the routing protocols exploited in the terrestrial Internet network at the IP layer are a further source of concern when extending IP links to the satellites. Performing the routing of the packets at the IP layer (*IP routing*) is in fact the most interesting solution in the terrestrial Internet network. IP routing has the advantages of being based on optimized and widely accepted routing mechanisms such as Distance Vector (DV) and the Link-State Algorithm (LSA) and is also easily augmentable to support multicast transmission by means of the Internet Group Management Protocol (IGMP) specified by the IETF documents. Nevertheless, in particular constellations in which LEO satellites equipped with ISL and OBP functionality are used to interact with the terrestrial IP segments, the routing issue still presents many challenges. The efficiency or even the correctness of the algorithms extensively proved in a terrestrial-only scenario could not be guaranteed any longer in satellite constellations because of both frequent topological changes and unbearable jitter (in LEO constellations) or round-trip (in GEO constellations) delays. The main problem is that Internet routing protocols, such as Open Shortest Path (OSPF) and the Routing Information Protocol (RIP), exploit topological information that, due to the satellite movements, could rapidly become obsolete. A further point that deserves a thorough investigation is understanding "where" the routing function has to be implemented. Because of the deployment of a new generation of satellites with OBP capabilities, investigations should be conducted to decide whether to implement that function in the satellite or on the ground. Obviously, different problems arise when the choice falls on each of the two alternatives.

A further aspect that attracts the interest of many researchers is represented by the *exterior routing protocol* typical of the Internet and derives from the concept of *Autonomous Systems (AS)*. To the typical protocols handling the internal routing, in fact, it is widely known that coupling suitable inter-AS (also called "exterior") routing solutions is necessary. If we assume that the satellite system embedded in an overall Internet network is an AS, there is a need to design effective border gateways implementing the Border Gateway Protocol (BGP)[24] adapted for the new scenario. Here, like above, there is the need to decide where to implement the BGP protocol adapted for the new heterogeneous scenario that can be composed of many satellites overlaying a terrestrial Internet area. The border gateway functionality can be implemented either on-board the satellites or in ground locations. The effect of implementing it on-board is the excessive computational and storage load. On the other hand, terrestrial gateways imply an extra round-trip delay.[3] To summarize, the research is currently investigating different IP routing schemes, some of which will be examined in upcoming sections, even if in the future the most probable solution will be that IP routing for satellite networks may be implemented via a combination of various techniques, such as tunneling, Network Address Translation (NAT), Border Gateway Protocol, IP/Asynchronous Transfer Mode (ATM) with Multiprotocol Label Switching, as well as proprietary routing techniques.[20]

Last but not least, the interest of the research community is currently focusing on the QoS issue in IP-satellite systems. You should not forget that IP is evolving into a protocol that differentiates the level of performance and the degree of QoS offered to the end user. The QoS guarantees need to be maintained also over the satellite for any conceived service class of the Internet (DiffServ and IntServ). IP QoS over the satellite is destined to be achieved not only through the use of IP over ATM satellite platforms, but also in an all native-IP terrestrial-satellite platform. The implementation of the former solution brings much overhead and complexity, thus strongly pushing toward the deployment of the second solution: the direct support of integrated and differentiated service models. We also discuss in detail how this problem can be currently addressed.

4.2.1 IP-QoS via Satellite: Achievement of IP QoS

To address the issue of IP QoS guarantees via satellite, we will focus on modern satellite platforms equipped with OBP technology. Many opportunities can be exploited in platforms for the QoS control, most of which are related to the switching and transport modality the designer decides to adopt on-board. Our attention will focus on satellite platforms based on three technologies:

- The ATM protocol, or at least partial aspects derived from the ATM protocol stack
- The DVB/DVB-RCS standard, coupled with either MPEG-2 or ATM transport streams
- Native IP exploiting IP switching (MPLS) coupled with IntServ/DiffServ models

In any case, the main objectives are the achievement of transparent interoperability with IP services, the need to differentiate the QoS offered to IP/non-IP traffic, and the maximization of the utilization of the platform's resources under heterogeneous (multimedia) offered traffic.

4.2.1.1 IP over DVB and DVB-RCS

The family of satellites that have been developed to jointly exploit digital transmission technology based on the MPEG-2 and digital video broadcasting (DVB) standards represent attractive platforms through which Internet traffic can also be conveyed in the direction of the Internet provider/end user and, more recently (with the advent of the new DVB-RCS standard), also in the opposite direction.

It is worth repeating that DVB stands for *digital video broadcasting*. In its name are summarized the points of strength of the standard. It is completely digital and thus able to transport any type of digitalized data both of the HDTV (High-Definition TV) and the SDTV (Standard Definition TV) type and of the multimedia/interactive data type. The fact that it was originally conceived for video transmission gives to the standard an intrinsic broadband nature, which is also a point of strength for supporting IP multimedia services and applications. Last, but not least, its broadcasting capability makes it suitable to support IP multicasting applications, which are recently gaining ground in the modern telecommunications market.

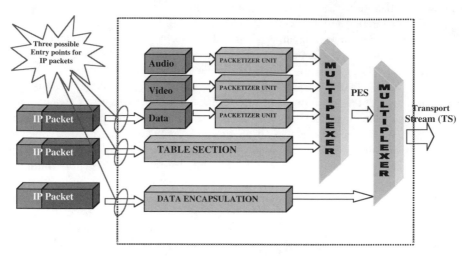

FIGURE 4.5 Entry points for IP datagrams over MPEG-2 TS/DVB.

One of the first European Telecommunications Standards Institute (ETSI) standards developed within the DVB project is called DVB-S (Digital Video Broadcasting via Satellite),[25] which has been designed with a modular organization, based on independent subsystems, in order to allow the standards developed later (DVB-T, DVB-C) to reuse portions of the DVB-S not related to the transmission medium. DVB-S is a powerful means for deploying broadcasting[26,27] of "data containers" into which any type of digital data can be put; IP datagrams can also exploit these data containers to transport Internet-based service information.

An interesting paper describing the DVB-S is reference 28. Refer to this paper if you are interested in gaining an in-depth understanding of this topic.

We need to understand how IP data information flows and is delivered through satellite IP platforms exploiting the DVB/MPEG-2 standard. For data broadcasting, the DVB standard foresees three modalities for data encapsulation into MPEG-2.[29–32] In Figure 4.5 the possible entry points for IP data traffic over MPEG-2 are shown:

1. IP packets can be encapsulated and transported within MPEG-2 PES, similarly to audio and video information flows. This is the *Data Streaming*[27] approach.
2. IP packets can be transported within section packets with a format specified for the transfer of system tables. In this case, these tables need an identification fixed at a predefined value that distinguishes when they are transporting IP packets and when they are transporting control messages. This approach is called *Multiprotocol Encapsulation (MPE)*[27] and is the one usually adopted because it suits the IP datagram transport optimization.
3. A protocol can segment the datagrams into a sequence of cells. This technique is called *Data Piping*.[27] Reference 5 offers more details on the methodology and overhead in various examples of IP mapping into MPEG-2 Transport Streams (TS).

In DVB-S systems for IP data broadcasting, the adopted end-to-end transmission approach could consist of the following. The (multimedia) IP content is received by particular satellite-transmitting modules of the IP-satellite network in the format of IP packets, coming from ISPs scattered across the external Internet. The IP flows can be delivered to these modules (such as uplink stations or satellite gateways) through high-speed dedicated lines or through the best-effort native Internet, according to the application requirements. These modules are responsible for the data transmission toward the DVB satellites, following a resource-assignment phase performed by appropriate terrestrial Network Control modules.

Before its delivery toward the satellite, data is processed and encapsulated into MPEG2 TS format (according to one of the rules addressed above). By assuming a satellite with OBP capability, the received IP-over-MPEG2 traffic is multiplexed and transmitted according to the DVB standard and received by the IP end users equipped with DVB receivers.

As for the request of the services in the return link direction, two different approaches are possible, depending on the exploitation of a DVB (return channel via terrestrial IP network) or DVB-RCS (return channel via satellite) solution. In fact, compared with the former standard, the latter also foresees a return link via satellite. Thus, if we envisage a DVB-RCS satellite platform, then the IP traffic in the direction *user-satellite-ISP* can travel encapsulated into the DVB standard and can be sent directly toward the satellite without interesting the terrestrial IP segment in the direction *user-ISP*. Otherwise, if the return channel via satellite is not foreseen (the DVB-S standard only), then the IP packets in the direction *user-ISP* travel across a terrestrial Internet platform and the only transmission exploiting IP satellite with DVB-S happens in the *ISP-satellite-user* direction.

Let us take a closer look at how the transport of IP traffic can be performed within DVB-RCS satellite platforms, according to the rules under way regarding standardization within ETSI standardization groups. The work is progressing very fast, and in Europe it is also strongly supported by the European Space Agency (ESA) through the activity of its interest group on Regenerative Satellite Access Terminal (RSAT). For our purposes, the most interesting aspect of such a standardization activity is the foreseeing of two different transport modalities to convey data traffic over DVB-RCS channels: One is based on the use of ATM cells and the other on MPEG-2 TSs. As clearly summarized in reference 33, the consequent transport of IP datagrams in DVB-RCS systems would be performed by mapping IP traffic either *over AAL5-over-ATM via DVB-RCS* or *over MPE/MPEG-2 Transport Streams via DVB-RCS*.

The various solutions imply a different overhead added to the IP packets and thus a different effectiveness in exploiting the satellite channel resources. As detailed in reference 33, the MPE/MPEG2-TS option may, on the average, have 5% less overhead compared with the AAL5/ATM solution. They generate a different number of Layer 2 fragments when an IP datagram has to be sent over the satellite link. However, what clearly emerges from a simple reasoning about the protocol encapsulation rules and the different PDU formats constraints is that the transport of single small IP packets through MPE/MPEG2-TS would result in a significant waste of resources.

4.2.1.2   DiffServ/IntServ over Satellites

A key question arises regarding the widespread use of satellites: Will the satellite influence the design choices of an IP-based Internet network with QoS guarantees? In other words, the issue to investigate is whether the IP QoS models (that is, *Integrated* Services[34] and *Differentiated* Services[35]) commonly used over wired networks are also able to ensure the target QoS over paths including satellite links. The main limiting factors are tied to the scarcity of the satellite bandwidth resource and the (long) latency over satellite links, which strongly influence any design decisions.

The most promising approach to guarantee differentiated QoS levels within the global IP network is the one assuming that the IP Integrated Services (IntServ) model is adopted in the satellite access system and is combined with an IP core network based on the Differentiated Services (DiffServ) framework, in which the *aggregate RSVP* protocol[36] is implemented. In this scenario, the design of the effective interworking between the terrestrial IP backbone and the satellite access platform becomes a key issue. The gateway design rules have to fulfill the following requirements:

- Seamless roaming between the wireless and the wired environments
- Efficient integration between the IP IntServ and the DiffServ models
- Suitable mapping of terrestrial onto satellite bearers for traffics with different profiles and QoS requirements

This section will focus on the extension of IP-based QoS mechanisms toward satellite networks and on the design of the gateway device acting as an interworking unit between terrestrial and satellite segments.

4.2.1.2.1   The IP-Based QoS Approaches

We will briefly recall some basics of the QoS model proposed for IP. This is done here to enable you to understand the concepts that we discuss in the remaining part of this section. Nevertheless, for a more in-depth description of these mechanisms refer to the specific chapter in this book.

The IETF has proposed several approaches to the issue of IP-based QoS guarantee; the two main proposals are the already addressed IntServ and DiffServ. The concept commonly agreed on by both approaches is the differentiated handling of each service class as a means of guaranteeing the target QoS to every type of IP application.

The IntServ basic concept is resource *reservation*: resources for each single traffic flow are reserved across the whole end-to-end path upon request and in advance. Therefore, "end-to-end" QoS control is provided on a "per-flow" basis; this means that every single flow is separately handled at each router along the data transmission path. Two IntServ service categories have been defined: *guaranteed services (GS)*[37] and *controlled load services (CLS)*,[38] characterized by strict and mild constraints on delay and loss probability, respectively. According to the IntServ approach, when each network node throughout a data transmission path implements one of the defined IntServ service categories (GS, CLS), then the network is able to deliver a controlled end-to-end QoS level, which is defined by the specific IntServ category in use.

To provide resource reservation throughout the end-to-end data path, many protocols have been proposed; among them, RSVP (Resource Reservation Protocol)[39,40] is the most commonly used. RSVP is based on some signaling messages (*Path, Resv, PathTear, ResvTear, ResvErr,* and *PathErr*) exchanged among sender, receiver, and intermediate network nodes. RSVP signaling is used to notify per-flow resource requirements (expressed in terms of *IntServ* parameters) to each network node. Then, each network node can apply admission control and traffic resource management policies to ensure that each admitted flow receives the requested service. The IntServ parameters used to describe per-flow resource requirements are the token bucket parameters (*token bucket rate, r, token bucket size, b, peak data rate, p, maximum packet size, M,* and *minimum policed unit, or m*). They all describe the source's generated flow. The traffic burstiness is defined as $\beta = p/r$.

The most critical IntServ class to convey over satellite links is the GS[37] class, which provides the user with guaranteed maximal packets' *queuing delay* and *delivery time.* Given the token bucket–based description of a traffic flow, a service element (for example, a router or a subnet) computes some parameters, which describe how it will handle the packets of this flow. By combining these parameters from the various service elements in a data path, we can compute the maximum delay experienced by a packet transmitted over that path. Then, by setting a bound on the target maximum end-to-end delay (*DB,* delay bound), the receiver is enabled to request a given bandwidth (or service rate) *R,* which is able to maintain the delay below the given bound. This is possible because the GS definition is based on the fluid model assumption, which ensures that the delay of a flow obeying a token bucket (*r,b*) and served by a line with bandwidth *R* is bounded by *b/R.* This holds on condition that *R* is higher than *r.*

R is computed based on the end-to-end delay bound DB, as described in the following equations:

$$DB = \frac{(b-M)}{R} \times \frac{p-R}{p-r} + \frac{M + C_{tot}}{R} + D_{tot} \quad \text{if } p > R >= r \tag{4.1a}$$

$$DB = \frac{M + C_{tot}}{R} + D_{tot} \quad \text{if } r <= p <= R \tag{4.1b}$$

C_{tot} and D_{tot} are error terms, which represent how the network elements' implementation of GS deviates from the fluid model. C_{tot} is the flow rate-dependent error term; D_{tot} is the rate-independent error term and represents the worst-case transit time variation through the service elements. For more details on Equation 4.1a and Equation 4.1b, refer to reference 37.

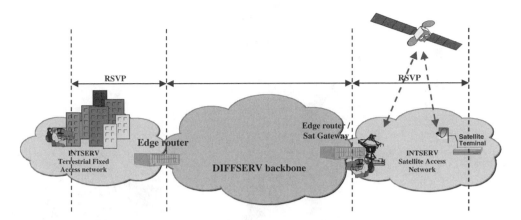

FIGURE 4.6 IntServ and DiffServ integration.

Each node on the route from source to destination updates the PATH message by adding the relevant values of C_i and D_i. On the reception of the PATH message, the end node will be enabled to know the C_{tot} and D_{tot} values relevant to the whole end-to-end path.

In contrast to the per-flow approach of RSVP, the DiffServ framework manages traffic at the "aggregate" level. The model defines the following mechanisms for differentiating traffic streams: differentiated Per-Hop Behaviors (PHB) for packets queuing and forwarding, traffic classification, metering, policing and shaping functions. These mechanisms are used at the edge of the DiffServ region. The internal routers in the DiffServ region do not distinguish the individual flows; they handle packets according to their PHB identifier, based on the DiffServ Codepoint (DSCP) in the IP packet header. Because DiffServ eliminates the need for per-flow state and per-flow processing, it scales well to large networks.

The IETF has standardized two types of DiffServ classes: *Uncontrolled* and *Controlled*. The former offers "qualitative" service guarantees; an example of this class is the *Assured Forwarding (AF)* PBH.[41] In contrast, the controlled traffic class uses per-flow admission control to provide end-to-end QoS guarantees. An example of this class is the *Expedited Forwarding (EF)* PBH.[42]

The combination of IntServ and DiffServ models can be an advantageous solution:[43,44] IntServ/RSVP and DiffServ can, in fact, be used as complementary technologies in the pursuit of end-to-end QoS. IntServ can be used effectively in the access network to request per-flow, quantifiable resources along an end-to-end data path, whereas the DiffServ scalability can be exploited in the core network. The integration of IntServ and DiffServ models can be based on the presence of an edge device, which acts like an IntServ-capable router on the access network and as a DiffServ router on the core network, as illustrated in reference 45. The advantage of this approach is that it proposes dynamic admission control in the DiffServ network region, without the burden of per-flow RSVP signaling.

4.2.1.2.2 The Extension of the IntServ-RSVP Paradigm to the Satellite Environment

Figure 4.6 illustrates a sample terrestrial-satellite network supporting end-to-end IntServ over a DiffServ core network. The main components are a *DiffServ Network Region* and some *IntServ Network Regions* outside the DiffServ region, and *Edge Routers (ER)*, which are adjacent to the DiffServ network region.

The *DiffServ Network Region* is a terrestrial core network that supports aggregate traffic control. This region provides two or more DSCP-based service levels. The *IntServ Network Region* may be a generic IntServ access network; for example, it could be an IntServ *satellite* access network on one side and an "*any*" IntServ terrestrial network on the other side.

ERs act as IntServ-capable routers in the access networks and as DiffServ-capable routers in the core network. The DiffServ network can be RSVP aware, and ERs also behave like border routers for the DiffServ region. This means that ERs participate in RSVP signaling and perform the admission control task for the DiffServ network. Therefore, changes in the capacity available in the DiffServ network region can be communicated to the IntServ-capable nodes outside the DiffServ region via RSVP. This feature

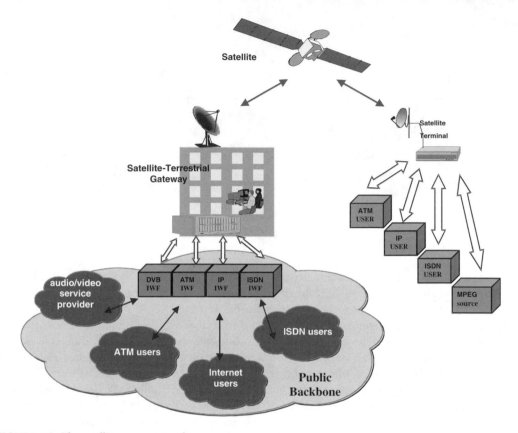

FIGURE 4.7   The satellite access network

gives the architecture proposed in reference 43 the further advantage of providing dynamic resource provisioning in the core network, in contrast to static provisioning. The accurate design of the ER/gateway is vital for effective interworking between the terrestrial backbone and the satellite access network.

As for the satellite access network, its main components are illustrated in Figure 4.7: a satellite with OBP capability; a gateway station, interconnecting satellite and terrestrial segments; satellite terminals of various types; and a Master Control Station that is responsible for the Call Admission Control (CAC). The satellite has OBP capability and implements traffic and resource management (TRM) functions.

The satellite network can be seen as an *underlying* network, aimed at interfacing *Overlying Networks (OLNs)* based on different protocols, such as IP, ATM, X.25, Frame Relay, N-ISDN, and MPEG based. A valid example of this type of system is that in reference 46. The transparency of the satellite network is guaranteed by the use of one *interworking function* (IWF) for each network protocol, both at the satellite terminal and at the gateway/provider terminals.[47]

To support dynamic resource provisioning within the DiffServ region, the aggregate RSVP mechanism[36] has been proposed by the IETF. Aggregate RSVP, an extension to RSVP, was developed in order to enable reservations to be made for an aggregation of flows between edges of a network region, rather than for individual flows as supported by the current version of RSVP. In other words, aggregate RSVP is a protocol proposed for the aggregation of individual RSVP reservations that cross an "aggregation region" and share common ingress and egress routers into one RSVP reservation from ingress to egress. Routers at the ingress and egress edges of an aggregation region are termed "Aggregator" and "Deaggregator," respectively. This is similar to the use of virtual paths in ATM networks. An aggregation region is a contiguous set of systems capable of performing RSVP aggregation. In accordance with reference 36, the aggregation region could be the one in which the DiffServ model is adopted. The ERs at the ingress and egress sides of the DiffServ core network act as Aggregator and Deaggregator. In our

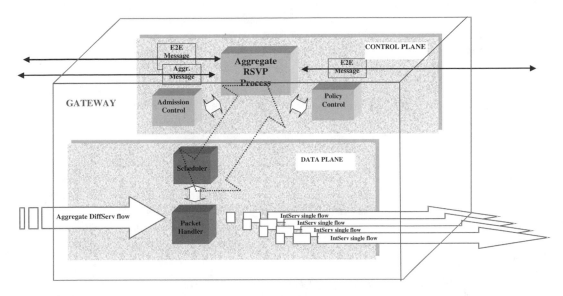

FIGURE 4.8 Gateway functional modules.

reference architecture, the ER in the terrestrial IntServ access network acts as an Aggregator, whereas the ER in the satellite IntServ destination network acts as a Deaggregator.

We will call "end-to-end" (E2E) reservations the reservation requests relevant to individual sessions, and *E2E Path/Resv messages* their respective messages. In a similar way, we will call "aggregate" reservations the request relevant to many E2E reservations and *aggregate Path/Resv* messages the relevant messages.

E2E RSVP messages must be hidden from RSVP-capable routers inside the aggregation region; this means that routers within the DiffServ region forward E2E RSVP messages transparently and with negligible performance impact. To this aim, the *IP Protocol Number* in E2E reservation messages is changed from its normal value (RSVP) to RSVP-E2E-IGNORE upon entering the aggregation region and is then restored at the egress point. This enables each router within the aggregation region to ignore E2E reservation messages, which are forwarded as normal IP datagrams. The aggregation of E2E reservations allows a reduction in the amount of states to be stored in the routers and of signaling messages exchanged within the aggregation region. *Aggregate Path* messages are sent from the Aggregator to the Deaggregator by using RSVP's normal *IP Protocol Number*.

There are numerous options for choosing which DiffServ PHBs might be used for different traffic classes crossing the aggregation region. This is the "service mapping" problem that will be addressed in the following section.

4.2.1.2.3   The ER/Gateway Specifications
The gateway station plays a fundamental role within the reference network architecture shown in Figure 4.7. As already outlined, it has a twofold functionality: *interworking* between the terrestrial and satellite network segments and the *aggregating/deaggregating* function. This functionality can be located in the IWF module of the gateway, which is, therefore, also considered an IP node.

The internal structure of the gateway device is shown in Figure 4.8; it is split into some building blocks that are included in the control plane or in the data plane.

4.2.1.2.3.1 The Aggregating/Deaggregating Function of the Gateway — The *control plane* takes care of setting up and tearing down data paths through the network. Because the gateway acts as an RSVP-capable router with functionality of Deaggregator at the egress of the DiffServ core network, it is responsible for managing *E2E Path* and *Resv* messages. Specifically, the gateway is involved in the reception of *E2E Path* messages from the Aggregator and in the handling of *E2E Resv* messages from the receiving terminal.

The control plane is also responsible for determining the mapping between DiffServ PHBs and IntServ classes. The "service mapping" problem is detailed in reference 48. Numerous options available as to which DiffServ PHBs might be used for various IntServ classes of traffic crossing the aggregation region. In any case, the rules for mapping E2E reservations onto aggregate reservations depend on the network administrator's objectives. An example of mapping would be the following:

- The IntServ *Controlled Load Service (CLS)* could be mapped onto the *Assured Forwarding (AF)* PHB, which allows the association of different loss priorities to the packets.
- The IntServ *Guaranteed Service (GS)* could be mapped onto the *Expedited Forwarding (EF)* PHB, which is the most suitable for real-time applications.

The *data plane* is responsible for the transmission of traffic generated by user applications. As shown in Figure 4.8, the data plane includes two functional blocks: the Packet Handler and the Scheduler. The *Packet Handler* is in charge of the management of the aggregated traffic in input; it changes the aggregated DiffServ traffic into individual IntServ flows.

The *Scheduler* is responsible for the transmission of packets queued in the *Packet Handler,* according to a defined scheduling policy. It decides the packet management at the network layer, based on the desired QoS.

4.2.1.2.3.2 The Terrestrial-Satellite Interworking Function of the Gateway — The gateway not only implements aggregation/deaggregation functions but also acts as an interconnecting unit at the border between the satellite and the fixed segment. The satellite and terrestrial environments differ in many aspects; they implement different protocols and use different transport modes (bearer services). Further complications come from the fact that IntServ is implemented in the access satellite network, whereas DiffServ is implemented on the fixed side.

4.2.1.2.3.3 Mapping of IntServ Classes over Satellite Connections — Two types of connections can be established in a generic GEO satellite system: *permanent* and *semipermanent.* Permanent connections assign a given number of channels to the terminal for the entire connection lifetime. Semipermanent connections have dynamically assigned resources on the basis of the source activity.

One of the main tasks of the gateway station that interfaces with the DiffServ core network and the satellite network is to decide the most effective mapping of the IntServ classes over satellite connections (permanent or semipermanent). In reference 49, the authors have investigated this issue in depth. Because the CLS class has no temporal constraints, it can be stated that its "natural" mapping is on *semipermanent* connections. This mapping allows a better utilization of the satellite bandwidth compared with the permanent one, even if it cannot guarantee any delay requirements. On the other hand, it is not possible to state *a priori* which is the best mapping of the GS class on satellite connections. In fact, the token bucket parameters of the GS source and the target end-to-end delay set by the receiver play a leading role in deciding the best mapping. Therefore, the gateway station makes use of lookup tables and functional modules for the monitoring of the satellite channel load to dynamically decide the mapping of the GS class over satellite links.

4.2.1.2.3.4 Supporting the RSVP Requests across the Satellite — In our view, the gateway station has two options to handle E2E RSVP messages in the satellite system segment. The simplest solution is to convey RSVP messages transparently on *best-effort* connections. The gateway is enabled to recognize IP datagrams carrying RSVP messages because the E2E RSVP *Path* and *PathTear* messages are forwarded across the DiffServ aggregation region like normal IP datagrams by setting the IP *Protocol* field to RSVP-E2E-IGNORE. Therefore, the gateway may deliver them over a newly generated or an already-existing best-effort connection that is identified by means of source and destination addresses. When receiving the *E2E Path* message, the satellite terminal replies with an *E2E Resv* message. Because satellite connections are typically unidirectional, each RSVP session needs two best-effort connections in opposite directions.

An alternative solution for managing RSVP messages is to deliver them on the already-active satellite connection assigned to the relevant application that opened the RSVP session. The information needed to identify the relevant satellite connection can be derived from the fields in the RSVP messages, which give information on the sender and receiver of the RSVP session. This second solution proves more difficult to implement than the first one but allows the use of fewer satellite connections.[47]

4.2.1.2.3.5 E2E Path ADSPEC Update at the Gateway — As already mentioned, E2E RSVP messages are hidden from the routers inside the aggregation region. This means that the ADSPECs of *E2E Path* messages are not updated as they travel through the aggregation region. Therefore, the task of updating the ADSPEC in the corresponding *E2E Path*, in order to reflect the impact of the aggregation region on the achievable end-to-end QoS, is left to the gateway. To do so, the gateway makes use of the information included in the ADSPEC from the *Aggregate Path* message because *Aggregate Path* messages are processed inside the aggregation region and routers update their ADSPEC.

Nevertheless, in a terrestrial-satellite system the gateway should update the ADSPEC including not only the impact of the aggregation region, but also the impact of the satellite path on the achievable end-to-end QoS. To perform this task, the gateway distinguishes two cases, according to the IntServ class involved in the reservation procedure. In the case of CLS, the ADSPEC includes only the *break bit* that is used to indicate the presence of a node not capable of managing the service along the data path. In this case, the gateway has to modify the break bit only if the satellite network does not support CLS. In the case of GS, the ADSPEC update depends on how this service is mapped over the satellite link. If GS is mapped over satellite permanent connections, the D term in Equation 4.2 of the *Delay Bound* (DB) includes only the worst-case waiting time before transmitting a burst — that is, the duration of a frame. If GS is mapped over semipermanent connections, the D term also includes the further delay due to the per-burst resource request.

In general, the following contribute to the D terms for a satellite connection:

- Time needed for the burst transmission request to reach the Traffic Resource Manager (TRM) and return (270 ms), namely the round-trip-delay
- Maximum waiting time of a request on board (TimeOut)
- One frame duration (that is, 26.5 ms), because the requests received during a frame are analyzed by the TRM during the subsequent frame
- One frame duration, due to the TDMA technique used on the satellite uplink
- Time needed for terminal configuration (100 ms) and on-board switching (54 ms)

The C term relevant to the satellite link is invariant. For a permanent connection, the D term is equal only to the frame time given by the TDMA scheme. For the semipermanent connection, all the terms listed are present. Thus, $D_{perm} = TDMA\_frame\_duration$, and $D_{semip} = round\_trip\_delay + TimeOut\_on\_board$. D_{semip} is variable, due to the presence of the term *TimeOut on board*. Thus, only by fixing the value of the *TimeOut* is it understood whether a call with a given maximum delay time can be handled by a semipermanent satellite connection. By fixing the minimum value of *TimeOut* (i.e., $TimeOut\_min = 0$), we obtain, in fact, the minimum *DB* that can be handled by a semipermanent connection.

To better understand the consequences of the difference between D_{perm} and D_{semip}, we use the most simplified expression of the delay-bound *DB*:

$$DB = \frac{b + C_{tot}}{R} + D_{tot} \qquad (4.2)$$

The bound on the delay is decided by the receiver, which requests an amount of bandwidth in the permanent and semipermanent case equal to, respectively:

$$R_{perm} = \frac{(b+C_{tot})}{(DB-D_{perm})} \; ; \quad R_{semip} = \frac{(b+C_{tot})}{(DB-D_{semip})}$$

where $D_{semip} > D_{perm}$ and thus $R_{semip} > R_{perm}$.

Because R_{semip} is always greater than R_{perm}, the mapping over permanent connection seems to be the best solution, as it can guarantee lower delay with smaller bandwidth. However, this is not always the case. Even if for the same guaranteed maximum delay semipermanent connections require a greater bandwidth, they could provide better utilization of satellite resources in some situations. This happens due to the release of resources performed by semipermanent connections during the source inactivity periods. The goal of the study in reference 49 was to discover in which situations the use of a semipermanent mapping is more effective than a permanent one and to identify the bounds delimiting the regions in which one mapping is preferred to the other one according to the involved traffic profile. As an example of this research study, which is still in progress, in a succeeding section we will report some results of a performance assessment campaign.

4.2.1.3 MPLS via Satellite

Multiprotocol Label Switching (MPLS) is a technology that is widely accepted in terrestrial IP environments as a means for guaranteeing differentiated levels of QoS to IP traffic. The question is: How feasible is an MPLS-based solution for IP QoS via satellite platforms? Literature about this issue is beginning to appear, and some typical features of MPLS via satellite are being addressed. An interesting work that could initiate you to this issue is undoubtedly reference 33, even if you could also gather a good understanding of MPLS and its virtues and limitations when extended to a satellite environment by reading the relevant recommendations.[50]

By mapping IP over ATM, a differentiation of the QoS guaranteed to IP traffic could be performed by means of the QoS-aware routing and switching that ATM offers. By means of the bearer services defined within the ATM network, in fact, a differentiation of the QoS can be easily achieved. This QoS awareness can also be guaranteed to IP traffic encapsulated into ATM cells across the satellite as long as most of today's broadband satellite platforms with OBP capability implement ATM switching and traffic engineering on board.

MPLS is an appealing alternative solution to the ATM transport solution for IP. In fact, it has been considered as means of providing efficient mechanisms for traffic management and, most important to our aims, of differentiating QoS support. As usually addressed in literature, we can refer to MPLS as a technique providing QoS guarantees to the user traffic by basing its behavior on IP switching and QoS routing techniques, without exploiting any feature offered by the ATM Layer 3. It only assumes the likely presence of an ATM layer 2. The exploitation of MPLS for allowing the use of IP in conjunction with any Layer 2 protocol is currently under investigation within the IETF.[50]

MPLS provides Layer 3 routing by using IP addressing and dynamic IP routing to set up Virtual Connections (VCs) that are called *label switched paths (LSP)*. Its strength lies in its use of a very light protocol called *label distribution protocol (LDP)* to perform its functionality.

Nevertheless, the research community still has to consider carefully the suitability of MPLS over a satellite platform. In reference 33, both advantages and problems arising when extending MPLS to a satellite platform are very clearly addressed. The main advantages seem to be the simplification of the signaling procedure when compared with a pure IP-over-ATM switching solution on board the satellites, a lower amount of information required for the forwarding tables when compared with the IP routing tables, and mainly the possibility of a differentiation of the IP QoS offered to the Internet-satellite terminal by means of the effective mapping of DiffServ classes into MPLS Forward Equivalence Classes (FEC), similar to what happens in the terrestrial segment. Logically, these advantages have to be paid in terms of optimization efforts required to decide how to distribute the different functionality (either on board or in the terrestrial segment) compatibly to the satellite payload constraints and how to reduce the signaling exchange among the great number of IP satellite and IP earth stations that exist, thanks to the wide-reaching nature of satellite coverage. In reference 33, for example, the authors show how to optimize the MPLS LDP for the satellite environment to minimize the signaling load.

4.2.2 Routing via IP Satellites

The exploitation of IP routing throughout a system made up of a constellation of satellites and an interconnected terrestrial platform is an interesting design solution, and many advantages can derive from its implementation. Therefore, it deserves some investigation aimed at highlighting the pros and cons of such a choice and a discussion of some algorithms and protocols relevant to this topic.

Several reasons for IP routing within a satellite system are clearly highlighted in the tutorial papers found in references 3 and 51, whereas in other interesting papers, such as references 21 and 52 to 56, proposals for the solution of the most relevant problems can be found. From literature, it emerges that the relevance of introducing techniques of IP routing is represented by aspects that involve characteristics of the IP satellite infrastructure, which go well beyond the mere correct delivery of IP packets across the satellite constellation. In fact, a correct IP routing implemented within the satellite constellation positively influences some choices relevant to the evolution of Internet services and the differentiation of the guaranteed QoS.

It is worth noting that the most relevant aspect favoring an IP-over-satellite solution is the inherent broadcast capability of such a telecommunications platform. This, in fact, enables the support of currently emerging multicast/broadcast IP services that are gaining ground in the terrestrial Internet infrastructure. The IGMP protocol[57] is widely exploited in the terrestrial Internet; thus, its use on the satellite constellation as well is straightforward. Nevertheless, IP multicast can be achieved by exploiting protocols such as IGMP only if IP routing is enabled in the satellite constellation. Furthermore, the evolution of the Internet toward a platform that enables differentiated QoS also supports the implementation of IP routing. The already addressed extension of QoS-IP models, namely IntServ and DiffServ, to the satellite platform can be achieved only if effective IP routing techniques are available within the satellite constellation. In this chapter, we address both issues relevant to routing in whole native-IP satellite platforms and issues relevant to the routing of IP packets in an IP-over ATM satellite solution.

Besides the highlighted advantages of IP routing in the satellite, several other problems accompany the introduction of this feature into the overall terrestrial-satellite IP network. Our interest is mainly on the features relevant to the satellite constellation, which could be considered as an Internet Autonomous System (AS), according to the terminology used in the routing area. If the satellite constellation and its terrestrial subsystem (Network Control Center, Interworking Units, etc.) are considered to be an AS, then the issue of IP routing can be split for illustration purposes into internal routing (relevant to the satellite terrestrial system and to the on-board switching and routing) and external routing (by considering some of the gateways of the IP satellite system operating as border gateways running exterior routing protocols). Before addressing some sample solutions that have been presented in literature for the issue of internal and external IP routing, it is worth pointing out some typical problems that are associated with IP routing within the satellite system.

4.2.2.1 Problems Associated with IP Routing in the Satellite Constellation

In an interesting tutorial paper, Wood et al.[51] highlight some common problems, which we need to observe as they apparently represent an impediment to IP routing via the satellite constellation. A first source of worrying is usually considered the impossibility of fitting variable-length IP packets into a satellite air interface that allocates channel capacity through the use of fixed-length time slots in a well-ordered frame structure. Nevertheless, the standard for IP networks foresees some countermeasures for this kind of impairment. IP packets can be split into fragments by means of explicit IP-level fragmentation or implicit lower-layer fragmentation, exploiting MAC layer or tunneling protocols.

Other features that apparently rouse some complications during the introduction of IP routing are related to the size of the routing table and the complexity of the protocol. We must recall that the weight is one of the most sensitive parameters to be kept under control when dimensioning the payload of the satellite, thus preventing the engineers from introducing too much complexity on board the satellite.

This implies that most of the complexity has to move to the ground segment and that the information exchange and update from terrestrial segment to satellite is destined to increase to unbearable levels. Undoubtedly, it is difficult to think that a LEO satellite would have so much space on board to hold the information necessary to route the traffic to every terrestrial network it is connected to. Thus, scalability problems for on-board routing and processing are straightforward. Nevertheless, in reference 51 it is noted that the impact of the highlighted problems on the system performance could be reduced by separating and isolating satellite and Internet routing updates to their respective routing domains.

Also, the reduced OBP capability represents a limiting feature to the introduction of IP routing in the satellite constellation. Routing an IP packet is more complicated and consumes more processing power than simple switching, because the IP routing elements have to read the global destination address, access lookup tables, and then perform the correct forwarding. Hope in this field is given by the continual enhancements in the database access procedures, processing power, and miniaturization of the hardware, which will permit us to have a more powerful and effective on-board functionality in the next generation of satellites.

Further features, also reported in reference 3, are a source of trouble when addressing the issue of generic routing across a LEO satellite platform. Logically, they adversely affect the routing of IP packets across such a kind of platform and therefore deserve to be investigated. The on-board capability in LEO systems, along with the possibility of establishing ISLs among satellites, implies a full connectivity in the sky but raises severe routing problems. They are caused by the fast dynamic changes in the network topology in the sky (where we can refer to a full meshed network made up of satellites, analogous to the terrestrial nodes, and ISLs, analogous to the terrestrial links), due to the movement of the satellites belonging to the constellation. This means that each routing algorithm to be exploited in such a scenario must be able to effectively cope with this high variability. This is not the case, unfortunately, of the most known interior routing protocols conceived for the Internet, such as Routing Information Protocol (RIP)[58] and Open Shortest Path First (OSPF),[59] which rely on distance vector and link-state routing techniques, respectively, and thus need to have updated information on the topology and status of the links among LEO satellites. Therefore, new approaches to the effective handling of such a dynamic topology in the sky for routing purposes turn out to be vital to enable IP routing as well and to consider the satellite as a standard Internet *router*. The most interesting approaches are undoubtedly the one conceived by Mauger and Rosenberg and described in reference 54 and the one proposed and studied by Werner and Werner et al. in references 52 and 53.

In reference 54, the authors assume that the satellites are connected to the neighbors in the four directions and that the constellation is regular and symmetrical. This allows them to create a virtual model of the network that makes transparent the movement of the satellites to the terrestrial segment and introduces the new concept of the fixed Virtual Nodes (VNs). When the satellites in the sky move, the set of VNs remain unchanged in their positions. Thus, the virtual topology remains fixed because VNs are just embodied by different satellites. This is logically due to the symmetry and the predictability of the movement of the LEO satellites in their relevant constellation orbits. For some periods of time, the routing information of a VN is included in a LEO satellite embodying that node; as a new LEO satellite replaces the previous LEO, it also receives the routing information of the previous LEO. Thus, the routing decision always assumes a regular topology that is virtual and physically implemented by different satellites as time goes by.

The second method,[52,53] called *discrete-time dynamic virtual topology routing (DT-DVTR)*, bases its behavior on the assumption that in some intervals of time the topology of the satellite network does not change, and that during these intervals the cost associated with each link connecting two LEO satellites is assumed to remain unchanged. This assumption implies that the typical routing algorithms exploited by the Internet can still be effective, because they rely on a fixed topology. The paths can be computed and then the set of routing tables associated with each of the states can be set to the satellites and stored on board. Some optimizations of this procedure (reducing the handoff ratio among satellites and minimizing the required storage capacity of the satellite) are also proposed.

For more details on such methodologies enabling the implementation of typical IP routing algorithms by trying to hide the satellite mobility, refer to the cited references.

4.2.2.2 IP Interior Routing Issues

A widely accepted approach to the problem of IP routing via satellite is based on the concept that the routing implemented within the satellite constellation has to be kept isolated from the routing implemented within the terrestrial segment. This, logically, has the main consequence of reducing the complexity of the overall IP routing procedure and enabling, through the logical separation of the two routing realms, the independent optimization of the different procedures to the specific environment. Two main modalities for decoupling the terrestrial and the satellite segments of the IP network are addressed in literature: *tunneling* and *Network Address Translation (NAT)*. They are widely used in terrestrial-only IP networks, but their use can be effectively extended to the heterogeneous environment we are discussing.

IP tunnels can be created for satellite-terrestrial segment isolation purposes across the satellites links between pairs of single IP hosts scattered across the terrestrial backbones, or even routers connecting two terrestrial IP local area networks (LAN)s across the satellite, or border gateway routers isolating the satellite IP AS from the rest of the Internet world.

In so doing, the satellite constellation is seen as a single hop connecting two terrestrial IP entities that perform IP packet encapsulation/decapsulation. Two advantages of this approach, highlighted in the literature,[51] are the following:

- Adapt the tunneling network layer and routing protocols inside the constellation network to the needs and constraints of that network.
- Separate routing updates and addressing in the constellation network from routing updates and addressing in the Internet.

In addition, tunneling is a very interesting solution for a further problem present in asymmetrical satellite platforms exploiting IP. It is, in fact, an effective way to make the asymmetry transparent to the routing algorithms, which assume the presence of symmetric two-way links with similar properties. In reference 3, the authors point out that dynamic routing algorithms (such as link-state routing) would not be able to work as long as they assume that receiving a cost notification from a given neighboring node means that they can reach a destination through that node. In asymmetrical satellite platforms, receiving information across the satellite forward link does not mean that a destination can be reached by sending the packets across the same link directly (the reverse link is in fact terrestrial). Thus, only the use of the tunneling can give the possibility of relying on a "virtual" bi-directional IP link between the satellite and the end user. Details on the modality of implementing this can be found in reference 3. The authors also note, however, that the scheme described is based on point-to-point unidirectional links, although satellites have an intrinsic point-to-multipoint broadcast capability. This means that additional research is required to identify an effective way to optimize the procedure in such a system.

An alternative method to tunneling is the translation of the IP address, which is performed in the traditional IP network by means of the NAT protocol.[60] Using NAT, the designer can separate the IP addressing within the constellation of satellites from the global Internet network addressing. The consequence of this separation would be the avoidance of propagating routing update information from terrestrial to satellite segment, and vice versa. This update information exchange would otherwise occur with too high of a frequency, due to the movements of LEO satellites and to the consequent changes in the satellite network topology. NAT can be implemented at the border between terrestrial and satellite segments, where the translation of addresses (internally/externally visible/invisible) can take place efficiently.

4.2.2.3 IP-over-ATM Routing in Satellite Constellations

In a previous section, we introduced some techniques that try to make the movement of ATM-based LEO constellations transparent to the terrestrial segment in order to enable the implementation of routing algorithms basing their behavior on the continuous knowledge of the topology status. This, indirectly, introduces a topic that deserves a brief discussion as well: IP routing when exploiting the IP-over-ATM transport solution. Note, in fact, that most of the new-conception satellite platforms for broadband multimedia services are conceived as ATM-satellites. This means that ATM switching and routing is implemented on board. In these cases, the routing of IP-over-ATM traffic across the satellite constellation can be achieved easily by means of the MPLS protocol,[51] strongly supported for the terrestrial backbones, also thanks to its compatibility with the DiffServ model of the new Internet with QoS capability. The discussion of IP MPLS over satellite has already been addressed with more details in Section 4.2.1.3. You can refer to reference 51 to investigate reasons why the use of MPLS for IP-over-ATM can be beneficial for IP traffic in satellite networks aimed at supporting the ATM protocol.

4.2.2.4 IP Exterior Routing and BGP in Satellite Autonomous Systems

We have already observed that the isolation of the satellite segment belonging to a common TCP/IP infrastructure from the terrestrial Internet segment can greatly simplify the routing of IP packets. The isolation can be easily achieved by considering the satellite subsystem as an AS. This means that besides the already addressed solution to the problems of interior routing, the issue of exterior — that is, inter-AS routing — needs some study.

When we consider the satellite system as an AS, we assume that some gateways are present, which implement the functionality of the Border Gateway Protocol (BGP).[24] The latter is, in fact, the protocol usually exploited for exterior routing across the terrestrial infrastructure. The most natural solution to BGP routing consists of implementing such a protocol in terrestrial gateways (those that are positioned at the border of the satellite IP subsystem). In fact, the functionality of the border gateway is such that its implementation on board the satellites, although feasible, is not advisable because of the computational complexity it would bring on board the satellite payload (also, address translation should be performed). The border gateways have to exchange a large amount of information relevant to routes across the AS outside that include the satellite constellation. This suggests that we should avoid involving this routing update information exchange (usually very heavy) for satellite links; it suggests we should exploit only terrestrial links between terrestrial BGP gateways. This choice solves many problems but does not mean that nothing has to be changed in the way BGP is executed. The reason is that when BGP routing is used across many terrestrial ASs, only the cost of external links must be evaluated, because the cost of internal routes can be considered negligible. Recall that in terrestrial-only Internet networks, the ASs have a reduced extension that makes the cost of internal paths very low when compared with the inter-AS links. This is not the case with satellite ASs, however. The satellite coverage has, in fact, an extension that makes the cost of internal routes comparable with or even greater than the cost of terrestrial inter-AS links. This is why the mere extension of the BGP routing to the case of terrestrial-satellite Internet networks is not sufficient to guarantee an effective overall behavior. The problem results mainly from the relative weight of internal and external metrics, which is different than in the terrestrial-only case.

An example of a proposed solution to the highlighted problems relevant to BGP routing is illustrated in reference 21. The authors propose a so-called *Border Gateway Protocol — Satellite version (BGP-S)*, whose behavior is implemented by considering the satellite network as an AS with special properties. BGP-S, which is designed to support automated discovery of paths that include the satellite hops network, is to be adopted jointly with the traditional BGP-4.[24] BGP-S is implemented in only one terrestrial gateway for each terrestrial AS. This reduces the complexity. Furthermore, BGP-S uses TCP connections between two peer gateways and uses BGP-4's *LocalPref* value to propagate the paths it has discovered through the satellite network. Details on the way paths are discovered through the BGP-S protocol can be found in reference 21.

4.2.3   Unicast vs. Multicast IP over Satellites

Multicast is a another feature that is becoming increasingly important in the telecommunication scenario thanks to the new multicast services that many service providers are offering over both terrestrial and satellite platforms. Nevertheless, due to their intrinsic broadcast capability, the satellites are the winning alternative approach to high-cost terrestrial solutions. The satellite, in fact, because of its great capacity to serve wide coverage areas, can avoid overloading terrestrial IP routers and links, usually underdimensioned when compared with the typical unicast services. Examples of satellite platforms offering multicast

IP services over DVB have emerged in the past few years. Most of them are asymmetric, meaning that return links through the satellite are missing.

A problem that has clearly emerged in literature[61] is that, although the traditional multicast service over satellites does not require transmission over the return link, this is not always the case with IP multicast over satellites. If IP multicast is implemented, in fact, the scenario we must consider involves a terrestrial IP network and a satellite IP segment. Thus, the satellite portion of the IP network has to follow the typical rules of IP multicasting over traditional terrestrial links. The point is that IP multicast is usually receiver driven. The main consequence of this feature is that the typical requirement in these kinds of systems is the presence of a return channel for full-duplex communications.

The presence of a return channel is not a problem either in the case of asymmetrical platforms (a return channel via terrestrial IP links) or in the case of symmetrical platform (a return channel exploiting low-cost shared return links). The problem could be the typical difference in capacity (especially in the former type of satellites platforms) associated with the forward and the return link. A partial solution to this problem could be achieved by means of the protocol currently under standardization, the topic of massive research that goes under the name of DVB-RCS (DVB with return channel via satellite), already mentioned in Section 4.2.1.1.

An interesting survey of various protocols for multicast transmission is given in reference 61. Here you can also find interesting comments on the suitability of each category of multicast protocols to the satellite IP scenario. We will briefly report here some considerations and invite you to consult the cited work for more details.

An insightful classification of routing algorithm foresees two main families of algorithms: *one-to-many* (star based, ring based, tree based, network assisted, and asynchronous layer coding) and *many-to-many* (cloud based). What emerges is that the *one-to-many* algorithms, based on star, tree, and asynchronous layer coding, do not require the presence of a return path, thus making them suitable for a satellite platform. The problem with some of them is mainly related to delay issues, which are intrinsic in GEO satellites. In contrast, the performance of ring-based *one-to-many* multicast protocols is adversely affected by the long propagation delay caused by the presence of GEO satellites and, furthermore, assumes the presence of a return channel to send back the General Acknowledgment (ACK) and the Negative Acknowledgment (NACK) messages. A slow-rate return channel is usually not suitable for these types of protocols. The problems are even worse in the case of *many-to-many* (NACK-based) protocols, which explicitly assume the presence of forward and return links with the same transmission capacity and suffer from the GEO satellite delay to a larger extent. Therefore, these protocols are not suitable at all for the GEO asymmetric IP satellite platforms.

Besides the drawbacks related to the typically high round-trip delay of the IP satellite links, a further drawback relates to the channel error characteristics of the satellites. Via satellite, the errors are bursty, and the presence of these bursts result in a higher IP packet loss rate than would be expected for the same bit error rate (BER) on terrestrial (wireline) links. The problem is that for reliable multicast applications, the IP datagram loss can, therefore, have a significant impact if retransmission is required by the multicast protocol. This aspect has also been addressed in literature,[62] and some performance evaluations are available.

A great research effort is under way within many European Community research frameworks on several aspects related to the multicast via satellite issue. The future results will be beneficial for the deployment of IP satellite platforms, which are the focus of this chapter.

4.2.4 TCP Performance over Satellites: Technical Challenges, Solutions, and Performance

In this section we will address the issue of TCP extension to a satellite scenario. This is a very hot topic widely addressed in the literature relevant to the topic of IP extension to a satellite-terrestrial integrated platform. Thus, what we aim at giving in this section is just a flavor of the many challenging aspects related to this issue, the intrinsic limitations of TCP when performing across satellite links, and the technological solutions that are under way regarding standardizations (or that are still at their infancy).

For an in-depth understanding of this issue, refer to the standard documentations and documents in literature that will be mentioned in this section.

First, let us try to identify the problems that a transport protocol like TCP experiences over satellite links. Many tutorial literature (see references 3, 15 and 63 to 67), is available as well as several RFC documents from the IETF group.[22,23] From their reading it is clear that the extension of TCP to the satellite environment is not effortless. The main problems arising in the satellite scenario derive from some peculiarities shown by the satellite links, well described in reference 22 and briefly reported here.

Usually the feedback loop in a satellite system is very long. This is due to the propagation delay of satellite channels, which could increase up to 270 ms for GEO satellite platforms. The impact of this feature is mostly on the TCP transmitter, which cannot be sure about the success of its transmission before a long time interval has passed. This delay also causes TCP to be compelled to keep a large number of packets sent but not yet acknowledged and, thus, to require expensive buffers (this is mostly related to the large *delay*bandwidth* product typical of GEO satellite systems).

Besides the delay, a further "killer" of the QoS in TCP connections over satellite links is the high error-prone (highness BER) of these links. As widely known, the reaction of TCP to a loss experienced over a connection consists of assuming that the loss is caused by a strong congestion level over the link, and consequently the window size is reduced to limit such an assumed congestion.

In particular satellite platforms, other typical problems could arise. One is the adverse effect on the performance of TCP caused by the presence of uplinks and downlinks with different capacities (usually the uplink is terrestrial and the downlink is via satellite) in asymmetric satellite networks. In fact, in literature it has been demonstrated that slower reverse channels can induce deleterious effects, such as ACK losses.[64] Another issue could arise in LEO platforms, in which the problems are represented by the *increased loss probability* related to the necessity of performing handovers, as well as by the *high fluctuation of the terrestrial-satellite propagation delay* (the amount of variability depends on the number of satellites, orbital dynamics, intersatellite routing algorithms, and so on; it is often assumed that the variability can range from a few milliseconds up to 80 ms). The negative influence of the handover feature is proved, whereas the actual negative impact of the delay fluctuation is currently under investigation. Anyway, it is easy to understand that a high variability in the round-trip delay implies a great inaccuracy when choosing the exact values of the TCP retransmission timeout. Studies performed for Teledesic show that the impact of this feature on TCP is not negligible. An example of test study over a LEO simulated platform, in which it is concluded that TCP performs better over LEO satellites than over GEO satellites, is described in reference 68. Nevertheless, the authors of this chapter think that more research is required before the assertion in reference 68 is fully assessed and accepted.

At this point, it is clear that the adverse impact of BER and large propagation delays on TCP is relevant, as these conditions usually cause excessive TCP timeouts and retransmissions and large bandwidth inefficiency. Nonetheless, this is not the only concern in a TCP-over-satellite scenario. Additional concerns are posed by other typical features of TCP whose limitation usually emerges in long-latency environments, such as the satellite: small window sizes, round-trip timer inaccuracies, startup windows, and so forth. The strategy usually followed to face these problems consists of operating at different protocol levels and intervening both on so-called *non-TCP mechanisms* (usually they are Layer 2 mechanisms) and on *standard-TCP mechanisms*. Furthermore, a strong research effort is currently trying to develop some *satellite-TCP specific mechanisms* aimed at improving TCP performance via the satellite. These are not yet standardized and accepted by the TCP/IP community because some research work is still required to assess their performance and effectiveness. Nevertheless, some of them are developed in TCP/IP satellite platforms, which are beginning to appear in the telecommunications scenario. We will go through some of the mechanisms belonging to each of the three addressed categories by pointing out their impact on TCP.

The mechanisms, which are not specific of TCP, are better highlighted in references 22 and 64 and are briefly outlined here: *Path MTU Discovery* and *forward error correction (FEC)* techniques.

Path MTU discovery[69] is exploited to establish the maximum packet size that can be conveyed over a given path of the network without being compelled to perform the IP fragmentation. Its adoption can allow TCP to use the largest possible MTU over the satellite channel. The advantage is related to the fact that the increase in the congestion window is segment based; therefore, larger segments enable TCP senders to increase the congestion window more rapidly in terms of bytes rather than smaller segments. In a high-latency environment, such as with the satellite, this corresponds to increasing the TCP performance while reducing the fragmentation overhead. By implementing such a technique, the only negative effect that needs to be controlled is the likely increase in the delay after the path MTU discovery procedure.

When a loss occurs, TCP interprets it as the evidence that the network is experiencing congestion and reduces the congestion window even if the loss is due to the errors occurring only on the satellite link. Recovering from this false alarm is a very slow procedure, with a resulting deterioration of TCP performance. Therefore, some techniques reducing the losses due to the channel can have a beneficial effect on TCP. The use of FEC techniques on a satellite link should achieve the main task of reducing the BER over satellite links.

Further mechanisms, more strictly related to the specificity of the TCP protocol, are briefly described in addition to the non-TCP mechanisms already discussed. More details can be found in references 3, 23, and 64.

As clearly stated in official IETF documentations,[23] the techniques we are going to present in the remaining part of this section have to be considered as just a brief presentation of the output of research that is currently conducted in the TCP-over-satellite issue. The protocol and algorithms presented are not standards (yet), neither are they recommendations. They are "possible TCP enhancements that may allow TCP to better utilize the available bandwidth provided by networks containing satellite links."[23] Perhaps they will become standards, but some research is still required. We will address just a few techniques. For a complete list of possible TCP-satellite enhancement measures, please refer to the document cited earlier.

TCP for Transaction (T/TCP), first studied in references 70 and 71, is an extension of the traditional TCP protocol, which foresees that the three-way handshake procedure required to set up a connection is avoided. Satellite Internet systems can, logically, derive a great beneficial from the adoption of this technique. It is, in fact, very effective in enhancing the performance of TCP over long-latency links by allowing data transfer to start more rapidly.

Alternative techniques aimed at reducing the duration of the slow-start phase, and thus making its use more efficient in long-delay networks, are also under investigation. The most relevant for our environment are the *use of larger initial windows* and the technique of *byte counting*. For satellite networks, it is strongly recommended that the initial size of the congestion window (*cwdn*) be increased from 1 to 2 or according to some equations that can be easily found in RFC documents or in literature. This will, in fact, reduce the duration of the slow-start phase by reducing the time required by a TCP source to increase the size of its congestion window during slow start. The same beneficial effect is achieved by means of *byte counting* instead of the traditional "ACK counting." According to this technique, increased throughput is achieved by incrementing the window based on the number of newly acknowledged bytes, independent of the receiver's ACK generation rate. However, some side effects resulting from this technique require further research to prevent large line-rate bursts[64] (i.e., *limited byte counting*[72]). The cited technique is expected to be especially beneficial to asymmetric networks.[23]

We have already addressed that by intervening into the mechanisms related to the "loss prevention" across the satellite TCP link (e.g., by using FEC coding) we can increase TCP performance. A further beneficial effect could derive from the adoption of new techniques at the "loss recovery" level, such as mechanisms based on the use of the *fast recovery* algorithm, which takes into account information provided by Selective Acknowledgments (SACKs)[73] sent by the receiver. It consists of allowing the receiver to specify the correctly received segments, thus enabling the sender to retransmit only the lost packets.

In the literature,[64] TCP SACK performed very well in long-delay environments with moderate losses when retransmitting all lost segments within the first round-trip time after fast recovery is triggered. It is also proved that timeouts can be reduced and that the SACK technique can be used jointly with other traditional techniques without problems. Nevertheless, under highly errored channel conditions (about 10^{-6}), TCP SACK required significant modifications to be actually effective. An interesting alternative proposal goes under the name of *forward* ACK (FACK) and still uses the information provided by the SACK option. Although additional research is required in highly errored satellite environments, FACK is expected to provide good performance gains. Please refer to reference 23 for an in-depth treatment.

A further approach proposed in literature for overcoming the TCP performance reduction in a satellite system consists of supporting a data transfer over satellite links by establishing *multiple TCP flows*. The beneficial effect of this approach is represented by the possibility of starting a transmission with a congestion window of N segments, which is more effective when compared with the single-segment *cwnd* size in traditional TCP flows.[22] The use of multiple connections is equivalent to the use of TCP based on a single connection but exploiting a larger congestion window. Thus, the beneficial effects already addressed for that case still hold in the present multiple-segment case. In reference 3, three different approaches for performing the TCP connection splitting over the satellite are reported: TCP spoofing, TCP splitting, and TCP caching. Consult the referenced article, as well as references 66 and 74 to 76, to have a clearer vision of the performance and effectiveness of some of these splitting techniques.

A further aspect worth highlighting is the possibility in new-generation OBP-equipped satellite platforms of supporting not only TCP over a best-effort service framework but also TCP over frameworks guaranteeing a minimum rate to services. The latter feature requires the study of TCP performance when in the presence of different buffer management and QoS scheduling techniques. Most of the research in this field has assumed that the switching techniques implemented on board are based on ATM. Thus, several buffer management techniques have been proposed to improve TCP performance over ATM networks. We will not go too much into detail with these techniques; often, they are strictly related to particular IP-over-ATM–based satellite platforms. A good reference paper in this field is reference 77, where the authors study buffer management techniques in satellite networks for TCP transport and present experimental results and guidelines to design satellite-ATM network architectures for the efficient transport of TCP data.

4.3   Implementation of IP-Satellite Services and Platforms

4.3.1   Examples of IP-Satellite Architectures and Scenarios for Internet Services

In this section we will briefly describe some examples of satellite architectures suitably conceived for matching the increasing needs of users who want to access multimedia services made available by ISPs through satellite links. We will focus on architectural solutions actually available either as prototypes or as demonstrators, representing the output of research projects involving partnerships made up of industrial, university, and research institutions. Several further solutions are available on the market; thus, those presented have to be considered as mere examples that should, it is hoped, help you understand how some of the concepts addressed so far could be implemented in real IP-satellite infrastructures.

4.3.1.1   The ASI-CNIT Experience

As a first example we consider the output of the project "Integration of Multimedia Services on Heterogeneous Satellite Networks," supported by the Italian Space Agency (ASI) under Computer Networking and Information Technology (CNIT) contract ASI-ARS-00-205.[78] It represents a reference solution for the interconnection of IP network segments in which the need for QoS awareness and QoS differentiation is strongly felt. The two main characteristics making it attractive for the end user are its simplicity and private nature.

The literature[78,79] shows that the system uses an ITALSAT II (13°EST) satellite in the Ka band (20–30 GHz), which is a platform currently exploited to provide new services. Each satellite station can be assigned a full-duplex channel with a bit rate ranging from 32 Kbps to 2 Mbps.

Different LANs, composed of PCs (a source of the services under test: TCP/IP videoconferencing tools, TCP/IP file transfer, and Web browsing), are connected either through a satellite link at 2 Mbps or with each other through ISDN links. The interconnecting devices utilized between terrestrial and satellite segments are routers operating at the IP level.

Although private, such an infrastructure presents typical management problems common to larger TCP/IP satellite-based networks. To reserve bandwidth for the sessions requiring a quantitative guaranteed QoS level, the IntServ paradigm is adopted within the integrated platform. Furthermore, an interesting modification of the TCP protocol has been introduced to improve the performance of the additional traffic crossing the network: file transfers and Web browsing. As for the tools exploited at the application level and exploiting the TCP/UDP and IP features offered by the network, the Mbone tools (session directory, SDR, for multicast session announcement; videoconferencing, VIC, for video; and robust audio tool, RAT, for audio) are used to transmit video and voice.[80] If you would like a more detailed look at the protocol and infrastructure characteristic of this platform, refer to references 78 and 79.

4.3.1.2   The SUITED Project

The second example of satellite systems is the one deployed within the *multisegment System for broadband Ubiquitous access to InTEr-net services and Demonstrator* (SUITED) IST1999-10469 European Union project,[2] whose goal is to meet the expectations of multimedia mobile users. The resulting infrastructure is very different from the one described previously in that it shows the typical features of a *Global Mobile Broadband System (GMBS)* — that is, an integrated satellite/terrestrial infrastructure including a *GMBS multisegment access network* and a *federated Internet service provider (ISP) network* (an Internet segment implementing appropriate functionality).

Undoubtedly, the most interesting aspect of this infrastructure is the user perception of a single network supporting mobile Internet services with QoS guarantees, whereas the multisegment access network is actually made up of the following components:

- The Ka band OBP-based satellite system
- The General Packet Radio Service (GPRS) system
- The Universal Mobile Telecommunications System (UMTS)
- The wireless LAN (WLAN) system

This means that the user is able to access the multimedia services offered by an ISP by exploiting a wide range of alternative access networks. In public documents it is stated that the approach to the deployment of the GBMS is of the evolutionary kind: "...in GMBS Phase I (ready to be operational by 2004), the satellite segment will be complemented by the GPRS system, while the GMBS Phase II (ready to be operational by 2006) will be characterized by the presence of the UMTS segment complementing/replacing the GPRS.... The WLAN will be present in both phases."[2]

The GMBS is composed of two main portions: *edge* and *core* segments. The former consists of a wireless segment and Internet subnetwork, whereas the latter consists of one or more subnetworks and the satellite segment (acting as a transit IP).

The IntServ and DiffServ models are adopted in the edge and in the core segment, respectively. These paradigms are enhanced by an admission control mechanism, namely Gauge&Gate Reservation with Independent Probing (GRIP),[81] suitably designed for the envisaged platform. To handle IP mobility, the federated ISP network implements the Mobile IPv6 protocol.

4.3.1.3   SkyplexNet: A Platform Utilizing DVB and DVB-RCS Standards to Support IP Traffic

The last platform we will focus our attention on is the SkyplexNet[82] system, representing an innovative ground-segment solution developed within the framework of an ARTES-3 line 1 ESA project, based on the "Skyplex on-board multiplexer," and currently operating on the HotBird 5 satellite. It is specifically thought to provide the end user connected to low-cost satellite terminals with a new generation of high bit-rate services.

The way this platform acts is typical of the asymmetrical IP satellite platform described in previous sections of this chapter. The user terminals are both connected to the IP terrestrial network and to the receiving DVD-S system and antenna.

The multimedia contents are requested by the user through the terrestrial return link from the *Content Providers* as IP datagrams. The IP flows are delivered to the *Service Center,* through high-speed dedicated lines or through the Internet, according to the application requirements. The Service Center collects the user requests and, after a suitable scheduling policy, exploits IP packets to send the contents to the *Uplink Station.* The so-called *IP Gateway* in the station multiplexes, scrambles, encodes into MPEG2 Transport Stream format, QPSK-modulates, and sends the data to the assigned satellite RF channel. In uplink, the system uses high-capacity satellite links (from 2 Mbps to 6 Mbps).

All flows that arrive are multiplexed on board the satellite and transmitted like a single DVB signal to the receivers in TDM modality at a rate of 55 Mbits/sec (27.5 Msymbols/sec).

SkyplexNet is particularly suited to the transmission of multimedia applications, Internet data, and broadcast television in MPEG2 Transport Stream format and coming from different remote locations. The features that make this platform a cost-effective asymmetric satellite communications solution with terrestrial return link include the characteristic SkyplexNet networking technology, the *SCPC/TDMA* (Single Channel Per Carrier/Time Division Multiple Access) modality exploited by the uplink stations, the *bandwidth-on-demand* resource allocation mechanism, and, last but not least, compliance with the DVB standard.

Note that this platform is currently evolving toward the SkyplexNet/HB6 architecture, which is enhanced through the deployment of suitable satellite access techniques based on the DVB-RCS Standard[83] and will be operative over the HotBird 6 satellite platform. For this system different types of terminals are foreseen, namely, two-way end-user SaTellite Terminals (SaTs) and provider, gateway, and NOC (Network Operation Center) terminals. They access the satellite resources either in the Ka or Ku band for the uplink and downlink channels. The access method is either SCPC or TDMA, and the transmission rate can vary according to the access scheme and bandwidth assigned.

4.3.2   An Example of Performance in Specific IP-Satellite Platforms

In this section we will report some performance results on the study of an efficient introduction of IP-with-QoS models (DiffServ and IntServ) into a real terrestrial-satellite platform that has been developed within projects partially funded by the European Space Agency and carried out by the EuroSkyWay (ESW) partnership.

4.3.2.1   A Brief Description of the Multimedia Satellite ESW Platform

The objective of the ESW project is the design and development of a new generation of broadband satellite networks operating in the Ka band (20–30 GHz) and equipped with a fully digital OBP payload. The OBP technology allows direct connectivity among any terminals, thus eliminating the need for expensive gateway stations.[49,84]

The ESW development plan is based on the realization of the SkyplexNet system and involves three steps.[84] The first step (1998–2001) provided the launch in orbit of OBP technology and the setup of the SkyplexNet-HotBirdTM5 communication platform for the service provider. The second step (2001–2003) foresees the development of the Ka band and interactive satellite user terminals. The last step, planned for 2004–2005, will provide for the launch of the first "dedicated" satellite with 1-Gbps capacity.

The main features of the ESW platform are the following:

- Provision of point-to-point and point-to-multipoint connectivity
- Direct access to the satellite
- Two-way, high-speed communications through small, flexible, and low-cost terminals operating at different transmission rates
- Bi-directional bandwidth on demand, dynamic bandwidth resources allocation (up to 2 Mbps in the uplink and 32.8 Mbps in the downlink)
- Full compatibility with the DVB/MPEG and DVB-RCS standards
- Small satellite dish (about 80 cm)
- Europe and Mediterranean Basin coverage

4.3.2.2 Performance Study

In this section we report the main results of a performance analysis that the authors of this chapter have carried out[49,85] with the purpose of finding the best mapping of IntServ traffic classes on the satellite bearer services offered by the reference ESW platform. Note that the results can be extended to any GEO satellite platform and are not related only to the specific platform under study. The reference network architecture is the one depicted in Figure 4.7.

The relevance of the reported results relates to the fact that the IETF suggestions for the mapping of Constant Bit Rate (CBR) and Variable Bit Rate (VBR), and best-effort traffic profiles over IntServ traffic categories valid for a terrestrial IP platform are not always suitable for a satellite platform. The main reason for this behavior is the long latency of GEO satellite links. To investigate this issue further, we refer to the detailed expression of the Delay Bound (DB) already shown in Equation 4.1a and Equation 4.1b, which are reported here for the sake of clarity. DB is the maximum delay assured to the Guaranteed Service (GS) class within the IntServ/RSVP framework. The DB value is computed by accounting for the source peak rate p, the token generation r, and the maximum packet size M:

$$DB = \frac{(b-M)}{R} \times \frac{p-R}{p-r} + \frac{M+C_{tot}}{R} + D_{tot} \quad \text{if } r =< R < p \tag{4.3}$$

$$DB = \frac{M+C_{tot}}{R} + D_{tot} \quad \text{if } r < p <= R \tag{4.4}$$

The DB value is different depending on whether the requested rate R is greater or lower than p. We can distinguish two cases based on the value of the delay DB corresponding to the condition $R=p$. We call this delay value *DBS*, which can be computed as follows:

$$DBS = \frac{M+C_{tot}}{p} + D_{tot} \tag{4.5}$$

If the requested delay DB is lower than *DBS*, the requested bandwidth R over the satellite is greater than p. Otherwise, if a delay greater than *DBS* is sufficient, then a smaller bandwidth is requested. Analytically, we obtain the following:

$$R = \frac{\left[\dfrac{b-M}{p-R} + (M+C_{tot}) \right]}{\left[\dfrac{b-M}{p-r} + (DB-D_{tot}) \right]} \quad \text{if } DB > DBS \tag{4.6}$$

$$R = \frac{M+C_{tot}}{DB - D_{tot}} \quad \text{if } DB <= DBS \tag{4.7}$$

Time delay is a major QoS issue, especially in satellite networks; thus, it is interesting to give some details on the end-to-end time delay that the illustrated IntServ-DiffServ platform can offer to the applications and to examine the effect that this delay has on the overall integrated system performance. Because the reference architecture includes a GEO satellite network, it is clear that it is unsuitable for voice and highly interactive real-time traffic because of long latency (including processing, path establishment, service mapping, and propagation time delay). Nevertheless, a wide range of low-interactive, real-time packet-based applications, which tolerate some end-to-end delay, can be successfully supported.

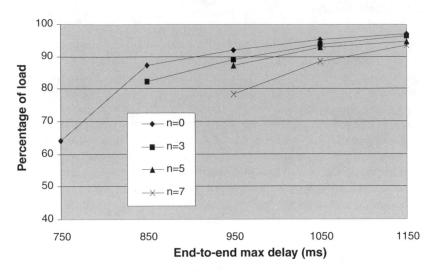

FIGURE 4.9 Percentage of traffic load vs. end-to-end max delay. (From Iera, A. and Molinaro, A.[85] With permission, © 2003 IEEE.)

We will show some curves illustrating the relationship between the end-to-end maximum delay allowed by the applications and the following main system performance indexes: the *system loading percentage* (defined as loading_percentage = Avg._num_of_busy_channels/Total_num_of_channels × 100), which predicts some idea of the achievable system resources utilization, and the *burst blocking probability*, which predicts the QoS performance guaranteed to the packet-based applications.

The curves in Figure 4.9, which are relevant to the system load, refer to GS traffic only. They are sketched by fixing the token bucket size $b = 128$ Kbit, the token generation rate $r = 256$ Kbps, and the source burstiness $\beta = 3$. The curves are shown for different values of the *TimeOut* expressed in terms of the number n of TDMA frames a resource request from a GS burst tolerates to wait on board before being satisfied.

The sustained load increases with the requested end-to-end delay; the same is true for the number of accepted sources (curves are not shown). In fact, when the maximum end-to-end allowed delay increases, the requested bandwidth R decreases; therefore, the number of accepted sources and the total exploited satellite channels increase. As expected, a worse behavior corresponds to a higher source burstiness. An increase in the burstiness, in fact, implies an increase in the requested bandwidth and a resultant decrease in the total exploited channels. Nevertheless, the system performance still remains satisfactory for the end-to-end delay range shown in Figure 4.9.

A further important issue to investigate derives from the observation that when conveying the GS class over semipermanent satellite connections, burst losses may occur due to the statistical multiplexing performed by the CAC in the Master Control Station (MCS) of the ESW system. Therefore, an interesting performance index is the *burst blocking probability (BBP)*, which measures the probability that a burst waiting for resources must be blocked (and then lost) due to unavailability of satellite channels. Figure 4.10 shows BBP curves for $b = 1024$ Kbit, $r = 256$ Kbps, $\beta = 3$, and for different values of *TimeOut*. These curves demonstrate that the loss caused by the mapping of GS flows on semipermanent satellite connections can be kept below the bound of 0.01 established by the CAC mechanism.[86] Furthermore, the BBP decreases when the *TimeOut* increases; in fact, the longer a burst is disposed to wait on board, the smaller the probability that a buffered request is discarded. In general, the BBP remains below the bound (0.01) established by the CAC unless the *TimeOut* is zero, independently from the burstiness value. Also the BBP behavior for different values of the b/r ratio, with fixed burstiness and $r = 256$ Kbps, can be analyzed. In this case as well, the BBP remains below the bound established by the CAC unless the *TimeOut* is zero. A similar behavior can always be found under any traffic profile and loading condition.

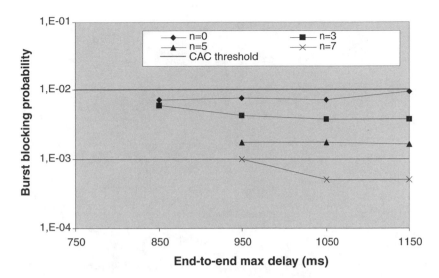

FIGURE 4.10 Burst blocking probability vs. maximum request delay, by varying the *TimeOut.* (From Iera, A. and Molinaro, A.[85] With permission, © 2003 IEEE.)

4.3.2.2.1 The Service Mapping over Satellite Connections

In references 87 and 88, IETF defines some guidelines for mapping the IntServ classes over the traffic classes of a high-speed fixed backbone, such as ATM, in order to provide effective end-to-end QoS.

The GS class has stringent temporal constraints; therefore, CBR and real-time VBR (rt-VBR) ATM service classes represent two feasible mappings. Between them, IETF recommends the mapping of GS over rt-VBR connections. The main reason lies in the fact that system efficiency can be improved as a result of the exploitation of statistical multiplexing of rt-VBR traffic sources. But things are supposed to change when the terrestrial connection is extended toward a satellite link or the wired backbone is accessed through the satellite.

For the sake of simplicity, but without losing generality, we consider a sender, *Tx,* communicating across the reference IntServ-DiffServ network with a receiver, *Rx. Tx* is a host in the terrestrial IntServ access network, and *Rx* is a mobile terminal of the IntServ satellite system.

To assess the performance of each type of mapping for the IntServ service classes over satellite connections, two metrics related to the satellite bandwidth utilization can be introduced: *Static_Efficiency* and *Utilization_Factor,* defined as follows:[49,85]

$$Static\_Efficiency\ (SE) = \frac{r}{R}$$

$$Utilisation\_Factor\ (UF) = \frac{r}{R} \times \frac{T_{conn}}{T_{poss}} = Static\_Efficiency \times \frac{T_{conn}}{T_{poss}}$$

The *SE* parameter is computed as the source *average data rate r* over the *allocated bandwidth R.* The *UF* parameter is introduced to take into account the average amount of data generated by a source during a session ($r \cdot T_{conn}$), compared with the total amount of data actually transportable when the bandwidth is not continuously allocated ($R \cdot T_{poss}$). T_{conn} is the actual duration of the connection, and T_{poss} is the fraction of T_{conn} during which the source actually owns the resources allocated to it. For the case of a permanent connection, T_{poss} coincides with T_{conn}, as the resources are allocated to the requesting source for the whole duration of the connection. Consequently, *SE* and *UF* coincide. For the case of semipermanent connections, the allocated resources (*R*) are released during the source inactivity (OFF) periods; thus, T_{poss} is shorter than T_{conn} and *UF* is higher than *SE.*

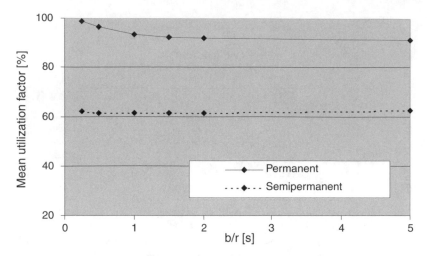

FIGURE 4.11   Mean utilization factor vs. b/r; β small. (From Iera, A. et al.[49] With permission, © 2003 IEEE.)

It is clear that the mere computation of *SE* does not directly allow us to conclude that a better resources utilization corresponds to a given mapping for any traffic profile and loading condition. In fact, *UF* must also be computed to determine the degree of system effectiveness. By measuring the *SE* and *UF* values it will be clear in the remainder of this section that the best mapping of the GS class over a satellite connection (permanent or semipermanent) strongly depends on the values of its traffic parameters. There is no unique optimal solution for each GS source type.

4.3.2.2.2 The Effect of the b/r Ratio and the Burstiness β on the System Performance

The following curves show the influence of β and b/r on the choice of the best mapping of the GS class over satellite connections. The ratio b/r represents the time needed for an empty bucket to be filled with b tokens. Therefore, given a value of traffic burstiness $\beta = p/r$, b/r determines the mean length of the transmission inactivity periods (Toff).

The first result is that, for *small* values of burstiness β, the most convenient mapping is always the permanent one. Figure 4.11 confirms this assessment; it shows the mean value of the *UF* parameter. For sources with *average* or *high* burstiness, this is not always true.

Figure 4.12 shows curves of the mean *UF* for average and high burstiness β values (2 and 50). They clearly show that higher burstiness corresponds to a lower utilization factor for both mappings. It is important to notice that the relative behavior between permanent and semipermanent mappings remains the same.

The curves in Figure 4.12 confirm that, in some situations, the system resource utilization can be higher if the GS class is mapped over semipermanent satellite connections. This is the case when b/r is higher than a given threshold. This supports our view that a comparison performed in terms of the *SE* index is not enough to state that the permanent mapping is better than the semipermanent one. *UF* is a more complete index, which gives information about the actual utilization of the resources allocated to the connections. By analyzing curves in Figure 4.12, when b/r increases, the *UF* of semipermanent connections increases as well, in contrast to what happens for the permanent case. This is a consequence of the possibility of sending longer bursts and of the increasing of sources inactivity periods after the increase of b/r. The latter point represents an advantage for semipermanent connections, because they release the allocated resources during inactivity periods.

Additional analyses of the system behavior conducted by varying the values of the parameters b/r and β provide some interesting hints regarding the impact of the cited parameters on the performance of each mapping solution. An expected result is that the permanent mapping is the best for very low burstiness values. A more significant result is that for burstiness values above a threshold the mapping decision has to always depend on the value of b/r.

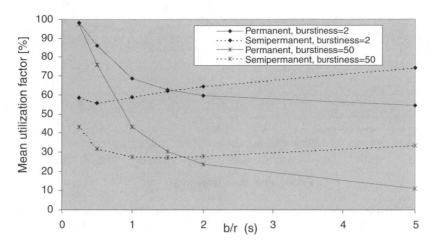

FIGURE 4.12 Mean utilization factor vs. *b/r*; average or high β. (From Iera, A. et al.[49] With permission, © 2003 IEEE.)

TABLE 4.1 Ideal Mapping

	b/r < 1.8 sec	*b/r* > 1.8 sec
β < 1.5	Permanent	Permanent
β > 1.5	Permanent	Semipermanent

Source: From Iera, A. et al.[49] With permission.

For average or high β, a threshold is *always* detected in the value of *b/r*, corresponding to the case where the *UF* of both mappings is almost similar. For values of *b/r* below the threshold, the permanent mapping is the most advantageous. For *b/r* values higher than the threshold, the semipermanent mapping is more effective, regardless of the requested delay. In these situations, in fact, *UF* for permanent connections is very low, because it is adversely affected by the increase in the inactivity periods of the source.

Table 4.1 summarizes the results achieved for the sample case shown here. For values of the burstiness lower than 1.5, the permanent mapping always proves to be the best. In contrast, for average or high (higher than 1.5) burstiness, *b/r* values equal to about 1.8 represents a threshold value; below this threshold the permanent mapping is the best, whereas above this threshold the semipermanent mapping performs better. The same behavior can *always* be observed for any traffic profile and system condition.

4.3.2.2.3 The Effect of the Target End-to-End Delay

Another interesting aspect to investigate is the effect of the target end-to-end delay desired by the receiver on the behavior of the GS mapping on permanent and semipermanent satellite connections. The target end-to-end delay has to include both queuing and propagation delays, and its value determines the amount of the requested bandwidth *R*.

The curves in Figure 4.13 illustrate the changes in the UF and the SE of both mappings (in the case of permanent mapping, they coincide). They show that by increasing the value of the target delay, the *UF* of both permanent and semipermanent connections increases. This is an expected behavior; in fact, an increase in the end-to-end-delay requested by the receiver corresponds to a reduction in the bandwidth *R* allocated to the connection. This smaller bandwidth results in less resource waste, which is observed when the source is momentarily silent and the satellite channels reserved to it remain unused. For the permanent case, the values on the right side of the curve correspond to a value of *R*, which is about equal to the rate *r*.

As for the slope of the curves, we note that the slope of the permanent curve remains almost unchanged, whereas a change in the slope of the semipermanent curve can be noted for a value of the target delay equal to 655 ms. According to Equation 4.3 and Equation 4.4, for delay values lower than this point, the required bandwidth *R* is higher than the peak data rate *p*. For higher delay values, the required *R* is lower than *p*.

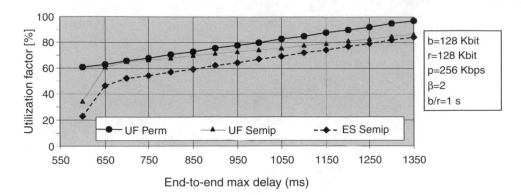

FIGURE 4.13 Utilization factor (UF) and static efficiency (SE) vs. end-to-end requested delay. (From Iera, A. et al.[49] With permission, © 2003 IEEE.)

It is worth noticing that the increase in the UF can always be observed for any traffic condition, as well as the discontinuity in the slope of the curve relevant to the semipermanent mapping.

Furthermore, Figure 4.13 shows that the most convenient mapping is always the permanent one. It is proven by the fact that the sources are characterized by $\beta > 1.5$ and $b/r < 1.8$ (see Table 4.1); regardless of the delay, the situation does not change. It can be concluded that by varying the end-to-end requested delay, the values of the UF change, but the relative behavior does not change if b/r and burstiness values are kept constant.

References

1. Mathy, L., Edwards, C., and Hutchison, D., The Internet: a global telecommunications solution?, *IEEE Network Magazine*, 14 (4), 46, 2000.
2. Conforto, P. et al., Ubiquitous Internet in an integrated satellite-terrestrial environment: the SUITED solution, *IEEE Communications Magazine*, 40 (1), 98, 2002.
3. Hu, Y. and Li, V.O.K., Satellite-based Internet: a tutorial, *IEEE Communications Magazine*, (39) 3, 154, 2001.
4. Cooper, P.W. and Bradley, J.F., A space-borne satellite-dedicated gateway to the Internet, *IEEE Communications Magazine*, 37 (10), 122, 1999.
5. Clausen, H.D., Linder, H., and Collini-Nocker, B., Internet over direct broadcast satellites, *IEEE Communications Magazine*, 37 (6), 146, 1999.
6. Ohmori, S., Wakana, H., and Kawase, S., *Mobile Satellite Communications*, Artech House, Boston-London, 1998.
7. Iera, A., Molinaro, A., and Marano, S., Geostationary satellite platforms for multimedia communications: signalling issues, *ASSI Satellite Communications Letters*, II (2), 1, 2002.
8. Evans, B.G. and Tafazzoli, R., Future multimedia communications via satellite, *International Journal of Satellite Communications*, 14 (6), 467, 1996.
9. Connors, D.P., Ryu, B., and Dao, S., Modelling and simulation of broadband satellite networks. Part I: medium access control for QoS provisioning, *IEEE Communications Magazine*, 37 (3), 72, 1999.
10. Mertzanis, I. et al., Protocol architectures for satellite ATM broadband networks, *IEEE Communications Magazine*, 37 (3), 46, 1999.
11. Grace, D. et al., Providing multimedia communications services from high altitude platforms, *International Journal of Satellite Communications*, 19 (6), 559, 2001.
12. Hase, Y., A novel broadband all-wireless access network using stratospheric platforms, in *Proceedings of the IEEE Vehicular Technology Conference*, Ottawa, Canada, 1998, 1191.
13. The Halo Network's Web site: http://www.angeltechnologies.com.

14. Sengupta, J.R., Evolution of the INMARSAT aeronautical system: service, system, and business considerations, in *Proceedings of the International Mobile Satellite Conference*, Ottawa, Canada, 1995, 245.

15. Jamalipour, A., Broad-band satellite networks — the global IT bridge, *Proceedings of the IEEE*, 89 (1), 88, 2001.

16. Pratt S.R. et al., An operational performance overview of the Iridium low earth orbit satellite systems, *IEEE Communications Surveys*, 2 (2), 2, 1999.

17. Hirshfield, E., The Globalstar system: breakthroughs in efficiency in microwave and signal processing technology, *Space Communications*, 14, 69, 1996.

18. Schindall, J., Concept and implementation of the Globalstar mobile satellite system, in *Proceedings of the International Mobile Satellite Conference*, Ottawa, Canada, 1995, A-11.

19. Sturza, M.A., Architecture of the Teledesic satellite system, in *Proceedings of the International Mobile Satellite Conference*, Ottawa, Canada, 1995, 212.

20. Farserotu, J. and Prasad, R., A survey of future broadband multimedia satellite systems, issues and trends, *IEEE Communications Magazine*, 38 (6), 128, 2000.

21. Ekici, E., Akyildiz, I.F., and Bender, M.D., Network layer integration of terrestrial and satellite IP networks over BGP-S, in *Proceedings of the IEEE Globecom*, San Antonio, 2001, 2698.

22. Allman, M. and Glover, D., Enhancing TCP over satellite channels using standard mechanisms, IETF RFC 2488, 1999.

23. Allman, M. et al., Ongoing TCP research related to satellites, IETF RFC 2760, 2000.

24. Rekhter, Y. and Li, T., A Border Gateway Protocol (BGP-4), IETF RFC 1771, 1995.

25. Digital Broadcasting Systems for television, sound and data services; framing structure, channel coding and modulation for 11/12 GHz satellite services, ETSI EN 300 421 v.1.1.2, 1997.

26. Digital Video Broadcasting (DVB): DVB specification for data broadcasting, ETSI EN 301 192 v. 1.1.1, 1997.

27. Digital Video Broadcasting (DVB): DVB implementation guidelines for data broadcasting, ETSI TR 101 202 v.1.1.1, 1999.

28. Cominetti, M. and Morello, A., Digital video broadcasting over satellite (DVB-S): a system for broadcasting and contribution applications, *International Journal of Satellite Communications*, 18 (6), 393, 2000.

29. MPEG-2 Information technology — generic coding of moving pictures and associated audio information — part 1: systems, ISO/IEC 13818-1, 1996.

30. MPEG-2 Information technology — generic coding of moving pictures and associated audio information — part 2: Video, ISO/IEC 13818-2, 1996.

31. MPEG-2 Information technology — generic coding of moving pictures and associated audio information — part 3: Audio, ISO/IEC13818-3, 1998.

32. Implementation guidelines for the use of MPEG-2 Systems, Video and Audio in satellite, cable and terrestrial broadcasting applications, ETSI ETR 154 v.3, 1997.

33. Ors, T. and Rosenberg, C., Providing IP QoS over GEO satellite systems using MPLS, *International Journal of Satellite Communications*, 19 (5), 443, 2001.

34. Braden, R., Clark, D., and Shenker, S., Integrated services in the Internet architecture: an overview, IETF RFC 1633, 1994.

35. Blake, S. et al., An architecture for Differentiated Services, IETF RFC 2475, 1998.

36. Baker, F. et al., Aggregation of RSVP for IPv4 and IPv6 Reservations (draft-ietf-issll-rsvp-aggr-04.txt), IETF draft, 2001.

37. Shenker, S., Partridge, C., and Guerin, R., Specification of guaranteed quality of service, IETF RFC 2212, 1997.

38. Wroclawski, J., Specification of the controlled-load network element service, IETF RFC 2211, 1997.

39. Braden, R. et al., Resource ReSerVation Protocol (RSVP) — version 1 functional specification, IETF RFC 2205, 1997.

40. Wroclawski, J., The use of RSVP with IETF Integrated Services, IETF RFC 2210, 1997.
41. Heinanen, J. et al., Assured Forwarding PHB group, IETF RFC 2597, 1999.
42. Jacobson, V., Nichols, K., and Poduri, K., An Expedited Forwarding PHB, IETF RFC 2598, 1999.
43. Bernet, Y. et al., A framework for Integrated Services operation over DiffServ networks, IETF RFC 2998, 2000.
44. Bernet, Y., The complementary roles of RSVP and Differentiated Services in the full-service QoS network, *IEEE Communications Magazine*, 38 (2), 154, 2000.
45. Eichler, G. et al., Implementing Integrated and Differentiated Services for the Internet with ATM networks: a practical approach, *IEEE Communications Magazine*, 38 (1), 132, 2000.
46. Losquadro, G., EUROSKYWAY: satellite system for interactive multimedia services, in *Proc. Ka-Band Util. Conf.*, 1996, 13.
47. Losquadro, G. et al., EuroSkyWay/INTERNET V.4 interworking requirements, *Doc. No. SPE-ESW-0014-ALS*.
48. Wroclawski, J. and Charny, A., Integrated service mappings for differentiated services networks (draft-ietf-issll-ds-map-01.txt), IETF draft, 2001.
49. Iera, A., Molinaro, A., and Marano, S., IP with QoS guarantees via GEO satellite channels: performance issues, *IEEE Personal Communications Magazine*, 8 (3), 14, 2001.
50. IETF MPLS Working Group Web site: www.ietf.org/html.charters/MPLS-charter.html.
51. Wood, L. et al., IP routing issues in satellite constellation networks, *International Journal of Satellite Communications*, 19 (1), 69, 2001.
52. Werner, M., A dynamic routing concept for ATM-based satellite personal communication networks, *IEEE Journal on Selected Areas in Communications*, 15 (8), 1636, 1997.
53. Werner, M. et al., ATM-based routing in LEO/MEO satellite networks with intersatellite links, *IEEE Journal on Selected Areas in Communications*, 15 (1), 69, 1997.
54. Mauger, R. and Rosenberg, C., QoS guarantees for multimedia services on a TDMA-based satellite network, *IEEE Communications Magazine*, 35 (7), 56, 1997.
55. Narvaez, P., Clerget, A., and Dabbous, W., Internet routing over LEO satellite constellations, in *Proceedings of the ACM/IEEE International Workshop on Satellite-Based Information Services*, Dallas, 1998.
56. Yegenoglu, F., Alexander, R., and Gokhale, D., An IP transport and routing architecture for next-generation satellite networks, *IEEE Network Magazine*, 14 (5), 32, 2000.
57. Fenner, W., Internet Group Management Protocol, Version 2, IETF RFC 2236, 1997.
58. Malkin, G., RIP version 2, IETF RFC 2453, 1998.
59. Moy, J., OSPF version 2, IETF RFC 2328, 1998.
60. Srisuresh, P. and Holdrege, M., IP network address translator (NAT) terminology and considerations, IETF RFC 2663, 1999.
61. Koyabe, M.W. and Fairhurst, G., Reliable multicast via satellite: a comparison survey and taxonomy, *International Journal of Satellite Communications*, 19 (1), 3, 2001.
62. Howarth, M.P., Cruickshank, H., and Sun, Z., Unicast and multicast IP error performance over an ATM satellite link, *IEEE Communications Letters*, 5 (8), 340, 2001.
63. Jamalipour, A. and Tung, T., The role of satellites in global IP: trends and implications, *IEEE Personal Communications Magazine*, 8 (3), 5, 2001.
64. Ghani, N. and Dixit, S., TCP/IP enhancements for satellite networks, *IEEE Communications Magazine*, 37 (7), 64, 1999.
65. Akyildiz, I.F. et al., Research issues for transport protocols in satellite IP networks, *IEEE Personal Communications Magazine*, 8 (3), 44, 2001.
66. Henderson, T.R. and Katz, R.H., Transport protocols for Internet-compatible satellite networks, *IEEE Journal on Selected Areas in Communications*, 17 (2), 326, 1999.
67. Partridge, C. and Shepard, T.J., TCP/IP performance over satellite links, *IEEE Network Magazine*, 11 (5), 44, 1997.

68. Choticapong, Y., Cruickshank, H., and Sun, Z., Evaluation of TCP and Internet traffic via low earth orbit satellites, *IEEE Personal Communications Magazine*, 8 (3), 28, 2001.

69. Mogul, J. and Deering, S., Path MTU discovery, IETF RFC 1191, 1990.

70. Braden, R., Transaction TCP — concepts, IETF RFC 1379, 1992.

71. Braden, R., T/TCP — TCP extensions for transactions: functional specification, IETF RFC 1644, 1994.

72. Allman, M., On the generation and use of TCP acknowledgments, *ACM Computer Communications Review*, 28 (5), 1998.

73. Mathis, M. et al., TCP selective acknowledgement options, IETF RFC 2018, 1996.

74. Allman, M. et al., TCP performance over satellite links, in *Proceedings of the International Conference on Telecommunications Systems*, Nashville, 1997.

75. Bharadwaj, V.G., Baras, J.S., and Butts, N.P., An architecture for Internet service via broadband satellite networks, *International Journal of Satellite Communications*, 19 (1), 29, 2001.

76. Minei, I. and Cohen, R., High-speed Internet access through unidirectional geostationary satellite channels, *IEEE Journal on Selected Areas in Communications*, 17 (2), 345, 1999.

77. Goyal, R. et al., Buffer management and rate guarantees for TCP over satellite-ATM networks, *International Journal of Satellite Communications*, 19, 111, 2001.

78. Adami, D., Marchese, M, and Ronga, L.S., TCP/IP-based multimedia applications and services over satellite links: experience from an ASI/CNIT project, *IEEE Personal Communications Magazine*, 3, 20, 2001.

79. Marchese, M., TCP modifications over satellite channels: study and performance evaluation, *International Journal of Satellite Communications*, 19 (1), 93, 2001.

80. Wittmann, R. and Zitterbart, M., *Multicast Communication, Protocols and Applications*, Morgan Kaufmann, San Francisco, 2001.

81. Bianchi, G., Blefari-Melazzi, N., and Femminella, M., A migration path to provide end-to-end QoS over stateless networks by means of a probing-driven admission control (draft-bianchi_blefari-end-to-end-QoS-00.txt), Internet draft, 2000.

82. Des Dorides, C. and Tomasicchio, G., SkyplexNet system: an advanced platform for satellite multimedia services, in *Proceedings of the European Conference on Satellite Communications*, Toulouse, France, 1999.

83. Digital Video Broadcasting (DVB); interaction channel for satellite distribution systems, ETSI EN DVB-RCS 333, REV 6.0 (based on ETSI EN 301 790 V1.2.2 (2000-12) DVBRCS001r15), European Standard (Telecommunications series), 2002.

84. EuroSkyWay Web site: http://www.euroskyway.it.

85. Iera, A. and Molinaro, M., Designing the internetworking of terrestrial and satellite IP-based networks, *IEEE Communications Magazine*, 40 (2), 136, 2002.

86. Iera, A., Molinaro, A., and Marano, S., Call admission control and resource management issues for real time VBR traffic in ATM-satellite networks, *IEEE Journal on Selected Areas in Communications*, 18 (11), 2393, 2000.

87. Garrett, M. and Borden, M., Interoperation of Controlled-Load Service and Guaranteed Service with ATM, IETF RFC 2381, 1998.

88. Crawley, E. et al., A framework for Integrated Services and RSVP over ATM, IETF RFC 2382, 1998.

5

Mobility Management in Mobile IP Networks

Lila V. Dimopoulou
National Technical University of Athens

Iakovos S. Venieris
National Technical University of Athens

5.1   Introduction

Internet Protocol (IP) mobility management comprises the functionality located in specific places within a network that enables mobile hosts to move across different IP networks and to receive uninterrupted services. This concept should be distinguished from the ability of a host to detach from a network and attach to a new one, where it can start receiving services because it implies the portability of the mobile terminal and does not guarantee IP mobility. We are interested, however, in allowing mobile hosts to continue receiving services when moving to a new IP subnet without such a change becoming visible to upper layers and causing their possible interruption. This means that active TCP sessions, for example, should not be reset after the host changes its access point and therefore should continue addressing the mobile host as if it were stationary and always attached to its home network (i.e., address the host at its home IP address). To accomplish that, special mechanisms need to be deployed in the network for keeping track of the host's current location and being able to deliver packets to it.

One way to achieve this is through the establishment of special routes along the communication path between the correspondent host (CH) — that is, the mobile host's endpoint — and the mobile host (MH) for the routing of packets to the latter's current location. (For the rest of this chapter, we use the terms *mobile host* and *correspondent host* when referring to the mobile terminal and its communication endpoint, either mobile or stationary. The terms *mobile node* and *correspondent node* are also found in the literature.) These special routes are needed because the host's home address can no longer be used by conventional IP routing mechanisms for the forwarding of packets to its actual location. This is because as soon as the MH moves away from its home network, its home address is topologically incorrect and cannot be used for routing purposes. By *home address* we mean the IP address assigned to an MH from its home network, either statically or dynamically. Therefore, special routes in the core network need to be established whenever the host changes its point of attachment. This is, however, not a scalable solution; it requires a great amount of processing and memory resources by core routers, not to mention the increasing load placed on the network to establish these routes. Note that the introduction of a hierarchical structure in the IP addresses has been motivated in the first place by the evident problems of a flat model, which would require a per-host routing state in core routers.

A second way to achieve IP mobility is to have specific nodes in the core network maintain location information for the host. Packets addressed to the host's home address, which pass through these nodes, are forwarded to the correct location by means of tunneling techniques (mainly in IPv4 networks). Such specialized nodes know only the next-hop-specific node that these packets should visit. Tunneling is used because the home IP address carried in the packets cannot be consulted for their routing to the host's actual location. Therefore, these packets are encapsulated into IP packets that carry the correct routing information. Such functionality can be located at any place within a network, including the CH. In this way, either the CH or any intermediary special node is responsible for tunneling packets to the MH's current location.

All proposed mobility management schemes are based on either of these two mechanisms for addressing the mobility of hosts at the IP level. However, it is important to introduce here an additional parameter, which differentiates such schemes into the ones handling micromobility (or regional mobility) and the others handling macromobility. Micromobility protocols address the mobility of a host within small regions and do not allow the distribution of its location information across the boundaries of these regions. These schemes provide transparency to the host's communication endpoints, which remain unaware of the MH's movements within the regional network. On the other hand, macromobility protocols, represented by the Mobile IP protocol for IPv4 and IPv6 networks, require that the whole path be updated for each movement of a host and therefore are considered ideal for handling the coarse grain mobility of a host — that is, the movement across regional networks. It is, however, unacceptable for these schemes to use special routes due to the scalability constraints mentioned earlier, but such routes can more efficiently be established in small domains (where micromobility protocols can be deployed), serving a small number of users.

The rest of the chapter is structured as follows. Section 5.2 is devoted to the description of the most popular macromobility protocol for both IPv4 and IPv6 networks: the Mobile IP protocol. Then, in Section 5.3 we describe several micromobility protocols that, along with Mobile IP, achieve global IP mobility for mobile hosts. We focus on comparing these protocols and highlighting their pros and cons in performing *path update*, *handover control*, and *paging*, the three constituents of a complete mobility solution.

5.2 Macromobility Protocols

5.2.1 Mobile IPv4

Mobile IP (MIP) is the Internet Engineering Task Force (IETF) proposal for enabling IP mobility. Its goal is to offer a simple network-layer solution for mobility support without affecting higher layers. In particular, it is aimed at maintaining active TCP connections while the user is moving across IP networks. For obviating the problem, we should describe here the two contradictory issues that MIP attempts to address. The movement of mobile hosts across IP subnets requires that their IP address be changed in order that they can still receive packets. This is because the IP address of a host reveals routing information and is consulted for the routing of packets across the Internet. On the other hand, upper layers are affected when the host changes its IP address because, for example, TCP sessions need to be reset. This is attributed to the additional role of the IP address, which serves as an identifier for the hosts. However, maintaining the host's IP address while the latter moves across IP subnets will not allow the routing of packets to its actual location. Mobile IP seeks, therefore, to support mobility under the constraints imposed by the nature of the IP address.

The fundamental principle adopted by MIP is that a mobile host should use two IP addresses: a permanent address — the *home address*, assigned to the host and acting as its endpoint identifier — and a temporary address — the *Care-of Address* (CoA), providing the host's actual location. In MIP's simple operation, CHs — peers with which MHs are communicating — address the MH via its home address (the mobility of the host is hidden from them) whereas MIP handles the host's movement and enables the delivery of packets to its actual location. This operation requires the introduction of two additional entities in the IP network infrastructure: the Home Agent (HA) and the Foreign Agent (FA).

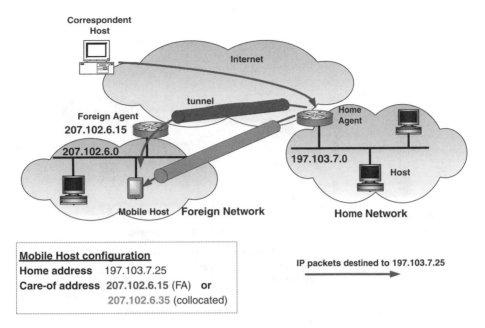

FIGURE 5.1 Mobile IP functional entities.

The HA is the entity that maintains location information for the host. It resides in the host's home network and is responsible (when the MH is away from home) for intercepting packets addressed to the host's home address and tunneling them to its CoA, which is its current point of attachment. The FA, on the other hand, resides in the visited network of the MH and provides to the latter routing services if needed — in other words, it detunnels the packets sent by the HA and delivers them to the host. Note that Mobile IP operates only when the host is away from its home network; otherwise, conventional IP routing mechanisms are used for the routing of packets to the MHs, as with any other stationary host. IP mobility mechanisms for a host are activated only when the latter attaches to a foreign network.

In any foreign network, the MH must acquire a new CoA, which reflects its current point of attachment. This CoA can be determined either by an FA present in the subnetwork (i.e., the CoA comprises one of the FA's addresses and is referred to as *foreign agent care-of address*) or by an alternative mechanism, such as the Dynamic Host Configuration Protocol (DHCP), in which case it is denoted as a collocated CoA. In the former case, the FA is the endpoint of the tunnel, as illustrated in Figure 5.1, and is responsible for decapsulating the data packets and delivering the inner data to the MH. Its main advantage is that it does not impose the need for extra IPv4 address space because one IP address can be used as the CoA of many mobile hosts. On the contrary, in the case that the host is assigned a collocated CoA, there is the disadvantage that each subnetwork should use a pool of addresses for serving visiting MHs. However, it does not require the presence of FAs — that is, MIP functionality to be deployed by visiting networks.

5.2.1.1 Overview

We can distinguish three phases in the Mobile IP operation: agent discovery, registration, and routing. When a mobile host moves within a new subnetwork, it needs the means to detect that it has changed its point of attachment, identify whether it is located in a foreign or its home network, and acquire a CoA on the new subnet. Agent Advertisements, transmitted periodically by mobility agents (FAs or HAs) on the link, facilitate the movement detection and IP address acquisition procedures. In particular, the Agent Advertisement constitutes an extension of the Internet Control Message Protocol (ICMP) Router Advertisement message, when augmenting the latter with a *Mobility Agent Advertisement Extension*. The Router Advertisement portion of the message enables movement detection, as we will see later, whereas the Mobility Agent Advertisement Extension reveals to the MH the routing services provided by the

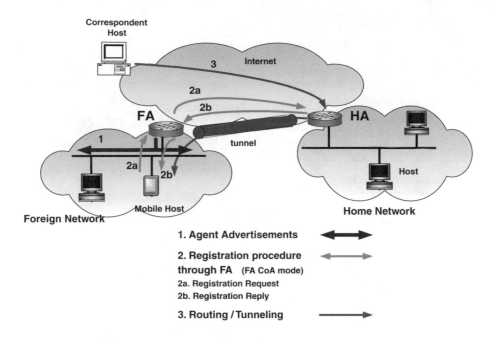

FIGURE 5.2   Operation of Mobile IPv4.

mobility agent (HA, FA, or both) on the link, as well as a list of CoAs in the FA case. The MH can possibly use an address from the list as its temporary CoA (FA CoA). Alternatively, it can use an external assignment mechanism such as DHCP for obtaining an address on the link (collocated CoA).

At the time the agent discovery and CoA acquisition phase has been completed, the MH should register with its HA. This is performed by sending to the latter a *Registration Request* message and by waiting for a *Registration Reply* message indicating the result of the registration. The HA binds in this way the host's home address with its new CoA. This procedure can be possibly carried out via the FA (see Figure 5.2), which receives the aforementioned messages and then relays them to the HA and the MH, respectively. Registration through the FA is always performed when the FA's services are to be used for the delivery of packets to the MH — that is, when the host operates in an FA CoA mode. The FA needs to extract information that will enable this delivery and will generally assist the provision of routing services to the MH. The same messages are exchanged (not through the FA this time) when the MH detects that it has returned to its home network. Here, it deregisters with its HA, which stops maintaining location information for the host and providing mobility services to the latter.

As soon as the HA is informed of the host's current point of attachment, it is able to deliver packets to it. Datagrams sent to the MH's home address are intercepted by the HA and tunneled by the latter to the MH's CoA. Note that the endpoint of the tunnel may either be the FA or the MH itself. In the former case, the FA detunnels the datagrams, checks in its visitor list to find a registered host with a home address matching the destination IP address of the inner packets, and delivers them to the host's link-layer address (both the FA and the host have an interface on the same link). In the latter case, the MH is responsible for decapsulating the packets. In the reverse path (for the packets being sent by the mobile host to its CHs), standard destination-based routing is used, unless otherwise requested. It is possible that reverse tunneling is requested during the registration procedure (refer to Section 5.2.1.3), in which case packets are first delivered to the HA by means of a tunnel and, after being decapsulated, are delivered to their actual destination via conventional IP routing.

The HA should be located on the same link indicated by the home address of its registered MHs for being able to intercept datagrams addressed to these hosts. It can reside either in the IP gateway router of the subnet or in a stand-alone host without routing functionality. This way, IP packets will be forwarded

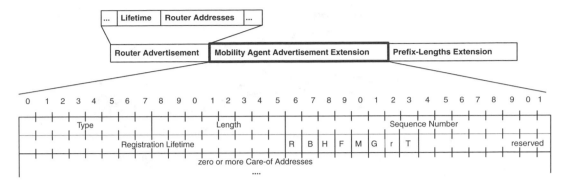

FIGURE 5.3 Mobility Agent Advertisement extension format.

to the subnetwork based on the host's IP address, and from there they will be delivered to the HA's link-layer address. This is because the HA will respond to broadcast ARP (Address Resolution Protocol) messages for the MH by providing its own link-layer address. This mechanism is referred to as *proxy ARP* because the HA responds to ARP messages on behalf of the MH.

In the following sections, we provide further details on the three main functions of Mobile IP and the protocol operation.

5.2.1.1.1 *Agent Discovery and Care-of Address Acquisition*

Mobility agents transmit Agent Advertisements in order to advertise their services on a link. Mobile hosts primarily use these advertisements to determine their current point of attachment and to identify the services provided by the agent. Advertisements are sent periodically by mobility agents or in response to Agent Solicitations transmitted by MHs for advancing the agent discovery procedure. Through these advertisements, MHs are able to detect whether they are located at the same subnetwork or have migrated to a new link (referred to as *move detection*) and whether this link constitutes a foreign network or their home network. If this is a new link, they need to acquire a temporary CoA and register with their HA. If they have returned home, they need to deregister with their HA and operate without mobility services.

It is interesting to examine how movement detection is performed. As already mentioned, an Agent Advertisement is an ICMP Router Advertisement message that is followed by a Mobility Agent Advertisement Extension and, optionally, by other extensions. The first method of move detection is based on the *Lifetime* field of the Router Advertisement part of the message. In particular, the MH records the lifetime value, and in the absence of another advertisement from the same agent until the lifetime expires, it deduces that it has moved to a new link. If in the meantime it receives an advertisement from another agent, it registers with the latter. Otherwise, it may transmit an Agent Solicitation for the timely discovery of an agent. The second method is based on the *Prefix-Lengths Extension* that may follow the Mobility Agent Advertisement Extension. These prefix lengths are applied to the *router addresses* that are found in the Router Advertisement part of the message and can serve as the identification of the subnetwork. Therefore, if the prefix advertised by an agent on a link differs from the prefix of the host's CoA (either FA or collocated CoA), then the MH assumes that it has moved. The procedure is the same as before: the host needs to register with the new agent. Last, the MH easily detects that it has returned home when it receives an Agent Advertisement transmitted by its HA.

Apart from facilitating move detection, the Agent Advertisement, through its Mobility Agent Advertisement Extension depicted in Figure 5.3, provides information for the agent's services. The *F* and *H* flags indicate the agent's role on the link (FA or HA). Note that a mobility agent can provide at the same time foreign and home agent services. An HA needs to advertise its presence on the link so that mobile hosts can detect when they return to their home network. The type of routing services (i.e., the encapsulation type) supported by an FA is indicated through the *M* and *G* flags, which correspond to Minimal and GRE (Generic Routing Encapsulation) encapsulation, respectively. However, these two encapsulation types are optional whereas IP in IP encapsulation provision is mandatory for all foreign and home agents.

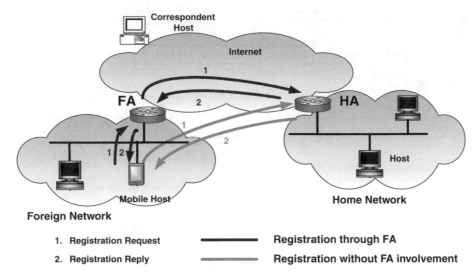

FIGURE 5.4 Registration procedure in Mobile IP.

When the *T* flag is set, the FA supports reverse tunneling (refer to Section 5.2.1.3) whereas the *B* flag, when set, shows that the FA is busy and cannot accept further registrations. The *R* flag indicates that registration with an FA is required. Although this is mandatory only for MHs using an FA CoA, such an indication obliges hosts with a collocated CoA to do so. This is needed in cases where the network enforces policies in visiting hosts requiring their authorization. Last, the *Care-of Addresses* field enables the advertisement of the agent's CoAs, which is used by the MHs when registering with their HAs.

5.2.1.1.2 Registration Procedure

The main purpose of the registration procedure is that the HA becomes aware of the mobile host's current point of attachment in order to be able to deliver to the latter packets addressed to its home address. In addition to that, a registration procedure takes place when the host needs to renew its soon-to-expire registration with its HA. Therefore, regardless of whether the MH migrates to a new network, it is engaged in sending renewal registrations to its HA for resetting the lifetime of its registration. The HA plays an active role in the whole procedure and in most cases requires the assistance of the FA, which, however, appears to have a passive role.

In particular, the registration procedure (see Figure 5.4) consists of two messages exchanged between the mobile host and the HA: a *Registration Request* (from MH to HA) and a *Registration Reply* (from HA to MH). Both messages are carried over User Datagram Protocol (UDP) datagrams at the well-known port number 434. They are sent to their endpoint, possibly through the FA, which processes the information of its interest and relays the messages to their actual destination. The FA is not involved in this procedure when the MH is using a collocated CoA unless it has set the *R* bit of its Agent Advertisement. In all other cases, the MH sends the Registration Request message to the FA, which relays it to the HA. In the reverse direction, the Registration Reply message is delivered to the FA, which relays it to the mobile host.

For the rest of our discussion, we will assume that the registration is performed through the FA unless otherwise stated. The Registration Request message is the one depicted in Figure 5.5. With the *Lifetime* field, the MH indicates the lifetime of its registration, taking also into consideration the *Registration Lifetime* value set by the FA in its Agent Advertisement. The HA does not always accept this value and may introduce a new one within the Registration Reply message, which in any case should not be greater than the one included in the request. Moreover, a lifetime set to zero is used for the deregistration of the host. The fields *Home Address* and *Care-of Address* are set to the host's home and CoA, respectively, whereas the *Home Agent* field indicates the IP address of the host's HA. The FA will use this information for setting the datagram's destination address and relaying the request to the HA.

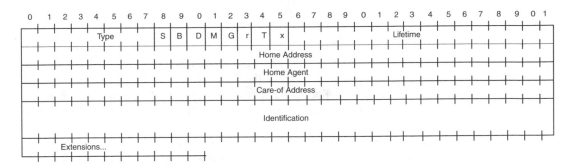

FIGURE 5.5   Registration Request message format.

The fixed portion of the Registration Request is followed by one or more extensions. Certainly, one extension should always be present in both Request and Reply messages, and that is the *Mobile-Home Authentication Extension*. This extension requires that a mobility security association exist between the MH and the HA and is used for authenticating registration data exchanged between these entities. Moreover, other authentication extensions can also follow for authenticating data exchanged between the MH and the FA or between the FA and the HA. However, these require that mobility security associations exist for the corresponding relationships. For example, a host, after securing the Registration Request message with a Mobile-Home Authentication Extension, may add other extensions to be processed by the FA and secures these extensions with a *Mobile-Foreign Authentication Extension*. The FA, when receiving the message, should process and remove any extension after the Mobile-Home Authentication Extension before relaying it to the HA. Likewise, it may add any information to be processed by the HA, which is secured with a *Foreign-Home Authentication Extension*. The same logic also applies to the Reply messages sent in the reverse direction. It is important, however, that the FA not modify any of the fields up to the Mobile-Home Authentication Extension because it is likely that the HA will not be able to authenticate this part of the message.

The remaining flags request forwarding services from the HA and communicate to the latter features supported by the FA (in the FA CoA mode). In particular, the *S* bit is set when the MH requests that the HA maintain simultaneous bindings for more than one CoA. In this case, the HA does not replace any previous mobility binding for the host with the new one but instead maintains multiple bindings. In such a case, the HA tunnels datagrams addressed to the host to all CoAs indicated in these bindings, which might result in the host receiving duplicate packets. Likewise, a mobile host may deregister one of its simultaneous CoAs (the *Care-of Address* field is set to this address and the *Lifetime* field is set to zero in the Registration Request) or all at once when returning home (*Care-of Address* is set to the home address and *Lifetime* is set to zero). This service is useful in cases where the host uses more than one wireless network interface in the overlapping transmission area of more than one FAs and when transmission links appear to be significantly error prone. Evidently in this case, packet loss experienced by the host can be significantly reduced when compared with a scenario where one FA is serving the MH in an error-prone wireless environment.

Moreover, by setting the *B* bit, the MH requests that broadcast packets destined to its home network are tunneled to its current location, whereas the *D* bit indicates how this tunneling should be performed. More specific, when the *D* flag is set, the MH declares to the HA that it is using a collocated CoA. In this case, the HA will tunnel broadcast packets to the host's CoA as with any other unicast packet. However, if an FA CoA is used, then the HA should first encapsulate the broadcast datagram in a unicast packet addressed to the MH's home address and then tunnel it to the MH's CoA. This is needed because the FA, upon decapsulating the packet, will detect its final destination by viewing the host's home address. However, if the first encapsulation were not carried out, then the FA would not know where to deliver a packet destined to an address (the home network's broadcast address) for which no mobility binding exists. The MH is responsible for further decapsulating these datagrams in order to receive the broadcast data. Last, the *M* and *G* bits allow the host to request alternative methods of encapsulation that are supported by the FA (in FA CoA mode), as indicated in its Agent Advertisement or the mobile host (in collocated CoA mode).

The FA, upon receipt of a Registration Request from a host, extracts and maintains information such as the host's link-layer address as well as the IP source address of the packet, which are used for delivering data to the MH. In particular, the IP source address of the Registration Request message is sent to the MH's home address or to its CoA if a collocated CoA is used. This way, the FA can deliver packets addressed to this IP address because it is aware of the associated link-layer address. The actual update of the FA's visitor list takes place when it receives a Registration Reply from the HA indicating the success of the procedure. The registration lifetime parameter carried in the Registration Request is important because it determines the duration of the registration. It is interesting to see when this timer is set at the three entities: MH, FA, and HA. Only the MH sets the timer when sending the Registration Request, whereas the HA and the FA set this timer when sending and relaying the Registration Reply, respectively. This will ensure that the registration lifetime expires at the MH before it does at the HA and FA. Consequently, a renewal registration will be transmitted early enough for updating the mobility bindings of these two entities. However, the registration lifetime not only is decided by the MH but needs the approval of the HA. Therefore, if the HA specifies a smaller lifetime value in the Registration Reply, the MH will simply decrease the registration's remaining lifetime by the difference of the two lifetime values.

The FA relays the Registration Request message to the HA, which updates its mobility binding for the host. It adds a new entry and either removes the existing ones (the *S* bit is not set) or maintains them in conjunction with the new one (simultaneous bindings have been requested). If the MH was previously located in its home network, a mobility binding is created because the HA did not maintain until that moment any location information for the host. Moreover, a lifetime set to zero will result in the deletion of all entries for the host or only of the one indicated in the *Care-of Address* field of the message, as explained earlier. The HA replies to the FA (i.e., the source of the Registration Request) with a Registration Reply and resets the timer for the registration lifetime. The FA relays the message to the MH, updates its visitor list with an entry for the host, and resets the timer of the registration lifetime to the value carried in the reply message.

5.2.1.1.2.1 Proxy and Gratuitous ARP — Apart from the registration and deregistration procedures performed when the mobile host leaves or returns to its home network, it is also necessary that appropriate ARP procedures be carried out for enabling other hosts in the home network to contact the MH. This is because hosts in the home network do not have valid ARP cache entries for the MH from the moment that the latter moves away. Consequently, as soon as the HA receives and accepts the Registration Request message, it performs gratuitous ARP for notifying hosts on its link that packets addressed to the MH's home address should be delivered to its own link-layer address — that is, the HA. This way, the HA will be able to intercept such packets and tunnel them to the MH. Accordingly, the HA starts using proxy ARP for replying to ARP requests for the MH. Note that the MH, while away from home, does not respond to or issue broadcast ARP requests in order to avoid creating ARP caches in hosts on the foreign link that can become stale. The exception to this is when the FA transmits unicast ARP requests to the MH for the latter's home address, in which case the MH replies to the FA with a unicast ARP reply message. This way, other hosts on the link are not able to extract information from these ARP messages.

Likewise, when the MH returns to its home network, it should perform gratuitous ARP for updating the ARP caches of hosts on the link. It therefore associates its own link-layer address with its home IP address for being able to receive packets addressed to its home address. Then, the MH sends the (de-) Registration Request to its HA, which, upon receipt and successful processing of the message, stops using proxy ARP for the host and once again performs gratuitous ARP. The gratuitous ARP is transmitted by both the MH and the HA because, in wireless environments, it is likely that their transmission area differs.

5.2.1.1.2.2 Dynamic HA Discovery and Home Address Assignment — Until now, we have assumed that the MH is configured with the HA's address and its home IP address. It is possible, however, that none of these two assumptions hold true. If the MH is not aware of the HA's address, it can use dynamic HA address resolution for discovering the IP address of an HA on its home network. This is performed by setting the destination address of the Registration Request message (if registering directly with the HA) or the *Home Agent* field of this message (if registering through the FA) to its home network's broadcast address. Home agents (more than one HA may serve on a link) that receive a request with a broadcast destination address must send a Registration Reply to the mobile host. This message should carry the HA's IP address and indicate that the request has been rejected. This will prevent the MH from possibly being simultaneously registered with more than one HAs. However, in this way, it becomes aware of at least one HA's address and can consequently proceed with registration.

Moreover, it is possible that the MH is not configured with a home IP address, in which case the *Mobile Node NAI Extension* can follow the Registration Request message for serving as the identification of the user. In this message, the home address is set to zero. The network access identifier (NAI) is of the form *user@realm* and is currently used for the authentication and authorization of dial-up users by AAA (Authentication Authorization Accounting) servers. However, it is expected that such services will be widely used in the future by MIP-aware MHs roaming to foreign domains with AAA servers, which should enable the assignment of suitable home IP addresses to the hosts. Most important, such integration of MIP with the AAA infrastructure will lead to the provision of MIP services by cellular infrastructures. This is attributed to the fact that cellular networks need to verify the identity of MHs and authorize connectivity in the absence of preconfigured security associations.

Using the NAI paves the way for dynamic IP address allocation. The NAI contains all the information needed by the AAA infrastructure (AAA servers in foreign and home domains) to authenticate the user and grant to the latter a dynamically assigned home address. This address is carried in the Registration Reply message, which is finally delivered to the host.

5.2.1.1.3 *Routing*

After the MH has registered with its HA, it is able to receive routing services from the latter in cooperation, if needed, with the FA for packets addressed to its home address. In particular, the HA intercepts packets addressed to the MH's home address (enabled through the use of proxy and gratuitous ARP) and tunnels them to the registered CoA. The end of the tunnel might be an FA or the mobile host itself. If the HA supports the *simultaneous bindings* feature and the MH has registered more than one CoA, the HA tunnels a copy to each one of them. The default mechanism used for tunneling is IP in IP encapsulation, which is supported by all MIP entities — HA, FA, and MH. It is also possible that an alternative type of encapsulation, such as Minimal or GRE encapsulation, has been requested by the MH and agreed upon during the registration procedure. These methods, however, are optional, and it is up to the HA to decide the encapsulation method to be used for delivering datagrams to an MH.

The encapsulated data reaches the FA (the end of tunnel in Figure 5.6), which checks its visitor list for an entry matching the inner packet's destination address — that is, the MH's home address. If an entry is found, the FA delivers the inner data to the MH by addressing the host's link-layer address. This address is known to the FA during the agent discovery or registration procedure, from either an Agent Solicitation or the Registration Request message transmitted by the MH, respectively. If the FA has no entry for the host in its visitor list, then it simply discards the message.

In the reverse direction, the host must be configured with a default router for serving the next hop for outgoing packets. If it has registered an FA CoA, it simply uses its FA as the first-hop router for outgoing packets. The FA's link-layer address is known from the agent's advertisements. Alternatively, the MH can choose its default router from the *Router Addresses* advertised in the ICMP Router Advertisement portion of the Agent Advertisement. These two methods are also used by MHs that use a collocated CoA. It should be noted, though, that unless a means other than broadcast ARP exists for the host to obtain the link-layer address of its default router, the FA should serve the default router because its link-layer address is always known from the Agent Advertisements.

Unless otherwise requested, standard IP routing is used for packets transmitted by the MH. The FA is responsible for forwarding packets coming from registered MHs with the use of conventional IP routing mechanisms. If reverse tunneling has been requested during the registration procedure (refer to Section

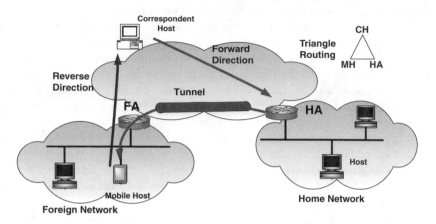

FIGURE 5.6   Routing in Mobile IP.

5.2.1.3), the FA should tunnel packets and deliver them to the host's HA. The tunneling mechanism to be used is the same as the one agreed on for the forward routing of packets. The HA should therefore be able to decapsulate packets addressed to itself and subsequently forward them based on the inner packet's destination address.

MIP routing services are used when the MH is away from home. However, if the MH is attached to its home network, it operates exactly as any other stationary host. Packets addressed to the host's home address reach the MH through standard IP routing. The HA maintains no mobility binding and thus is not involved in the routing of datagrams; it does not intercept packets addressed to the host's home address.

5.2.1.2   Route Optimization

The base Mobile IP protocol provides transparency to the operation of CHs by letting them believe that mobile hosts always reside in their home networks. In other words, mobility is hidden from CHs, which assume that the home address indicates the MH's current point of attachment. Although this property is desirable because it supports mobility without introducing any changes to the CH's protocol stack, it has the disadvantage that the triangle routing problem arises and that unnecessary load is placed on the network. The triangle routing problem and, more specifically, the indirect routing of packets in the forward direction may lead to undesirable effects to applications with stringent delay requirements. Moreover, congestion is likely to arise in the home network, and a great consumption of network resources will be performed in cases with distant home agents and communicating hosts that are situated nearby.

For the above reasons, route optimization extensions have been proposed to the Mobile IP operation so that the HA is bypassed and data packets are routed directly from the CH to the MH. Similar to the HA binding caches, CHs are then required to maintain binding caches for the hosts they communicate with so as to directly send packets (tunneled packets) to the latter's actual point of attachment. The fact that mobility information for the host is now available to all possible communicating CHs renders the timely update of this information more difficult when the MH changes its point of attachment. While in the standard protocol only the HA needed to be notified, now, in addition, all CHs need to update their binding caches. Until this happens, packets forwarded to the old FA are lost. For that reason, extensions are also provided to allow the forwarding of packets, already on their way to the old FA, to the MH's new CoA.

Route optimization defines four messages for handling this operation. The *Binding Update (BU)* message reveals the CoA of an MH and, when received by an entity (here, the CH), results in the update of the latter's binding cache for the specific host. The HA sends a BU message to the CH when it receives data packets from the latter destined to an MH away from home. It then deduces that the CH is not aware of the host's actual location because packets are addressed to the host's home address. The BU message, apart from the binding information, carries the binding's maximum lifetime, which should be set by the HA to a value not greater than the host's remaining registration lifetime. The binding's lifetime

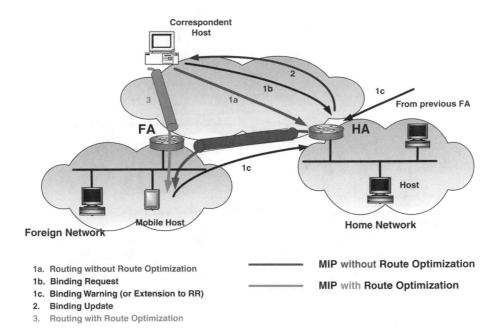

FIGURE 5.7 Route optimization.

as well as the IP address of the BU's originator (i.e., HA) is maintained by the CH for the latter to send a *Binding Request* message to the host's HA when it needs to update the corresponding entry in its binding cache. This might be the case when the CH is still in active communication with the MH and wishes to update the latter's binding information before it expires. The HA responds with a BU message providing this information, as described earlier. Note that the MH may have requested (via its registration to the HA) that the HA not provide any binding information to CHs. In this case, the host's CoA in the BU message is set equal to its home address and the lifetime is set to zero. There is no need for the CH to acknowledge the receipt of the BU in these two cases (i.e., by replying with a Binding Acknowledge message); the absence of packets sent to the HA from the CH will imply the correct reception of the BU message.

We have examined how data packets as well as the Binding Request message received by the HA trigger the Binding Update procedure. In addition to these cases, a BU message is sent by the HA in response to a *Binding Warning* message, as illustrated in Figure 5.7. This might occur when an FA receives tunneled packets for an MH that it is not currently serving but for which it maintains a binding cache (for the support of smooth handover functionality). This means that the FA was serving that host just before the latter moved to a new FA. However, the CH has not yet been informed of the MH's change of CoA and keeps sending packets to the old one. As expected, the FA understands that the CH needs to update its binding cache for the host and sends to the host's HA a Binding Warning message, which carries the CH(s) to be updated. Then, the HA issues a BU message to the indicated CHs to inform them of the MH's new CoA. A Binding Warning message (or a Binding Warning extension to the Registration Request message) can also be sent by the MH to its HA at the time it registers a new CoA. This way, the HA is informed in a timely manner of the CHs to be updated during the registration procedure. This is important, in particular when the host returns to its home network (where smooth handover is not performed) and needs to immediately notify its CHs to start sending packets to its home address.

Smooth handover has been defined for supporting the cases of data being sent based on out-of-date binding caches maintained at CHs or even of data packets in flight while the MH changes its point of attachment and before registering with its HA. In both cases, either the CHs or the HA maintain "wrong" information for the MH with respect to its CoA. For that reason, the old FA needs to be informed of the host's new location in order to redirect packets addressed to the host. This is performed by having

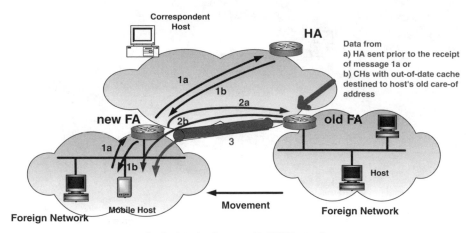

1a. **Registration Request with PFANE extension**
1b. **Registration Reply**
2a. **Binding Update**
2b. **Binding Acknowledge**
3. **Smooth Handoff**

FIGURE 5.8 Smooth handover.

the MH instruct its new FA (via a Previous Foreign Agent Notification, or PFANE, Extension to the Registration Request message) to send a BU message to its previous FA, as depicted in Figure 5.8. Upon receipt of this message, the old FA deletes the visitor list entry for the mobile host (because it is no longer serving this host) and adds a binding cache entry for it (for forwarding packets to its new location). Then, it replies to the MH with a *Binding Acknowledge* message, because the host needs to be sure that the update has been carried out. In the absence of this message, the host retransmits BUs to the previous FA. In the case that the MH returns to its home network, smooth handover functionality is not supported (the host does not have any CoA) and the BU message causes only the previous FA to delete its visitor list entry.

Note that security is of great importance here because routing information for mobile hosts is affected. CHs as well as FAs must be sure of the authenticity of the BUs, and for that reason, these messages must be accompanied by the *Route Optimization Authentication Extension*. Authentication is based on mobility security associations established in advance among the entities exchanging such messages and can be created using any registration key establishment methods.

Route optimization requires some modifications/extensions to the base Mobile IP protocol. In particular, a new flag has been added to the Registration Request message that allows the MH to declare whether it wishes its CoA to be kept private. In this case, the HA does not inform the CHs of the host's CoA, and route optimization is not performed. Moreover, a Binding Warning extension to the Registration Request message may be added by the MH when it wants to communicate in a timely manner the change of its location to its CH(s). Last, FAs need to advertise their ability to support smooth handover by setting an additional flag to the Mobility Agent Advertisement Extension.

Certainly, route optimization has its pros and cons. It eliminates the triangle routing problem but, on the other hand, distributes mobility information for a host in several CHs, which renders the update of this information more difficult. This problem is addressed by notifying the previous FA of the host's new location. Moreover, this update allows the immediate release of any resources reserved for the host by the previous FA because the latter no longer waits for the registration lifetime to expire. BUs being sent to the previous FA carry a lifetime (as is the case with BUs being sent to CHs), which can be viewed as the transition period for the HA and the CHs to be updated. However, route optimization requires changes in the IP stack of the CH, because the latter must be able to encapsulate IP packets and should be equipped with the Route Optimization protocol. In addition, it is not always certain that FAs will support smooth handover functionality, and in this case, more packets will be lost given that extra time is needed for the CHs to be updated with the host's new CoA (when compared with the time needed for notifying only the HA).

5.2.1.3 Reverse Tunneling

Security concerns have forced Internet service providers (ISPs) to apply security policies such as ingress filtering in their networks' border routers for interrupting packets with a source IP address that appears topologically incorrect (i.e., it does not belong to the network's pool of addresses). This way, malicious users are not able to impersonate other users when attached at ingress-filtering domains. However, ingress filtering prevents the standard operation of Mobile IP; mobile hosts residing in a foreign domain are likely to issue packets that have as the source IP address their home address.

Reverse tunneling seeks to address this problem by introducing the use of a reverse tunnel, having as a starting point the CoA of the host and as an endpoint the HA, hiding in this way the topologically incorrect home address of the MH. Mobile hosts are aware that FAs support reverse tunneling by detecting if the *T* flag is set in their Agent Advertisements. Consequently, they request this service during their registration through the selected FA by setting the corresponding flag (*T* bit) in the Registration Request message. Likewise, HAs must accept the reverse tunnel support and should be ready to decapsulate packets originating from the host's CoA.

Two delivery styles of reverse tunneling operation exist. The *Direct Delivery Style,* illustrated in Figure 5.9, constitutes the default operation where the MH specifies the FA as its default router and uses the latter as the first-hop router for outgoing IP packets. From the MH's point of view, packets to be reverse tunneled are processed in the same way as any other packet. The destination address of such packets is the CH's IP address. The FA, in turn, is responsible for detecting these packets by checking their source IP address and for tunneling them to the MH's HA. The benefit of this style, when compared with the second one, is that no extra processing is required at the MH and that packets consume less bandwidth on the link between the host and the FA because they are not encapsulated. However, it lacks flexibility because all packets will be directed to the HA and there is no way for the MH to avoid reverse tunneling at runtime — that is, while transmitting packets. This is a disadvantage because packets will always be tunneled to the HA even if they are addressed to local resources, in which case they would not reach the border router and no reverse tunneling would be required.

In the *Encapsulating Delivery Style,* the MH does not send packets directly to the CHs but rather encapsulates them to the FA. The latter is responsible for verifying that the packets have been sent by an MH that has requested this service. Then, the packets are decapsulated and finally re-encapsulated to the host's HA. The HA, in both schemes, will decapsulate the packets and forward them based on the destination address of the original packet — that is, the CH's address. The benefit of this style is that it allows the MH to indicate at runtime whether reverse tunneling should be used by the FA. This way, if the MH, which has requested the Encapsulating Delivery Style service, sends packets to its CH without encapsulating them first to the FA, the latter does not perform reverse tunneling to these packets. As a consequence, it will forward them based on standard IP routing mechanisms. An MH requests this service by including an Encapsulating Delivery Style Extension to the Registration Request message. This extension is added after the *Mobile-Home Authentication Extension* and is removed by the FA after being processed, because it affects only the latter's operation.

In all cases, encapsulation is performed in accordance with the forward tunnel configuration. This means that the same encapsulation type (IP in IP unless otherwise requested) and the same tunnel endpoints are used for both forward (HA to host's CoA) and reverse tunnel. This symmetry between the two tunnels cannot be maintained, however, when route optimization is used; the MH cannot make any assumption as to whether the CH is able to decapsulate packets. On the contrary, this is obligatory for the HA.

Until now, we have assumed that MHs use the FA's CoA. If this is not the case and the host is configured with a collocated CoA, it is the one that performs the tunneling of packets to its HA. This provides the MH with the flexibility to decide at runtime whether reverse tunneling is to be used. Recall that this flexibility is allowed only in the Encapsulating Delivery Style for the FA CoA mode, as described earlier.

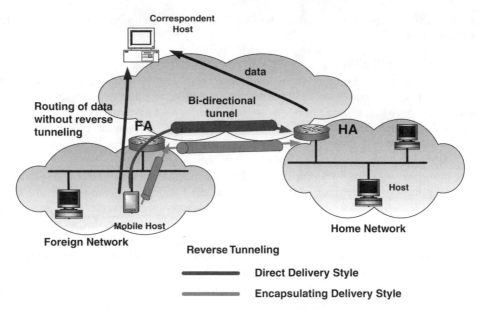

FIGURE 5.9   Reverse-tunneling delivery styles.

5.2.1.4   Discussion

Our earlier description of the Mobile IP functionality shows that this protocol is a simple and robust solution, which renders it the best candidate for addressing the IP mobility of hosts. We have highlighted, however, the main deficiencies of the protocol and examined the proposed solutions. The indirect routing is considered a serious drawback for traffic with stringent delay requirements, such as voice applications, because it increases the end-to-end delay of packets when compared with their delivery through an optimal route. Route optimization addresses this issue by imposing additional requirements on CHs that are no longer left intact by the mobility of hosts, preventing in this way its fast deployment.

The indirect routing may also hinder the scalability of the protocol in a scenario with the IETF Mobile IP widely deployed. This is because congestion may easily arise in the home network if a great number of MHs situated in visited networks is served by the HA. Route optimization is thus considered important for minimizing the network load and eliminating congestion at the home network, protecting the latter from single points of failure, such as the HA, and generally contributing to the overall health of the Internet.

In conjunction with the indirect routing, the scalability is also jeopardized by the high control overhead that MHs bring into the network each time they cross the boundaries of an IP subnet. Note that an MH needs to notify its HA each time it changes its point of attachment, which results in a change of the IP path to the host. This proves highly inefficient in the cases of the host's movement within very small regions that cause only a small change to the host's path. Mobile IP addresses such movements in the same way as any other movement because it requires that control traffic travels all the way to the HA. In other words, the path from the MH to the HA is always reestablished independently of the granularity of the host's movement. Numerous solutions have been proposed for addressing the micromobility of hosts while at the same time cooperating with Mobile IP for achieving global mobility. The idea is that the host's mobility within the regional domain should be handled locally, whereas Mobile IP should be used for handling mobility across such domains. Among these proposals, the MIPv4 Regional Registration (MIPv4 RR) solution has captured the IETF community's interest and is considered as the most dominant for addressing these needs. A detailed description of MIPv4 RR along with other candidate proposals is provided in Section 5.3.

Micromobility protocols also try to address another deficiency of Mobile IP — that is, the service disruption caused during IP handovers. Note that from the moment a host moves to a new network and until it notifies its HA of its new point of attachment, packets already on their way to the host's previous location are lost. This loss of packets during the IP handover causes a disruption of service experienced by users. Micromobility protocols alleviate this problem by reducing the time needed to restore the communication path to the host's new access point and thereby reducing the time interval during which packets are lost. However, it has been shown that IP handover is more efficiently addressed by protocols that are solely dedicated to handover control while at the same time cooperating with protocols performing path update functionality, such as MIP. To this aim, the IETF community has converged to the adoption of the *Low Latency Handovers in Mobile IPv4* for handling the IP handovers performed by mobile IP users. This proposal is based on the same principles adopted by *Fast Handovers in MIPv6*, both of which are briefly described in Section 5.3.

5.2.2   Mobile IPv6

Mobile IPv6 (MIPv6) is based on the same principles as Mobile IPv4 but appears to be much more powerful. It is well integrated into the new IPv6 protocol and exploits its facilities for overcoming several deficiencies encountered in MIPv4. The idea is always the same, which is that an MH needs to notify its HA of its current location in order that the latter route packets to the host's current point of attachment. Therefore, an MH maintains a home address, which serves as its identification to higher layers, whereas it is configured with a temporary CoA on each visiting link, which serves as a routable address for being able to receive packets.

One major difference with MIPv4 is that no FAs are needed to support the users. IPv6 functionality is sufficient for letting mobile hosts acquire their CoAs without the FA's assistance. In particular, an MH can acquire its temporary CoA through conventional IPv6 stateless or stateful autoconfiguration mechanisms. In the former case, the address can be generated easily by combining the network prefix of the visited network and an interface identifier of the host. In the latter case, a DHCPv6 server is required for making such address assignment to the host. Address exhaustion is not a problem in IPv6; its main driving force has been the elimination of this problem, and for that reason it has adopted 128-bit addresses instead of the 32-bit addresses used in IPv4.

The typical operation of Mobile IP — that is, where the CH addresses the MH at the latter's home address — does not require from the IPv6 correspondent hosts to implement the specific IPv6 extensions, which actually form MIPv6. In the opposite case — when the CHs are augmented with the MIPv6 functionality — an improved version of route optimization (as specified in MIPv4) can be used for the direct delivery of packets to the MH without the intervention of the HA. These CHs are able to associate the MH's home address with a CoA. However, such packets will not be encapsulated for delivery to the MH, as is the case in MIPv4, but instead an IPv6 Routing header will be used for this purpose. Keep in mind that route optimization is integrated into MIPv6 and does not constitute a set of optional extensions that might not be supported by all IP hosts. Mobile IPv6, on the other hand, comprises a protocol that extends the basic IPv6 functionality by means of header extensions rather than being built on top of it. Recall that this is the case with MIPv4, which is a separate protocol running on top of the IP/UDP layer.

Moreover, in MIPv6, a new feature is introduced that allows the direct routing of packets to CHs without the need for performing reverse tunneling to deal with ingress filtering routers. This can be accomplished because the host's CoA is used as the IP source address in the header of outgoing packets. For achieving transparency to higher layers, MIPv6 introduces the *Home Address* option, which is carried by the IPv6 Destination Option extension header, to inform the recipient (the CH) of the MH's home address. The CH, after being provided with this information, may process the packet accordingly and deliver it to higher layers to keep them unaware of the host's mobility.

Home-agent address discovery is also supported here by using a new IPv6 routing procedure called *anycast*. An MH, when it needs to discover an HA, makes an inquiry to its home network using IPv6 anycast. This mechanism returns a single reply to the MH, providing the latter with the requested information, instead of having separate replies from each HA on the home network. This makes the

MIPv6 HA address-discovery mechanism more efficient and reliable than the IPv4 directed broadcast used in MIPv4.

In addition, IPv6 Neighbor Discovery instead of ARP is used by the HA to intercept packets addressed to an MH's home address. This mechanism is used for determining the link-layer address of neighbors known to reside on attached links. This decouples the MIPv6 functionality from a particular link layer and therefore enhances the protocol's robustness. IPv6 Neighbor Discovery is also used in MIPv6, with some modifications when required, to allow move detection by MHs and to enable HA address discovery.

5.2.2.1 Overview

An MH, while attached to its home network, is able to receive packets addressed to its home address and that are forwarded by means of conventional IP routing mechanisms. When the host crosses the boundaries of its current serving network, movement detection is performed for the host to identify its new point of attachment and further acquire a new CoA. This CoA is acquired through stateful or stateless IPv6 autoconfiguration mechanisms and has the subnet prefix of the new link (for being globally routable).

Once configured with a CoA, the MH needs to send a *Binding Update (BU)* message to its HA to register this temporary address, which is referred to as the primary CoA. The BU to the HA is always acknowledged with a *Binding Acknowledgment* message. We refer to a primary CoA because the MH may have formed more than one CoA, possibly when it is reachable by more than one wireless link or when more than one prefix is advertised on its current link. Note that in wireless environments, a host may be attached simultaneously to two links. However, only one CoA should be registered with the HA at a time. This functionality — that is, multiple CoAs — is particularly important for the support of smooth handovers because the MH will be able to receive packets in its previously registered CoA while attaching at the same time on a new link and registering a new primary CoA with the HA.

The BU message is carried within an extension header to the IPv6 protocol, which is named the *Mobility header*. The latter is specified for the needs of the Binding Management functionality in MIPv6 and may carry more messages, as we will see below. By Binding Management functionality, we mean the procedures carried out for the registration of the MH's CoA with the HA and the CHs. Upon receipt of the BU message and after validating it, the HA updates its binding for the MH. This enables the forwarding of packets, addressed to the host's home address, to its current location with the use of IPv6 encapsulation. In the reverse direction, packets are reverse tunneled to the HA and finally delivered to the CH. The interception of packets is no longer performed with ARP mechanisms, as is the case with MIPv4, but instead IPv6 Neighbor Discovery is used. In particular, the HA will send a Neighbor Advertisement message to the *all-nodes multicast address* on the home link so it can associate its own link-layer address with the host's home IP address. Certainly, the HA needs to reply to any Neighbor Solicitations addressed to the MH after checking that it maintains a binding cache for it. This way, all packets destined to the host's home address will be delivered to the HA, which will finally tunnel them to the host's primary CoA.

This mode of operation allows the routing of packets, in both forward and reverse directions, through the HA and leaves the CHs unaware of the host's mobility. Therefore, it does not require from CHs that they implement the specific extensions to the IPv6 protocol that constitute the MIPv6 functionality. However, the main benefit of MIPv6 is that it allows route optimization for packets addressed to MHs in an efficient way without the need for tunneling techniques. As expected, the MH needs to inform its CH of its current location by sending to the latter a BU message (see Figure 5.10). This message is the same as the one used for registering with the HA. However, here a *return routability* test must be performed first, which will authorize the binding procedure and will provide to the CH an assurance that the MH is addressable at both its home and its CoA. Note that the messages exchanged for the needs of the return routability procedure are carried within the Mobility header.

The CH will now be able to route packets directly to the MH's CoA. In general, an MIPv6-equipped CH, when communicating with another host (mobile or stationary), always checks whether it maintains a binding for the host's home address, and if it does, it uses the *route optimization* mode for the delivery of these packets. In particular, it sets the destination address to the MH's CoA for the direct routing of

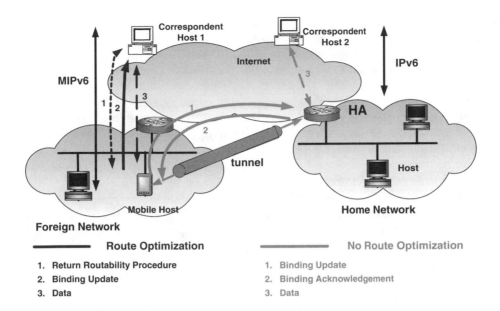

FIGURE 5.10 Mobile IPv6 operation.

packets to the host's current location. However, for maintaining transparency with upper layers, a new type of IPv6 Routing header defined for MIPv6 is added for carrying the host's home address. This information is retrieved by the MH, which accordingly sets the final destination address of the packets.

Likewise, in the reverse direction, the MH sets the source address of outgoing packets to its CoA, thus eliminating the ingress filtering problem. Here, a new destination option has been defined, namely the *Home Address* destination option, which communicates to the CH the home address of the MH. The CH uses this information for keeping higher layers unaware of the host's mobility. The MH will use this mode of communication only after having checked in its BU List that the addressed CH maintains a binding for its home address. This is because the CH will drop packets carrying *Home Address* information for a host not included in its binding cache.

In addition to the header extensions and options specified for MIPv6, some new ICMP messages have been introduced for enabling dynamic HA address discovery and MH configuration in case of prefix changes in the home network while the host resides away from home. If the MH is not configured with the HA's address, it can send an ICMP *Home Agent Address Discovery Request* message to its home network and, in particular, to the Home-Agents anycast address for acquiring this information. The HA that processes the message replies with an ICMP *Home Agent Address Discovery Reply* message, which entails a list of the link's available home agents. Note that possibly more than one HA provides services on the link. An HA is aware of the other nodes on the link providing similar services by means of Router Advertisements, which are extended accordingly in the context of MIPv6. These advertisements indicate that a router provides HA services on the link and reveal the HA's global address instead of the link local address along with a preference value. This *preference* information is used for constructing a hierarchy of the available HAs on the link, which is carried in the reply to the host.

Next, we discuss in more detail the main procedures performed in Mobile IPv6 and, more specifically, the movement detection, the binding management, and the routing of packets.

5.2.2.1.1 Movement Detection

In MIPv6, IPv6 Neighbor Discovery facilities are used for enabling the bidirectional confirmation of a MH's ability to communicate with its default router in its current location. By that, we mean that confirmation is needed for the forward direction — that is, packets from the router to the MH — as well as the reverse one. This way, "black hole" situations, where the link to the router does not work equally well in both directions, can be detected. This method is an enhancement to the one used for MIPv4, where only the forward direction is checked.

Movement detection is performed in cases where the MH loses communication with its default router. Most probably, this is because the host has moved to a new link and consequently needs to switch to a new default router for outgoing IP packets and proceed with the registration of a new CoA with an HA on its home link. However, note that these actions are not always the result of a loss of communication with the host's default router. It may be the case that an MH is attached at the same time to two links (for example, in wireless environments) and that it is configured with two CoAs corresponding to these links. For any reason, such as policy, the host may decide to switch its primary CoA and its default router to the currently nonprimary one. Therefore, although the host is not out of the reach of its currently default router, it decides to perform this switch. Keep in mind that only one CoA is registered as its primary CoA with the HA, although the host may use multiple CoAs. Therefore, the HA can address the MH at only its primary CoA. Recall that in MIPv4, multiple bindings are allowed. CHs, however, can possibly communicate with the MH through the nonprimary CoA(s) if this address is provided during the binding procedure.

In case no communication path exists in either direction with the host's default router, the MH decides to attach to a new link and register a new CoA, based on the link's advertised prefixes, carried in Router Advertisements. With respect to the reverse direction, the MH may determine through the *Neighbor Unreachability Detection* mechanism that there is no communication with its default router. In this case, the action to be performed is the configuration of the host with a new default router and possibly with a new primary CoA. For the opposite direction (the communication link to the MH), there are several ways that an MH can detect unreachability by its default router and thus proceed with the aforementioned actions.

The most common way that an MH detects reachability from its default router is when it keeps receiving IP packets from the latter. However, in the absence of data addressed to the MH, Router Advertisements periodically sent by its default router can alternatively be used for this purpose. Changes have been proposed with respect to the timing rules of Router Advertisements, and a new *Advertisement Interval* option has been introduced to these messages for enabling the movement-detection operation of the host. The latter is aware of the frequency that these messages are being sent through the Advertisement Interval option, if present in the advertisements. Therefore, when this time interval elapses without receiving any advertisement from its router, the host may proceed with switching to a new router and a new primary CoA.

Another way for the host to detect reachability from its default router is by setting its network interface to promiscuous mode so that it is able to receive packets on the link that are not necessarily addressed to its own link-layer address. Therefore, even if not actively participating in a session, packets addressed to other hosts may provide the indication that the forward communication path with the MH's router is successful. However, because it is likely that this option will result in a great consumption of the MH's power resources, it should be selected only when the load in the network is expected to be low. If the host is still unable to detect whether it has moved, it may send a Router Solicitation requesting from its router to respond with a Router Advertisement. Movement detection can also be performed with the help of Layer 2 mechanisms, which indicate a Layer 2 movement. This proves useful, however, when a Layer 2 movement results in an IP movement as well. The host can be timely aware of this fact by issuing a Router Solicitation when triggered by a Layer 2 indication.

In any case, the MH forms a new CoA on the new link with the use of stateful or stateless address autoconfiguration and registers with its HA. Before doing that, Duplicate Address Detection (DAD) needs to be performed for the new CoA on the link for verifying its uniqueness. When the host returns to its home network, a deregistration with its HA should be carried out. A host detects that it has returned home when its home subnet prefix is carried in the received Router Advertisements. Deregistration is performed by sending a BU message to the HA, requesting that the latter stop providing its mobility services. The HA, after processing the message, stops acting as a proxy for the host — in other words, it no longer responds to Neighbor Solicitations addressed to the host. In addition, because hosts on the home link need to update their neighbor caches, the MH multicasts a Neighbor Advertisement for associating its own link-layer address with its home address.

5.2.2.1.2 Binding Management

Mobile IPv6 uses four messages for binding management carried in the MIPv6 defined Mobility header. In addition, the Mobility header carries four other messages for the Return Routability procedure. A BU message is transmitted by the MH to the HA whenever the former wants to register a new primary CoA and consequently inform the HA of its new location or update the lifetime of an existing binding. The BU carries, in addition to the CoA information, the lifetime of the binding and always has the *acknowledge* bit set for requesting the acknowledgment from the HA. Particularly, the lifetime is set to a value not greater than the remaining lifetime of the home address and the CoA specified for the binding. The CoA information is carried in the source address of the BU message, in which case the home address information is carried in the Home Address destination option. Alternatively, the CoA can be carried within a new mobility option defined for MIPv6, namely, the *Alternate Care-of Address* option (see Figure 5.11). The CoA information should be extracted from this mobility option if present in the message. It is used in cases where the CoA to be registered is not topologically correct or when the applied security mechanism does not protect the IPv6 header and thus the CoA carried in the source address of the packet.

The HA, upon receipt of the message and after validating it, creates a new Binding Cache entry for the host or updates the existing one. In the former case, the HA must perform a DAD for the host's address on the home link to ensure that this address is not being used by any other host. DAD is also performed when the host registers a home address, which differs from the already registered one. This might be the result of prefix changes at the home network that will lead to a change on the host's home address. If DAD fails, then the host needs to be informed accordingly. The HA, in all cases, replies with a Binding Acknowledgment message where it sets the status field for indicating the success or failure of the binding procedure.

An MH performs deregistration when it returns home or when the lifetime of its CoA is due to expire. In this case, a BU is sent to the HA as before, where the *Lifetime* field is set to zero and the CoA is set to the host's home address. The HA will delete its entries for the MH and will stop providing HA services to the latter. A Binding Acknowledgement is also sent at this point for ensuring the success of the deregistration procedure.

Apart from informing the HA of its current point of attachment, the MH can also send BUs to CHs with which it communicates. This is for enabling route optimization in order to bypass the HA and avoid unnecessary delays to the delivery of packets as well as congestion in the home network. An MH sends a BU message to its CHs for the latter to either update their binding cache or create a new entry for the host. For this reason, the MH maintains in its BU List the CHs that hold such binding information and knows each time when a CH needs to update an invalid binding. Moreover, it can detect that a CH is not using route optimization when it receives encapsulated packets from the HA instead of packets carrying the Type 2 Routing header extension (refer to Section 5.2.2.1.3).

Note that the CoA carried in the BU and provided to CHs may not necessarily be the same as the one registered with the HA — that is, the primary CoA. Therefore, an MH may actively participate in sessions with CHs by using more than one CoA. The CoA is carried in either the Alternate CoA option, if present, or in the packet's source address. In the same way, the home address is indicated in the Home Address destination option, if present, or in the source address field of the IP packet. An acknowledgment is optionally requested here, although it is always sent to indicate the rejection of the BU, for example, when insufficient resources exist for maintaining a binding. Note that a return routability procedure must be performed before the transmission of the BU message by the MH. The latter may avoid performing a binding procedure with its CHs when it wants to keep its location private and therefore not expose its current CoA.

The CH might initiate the binding procedure by requesting that the MH provide binding information. This might be the case when the CH is actively communicating with the MH in route optimization mode

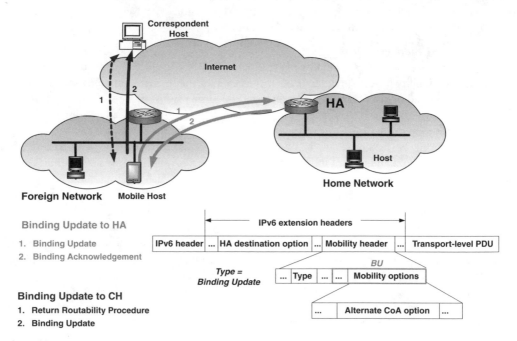

FIGURE 5.11 Binding management in MIPv6.

and the binding is nearing its expiration. To avoid any unnecessary change of path after the binding expires (to avoid the routing of packets through the home network), the CH may request a refresh of the binding through a *Binding Refresh Request* sent to the MH. Last, a *Binding Error* message is sent by the CH to the MH when a problem is detected in the MIPv6 operation. More specific, the CH might receive an MIPv6 control packet, with a Mobility header it is unable to recognize, or it might receive data packets that carry an invalid *Home Address destination option* (e.g., no binding entry is maintained for the indicated home address; for details, refer to the next section).

5.2.2.1.3 Routing

As already mentioned, two modes of communication exist between an MH and a CH. One does not require that the CH implement MIPv6 functionality (extensions to IPv6) and is based on the well-known MIP routing scheme. According to that scheme, all traffic passes through the home network, where the HA captures it and redirects it, through IPv6 encapsulation, to the host's actual location. In the reverse direction, the same path and the same mechanisms are used; a reverse tunnel is constructed for the routing of packets to the HA, which finally decapsulates them and delivers them to the CH. Here, the two endpoints of the tunnel are the HA's address and the host's primary CoA. It is well understood that this scheme creates single points of failure, particularly when the home network is equipped with a single HA; introduces unnecessary load to the core network when the communicating hosts are nearby situated; and causes delays to the delivery of packets, which may prove deleterious to delay-stringent applications.

As an alternative to this scheme, route optimization is proposed, based on the enhanced IPv6 facilities. In particular, the Type 2 Routing header has been introduced to address the needs of this communication mode. Packets sent by the CH to the MH are addressed to the host's CoA (the one maintained in the CH's binding) and are extended with a Type 2 Routing header, which carries the home address of the MH. Therefore, the routing of packets is performed based on the destination address field in the IPv6 header. The recipient of such packets (the MH) will proceed with processing the extension headers and will identify the Type 2 Routing header. The home address information, carried in this header, will be provided to the upper layers for maintaining transparency.

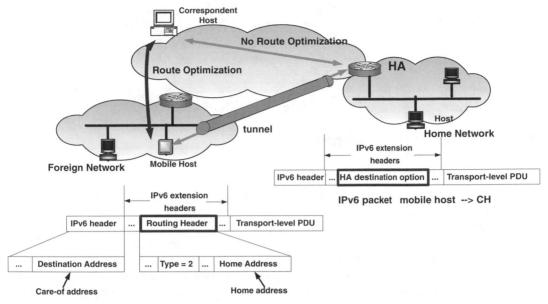

FIGURE 5.12 Routing in MIPv6.

We consider essential at this point to make a short reference to the functionality of such extension headers in IPv6. Extension headers, as specified in IPv6, follow the IPv6 header of a packet and are not examined or processed by any host along a packet's delivery path until the packet reaches the host identified in the *Destination Address* field of the IPv6 header. As far as the Routing header is concerned, it is added by the source host in order to list one or more intermediate nodes to be visited on the way to a packet's destination. The destination host is responsible, after processing the header, to set the destination address of the packet to the next node to be visited, and so forth. IPv6 defines the Type 0 Routing header for this purpose.

Likewise, MIPv6 has specified the Type 2 Routing header carrying the MH's home address. The MH, upon receipt of a packet addressed to it and after processing the Routing header, swaps the packet's IPv6 destination field with the Home Address field carried in the Routing header. This is interpreted by the host that the next node to be visited by the packet should be the one identified by the home address. Note that with a Type 2 Routing header, the next node is always the final hop. Therefore, this processing is internal to the MH because the CoA and home address are both addresses within the host. In this way, upper layers view only the home address as the destination address of the packets.

Route optimization is used by a CH when it maintains a binding for the MH. Otherwise, packets are addressed to the host's home address and encapsulation is performed by the HA. In the reverse direction, the MH includes the Home Address destination option (see Figure 5.12) in the packets addressed to the CH and routes them in an optimized way without using reverse-tunneling techniques. In particular, the source address of these packets is set to the host's CoA, whereas the Home Address destination option carries the host's home address. This option is added for allowing packets pass through routers implementing ingress filtering (their source address is topologically correct) and for providing transparency to upper layers at the CH. The latter will replace the packet's source address with the one carried in the Home Address destination option, and consequently, upper layers will process the packet without knowing its actual source. The CH, however, must maintain a binding for this home address; otherwise, such packets are dropped and a Binding Error message is sent to the MH. Note that route optimization introduces up to 1.5 round-trip delay before it can be turned on because a return routability procedure and a binding procedure should be done first.

An MH may choose to operate without the mobility services provided by MIPv6. In short-term communications, for example, or when the host's mobility is low and an IP handover is not expected, the MH may communicate with CHs by providing its CoA as the source address of outgoing packets, without the presence of a Home Address destination option. In this way, CHs regard the MH as stationary and address the latter at its CoA without the need for a Routing header. This scheme certainly reduces the overhead introduced by the extra options in packets. However, it does not allow the maintenance of a session while the host moves across IP subnets.

5.3   Micromobility Protocols

5.3.1   Introduction

Mobile IP has been widely accepted as the most appropriate protocol for addressing the needs of IP mobility management in future wireless mobile networks. However, it suffers from several well-known weaknesses, such as the increasing signaling load placed on the network and the latency in restoring the communication path to the host's new point of attachment. This latency further results in packets being dropped while the end-to-end path is being restored (from the MH to the HA and the CHs). Therefore, although providing a robust and simple solution, MIP does not guarantee seamless mobility, and thus MHs participating in active sessions experience disruption in their service when changing their point of attachment.

These deficiencies have led to the investigation of other solutions that complement Mobile IP and aim at handling locally the mobility of hosts within the boundaries of an access network for alleviating the core network from the increased signaling volume. This way, a hierarchical approach to mobility management is adopted in which MIP operates in the core network and micromobility protocols cover the region of an access network. The main functionality of these solutions consists of performing location management by locally establishing routing paths (*path update* functionality) for the hosts that move within the boundaries of their operating access network. Their cooperation with Mobile IP is essential for providing global mobility to hosts and enabling their movement across regional domains. Only interdomain movements bring MIP signaling into the network, whereas the intradomain movement of hosts causes paths to be reestablished locally without affecting MIP entities. This also reduces the latency of the path reestablishment and causes less disruption in active sessions.

Although the path update functionality suffices for a micromobility protocol to handle the mobility of hosts, some further requirements need to be met for rendering these solutions scalable in order to keep the network load low as the network grows and the number of MHs increases. Moreover, it is required that certain characteristics of the operating environment, such as the resource scarcity of the wireless access, as well as the capabilities of mobile terminals, be taken into account. Last, performance transparency should be provided for higher layers to be unaffected by the mobility management support. In other words, services and applications should not experience packet loss and should not be disrupted while a MH changes its access point.

The requirement for reducing the path update signaling within the access network and over the resource constraint wireless medium, as well as the power-limited lifetime of mobile devices, has obviated the need for an MH to operate in two modes: an idle and an active one. When in active mode, an MH notifies the network of each movement it performs (*path update* signaling) and keeps it aware of its exact location. This mode is used when an MH is engaged in communication with another host and the network needs to know the MH's location for forwarding packets to it. It is evident that most of the time a host will not participate to any session, though it will be active for a very small duration. Therefore, in this case, it is advantageous for the host to enter the idle mode, reducing in this way the battery consumption and relieving the network of the increasing signaling load. When in idle mode, there is no need for the host to notify the network of each movement it performs but only when entering a new paging area. This reduces the signaling overhead introduced to the network but at the expense of the latter being aware of the host's location in a coarse granularity of a paging area. An access network is organized in paging areas, which consist of a set of base stations and are configured by the network administrator for enabling paging support.

When data packets arrive for an idle MH, the network needs to page the host in order to become aware of its exact location and forward packets to it. Until this moment, the network knows only the paging area where the MH is located, which may span multiple access points. Paging is performed within the boundaries of the host's paging area in order that the MH wake up and notify the network of its whereabouts. It should be noted that the idle mode and the paging functionality have not been introduced just to meet the needs of IP mobility management but have already received great acceptance and practice in the wireless telecommunication world.

So far, we have discussed *path update* and *paging*, which together comprise the *location management* functionality of a micromobility protocol. Both mechanisms result in the establishment of location information within the network, enabling the delivery of packets to the hosts' current location. However, it is also essential that communication be restored in a seamless and efficient way when the hosts change their point of attachment. For this reason, *handover management* is required for dealing with the time-critical period of transition that precedes the path reestablishment and during which packets should be diverted to the host's new location.

Handover management is responsible for maintaining the active sessions of the MH as the latter moves across the coverage area of various access points. Here, we are concerned with handovers that result in a change of the network-layer (IP) connectivity of the host. Note, however, that this is not always the case because a change in the link-layer connectivity will not always result in an IP handover. A micromobility protocol should ensure that handovers are fast and smooth — that is, they should be performed without significant delays and without loss of packets. *Handover delay* is defined as the time between the delivery of the last packet to the host from the old access point and the delivery of the first packet from the new access point. Among the most adopted handover schemes are the establishment of temporary tunnels between the old and the new access point and the bicasting of packets to both access points. Both of these schemes might also use buffering techniques.

Regardless of the handover mechanism used within an IP access network, it should be noted that the handover performance is also highly dependent on the underlying radio technology and on the information that the latter provides to the IP layer. For example, a radio layer that provides indications to the IP layer of an impending handover enables the preparation and possibly the completion of the IP handover before the MH loses its Layer 2 (L2) connectivity. In this case, the Layer 3 (L3) handover delay is equal to the L2 handover delay because at the time the host regains the L2 connectivity it can start receiving IP packets. Moreover, the radio technology may allow for soft handovers — that is, the MH may simultaneously communicate with more than one access point — in which case the L2 handover delay is eliminated. Alternatively, a handover decision can be based on only L3 indications and be completely independent of the L2 technology, resulting in greater handover delay and a greater possibility of service disruption. It is evident that each approach carries its pros and cons; on this account, it remains to be decided within the standardization bodies to what degree these two layers should be coupled and synchronized.

Note that the *handover* functionality is not to be confused with the *path update* functionality. The former involves a time-critical operation that "locally" redirects packets to the host's new location, whereas the latter reestablishes the path after the handover has been performed and the IP connectivity has been regained. However, it might be the case that some protocols implement these functionalities in an integrated manner (e.g., HAWAII's path setup schemes) and do not separate the handover from the path update signaling.

You could expect that a micromobility protocol be broken down into three functions: path update, handover management, and paging. Although protocols exist that implement more than one of these functionalities, it is the current trend to have different protocols deal with each aspect while at the same time defining their interaction points for providing a complete solution. This is important because it facilitates the independent evolution of constituent functionalities.

Another aspect that should be kept in mind when designing a micromobility protocol is the need for robustness and scalability. It is beneficial if no single points of failure exist and if the protocol can respond in a timely manner to failures in nodes, links, or other entities with location-management functionality. Additionally, scalability should be of prime concern because the protocol should scale well if applied in any geographical area and for any number of users. Therefore, the amount of control traffic placed on the network (path updates, refreshes, paging) should be minimized and should be distributed in order not to burden specific nodes in the network.

Moreover, the address-management property of a micromobility protocol comprises a significant design choice, particularly in IPv4 networks where there is scarcity of addresses. It is possible that the access network uses no extra pool of addresses for serving visiting MHs, in which case the latter are addressed at their home address assigned by their home networks. The case where a collocated CoA is assigned to every host for identification and routing purposes is certainly less advantageous and may raise concerns if a great number of users have to be served.

Quality-of-service (QoS) support in IP mobile networks is another issue that has received great attention over the past years. A micromobility protocol should be able to interact with the QoS mechanisms used within the access network, if any. This is because a change in the routing path to the MH should be followed by a reestablishment of the QoS path. The QoS mechanisms, however, should operate locally and reserve resources for only the affected part of the path ensuring efficiency and scalability. It is desirable, therefore, that an effective integration of the micromobility and QoS protocols be facilitated for preserving the QoS of an active session and achieving the goal of performance transparency.

Several proposals have been made for handling the mobility of a host locally. Most of them consist of extensions to Mobile IP or interact with MIP signaling for providing global mobility to users. A classification of these protocols has been attempted by researchers based on the forwarding mechanisms that are used for the routing of packets to the MHs. Protocols may be categorized as those that use *hierarchical tunneling* techniques as their forwarding mechanism and those that use *host-based forwarding entries* along the path to the MHs. Their primary difference lies in where location information is maintained within the network. In the first category, agents, often placed in a hierarchical manner, hold this information and use the standard IP routing of the "mobile-unaware" fixed infrastructure for the forwarding of packets. In contrast, host-based forwarding schemes establish location information for the hosts in all nodes pertaining to the downstream forwarding path. In what follows, we have chosen to present two candidate protocols from each category. We begin with the description of two *host-based* protocols — Cellular IP and HAWAII — and continue with MIPv4 Regional Registration and Hierarchical MIPv6. The last two have received great attention within the IETF Mobile IP workgroup and are considered the most dominant to be applied in IPv4 and IPv6 networks, respectively.

5.3.2   Cellular IP

Cellular IP (CIP) is a micromobility protocol proposed by Columbia University and Ericsson for addressing the needs of local mobility and fast handover support, with the aim of preserving simplicity and scalability for operating from local to metropolitan area networks. Among its basic principles is that the location-management functionality is assisted by the data plane forwarding functionality on the uplink, and that its operation is differentiated for the case of idle and active MHs. Of interest is that CIP replaces IP routing in the wireless access network and uses its own routing mechanisms for the forwarding of packets to MHs.

CIP relies on Mobile IP for enabling global mobility across CIP access networks. In a global mobility scenario, Mobile IP is used for handling the movement of a host across adjacent CIP domains, whereas the CIP mechanisms handle mobility within the boundaries of an access network. The routing of packets in the backbone network is performed based on the host's CoA (more specific, the FA CoA) if the MH resides in a foreign domain, whereas the host's home address is used for routing purposes within the CIP domain. In particular, MHs use the IP address of the access network's gateway as their CoA, whereas the latter is responsible for decapsulating packets and forwarding them to the host based on its home address. By *gateway*, we mean a node with FA and HA functionality situated on the border between the CIP access network and the core IP network.

As already mentioned, in a CIP network MHs are identified by their home address, and for that reason CIP does not compound the address scarcity problem. The establishment of routes within the access network for the downstream routing of packets is accomplished by *route-update* messages transmitted by MHs to the gateway, where the shortest path is followed. All nodes traversed by these messages update their route caches with routing information for the host. These route entries point to the neighbor from which the route-update message was received. This neighbor is referred to as the *downlink neighbor*, and we explain below how it is differentiated from the *uplink neighbor*. In this way, a path is established, allowing the downstream routing of packets to MHs. Although explicit signaling is used for the establishment of these entries, data packets transmitted by MHs while participating in an active communication refresh this information and prevent it from being timed out. This is required because route caches are soft state and need to be refreshed regularly so they do not become invalid.

The upstream routing of packets is simple; it follows the shortest path to the gateway and is configured either manually or with the use of an *uplink neighbor selection algorithm*. In the latter case, the gateway broadcasts to the access network a packet carrying control information. The result is that all nodes are assigned one uplink neighbor, which is identified by (a) the interface from which the packet was received, and (b) the packet's source Medium Access Control (MAC) address. After this assignment, all remaining neighbors become *downlink* neighbors of the node. Packets coming from downlink neighbors are considered to be coming from MHs, whereas packets coming from the uplink neighbor are addressed to MHs. The routing of packets coming from the uplink neighbor is performed based on the established routing mappings, whereas packets coming from the downlink neighbors are simply forwarded to the uplink neighbor.

Note that route entries for MHs is the only routing information maintained at CIP nodes, and for that reason the latter are not aware of the network topology and of the routes to other nodes of the fixed-access network infrastructure. CIP nodes can forward packets only to the MHs based on their home address and to the gateway independent of the destination address carried in the packets. It can be said, therefore, that CIP nodes operate solely with their own routing mechanisms while trying to preserve the simplicity of Layer 2 switches; there is no need for them to be deployed on higher-cost IP routers.

However, the routing information maintained along the path to the MHs concerns only active hosts — that is, hosts that participate in an active session. Passive connectivity is supported for idle MHs, which inform the network of their position when changing paging areas. Mobile hosts send *paging-update* messages along the way to the gateway in order to update the paging caches residing in selected nodes within the network. The routing of packets addressed to idle MHs is performed based on the paging caches because route cache entries for these hosts have been cleared. The size of paging areas (the number of base stations they consist of) is determined by the network topology and the places in the network where paging caches are located.

In the following sections, we examine in more detail the functions of the CIP protocol: location management for both active and idle MHs, handover management, and routing.

5.3.2.1 Protocol Overview

The CIP access network, illustrated in Figure 5.13, mainly consists of base stations (BSs) that carry out routing and location management functionality and also provide a wireless interface to MHs. Nodes that perform only routing and location management, without serving as a wireless access point, are referred to as CIP nodes and can also be found in a CIP network. The CIP node that connects the CIP network to a regular IP network is referred to as the CIP gateway and hosts as well Mobile IP functionality (HA and FA) for enabling the global mobility of hosts.

The operation of the CIP network is unaffected by whether the host resides in its home or in a foreign network. Packets addressed to a host residing in a foreign CIP domain will always be decapsulated by the FA located at the CIP gateway. This way, the original packets, addressed to the host's home address will traverse the CIP domain. On the other hand, packets destined to a host residing in its home network will not undergo any encapsulation and will naturally carry the host's home address as the destination address. An MH is always identified by its home IP address within a CIP network.

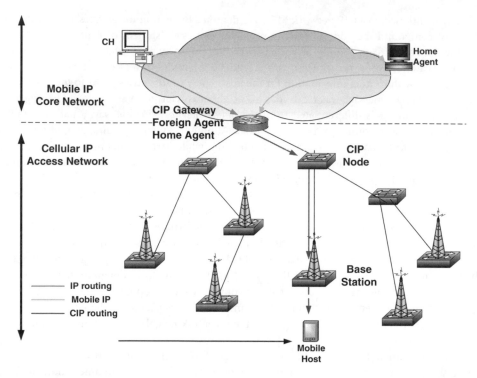

FIGURE 5.13 CIP architectural overview.

The host is responsible for registering with its HA whenever it moves to a foreign CIP network by communicating the FA's IP address as its CoA. It becomes aware of this interdomain movement by means of the beacon signals transmitted by BSs that carry, among other parameters, the CIP Network Identifier and the CIP gateway's IP address. Both parameters identify a CIP domain while the gateway's IP address is used by the host as its CoA. In this way, both CIP and Mobile IP mechanisms are activated for handling the mobility of the host within a CIP access network and across CIP domains, respectively.

5.3.2.1.1   Routing

Routing in the CIP network is performed based on specific CIP routing mechanisms. As we have explained, CIP nodes are aware of their uplink neighbor to which they relay packets originating from MHs and of their downlink neighbors to which they relay packets destined to MHs. Packets coming from MHs are forwarded hop-by-hop to each node's uplink neighbor and eventually reach the gateway, which is the first node performing regular IP routing. Until this point, packets are forwarded regardless of their destination IP address.

The gateway either discards packets, as is the case with CIP control packets addressed to the gateway, or forwards them to the external IP network. Packets pertaining to a mobile-to-mobile communication do not leave the CIP network but are forwarded from the gateway back to their destination — that is, an MH in the domain. This is feasible only if MIP Route Optimization (RO) is used, in which case the CH (here, an MH residing in the same domain with the MH) performs the tunneling instead of the HA. Therefore, packets transmitted by the CH have as their destination the gateway (FA CoA) and not the MH's home address. The gateway simply decapsulates packets and forwards them back to the access network based on the carried destination address (the host's home address). However, if RO is not used, packets are naturally forwarded to the MH's home network and from there are tunneled to the FA serving the host. In order to bypass the HA, the gateway may check if packets coming from the access network are addressed to an MH for which it maintains a route cache entry and then forward them back to the host, in this way eliminating the MIP operation.

The upstream routing of packets helps the establishment and maintenance of downstream paths to the BSs serving the MHs. This is achieved in the following way. Given that a host is in active mode, it sends a route-update message each time it is being served by a new BS in order to establish a routing path from the gateway to its current point of attachment. The route-update message, which constitutes an ICMP message addressed to the gateway, is processed or better monitored by all nodes pertaining to the path to the gateway. CIP nodes identify the type of message and record the carried information for updating their route caches. In particular, the nodes track the source IP address of the packet (the host's identifier), the interface where it was received, and its source MAC address (the last two parameters identify the downlink neighbor to the MH). In this way, route cache entries for the host are established and are consulted for the downstream forwarding of packets. Note that only route-update messages can establish new route caches entries because they carry authentication information.

On the other hand, regular IP packets transmitted by MHs and that follow the default route to the gateway are used for refreshing route entries. Certainly, this is possible only when the host has packets to send and is not just a receiver in an ongoing session. Nodes, in the same way, check the source IP address of the packet for identifying the MH, and if the packet is received from the same downlink neighbor as the one indicated in the route cache, they refresh the host's entry — that is, reset its expiration time. If the packet comes from a different neighbor, they discard it. While participating in a session, if the host has no data to send, it is responsible for sending route-update messages for refreshing the route cache entries in CIP nodes. This is to avoid the removal of the host's entries, which would result in having to page the host for newly coming packets.

In summary, the upstream routing of packets follows a preconfigured path where packets are forwarded to the uplink neighbor of each CIP node. At the same time, CIP nodes record information from the uplink packets in order to establish the reverse path to the MH. This path is established and updated only by CIP route-update messages, whereas it can be refreshed by regular data packets. On the other hand, packets coming from uplink neighbors are assumed to be addressed to MHs. Here, the route caches are consulted for forwarding the packets to the appropriate downlink neighbor.

5.3.2.1.2 *Handover*

CIP provides two types of handover schemes. In both schemes, handover is initiated by the MH, which sends a route-update message to the new BS for establishing a new path to the gateway. In *hard handover*, packets are redirected to the new BS at the crossover node after the host has regained its connectivity with the new BS. However, in *semi-soft handover*, the crossover node forwards packets to both the old and the new BS for a short time before handover takes place. When the host finally migrates, packets are already on their way to the new path.

5.3.2.1.2.1 Hard Handover — An MH that listens to beacon signals transmitted by BSs decides to switch to a new BS, based, for example, on signal strength measurements. After switching to the new BS, it sends a route-update message to the gateway to update its routing path (see Figure 5.14a). This message is processed by all CIP nodes, along the path to the gateway, that update their route caches. At the time that the route-update message reaches the crossover node (the first CIP node that also belongs to the old path) resulting in the update of its route cache, arriving packets are diverted to the new path. However, it is evident that until this moment, packets arriving at the crossover node follow the old path and are consequently lost because the MH has already switched to the new BS. The worst case for the hard-handover scheme is when the crossover node is the gateway, meaning that the route-update message must travel all the way to the gateway for establishing the new path and, as a result, more packets are lost. Route caches in CIP nodes belonging to the old path (except for the common part of the two paths) will eventually time out because they will not be refreshed anymore by the MH. It can be said that hard handover is based on a simple approach refraining from extra handover signaling, such as the activation of forwarding mechanisms from the old to the new BS, to the detriment of the user experiencing packet loss. Actually, the hard-handover scheme is simply the path update procedure that takes place when an MH changes its point of attachment, and no extra technique is used for alleviating the packet loss problem.

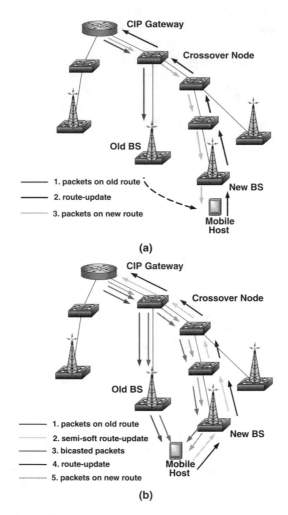

FIGURE 5.14 (a) Hard handover; (b) semi-soft handover.

5.3.2.1.2.2 Semi-soft handover — The semi-soft handover, depicted in Figure 5.14b, has been proposed as an alternative for minimizing packet loss. The idea is that the route cache entries along the path to the new BS must be created before the actual handover takes place. Here also, the MH transmits a route-update packet to the new BS but continues to listen to the old one to receive packets. The route-update message, which carries an indication of the type of handover, is processed as before by the CIP nodes, which update their route caches. However, the crossover node adds a new mapping for the host instead of replacing the old one. From this moment, packets arriving at the crossover node for the MH are forwarded to both downlink neighbors. The MH finally decides to switch to the new BS and only then sends a route-update message to complete the handover procedure. Actually, the procedure from this point is similar to the hard handover. The message will once again update the route caches in the CIP nodes along the path to the gateway and will force the crossover node to replace its old entry with the new one, pointing only to the new path. Therefore, apart from the path update procedure that finally takes place, bicasting techniques are used for handling packet loss. It is evident that at the time the host switches to the new BS, packets already follow the new path and the MH encounters no packet loss.

However, this is not always true if we take into account that the path from the crossover node to the new BS might have a shorter transmission delay than the path to the old BS. In this case, imagine that it takes longer for the same packet to reach the old BS than to reach the new one. This means that when the host migrates to the new BS and starts receiving packets, there may exist packets that have not been delivered to it. Certainly, this is not the only parameter that may contribute to packet loss; in addition, the time it takes the MH to switch to the new BS or the load that the two paths experience at that time are parameters to be taken into account. In any case, packet loss in semi-soft handover is negligible when compared with that experienced in the hard-handover scheme.

For alleviating this problem, Campbell et al. (1999) have proposed that some delay be introduced to the packets being forwarded along the new path in order to compensate for the difference between the propagation distance of the old and new paths. This buffering is naturally placed on the crossover node because the latter is aware of the semi-soft handover due to its existing entry for the host pointing to a different downlink neighbor. Although this technique may cause packets to be delivered twice to the MH, this is preferred to losing packets. If flows prefer to experience some packet loss but cannot tolerate delay, the buffering techniques may not be applied. Differentiation can be achieved, for example, based on the used transport protocols (TCP, UDP, RTP) that permit the identification of the flow's QoS needs.

5.3.2.1.3 *Paging*

CIP allows idle MHs to move within the boundaries of a paging area without having to notify the network of their exact location — that is, their serving BS. Paging areas comprise groups of BSs, which are assigned a Paging Area Identifier. An MH is always aware of the paging area it resides in from the beacon signals transmitted by BSs, carrying this identifier among other information. Paging areas are determined by paging caches, which are located in selected nodes in the network (see Figure 5.15). Paging caches have a longer expiration time than do route caches and are created by *paging-update* messages: authenticated ICMP packets transmitted by MHs to the gateway while in the idle state. A host sends paging update messages when it enters a new paging area or to refresh its paging entries periodically. In a similar way, the MH might send a *paging-teardown* packet to explicitly remove paging and route entries when leaving a service area instead of letting them time out.

Note that paging caches are also created and refreshed when the host is in the active state. In this case, route-update messages are used for establishing and refreshing both route and paging caches, whereas upstream data packets only refresh them. This way, the network maintains at the same time paging and routing information for active MHs; in the case of idle hosts, only paging information is preserved, whereas route caches time out. A paging cache carries the same information as a route cache: the host's home address and the downlink neighbor for paging an MH.

When packets addressed to an idle MH arrive at the CIP network, the CIP nodes normally search their route caches for identifying the downlink neighbor to route packets. When no entry is found for the host, the network is implicitly informed that the host is in the idle state and that it needs to be paged. CIP does not define an explicit paging message; therefore, the first packet is used to page the host (implicit paging). The paging cache is consulted and the packet is forwarded to the indicated downlink neighbor. If there is no entry for the host, the packet is simply discarded; it is assumed that the host is not located at the paging area(s) covered by this node. As explained earlier, not all CIP nodes are equipped with a paging cache. If this is the case, the packet is broadcasted to all its downlink neighbors. This way, paging caches "route" the first packet to the indicated paging area, whereas CIP nodes with no paging cache perform the broadcasting of the packet within the paging area. The configuration of paging areas may vary from having each BS comprise a paging area, in which case all nodes but the BSs are provided with paging caches, to having the whole network constitute a paging area, in which case no CIP node is equipped with a paging cache. It is up to the operator to determine the trade-off between the *paging load* brought on by the broadcasting of packets and the *computational burden* in the CIP network affecting nodes augmented with paging caches.

When the MH receives the packet, it switches to the active state and sends a route-update message to the gateway for establishing route cache entries in the network. From this moment, data packets addressed to the MH are delivered to it following the established route cache entries, which should be kept updated regularly. Consequently, paging caches are no longer consulted until the host falls again to the idle state.

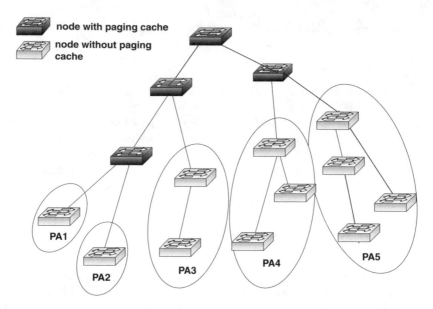

FIGURE 5.15   CIP paging architecture.

5.3.3   HAWAII

The HAWAII (Handover-Aware Wireless Access Internet Infrastructure) protocol has been proposed by Lucent Technologies for addressing the deficiencies of Mobile IP regarding the overhead brought on backbone routers due to the frequent registrations to HAs and the delays caused to the path update procedure resulting from the potential long-traveling registration messages. As is the case with every micromobility protocol, HAWAII seeks to handle in a local manner the movement of an MH, when the latter moves within a region, by leaving the backbone IP network unaffected. However, its operation has also been determined by the need to meet some additional design goals. These include the following:

- The efficient use of network resources, which implies avoiding the tunneling overhead and the triangle routing introduced by the MIP operation
- The per-flow QoS provision
- The local reestablishment of QoS reservations
- The reliability support against the occurrence of failures in the network

These goals have dictated the adoption of specific design choices for the protocol operation that are described next.

To begin with, it is useful to place the operation of the HAWAII protocol in the global infrastructure and examine its interoperation with a macromobility protocol: Mobile IP. A two-layer hierarchy is introduced; Mobile IP is assumed to operate at the first level for handling macromobility among HAWAII administrative domains, whereas HAWAII operates at the second level and runs exclusively within an administrative domain for handling mobility within the domain's boundaries. The domain root router (the gateway of a HAWAII domain) is the first router of the domain that routes downstream packets based on the HAWAII routing scheme.

Mobile hosts retain their IP address while moving within a HAWAII domain. In particular, when residing in their home network, they are identified by their home IP address; when attaching to a foreign network, they are identified by the collocated CoA assigned by the visiting domain. Therefore, a global unique address corresponds to each MH that is used for routing purposes across the backbone network and within the HAWAII network. The collocated CoA mode facilitates the per-flow QoS support in the backbone network, albeit aggravating the problem of limited IP addresses. Moreover, by maintaining the host's IP address, scalability is provided because there is no need for the HA to be notified for every local

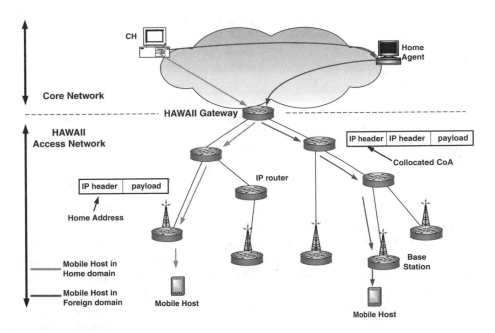

FIGURE 5.16 HAWAII architectural overview.

movement of the host. The path to the domain root router remains unaffected, and the change is performed on only the local path, which is managed by the HAWAII forwarding mechanisms. At the same time, this choice also enables the local reestablishment of QoS reservations, as we discuss below.

The routing of packets addressed to an MH located in the home domain evidently differs from the case of a host located in a foreign domain. In the former case, the packets are routed to the root domain router via standard IP routing and are then forwarded to the MH by means of the HAWAII routing scheme and the established per-host routes. Mobile IP is not activated here because the host's home address suffices for the routing of packets to the HAWAII gateway. This allows for efficient use of network resources; optimized routing is performed and packets are not encapsulated. On the other hand, packets addressed to an MH residing in a foreign domain may experience a non-optimized routing due to the intervention of the HA. Moreover, the packets forwarded within the HAWAII domain are encapsulated IP packets and are routed based on the collocated CoA of the MH, as illustrated in Figure 5.16. It is therefore evident that having the MH served in its home domain is considered the desired operation of the HAWAII protocol. For that reason — and given that an MH, after having powered up, will possibly remain in the same domain — Ramjee et al. (2000) have proposed an address-assignment scheme in which the MH is dynamically assigned the home domain based on the location where it powers up.

As a result, when the MH powers up, it dynamically acquires an IP address from the available pool of the domain's addresses that serves its home address. This dynamic assignment of the home address is acceptable if we consider that MHs usually act as clients and therefore do not need to have statically configured home addresses. This operation has the benefits we explained earlier and also allows for an efficient assignment of IP addresses in a HAWAII domain because, in most cases, hosts will be configured with only one IP address — their home address.

After having examined the protocol's basic design choices regarding the address assignment scheme, the use of the collocated CoA mode, and the host's IP address maintenance, let us delve into the main functions of the protocol: path update, handover, and paging.

5.3.3.1 Protocol Overview

The HAWAII network consists of the routers executing a generic IP routing protocol, the BSs that provide the wireless interface to MHs, and the domain gateway (domain root router) that serves the border with the backbone network infrastructure and demarcates the HAWAII protocol operation. For the rest of our discussion, we assume that the BSs run the IP protocol and act as IP routers, although this may not be the case in a cellular environment. Therefore, IP handover takes place whenever the MH changes its serving BS. The topology adopted by a HAWAII network is preferably hierarchical because the path-setup functionality of the protocol is optimized for such topologies. However, any arbitrary topology may be used.

The HAWAII protocol can be viewed as operating on top of this infrastructure, utilizing its routing and fault management mechanisms. Its main functionality consists of establishing and maintaining special host-based entries in the IP routers of the domain that will aid the forwarding of packets from the domain root router to the MHs. The selected routers that are augmented with this information and the point in time that these entries are established are determined by the path-setup scheme that operates in the HAWAII domain.

A path is initially established when the MH powers up and is updated each time the host migrates to a new BS. After power-up, the establishment of the path is quite straightforward; the relative control message transmitted by the MH simply follows the default route to the domain root router and establishes downstream forwarding entries in all traversed nodes. Each router is configured with the default route that is used for the forwarding of upstream packets to the domain gateway. As for the update of a path, the HAWAII protocol hinges on the assumption that there will always exist a common part between the old and the new path, and for that reason it refrains from a full reestablishment of the new path — that is, signaling messages do not reach the gateway. Four path-setup schemes are defined that are classified as forwarding and nonforwarding. These schemes perform the *path update* and *handover* functionalities in an integrated manner. Their classification is based on whether packets are forwarded from the old BS to the new one during handover.

As long as a path is set up and forwarding entries have been established in the routers pertaining to this path, the HAWAII protocol will ensure that the path is maintained. This is because HAWAII forwarding entries are soft state, which implies that they should be refreshed periodically, or otherwise they are removed. This property shields the operation of the protocol from potential link and router failures and also means it does not have to rely on MHs for the removal of path state from routers. The refresh mechanism is accomplished in a hop-by-hop manner to the domain root, and aggregate refreshes for a number of hosts with overlapping paths is also enabled.

Of great importance is the transparency that HAWAII provides to MHs. The latter are Mobile IP clients with NAI, route optimization, and challenge/response extensions. Therefore, there is no need for a host to run the HAWAII protocol in order to be served by a HAWAII network. This property places some extra requirements on the BSs, though, which are responsible for processing MIP messages, generating the appropriate HAWAII messages, and deciding on MIP registrations with HAs on behalf of the hosts.

5.3.3.1.1 *Path Setup*

In this section, we further elaborate on how HAWAII performs path update and handover management. We initially describe the path establishment procedure when the MH powers up, and then we present the operation of the path-setup schemes that reestablish the path when the host is handed over to new BSs. Forwarding and nonforwarding schemes have been defined, enabling lossless and fast handover being optimized for certain wireless link capabilities. A hierarchical topology is used for demonstrating the path setup functionality; however, meshed topologies are also examined for showing how they may lead to nonoptimal downstream paths.

5.3.3.1.1.1 Path Setup after Power-up — In Figure 5.17, the path-setup procedure is illustrated when the MH powers up. Let us assume that the host is not configured with a static home IP address and decides to make the current network its home network. It acquires a home IP address, for example, by means of DHCP and sends a MIP Registration Request to the nearest BS. When the BS receives the message, it detects that this is the host's first registration after power-up because the message has no PFANE extension (specified in MIP Route Optimization) and then by comparing the host's NAI with the current domain's NAI, checks whether this is the host's home domain. If this is not the case, it registers the MH with its

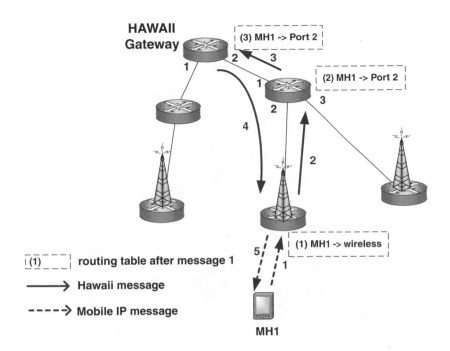

FIGURE 5.17 Path setup after power-up.

HA. In our example, because the host is in its home domain, the BS simply adds a forwarding entry for the host and then sends a HAWAII *Power Up update* message to the next hop router along the default route to the domain root router. The same processing is performed here, and the message is forwarded until it reaches the domain root router. Each router in the path adds a forwarding entry for the host. The domain root router sends an acknowledgment back to the BS, which finally replies to the MH with an MIP Registration Reply. Note here that if the MH was located in a foreign domain, the BS would also send an MIP Registration Request to the HA on the host's behalf for establishing the path between the HA and the HAWAII gateway.

At this point, packets addressed to the MH are routed within the HAWAII network based on the established forwarding entries. Routing outside the HAWAII domain is based on the subnet portion of the host's home address. If the host was in a foreign domain, then MIP would be used for the routing packets to the domain root router, and from there HAWAII routing would be performed for the encapsulated packets. Entries in the HAWAII domain are soft state and therefore should be refreshed regularly. MIP renewal registrations transmitted by the MH to the BS cause aggregate *refresh* messages to be sent periodically along the path to the domain gateway on a hop-by-hop basis. Moreover, in the case of a host located in a foreign network, the BS is responsible for keeping alive the MH's registration with the HA and therefore periodically generates Registration Requests on behalf of the host. However, these surrogate requests require modifications on the HA functionality because they do not contain a valid *Mobile-Home Authentication Extension*.

Although the per-host forwarding entries as well as the soft-state property may raise great concerns for the scalability of this approach, HAWAII's strength lies in the fact that only selected routers in the domain process HAWAII messages, as you will see next when we discuss the path-setup schemes. Certainly, the overhead in memory needs at the domain root router cannot be reduced because the latter maintains entries for all MHs served by the domain. This, however, ensures that packets belonging to a mobile-to-mobile communication will, in the worst case, reach the domain gateway by following the default route and then will be forwarded to the MH.

5.3.3.1.1.2 Forwarding Path-Setup Schemes — The main goal of the forwarding path-setup schemes is to achieve lossless handover by means of forwarding packets from the old BS to the one currently serving

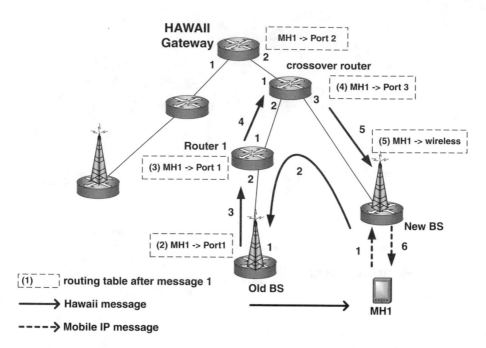

FIGURE 5.18 Multiple Stream Forwarding scheme.

the MH. When path update is eventually accomplished, packets stop being routed to the old BS and follow the new path to the MH's current BS. Assuming that we have a tree-like topology, the two paths (old and new) will always have a common part, which is the path between the domain root router and the crossover router. We define *crossover router* as the router closest to the MH that is located at the intersection of two paths: one between the domain root router and the old BS, and the second between the old BS and the new BS. HAWAII setup schemes base their functionality on the fact that signaling from the current BS to the old one will always pass from the crossover router, establishing in this way the new part of the path. Consequently, there is no need for signaling to be sent to the gateway.

As illustrated in the Multiple Stream Forwarding (MSF) scheme depicted in Figure 5.18, the MH first sends an MIP Registration Request to the new BS. The host must be configured in such a way (e.g., a netmask consisting of all 1's) that it always regards the new BS as belonging to a foreign domain and that it sends MIP registration messages on every movement. On the other hand, by comparing its NAI with the domain's NAI advertised by the BS, the host can identify whether it resides in its home domain and set the appropriate CoA in the Registration Request message. The CoA is set to the already-assigned collocated CoA of the host (when in a foreign domain) or to the new BS's advertised address (when in the home domain). Note that the host must retain its address as long as it moves within a HAWAII network.

Upon receiving the Registration Request, the new BS identifies that this is a *handover update* procedure, and not a power-up because of the PFANE extension of the message. This extension points to the old BS that was serving the user. To activate the forwarding procedure, the new BS sends a *handover update* message carrying its address to the old BS. The latter performs a routing table lookup for the new BS and determines the outgoing interface and the next hop router. Then, it sets the forwarding entry for the MH to this interface and forwards the message to the next hop router. The same processing takes place at all routers between the old and the new BS. Finally, the latter sends an MIP Registration Reply to the MH. Note that this reply does not contain authentication information from the HA, and this requires changes in the MIP client operation at the host.

Let us see what happens to the packets on their way to the MH. Packets that have reached the crossover router before the latter has updated its forwarding entry for the host (with message 4) are forwarded to

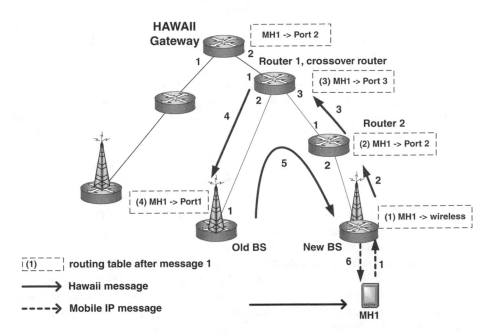

FIGURE 5.19 Unicast Nonforwarding scheme.

the old BS. These packets will finally be forwarded to the new BS, possibly out of order. Keep in mind that the forwarding of packets to the new BS is performed through the established forwarding entries and that there is no need for using a tunneling mechanism (which would certainly cause extra overhead). The out-of-order problem results from the fact that, for example, newer packets forwarded by Router 1 will reach the MH before older packets forwarded by the old BS. Likewise, packets that follow the path to the old BS may arrive at the MH later than newer packets diverted at the crossover router to the new path. The benefit of this scheme is that it reduces packet loss, although it incurs delay due to the buffering and forwarding mechanisms. However, packet loss will always exist because packets arriving at the old BS before message 2 are lost.

As an alternative, Ramjee et al. (2002) have proposed the Single Stream Forwarding (SSF) scheme for dealing with the problem of out-of-order packets. This scheme uses additional information — the incoming interface of the packets — in order to perform forwarding. As a result, packets following the old path always reach the old BS and are forwarded as a single stream to the new BS. The old BS informs the crossover router that it has cleared its buffers for the host in order for the router to divert newly coming packets to the new BS. The procedure ends with a Registration Reply message being sent to the MH. The entries along the old path will eventually time out. However, in most common handovers where the crossover router is close to the BSs, SSF does not significantly improve performance when compared with MSF.

It is interesting to see what happens in cases where the topology is not hierarchical. The crossover router may be the old BS if the path between the old and the new BS has no common part with the old path between the gateway and the old BS. In this case, the path-update procedure will lead to a suboptimal routing of packets, which represents the trade-off for not having the handover signaling reach the domain root router, as is the case with the power-up update procedure. Only the new and the old BSs including the routers connecting them are involved in processing the path-update messages. However, by keeping these messages close to the BSs, the HAWAII protocol protects the routers close to the gateway from dealing with a high volume of control messages that could potentially cause bottlenecks.

5.3.3.1.1.3 Nonforwarding Path-Setup Schemes — In the nonforwarding schemes, data packets are diverted at the crossover router to the new BS without any forwarding of packets from the old BS. In the Unicast Nonforwarding (UNF) scheme depicted in Figure 5.19, the new BS, upon receipt of the MIP Registration Request, adds a forwarding entry for the host by setting the outgoing interface to the one where the message was received. Then, it performs a routing table lookup for the old BS indicated in the PFANE extension for determining the next hop router (Router 2) and sends to the latter a HAWAII *handover update* message. Router 2, in turn, performs the same actions and forwards the message to Router 1. Packets arriving at Router 1, which also happens to be the crossover router, are now diverted to the new BS serving the host. Finally, the message reaches the old BS, which changes its forwarding entry and sends an acknowledgment to the new BS. The latter replies to the MH with a Registration Reply message.

Note that the UNF scheme resembles the CIP hard-handover scheme, although the destination of the handover signaling differs. UNF has better performance in networks where the MH is able to listen/transmit to more than one BS. This way, packets that arrive at the crossover router before the latter updates the host's forwarding entry (message 3) follow the old path and are delivered to the MH by the old BS, whereas newer packets are delivered by the new BS. As an alternative, the Multicast Nonforwarding (MNF) scheme is proposed; the main difference is that packets are bicasted from the crossover router to both paths for a short time until they are diverted to the new BS. It is evident that the selection of the appropriate path-setup scheme depends on the operator's priorities for eliminating packet loss, minimizing handover latency, and maintaining packet delivery order.

The path-setup schemes that have been described here operate within a HAWAII domain and are thus triggered for the intradomain movement of a host. If the latter enters a new HAWAII domain, then the procedure is similar to the one followed when the MH powers up. The host detects that it has entered a new foreign HAWAII domain, acquires a new collocated CoA, and sends an MIP Registration Request to the new BS. The latter identifies that this is an interdomain handover and initiates the power-up procedure for establishing a path for the host within the domain. After receiving the acknowledgment from the domain root router, it sends an MIP Registration Request to the host's HA. When the HA replies, an acknowledgment is finally sent to the MH. In this way, packets routed from the HA to the host's CoA reach the domain root router and, from there, the HAWAII mechanisms are used for their forwarding to the MH.

5.3.3.1.2 *Paging*

The HAWAII protocol is extended to distinguish active from idle users. When the MH is in the active state, the network knows its current BS, whereas when in the idle state, the network is aware of its location in the granularity of a set of BSs. This set of BSs comprises a paging area (PA) and is identified by an IP Multicast Group Address (MGA). Paging is accomplished by augmenting the forwarding functionality with paging entries. The protocol ensures that paging entries will always exist along the path from the domain root router to one BS in the PA of the MH. An MH can be paged by any router that belongs to this path and that can multicast a paging request, protecting in this way the paging functionality from single points of failure. However, certain "paging" rules aim at placing the paging load (or, in other words, the paging initiator) as close to the BSs as possible in order to protect routers higher in the hierarchy from extra overhead. Note that these routers are the ones experiencing the most processing and communication load in a HAWAII network.

When the MH stops being active for a certain time, it enters the idle mode and stops notifying the network of each movement it performs. It simply sends *paging updates* when entering a new PA and paging refreshes for refreshing its paging entries along the path to the domain root router. The network can page an MH by sending a *paging request* addressed to the corresponding PA's IP multicast address, resulting in all BSs belonging to this MGA to receive the request. The BSs, in turn, page the host by using L2 paging. Therefore, HAWAII paging relies on the underlying link layer's paging functionality. In particular, it is assumed that the MH can identify the PA it resides in, for example, by listening to broadcast messages carrying the PA identity and can detect paging requests.

In the network illustrated in Figure 5.20, we assume that BSs 1 and 2 comprise one PA and that Router 2 (R2) is able to route packets to this multicast tree (it has an entry for MGA1). The MH powers up and establishes forwarding and paging entries along the path to the domain root router (BS1-R2-R1). Note that the HAWAII *path update* message has been extended with the MGA field that corresponds to the PA of the host, enabling paging entries to be set at the same time as route entries. Let us assume that the

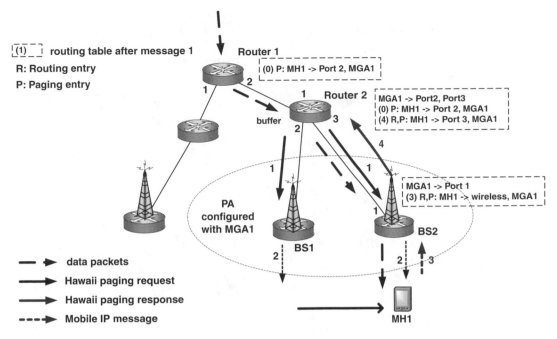

FIGURE 5.20 HAWAII paging.

MH enters the idle mode, resulting in forwarding entries in routers being deleted for this host after a short time while paging entries remain (state 0 in the routing table). The network therefore knows that the host is located in the PA identified by MGA1 but is not aware of the movement of the host to BS2. Note that there is no need for the host to send a paging update message because BS2 belongs to the same PA.

Now consider that data packets arrive for the host at Router 1. The latter detects that only a paging entry exists for the MH; still, it has no multicast entry for the MGA of the host's PA. Consequently, it forwards the packets to Router 2. The latter performs the same actions and identifies that it is part of the multicast tree for this PA — that is, it has a multicast entry for this MGA. Consequently, it buffers packets and initiates a paging request addressed to MGA1 (message 1). Base stations BS1 and BS2 receive this message and broadcast a paging message (message 2) utilizing the wireless link's paging mechanism. When the host receives the paging message, it responds with an MIP Registration Request to BS2, which updates its routing and paging entries for the host. In its turn, BS2 sends a *paging response* (message 4) to the paging initiator, Router 2, which also updates its entries. From this moment, the buffered packets can be forwarded to the MH based on the newly established routing entries.

The main benefit of the HAWAII paging approach is its distributed nature because paging initiators are not fixed nodes in the network. This, along with the soft-state nature of the paging entries, renders the protocol robust against potential failures. Last, HAWAII paging achieves load balancing by requiring that paging initiators not be heavily loaded, for example, with paging requests to other hosts.

5.3.3.2 QoS Support

HAWAII interacts in an efficient way with QoS provisioning mechanisms such as the Resource Reservation Protocol (RSVP). In particular, it facilitates the flow-based QoS provision of RSVP and enables the local restoration of a QoS path. The former is achieved due to the collocated CoA mode of MHs, whereas the latter is due to the maintenance of the host's IP address within a HAWAII domain. The collocated CoA mode enables RSVP to identify and treat separately the flows of the hosts residing in a HAWAII domain (this would not be feasible if hosts were using the same FA CoA). Recall that the treatment of packets is based on the packet's destination address, among other parameters. Moreover, if the hosts were assigned a new IP address whenever changing their point of attachment in a HAWAII domain, a new QoS

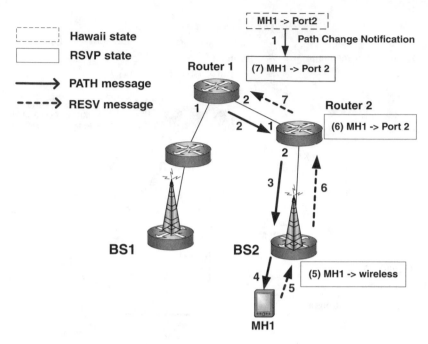

FIGURE 5.21 HAWAII-RSVP integration, with a mobile host as a receiver.

reservation would have to be performed each time along the end-to-end path, although only a part of the path would be affected. By maintaining the IP address of the host, it is possible to locally restore the QoS path.

In Figure 5.21, we illustrate the interactions between HAWAII and RSVP when the MH is the receiver of a data flow. Imagine that the MH is handed off from BS1 to BS2, which results in the path-update procedure taking place. We assume that the nonforwarding scheme is used, in which case forwarding entries for the host are established up to Router 1 (the crossover router). After Router 1 processes the handover-update message, it changes its forwarding entry for the host from port 1 to port 2. This results in a path-change notification (message 1) being sent to the router's RSVP daemon. The latter is configured to immediately trigger an RSVP PATH message, which follows the already established path to the new BS. The PATH message is processed by all routers to the new BS and installs the PATH state for the host. The MH responds with an RESV message, which follows the reverse path to Router 1. The latter stops forwarding the RESV messages because the reservation state in upstream routers remains as it is. This way, the reservation path is established locally and in a timely manner without affecting the end-to-end path.

The procedure is simplified when the MH is the sender because it is also the one responsible for sending the RSVP PATH message along the new path after the handover procedure is completed. The natural integration of HAWAII with RSVP is because RSVP responds to routing changes and follows the established paths regardless of the entity that performs these changes and maintains the routing entries (the HAWAII protocol in our case).

5.3.4   MIPv4 Regional Registration

The MIPv4 Regional Registration (MIPv4 RR) proposal from Ericsson and Nokia uses a hierarchy of FAs, located in a visited domain, with the aim of handling locally the movement of MHs and disengaging the latter from frequent registrations with the HA. The idea here is that the movement of a host within the visited domain and, in particular, under one globally routable entity, denoted as the Gateway Foreign Agent (GFA), is hidden from the HA. This means that the number of signaling messages to the home network, in conjunction with the time needed for the host to update the path to its current location, is

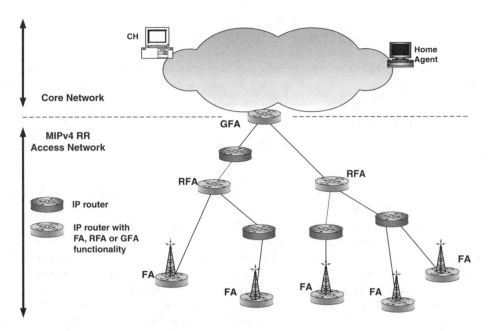

FIGURE 5.22 MIPv4 Regional Registration architectural overview.

reduced. This is because a change on the host's path within a visited network is not propagated to the HA or the CHs (in Route Optimization) but is handled locally.

The Regional Registration proposal defines a tree-like hierarchy of FAs where the FA located at the root of the tree is referred to as the GFA. It is possible that several levels of regional foreign agents (RFA) are supported between the GFA and the lowest-level FAs (alternatively local FAs). This hierarchy of FAs constitutes a visited domain with one or more GFAs providing their globally routable addresses to the public network (see Figure 5.22).

When the MH attaches to a foreign domain, it registers with its home network where it uses the IP address of the GFA as its CoA. In other words, the GFA is, from the HA's point of view, the tunnel's endpoint for the packets being addressed to the host's home address. However, the MIP Registration Request message sent to the HA, and augmented with appropriate "regional" extensions, is used at the same time for the establishment of tunnels along the FAs of the visited domain. This way, a path composed of sequential tunnels is established in the regional network to the MH. The location information resides only at the FAs that form the path from the GFA to the local FA, whereas routers also pertaining to this path have no routing information for the host (in contrast to HAWAII and CIP). It can be said, therefore, that MIP RR operates on top of the IP layer and uses the IP routing service of the fixed network for building an overlay network able to route packets to MHs.

After the registration with the HA, the mobility of the host is handled locally, given that the latter moves within the boundaries of the visited domain and under the same GFA. If, however, the host changes its GFA even within the same domain, it has to register with its HA so that the latter becomes aware of the new tunnel endpoint for packets addressed to the host. With respect to the RR operation, the host sends a Regional Registration Request message to the GFA through the FA hierarchy for updating its location information within the domain. The message makes it all the way to the GFA or to an intermediate-level RFA (the crossover RFA) in a multilevel hierarchy if part of the path is affected by the host's movement. In any case, the HA is not informed of the path-update procedure within the regional domain. Note that the MH is able to perform Regional Registration with any RFA (instead of the GFA) as long as it provides the latter's publicly routable address for registration with the HA.

Paging extensions to MIP RR allow the idle mode operation of MHs (for power-saving reasons) and reduce the routing state information maintained within the visited domain. The location of idle MHs is known to the granularity of a PA, which is defined as an area containing multiple FAs. The root FA of a PA is referred to as *paging FA* and assumes the role of initiating paging for an idle MH so that the latter establishes a routing path at its current point of attachment.

In the following sections, we examine in more detail the protocol operation — that is, how mobility is handled within a regional network when the MH enters a domain (Home Registration) or moves within the boundaries of a domain (Regional Registration) and how paging is supported.

5.3.4.1   Protocol Overview

As we mentioned earlier, a visited domain that supports Regional Registration is composed of a hierarchy of FAs — the GFA, RFAs, and FAs — which maintain location information for MHs and utilize the IP routing service of the fixed network infrastructure. Mobile hosts within a regional domain are identified by their home address. For the rest of our discussion, we assume that a two-level hierarchy of FAs is used, consisting of the GFA and the local FAs unless otherwise stated.

MIP RR proposes two new messages for handling the regional mobility of the user as well as extensions to the conventional MIP messages. However, the HA's functionality remains unaffected by the operation of MIP RR, whereas MHs can simply run standard MIP if they do not support these extensions. This is because, except for some cases where backward compatibility is not enabled, the communication between the MHs and the local FAs, as well as between the GFA and the HA for the Home Registration procedure, is performed via standard MIP messages. This way, the HA may be unaware of whether MIP RR is used by the MH. The MH, on the other hand, is responsible for determining (through the selection of the CoA registered with the HA) whether it will operate in standard MIP or MIP RR. There are, however, some exceptions to this backward compatibility, as we will see later, that require that the HA support an extension to the MIP Registration Request message and that do not allow for the host to operate in standard MIP.

In an MIP RR domain, the FAs typically announce their presence via Agent Advertisements that are modified accordingly to indicate the support of regional tunnel management. In particular, the *I* flag is inserted in the Agent Advertisement for that reason. Moreover, the FAs should also include their NAI (FA-NAI) to allow the MH to determine whether it is located in its home domain and whether it has changed domains since it last registered. The *Care-of Addresses* field in the advertisement indicates the addresses of the FAs hierarchy, with the first CoA being the local FA's address and the last one being the GFA's address. This way, the MH is aware of the GFA for performing Regional Registration with it and for using its address in registrations with the HA. If only one CoA is present in the advertisement, it can be either the local FA's address or an address set to *all ones*. Here, the MH is obliged to request for a GFA to be dynamically assigned to it. Note that the *all ones* address does not support backward compatibility because hosts operating in standard MIP cannot register the advertised CoA with their HA.

Depending on the MH's movement — either entering a visited domain or moving within the same domain — a registration needs to be performed with the HA to establish the path from the HA to the host or with the GFA for updating the path from the GFA to the host, respectively. In what follows, we examine both registration procedures and the protocol's paging extensions, as well as the differences between MIP RR operations in a multilevel and a two-level domain.

5.3.4.1.1   *Home Registration*

An MH performs Home Registration when it is assigned a new GFA — that is, a new CoA — either within the same visited domain or when it enters a new visited domain. The host identifies from the local FA's advertisement that it resides in a Regional Registration domain and becomes aware of the GFA's IP address. Moreover, by comparing the domain part of its own NAI with the FA-NAI carried in the advertisement's FA-NAI extension, it detects that it is located in a foreign domain and therefore proceeds with the registration. It transmits an MIP Registration Request in order to register with the HA and sets the CoA field to the GFA's address. This results in the establishment of a tunnel between the HA and the GFA for the routing of packets addressed to the host's home address. The FAs within the visited domain should also process the registration message to establish the path between the GFA and the MH. Therefore,

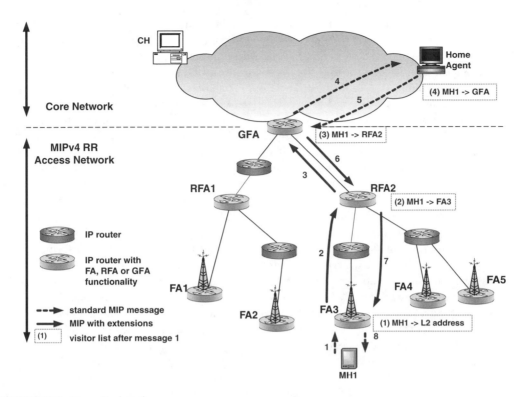

FIGURE 5.23 Home Registration.

this message is not sent directly to the GFA but is processed by all intermediate-level FAs. In a two-level hierarchy, only the local FA and the GFA process the message.

Let us look at the procedure depicted in Figure 5.23. The MH sends the Registration Request message to the local FA. The local FA identifies from the *Care-of Address* field of the message the GFA to which the message should be relayed. If a multilayer hierarchy is considered, it relays the message to the next RFA in the hierarchy (message 2). Note that the FAs are preconfigured with this information. The local FA, before relaying the message to the GFA, adds a Hierarchical Foreign Agent (HFA) extension carrying its own IP address. This information is used by the GFA to establish a tunnel with the local FA for the packets addressed to the MH. Likewise, in the multilayer hierarchy, each RFA removes the HFA extension received by the previous FA/RFA and adds its own extension to inform accordingly the next RFA/GFA (message 3). When the GFA finally receives the message, it removes the HFA extension, updates its visitor list with the current CoA of the host (the local FA or the next low-level RFA), and sends the request to the HA. Upon receiving the Registration Reply from the HA, the GFA consults its visitor list for the host's CoA in order to send the reply to this address. Note that the host's CoA for the HA is the address of the GFA, whereas the host's CoA for the GFA is the address of the local FA or the next low-level RFA. The local FA finally delivers the message to the MH. At this point, a path composed of sequential tunnels between the HA and the hierarchy of FAs to the host's actual location has been established. These tunnels are used for the downstream routing of packets to the host.

The upstream routing of packets to the CH is performed via standard IP routing based on the CH's destination address unless the MH has requested reverse tunneling during the registration procedure. If this is the case, then the local FA should include in its visitor list a field for the GFA's address. When packets arrive from an MH in its visitor list, the FA checks whether reverse tunneling has been requested and, if so, tunnels packets to the GFA or the next-level RFA. Likewise, the GFA maintains the HA's address and is responsible for decapsulating packets received by the local FA and reencapsulating them for delivery to the HA.

Until now, we have assumed that the host is aware of the GFA serving a domain from the FA's advertisements. If, however, this is not the case — that is, the FA advertises an *all ones* CoA or only its own address — a GFA should be dynamically assigned to the MH. The latter sets the *Care-of Address* field of the Registration Request message to zero and sends the message to the FA. The local FA is now responsible for assigning a GFA to the host. There is no established way for selecting a GFA, but in the simplest case, a default GFA is assigned to all MHs. Here, as in any network where one node maintains location information for all served hosts, scalability concerns may arise. On top of that, this node also comprises a single point of failure. After selecting a GFA for the MH, the FA adds a HFA extension to the message and forwards it to the GFA. The GFA processes the message as before and adds a GFA IP Address extension carrying its own address before relaying it to the HA. This extension informs the HA of the host's CoA when this is set to zero in the Registration Request message. It is not possible, however, for the GFA to change the value of this field because it will cause the Mobile-Home authentication to fail. The HA generates a Registration Reply including the GFA IP Address extension and sends it to the GFA. This message finally reaches the MH, which is now informed of its assigned GFA. Note that the host needs this information in order to identify whether subsequent movements are performed under the same GFA and to determine the type of registration to be carried out (Home or Regional).

An MH with a collocated CoA can also use Regional Registration. In this case, the host sends the Registration Request message directly to the GFA and not via the local FA. It is responsible, however, for adding the HFA extension to the message with its collocated address. This extension notifies the GFA of the host's CoA or, in other words, of the end of the tunnel for packets addressed to the MH.

Registrations to the HA need to be renewed when they are about to expire. The MH is responsible for sending registration renewals in order to refresh the bindings kept at the HA. The message is processed, as we explained earlier, by the entities involved in the registration procedure, which update their entries accordingly. The GFA identifies that this is a renewal message because it already has an entry for the host and thus updates only the lifetime of the host's registration. If it happens that the host has moved to a new FA when sending the message, the GFA will also update the host's *Care-of Address* field in the visitor-list entry. Note that here a Regional Registration for updating the local path (GFA-MH) is not needed because the preceded Home Registration has updated the whole path (HA-GFA-MH). An MH performs Home Registration instead of Regional Registration only when its Home Registration is due to expire.

5.3.4.1.2 *Regional Registration*

After having registered with a GFA within the visited domain, and as long as it moves under the same GFA, the MH can perform Regional Registrations. The path from the HA to the GFA is kept intact, and therefore Regional Registration aims at operating in a localized manner without involving the HA. The MH receives Agent Advertisements from the new FA and detects that the latter is located under the GFA with which it is already registered. This is performed by comparing the advertised GFA address with the one registered with the HA or the domain part of the advertised FA-NAI with the FA-NAI of the previous FA. Note, though, that the last method does not always ensure that the GFA is the same because the visited domain may have more than one GFA. In this case, the MH will be notified appropriately and will perform a Home Registration.

Otherwise, the MH sends a *Regional Registration Request* to the GFA via the new FA. Note that if the collocated CoA is used, the message is sent directly to the GFA. The message includes the GFA's address and the host's CoA (either the collocated one or the local FA's address). In this way, the GFA is informed of where to tunnel packets addressed to the MH and updates its visitor list accordingly. If the MH is not aware of the FA's address, it sets the *Care-of Address* field to zero. The FA is now responsible for adding a HFA extension to the message for informing the GFA of the host's CoA. Finally, the GFA replies to the host via the FA with a *Regional Registration Reply* message. This way, the path from the GFA to the MH is updated locally by reestablishing the tunnel from the GFA to the local FA or the host's collocated CoA. The delay caused by waiting for this update to take place is certainly smaller when compared with a path update to the HA because it is proportional to the round-trip time from the MH to the GFA (in a two-level hierarchy).

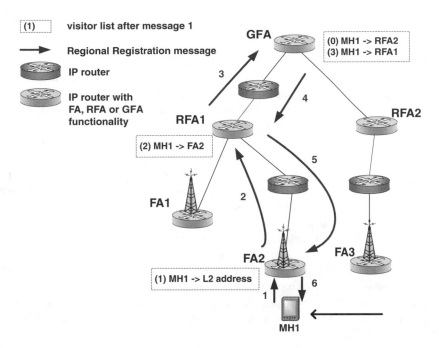

FIGURE 5.24 Regional Registration.

Note that the Regional Registration of an MH with the GFA should not have a greater lifetime than the remaining lifetime of the Home Registration. This is evident because there is no reason in maintaining the path between the GFA and the MH if the path between the HA and the GFA has expired. Therefore, upon receiving a Regional Registration Request, the GFA should set the lifetime (also in the reply message) to a value not greater than the remaining lifetime of the Home Registration.

In a multilevel hierarchy of FAs, the Regional Registration Request is answered by the crossover RFA (or the GFA; see Figure 5.24) residing at the intersection of the old and the new path. Therefore, the first RFA that already has an entry for the host is responsible for transmitting the Regional Registration Reply message. The crossover RFA will forward the request to the next RFA to the GFA only if the lifetime of the host's binding has expired and, consequently, the whole *regional* path needs to be updated.

Here, apart from the establishment of the new path to the crossover RFA, the protocol requires that smooth handover, as specified in the Route Optimization operation of MIP, be performed. This way, the host must include the PFANE extension in the request message in order to inform the new FA of the previous FA serving the host. This will result in a BU being sent from the current to the previous FA in order that the latter redirect packets already on their way along the old path, to the host's new location. Handover delay is significantly reduced; the host starts receiving packets at its new location when the old FA is informed of it. However, some extensions to smooth handover as specified in Route Optimization are needed so that all RFAs pertaining to the old path are notified of the MH's movement. As such, the BU is forwarded along the old path to the GFA until it reaches the crossover RFA. The latter is responsible for replying with a Binding Acknowledgment (BACK) to the MH via the old route. The processing of the BU that takes place at the RFAs involves the updating of the host's binding lifetime according to the value indicated in the BU.

In smooth handover, the crossover RFA knows that it is the one to reply with a BACK because earlier it has received a Registration Request message on the new path from the MH. If this is not the case and the BU has reached the crossover RFA before the Registration Request, then it simply forwards the BU to the next RFA and replies with a BACK to the MH only when it finally receives the Registration Request message. However, here, it forwards the Registration Request to the higher-level RFA so that the latter's visitor list is updated according to this message and that the previously received BU is ignored. The same forwarding procedure and processing takes place at all higher-level RFAs that have earlier forwarded a BU for the host.

5.3.4.1.3 *Paging*

Paging extensions to Mobile IP Regional Registration, referred to as Mobile IP Regional Paging (MIRP), have been proposed with the aim of supporting the power-constrained operation of MHs. A regional network is organized in PAs that comprise a subtree of FAs within the visited domain. The root FA of the PA is called the *Paging Foreign Agent (PFA)* and all FAs under the same PFA belong to the same PA. The latter is identified by a Paging Area ID (PAI). Paging information for an MH is located only at the PFA of the host's PA. All FAs under the PFA have no location information for the host and are unable to route packets to it. Paging is triggered by the PFA, when the latter receives packets for an idle MH within its PA, whereas the lowest FAs are responsible for transmitting paging Agent Advertisements for a specific host to a multicast IP address.

The lowest FAs advertise paging support by including a PAI extension in the Agent Advertisements. Mobile hosts identify a PA by the pair *paging area ID-realm* part of the FA-NAI and perform paging updates when entering the idle mode, when they need to reset the lifetime of the paging entries or when they detect movement to a new PA. One of the main features of this scheme is that it enables MHs and FAs within a PA to agree on the time instants that FAs transmit Agent Advertisements and paging Agent Advertisements for a host. Note that Agent Advertisements advertise paging support whereas paging Agent Advertisements are used for paging an MH. This way, the MH may power on its receiver only for the agreed slots that advertisements are expected (referred to as *time slot based paging*).

Mobile hosts perform paging updates by using the Registration Request messages (either Home or Regional). These messages carry an *Idle Mode Request* extension. Imagine in Figure 5.25 that the MH decides to enter the idle mode and performs an idle-mode registration. If an *idle mode Regional Registration Request* message is sent, it will be routed up the hierarchy until it reaches PFA1. Note that Regional Registration messages being sent while the host remains within the same PA (movement 1) will never cross the boundaries of the PA because the crossover FA will always reside in the PA. PFA1 marks the MH in its visitor list as idle and replies to the MH with a Registration Reply message including an Idle Mode Reply extension. This message communicates to the host the multicast IP address that will be used for paging purposes, if needed. FAs under PFA1 remove the host's entry after they forward the reply message to the host. As a result, routing information is removed from FAs under PFA1. However, it is important that routing information be maintained at FAs above the PFA as well as at the HA. In our example, because the MH remains under the same PFA, its routing path from the HA to PFA1 is unaffected. In this case, the host is responsible for refreshing it only when the binding to the HA is due to expire. An *idle mode Home Registration Request* is sent to the HA, which is processed by all entities as a normal Registration Request message and is used by PFA1 to refresh its paging entry for the host.

Likewise, when the host moves to a new PA (Paging Area 2), its path above the new PFA will always need to be updated (RFA1-PFA2 in our example). The host transmits an idle mode Regional Registration Request, which reaches the crossover RFA and results in the PFA2 replying with an idle mode Registration Reply. The crossover RFA (RFA1) updates its entry as in the normal Regional Registration procedure. The host could alternatively perform an idle mode Home Registration if its home binding was due to expire. In both cases, the routing information for the host from the HA to the new PFA will be updated. At any time, the MH may refresh its paging entries by performing the same idle mode registration procedure.

Data packets for an idle MH arrive at PFA2 by using the established tunnels along the path from the HA to PFA2. At this point, PFA2 buffers the packets and pages the MH. This is performed by sending a *Paging Request* for the MH to the lower-level FAs, and the message is forwarded down the hierarchy of FAs. In Figure 5.25, there is only one level of FAs below PFA2: FA3 and FA4. The leaf FAs send a *paging Agent Advertisement* to the MH's paging multicast address. The MH, upon receiving the paging message, performs an ordinary Home or Regional Registration, in this way updating the last part of its path within the PA. PFA2 is now able to deliver the buffered packets to the host. Alternatively, the host may enter the active mode when it needs to send packets without being paged earlier.

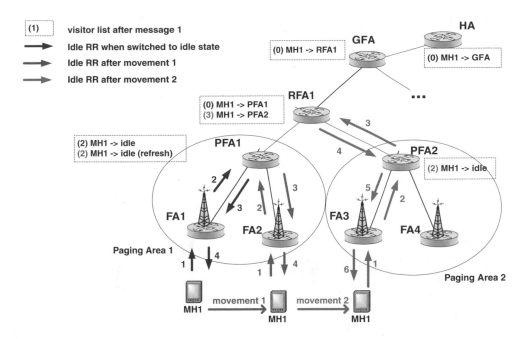

FIGURE 5.25 MIP Regional Paging.

5.3.5 Hierarchical MIPv6

Hierarchical MIPv6 (HMIPv6) is a proposal from Ericsson and the French National Institute for Research in Computer Science and Control (INRIA) whose purpose is to handle the mobility of hosts locally in order to reduce the signaling placed on the core network and to optimize the handover speed performance. Recall that in MIPv6, the MH is obliged to send a BU to its CHs and the HA each time it changes its point of attachment. This way, regardless of the granularity of the user's movement (micro- or macro-mobility) and consequently of the change made to the path, signaling messages travel all the way to the CHs and the HA, causing significant communication and processing load as the number of MHs increases. In addition, handover-speed performance is aggravated because the MH waits for the end-to-end path to be established so that it can receive packets on the new access router (AR).

For these reasons, HMIPv6 introduces a local entity within the access network, the Mobility Anchor Point (MAP), see Figure 5.26, which can be located at any level in a hierarchy of routers, including the AR. The idea is that the movement of the user within a MAP domain — that is, across the ARs that are under the MAP — is not visible to the CHs and the HA, which means there is no need for the latter to be notified of such movements. In particular, the CHs and the HA consider the MH as being located in the MAP's subnet. The MAP acts exactly as a HA; it intercepts the MH's packets and tunnels them to the actual location of the host. Note that HMIPv6 does not function in a similar way to MIPv4 RR, where packets are actually addressed to a gateway point (the GFA) and then tunneled within the access network. Here, packets are destined to an address within the MAP's subnet and not the MAP itself, and for that reason the latter needs to function as a typical HA for intercepting data packets and tunneling them to the host's location.

An MH residing in a MAP's domain is configured with two IP addresses: a Regional Care-of Address (RCoA) on the MAP's subnet and an on-Link Care-of Address (LCoA) that corresponds to the actual location of the host. The latter registers with the local MAP to provide binding information for its RCoA and LCoA. As long as the MH moves under the same MAP, it does not need to transmit BUs to its CHs and the HA. Instead, it must locally notify the MAP of its movements, resulting in a change of its LCoA. When the host crosses a new MAP's domain, then apart from registering with the new MAP, a BU needs to be sent by the MH to its CHs and the HA to notify them of its virtually new location (on the new

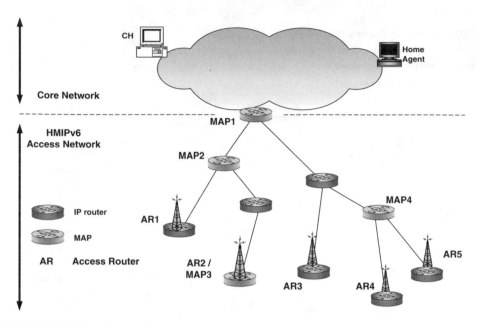

FIGURE 5.26   HMIPv6 architectural overview.

MAP's subnet). Therefore, the CHs and the HA maintain the binding *home address – RCoA* whereas the MAP maintains the binding *RCoA – LCoA* and is unaware of the host's home address.

The registration procedures we described are preceded by a MAP-discovery procedure that is based on the IPv6 Neighbor Discovery protocol. In particular, it is essential for an MH to detect the presence of the MAP(s) it can register with when it moves to a new AR. Being aware of the available MAP(s), the host can determine if only a local registration (with the current MAP) or a global registration (involving the CHs and the HA) is needed. ARs communicate to the hosts via router advertisements (RAs) the *global address* of the MAP(s), their *distance* from the host, and a *preference* value. This information helps the MH select a MAP according to, for example, its mobility pattern, or even select several MAPs simultaneously for handling different sessions.

Optimizing the handover-speed performance is another feature of the protocol that is attributed to the localized handling of the host's mobility. More specific, several techniques, such as MAP anchoring or the integration of HMIPv6 with Fast MIPv6 handovers, can be considered for achieving satisfactory handover delays. In what follows, we detail the protocol operation with respect to registration, handover, and paging functionality.

5.3.5.1   Protocol Overview

HMIPv6 introduces a new entity, namely the MAP, for undertaking the role of locally handling the mobility of hosts. The flexibility of the protocol operation lies in that a MAP can be located at any level in a hierarchical network of routers, from the gateway to the ARs, and in that several MAPs may exist in a domain, possibly covering the same area of routers. Therefore, no constraints exist on the location and number of MAPs, allowing in this way their gradual deployment in an HMIP domain. Moreover, unlike FAs, MAPs are not required to reside in each subnet of the domain, which would be essential if, for example, Agent Advertisement functionality were implemented in this entity. On the contrary, the IPv6 Neighbor Discovery protocol used for this purpose exempts the MAP entity from such constraints.

HMIPv6 proposes an extension to the IPv6 Neighbor Discovery protocol for enabling MAP discovery within a domain. The MAP option for RAs is introduced for communicating to MHs the required information for selecting a MAP. The *preference* field of the message allows the administrator to perform load balancing within the domain because this field is taken into account by hosts when selecting a MAP. The global IP address of the MAP is used for registration with the selected MAP, whereas some additional fields provide indications for the address to be used as source for packets originated by the MH as well as for the use of reverse tunneling. Note that the MAP's address is also used by the host for composing its RCoA by assuming a static prefix length of value 64.

At this point, let us describe how MAP discovery is performed in an HMIPv6 domain. Currently, there are two methods that enable ARs in a domain to discover MAPs and to let MHs obtain this information through RAs. In the *Dynamic MAP Discovery* method, MAPs and routers propagate the MAP option in RAs through certain manually configured interfaces, depending on the network topology (routers listen explicitly for RAs with the MAP option). A MAP can change its *preference* value at any time and propagate this change dynamically into the network, which is subsequently viewed by MHs. Alternatively, the *Router Renumbering* method may be used, which is best suited for large networks where the manual configuration of all routers is not a simple task. Here, the MAP option is sent from a central node to all ARs within a MAP domain without requiring that each router listen to the MAP option. Last, an administrator can resort to the manual configuration of the MAP option in the ARs and the MAPs without using any of the above methods.

Along with the extension to the Neighbor Discovery protocol, an extension to the MIPv6 protocol is also proposed. The latter involves the MIPv6 BU message that is transmitted by the MH to the local MAP for providing its RCoA and on-link CoA. In particular, a new flag is added to the BU message that indicates a *MAP registration* as opposed to a registration with the HA or the CHs. A local BU sent to the MAP also triggers the DAD procedure for the host's RCoA — that is, an address on the MAP's subnet. A Binding Acknowledgment indicating the result of the procedure is sent to the host.

In brief, the protocol has minimal impacts on IPv6 and MIPv6, and moreover, these extensions are applied only to the MH's protocol operation and do not affect the operation of HAs and CHs. If an MH runs the standard MIPv6 protocol, then it can simply operate as an MIP client without certainly taking advantage of the HMIPv6 domain's capabilities. The host will not identify the MAP option piggybacked on the RAs and will simply not proceed with MAP registrations.

5.3.5.1.1 *Registration Procedures*

Two registration procedures must be performed for an MH to be able to receive packets while moving within an HMIPv6 domain: a Home Registration and a Local Registration. The MH registers with the HA (and possibly with its CHs) when it enters a MAP domain (or leaves a MAP domain) and consequently needs to register with a new MAP. Note that MAP domains may overlap, and therefore a host may be registered with a MAP while at the same time residing at the overlapping area of two or more MAPs. In this way, MHs are allowed to bypass some levels of hierarchy and register with the most appropriate MAP among the advertised ones for optimizing their operation. The MH needs to establish a path from its HA and CHs to the new MAP as well as from the new MAP to its current location on the new AR's subnet. Before establishing the path, it must configure two CoAs: the RCoA on the MAP's subnet and the LCoA on the current AR's subnet. Both addresses are formed in a stateless manner.

After having configured the two addresses, the HMIPv6-aware MH performs MAP registration with the selected MAP (MAP4). As illustrated in Figure 5.27, it sends a local BU message to the MAP (message 1) in order to bind its RCoA with the on-link CoA. Here, the Home Address option of the BU message is set to the RCoA, which is reasonable if we consider and the MAP operates like an HA to the MH. The latter is assumed to be located on the MAP's subnet and has the RCoA serve its home address. The MAP checks by means of DAD whether another host on its subnet uses the MH's RCoA, and if not, it creates a new entry in its binding cache for the host's RCoA (or updates its existing one in the case of intra-MAP mobility). Then, it replies with a Binding Acknowledgment to the MH (message 2).

After having registered with the MAP, the MH needs to register with its HA for establishing the path between the HA and the MAP. It therefore sends a BU (message 3), where the Home Address option is set to the host's home address and the *Care-of Address* field is set to the RCoA. This way, the HA considers the host to be located on the MAP's subnet. The CoA (i.e., the RCoA) is either identified by the source address of the packet or can be found in the alternate-CoA option of the BU message. The BU message can also be sent to the host's CHs, in this way enabling route optimization. At this point, the whole path

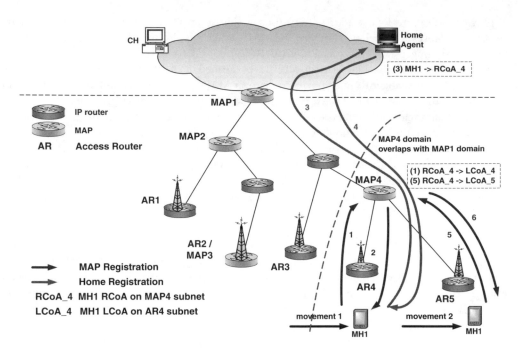

FIGURE 5.27 MAP and Home Registration procedures.

has been established and packets can be routed to the host either through an optimized route or through the HA. The MAP intercepts packets destined to the host's RCoA and tunnels them to the LCoA — that is, the on-link CoA of the host. The latter decapsulates the packets and processes them as in standard MIPv6 for MHs away from home. The MH is able to identify whether direct delivery to the CHs or reverse tunneling through the HA needs to be performed because the original packets are left intact by the MAP.

As mentioned earlier, the care-of address (RCoA) that the host registers with its HA can be found either at the source address of the BU packet or at the alternate CoA option. Whether the MH uses the RCoA instead of the LCoA as the source address of the BU message depends on the operating mode of the MAP that is known to the MH from the MAP option (in RAs). A MAP may request that (a) only the RCoA be used as the source address of the packets originated by the host, for example, when the operator does not want to expose the host's LCoA to nodes outside the domain and (b) reverse tunneling be used for the path between the MH and the MAP if ARs perform ingress filtering and operation a is also applied.

If operation a is requested, then BU packets should have the RCoA as their source address and there is no need for the alternate CoA to be set. If operation b is requested as well, then packets should also be tunneled to the MAP by having the outer packet's source address set to the LCoA and the destination address set to the MAP's address. If none of the operations is requested, then the source address of the BU message can be set to the LCoA whereas the alternate CoA should be set to the RCoA. However, if an MH wishes to hide its location from its CHs, it can choose to operate in mode a. In the case of data packets addressed to CHs either through the HA or with direct communication, the source address should always be set to the one registered at the HA and the CHs as the host's CoA — that is, the RCoA. Therefore, the MH should check only whether reverse tunneling is required.

The procedure that we have described involves an MH that changes its MAP and notifies the HA and CHs of this change. On the other hand, when the host moves locally within a MAP domain (movement 2) and changes its point of attachment (i.e., change of LCoA), then it needs to perform a MAP registration with only the current MAP to register its new LCoA. As a result, packets addressed to the host's RCoA will be intercepted by the MAP and tunneled to the host's new location (its new LCoA). The movement of the host under the same MAP is hidden from its CHs and HA.

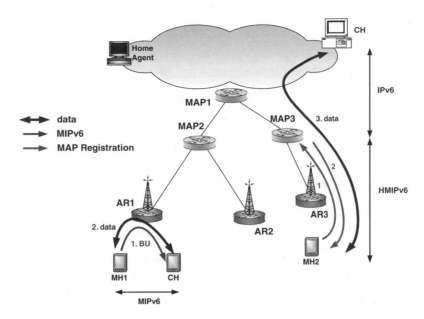

FIGURE 5.28 Mobile host operations in an HMIPv6 network.

It is therefore evident that the selection of a MAP for serving an MH determines significantly the *frequency* with which the CHs and HA are notified of the host's movement. If, for example, the selected MAP is high in the hierarchy of routers, it will cover a large area, and every movement of the host within this area will be hidden from the CHs and the HA (low frequency). If, however, the selected MAP is very low in the hierarchy or even located at the AR, it will cover a small area, and there is a greater possibility that the host will cross the boundaries of this area, resulting in a change of MAP (high frequency).

The MAP selection is certainly dependent on the MAP architecture, which is configured by the network operator. In particular, the placement of MAPs in the network is highly related to the mobility scenarios followed by mobile users in the covered area. A distributed placement of MAPs, with one MAP above the other, enables the user to choose the appropriate MAP optimized to its mobility pattern (e.g., speed of the user). On the other hand, a flat architecture in which the MAPs are located at the ARs is a better choice for cases where an inter-AR movement is considered a rare event. By deploying the MAPs at any level of the hierarchy, scalability is also provided because the processing load is being distributed to several MAPs, in contrast, for example, to an MIPv4 RR domain where the GFA maintains one entry for each MH in the region.

Moving a step further, it is possible for an MH to bypass the HMIPv6 operation and either revert to standard MIPv6 or not even use MIPv6 at all when global mobility support is not desired. In particular, when the MH (MH1 in Figure 5.28) communicates with CHs attached on the same link, it is desirable that data not be routed via the MAP in both directions but follow the shortest path. In this case, the MH may decide to use standard MIPv6 and therefore sends a BU to the CHs where it sets the CoA to its LCoA and not to its RCoA. Moreover, if the host (MH2 in Figure 5.28) is engaged in short-term sessions or if the mobility of the user is low, it may decide not to use MIPv6 for the communication with its CHs. Here, the MH pretends that the RCoA is its home address and packets addressed to the CHs have the RCoA as the source address without carrying the Home Address option. The host has previously registered with the MAP, indicating that packets addressed to its RCoA are tunneled to its LCoA, while it performs this update each time it changes its LCoA. This scheme, however, provides local mobility support to the host (i.e., mobility under the same MAP) and does not allow global mobility, forcing communication to resume when the host crosses the MAP domain's boundaries. Nevertheless, it reduces the overhead introduced to packets by the MIPv6 options (note that the packets that reach the MAP are IPv6 and not MIPv6 packets) and enables route optimization even in the case that CHs do not run MIPv6.

5.3.5.1.2 *Handover*

One of the aims of the HMIPv6 proposal has been the improvement of the handover-speed performance. The handover delay is indeed reduced when compared with standard MIPv6. In this case, the MH must register with the HA and its CHs, which may be located in a distant external network. It is evident that it will take significant time for these entities to be notified of the path change and to redirect the traffic to the MH's new location. On the other hand, when HMIPv6 is used, the MH performs only a local BU when handed over to ARs within a MAP domain, which causes a handover latency proportional to the round-trip time between the MH and the MAP.

When inter-MAP movement is performed, it is logical that the MH will start receiving packets from the new AR (NAR) only when the HA/CHs become aware of the new MAP serving the user. Registration with the new MAP is certainly needed before registering with the HA or the CHs. In order to speed up the handover between MAPs, the MH, after attaching to the new AR, may send a local BU to its previous MAP in order to inform the latter of its new LCoA. This way, packets already on their way to the old MAP will be forwarded to the new location of the user. However, this does not ensure that packet loss is eliminated because packets arriving at the old MAP before the latter updates its entry for the host are lost. Here, the handover latency is proportional to the round-trip time between the MH and the old MAP. Certainly, after sending the local BU to the old MAP, the host may register with its new MAP and the HA/CHs to update the whole path. From this moment, packets will stop being forwarded to the old MAP and will follow the new path. The idea here is that a localized handover procedure (i.e., involving only the MAPs) takes place to reduce packet loss as much as possible, which is later followed by the update of the whole path.

Going one step further, Soliman et al. (2002) propose that HMIPv6 be integrated with Fast Handovers for MIPv6 in order to attain even better results. The *Fast Handovers for Mobile IPv6* protocol has attracted great interest within IETF and in particular within the Mobile IP working group and is considered a promising handover solution to be applied to MIPv6 networks. It is based on the availability of L2 triggers that enable Layer 3, either at the MH or at the old AR, to anticipate a L3 handover. The result is that a tunnel is established between the old AR (OAR) and the new AR (NAR) before the host moves there. At the time the host attaches to the NAR, it is able to receive packets forwarded by the OAR. In particular, the OAR intercepts packets addressed to the host's previous CoA and tunnels them to the NAR. As described, the whole handover procedure is completed before the MH attaches to the NAR, minimizing significantly the handover delay. One issue that has arisen here is that there is no synchronization regarding the time that the host moves to the NAR in order that the OAR begin the forwarding of packets. Packet loss will occur if the forwarding is performed too late or too early with respect to the time the host detaches from the OAR and attaches to the NAR. Buffering techniques can be used at both the OAR and the NAR to alleviate this problem.

The integration of these two schemes comes natural if we consider that the AR's functionality in Fast Handovers resembles the functionality of MAPs (i.e., ARs act as local HAs), which maintain binding caches for the MHs and receive BUs (more specific Fast BUs) for establishing these entries. The first method of integration proposes the placement of MAPs at the ARs for handling the forwarding of packets between OARs and NARs. However, this architecture may prove inefficient, especially in the case where ARs are not connected directly and traffic between them is routed higher in the hierarchy of routers until it reaches an aggregation router. Here, during the handover procedure, packets that reach the OAR are tunneled to the NAR through the aggregation router. In the meantime, the MH updates its path and starts receiving packets from the new as well as from the old path, possibly out of order. This is because newer packets reaching the aggregation router and following the new path will be delivered to the MH earlier than older packets following the old path. This also results in the inefficient use of resources at the link between the aggregation router and the OAR. For that reason, a second way of integration is proposed in which the forwarding entity for such packets becomes the aggregation router (i.e., MAP functionality is placed at his router). According to this scheme, packets following the old path are redirected to the aggregation router and do not reach the OAR. In any case, the handover delay is minimized because the L3 handover is anticipated and the MAP is aware of the host's new location before

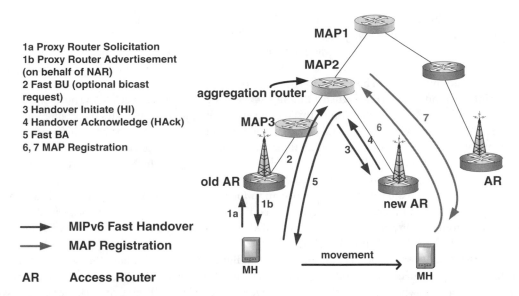

1a Proxy Router Solicitation
1b Proxy Router Advertisement
(on behalf of NAR)
2 Fast BU (optional bicast
request)
3 Handover Initiate (HI)
4 Handover Acknowledge (HAck)
5 Fast BA
6, 7 MAP Registration

MAP1
MAP2
aggregation router
MAP3
old AR
new AR
AR
MIPv6 Fast Handover
MAP Registration
AR   Access Router
MH   movement   MH

FIGURE 5.29 HMIPv6 — Fast Handovers integration.

the host moves to the NAR. Because the exact time to start forwarding is not known to the MAP, it is proposed that simultaneous bindings be maintained at the MAP in order that traffic addressed to the host be sent to both the OAR and the NAR for a short time.

In Figure 5.29, the MH, which is served by MAP3, becomes aware of an impending handover and sends a Proxy Router Solicitation message to its current AR (the OAR), including the link-layer identifier of its potential attachment point. In response, the OAR sends a Proxy Router Advertisement message, which carries the link-layer address and network prefix information of the NAR. Alternatively, the whole procedure can start from the network side by means of a trigger at the OAR, in which case the Proxy Router Solicitation message is not sent to the OAR. The MH uses the above information for obtaining an LCoA on the new link. Then, it sends a Fast Binding Update message to MAP2, with an optional bicast request, which associates the host's current RCoA with the NAR's IP address so that packets arriving at the MAP2 are tunneled to the NAR. MAP2 sends a Handover Initiate (HI) message to the NAR in order to establish a bidirectional tunnel between the two routers. This way, the MH can continue to use its current RCoA for its existing sessions. The NAR sets up a host route for the host's current RCoA, checks if the host's new LCoA can be used on its link, and responds with a Handover Acknowledge message. MAP2 sends a Fast Binding Acknowledgment message to the host, indicating whether the new LCoA can be used. As soon as the host attaches to the new link, it can receive packets addressed to its current RCoA and that are now tunneled to the NAR. At this point, the host can use its new LCoA (either the one already configured or a new one) for performing MAP registration with MAP2, which will subsequently tunnel packets to the host's new LCoA. Note that the host registers a new RCoA with MAP2. Here, a Home Registration is also needed because a different MAP will serve the user.

5.3.5.1.3 Paging

A paging extension to HMIPv6, referred to as HMIPv6 Regional Paging (HMIPv6RP), has been proposed that adopts many aspects of Mobile IP Regional Paging (see Section 5.3.4.1.3). This scheme also does not necessitate any L2 paging support for its operation. The entity initiating paging requests is the *Paging Mobility Anchor Point (PMAP)*, which comprises the highest MAP in an HMIPv6 domain and is in charge of the PAs organized in the domain. At least one PA is required. PAs are identified by a Paging Area ID (PAI), which is carried in the Router Advertisements of the ARs and indicates paging support within the domain. The routers within a PA are members of the PA multicast address, which is a permanently assigned IPv6 multicast address and which uniquely identifies a PA. In particular, the PA multicast address is constructed by the PAI. Time slot–based paging can be supported by having the ARs of a PA advertise simultaneously and by paging MHs at specific slots. This way, MHs are able to power on their receiver at agreed time slots and thus save power.

When an MH wishes to switch to idle mode, it sends a Regional or a Home BU with an *Idle Mode Request* extension to the PMAP or the HA, respectively. This extension carries the PAI advertised by the current AR serving the host. The PMAP or the HA processes the registration message as a standard BU. Paging information is maintained when a Binding Acknowledgement is sent back to the host. The PMAP includes an *Idle Mode Reply* extension to this message where it optionally indicates the Paging Multicast Address that will be used for paging the MH. This address is included only if paging will be performed simultaneously for other hosts as well within the same PA. Otherwise, the solicited-node multicast address (SNMA) obtained from the MH's RCoA is used by the ARs for paging the MH. The PMAP updates its cache for the host by adding the PA at which it is located and the Paging Multicast Address, if any. Note that only the PMAP maintains location information for idle MHs. Idle-mode registration is also performed by MHs when they enter a new PA or when they need to renew the lifetime of their idle-mode operation (Home Registration). This way, the PMAP is informed of the host's new PA and the path between the HA and the PMAP is refreshed.

When the PMAP receives packets for an idle MH, it buffers them and initiates the paging procedure for the host. In particular, it sends a *Paging Request* (i.e., an IPv6 packet with a Paging Request destination option) to all routers within the host's PA (the message is addressed to the Paging Area Multicast Address). The message indicates the Paging Multicast Address, if more hosts are to be paged at the same time, along with their RCoAs. The ARs of the PA, after receiving the Paging Request, send a *Paging Router Advertisement* to the MH(s), where the destination address of the packet is set to the host's SNMA or to the Paging Multicast Address. The MH, upon receiving the paging advertisement, enters the active mode and responds with a BU, either regional or to the HA, for establishing routing information within the domain. When the PMAP receives the response from the host, it may start forwarding the buffered packets. Certainly, a host can enter the active mode when it needs to send packets — that is, without being paged first.

5.3.6   Comparison

In this section, we carry out a qualitative comparison of the four described protocols, with respect to their architectural design principles and their mechanisms in performing micromobility management. Although it is possible to assess their effectiveness and scalability based on their design properties, a quantitative approach based on numerical data and simulations is also useful for their evaluation in a realistic network model. Such comparisons will not be performed here. However, the reader may refer to Campbell et al. (October 1999), Campbell et al. (August 2000), Ramjee et al. (July 2000), Ramjee et al. (June 2002), and Valko et al. for such performance studies.

As you can see in Table 5.1, all protocols but MIP RR operate at L3 — that is, the forwarding decision is taken based on the IP address information found in the L3 header. MIP RR handles mobility at L3.5 because the destination address of the inner IP packet is used for pointing to the next FA for the packets to be forwarded. In CIP and MIP RR, a host is identified by its home address, in HAWAII by its CoA (unless the host is located in its home network), and in HMIPv6 by its Regional CoA. Address management is of particular importance in IPv4 networks where there is a scarcity of IP addresses. Here, we note that CIP does not use any extra pool of addresses (FA CoA is used for all hosts in the domain) whereas HAWAII needs a CoA for each visiting host in the domain. MIP RR, on the other hand, does not impose extra address space requirements because the FA's address serves the CoA for visiting hosts.

As for the network topology, no restrictions are imposed on CIP and HAWAII networks. Nonetheless, the HAWAII protocol has been designed for optimally operating in a hierarchical topology. Nodes in these networks are CIP and HAWAII aware, respectively. In MIP RR, enhanced FAs (FAs, RFAs, GFA) form a tree-like topology, whereas in HMIPv6, MAPs may be located at any place within the domain. The first two protocols use host-based entries for forwarding packets to MHs, whereas the others use tunneling mechanisms instead. Tunneling is used when location information for a host is maintained at

TABLE 5.1   General Design Issues

	CIP	HAWAII	MIPv4 RR	HMIPv6
Layer handling mobility	L3	L3	L3.5	L3
Host identifier	Home address	Care-of address in foreign network/ home address in home network	Home address	RCoA
Routing topology	Nodes may form any topology	Routers may form any topology (although optimal functionality is with hierarchical topology)	FAs extended with RR functionality form a tree topology	MAPs may be located at any level in a hierarchical network of routers
Nodes involved	All CIP nodes replacing standard IP routing	All enhanced IP routers with HAWAII functionality	FAs, RFAs, GFA	MAPs
Routing	Host-based entries	Host-based entries	Tunneling	Tunneling
Address management	Home address	Static collocated CoA in foreign domain; home address in home domain	Collocated CoA or FA-CoA	RCoA and LCoA
Extensions/ modifications to the MIP/IP protocol running at network entities	No extensions	HAs must accept registrations without a Mobile-Home Authentication Extension	HA must handle the GFA IP address extension	MAP option in RAs
Extensions to the MIP protocol running at MHs	No extensions	Host's NAI, FA NAI, Route optimization, Challenge/Response extensions	*I* flag in Agent Advertisements, GFA IP address extension, registration with RFAs/GFA	*M* flag in BU indicating MAP registration

specific entities in the network while the underlying IP routing infrastructure remains unaffected. Packets are forwarded to the next location-aware entity pertaining to the downstream path, for example, the RFAs or FAs in MIP RR. In CIP and HAWAII, special route entries replace the IP routing functionality and serve the forwarding of packets to MHs. IP routing, however, is used in HAWAII when the network's fixed nodes are addressed, for example, the BSs in the path-setup schemes.

All protocols but CIP propose extensions to the standard MIP operation in MHs and HAs. The most significant modification required by the HAWAII protocol is that HAs must accept registrations without the *Mobile-Home Authentication Extension*. In MIP RR, the *GFA IP address* extension must be supported by both HAs and MHs (although not always needed), and the MH should identify the *I* flag in Agent Advertisements and be able to register the address of the GFA or any RFA with its HA. Likewise, HMIPv6 extends the RAs of IPv6 with the *MAP option* and introduces the *M* flag in MIPv6 BUs for MAP registrations.

In Table 5.2, the properties of the *path update* and *routing* functionalities are presented. All protocols use signaling messages for establishing routing paths to the MHs (i.e., for installing location information in the network). Their difference lies in the way that location information is propagated and the place at which it is maintained within the network. In CIP, *route-update* messages are propagated all the way to the gateway, although part of the path may be unaffected. HAWAII transmits *path update* messages to the old BS with the aim of localizing the path update procedure and integrating it with the handover control functionality. This way, it refrains from loading the gateway (communication as well as processing load) but may lead to a nonoptimal routing of packets. In MIP RR, *Regional Registration* messages are destined to the gateway (GFA), although they can be answered earlier at the crossover RFA, whereas in

TABLE 5.2   Path Update and Routing

	CIP	HAWAII	MIPv4 RR	HMIPv6
Path Update message destination	To gateway	To old BS	To GFA or crossover RFA	To MAP
Means for Path Update	*route-update* ICMP messages	*path update* UDP messages	MIP home registration messages, RR messages	MAP BUs
How paths are refreshed	IP data packets sent by MH or route-update messages in absence of data packets	Explicit aggregate refresh messages	MIP home registration messages	MAP BUs
Load balancing for path update signaling?	No, more loaded at the root	Yes, path updates are kept localized	No, more loaded at GFA	Yes, by setting the MAP preference value
Location information for a host resides where?	Selected nodes in the network	Selected nodes in the network	GFA, RFAs, FAs	MAPs
Forwarding cache sizes	At the gateway, an entry for each MH	At the domain root router, an entry for each MH	At the GFA, an entry for each MH unless hosts also register with intermediate RFAs	Each MAP serves its own MHs
Downlink routing	Host routes for original packets; routing based on home address	Host routes for encapsulated packets/original packets in home network	Sequential tunnels	Tunneling
Uplink routing	To the gateway; independent of destination address	Based on destination address, default path to the gateway is followed if no entry exists for the host	Standard routing based on destination address, or reverse tunneling	Standard routing based on destination address, or reverse tunneling to MAP or to HA
Optimal routing in mobile-to-mobile communication	Through gateway	Crossover router if route optimization is used or MH is in the home domain; otherwise, through HA	Through GFA when route optimization is used; otherwise, through HA	Through MAP if BU is sent to the CH as well; otherwise, through HA

HMIPv6, *MAP BUs* are addressed to the MAP serving the user. A MAP entity operates in a stand-alone manner without being affected by other MAPs possibly operating in the same domain. By that we mean that location information for a host is maintained in only one MAP at a time. As a consequence, the load induced by BUs is distributed to the MAPs of the domain. Load balancing can be performed by setting the *preference* attribute of a MAP, in this way mitigating overloaded conditions.

Among the four protocols, only CIP uses data packets for refreshing the host's routing entries along the already established forwarding path. Location information is maintained at specific nodes in MIP RR and HMIPv6 (FAs and MAPs, respectively), whereas all nodes pertaining to the path between the gateway and the MH are location aware in CIP and HAWAII. It is evident that the gateways in all protocols but HMIPv6 maintain a routing entry for all hosts within their domain. However, this might not be true in an MIP RR network if MHs choose to register with intermediate RFAs instead of the GFA.

The scalability property of these schemes is of prime importance when the network grows and the number of MHs increases. Scalability is mainly affected by the induced signaling load (i.e., the bandwidth consumed by the protocol's control messages), the processing load due to the processing of *path update* and *refresh* messages, as well as memory requirements for routing state maintenance.

TABLE 5.3 Handover Control

	CIP	HAWAII	MIPv4 RR	HMIPv6
Handover management	Hard/semi-soft handover	Forwarding/ nonforwarding schemes	Standard MIP, Route Optimization	Fast Handovers for MIPv6
Handover initiator	Mobile host	Mobile host	Mobile host	Either MH or OAR
Requirements from L2	No	No	No	Yes, L2 triggers indicating an impending L3 handover
Tunneling/ forwarding	No	Forwarding from old to new BS in forwarding schemes (MSF, SSF)	Tunneling from previous to new FA if smooth handover is used	Tunneling from OAR/MAP to NAR
Bicasting	Yes, at crossover node in semi-soft handover	Yes, at crossover router in MNF scheme	No	Yes at MAP acting as the forwarding entity in Fast Handovers
Duplication/out-of-order packets	Duplication may appear in semi-soft handover	Out-of-order packets in MSF Duplication may appear in MNF	Possibly out-of-order packets in smooth handover	Out-of-order packets when forwarding takes place at OAR

Downstream routing in MIP RR and HMIPv6 is performed with the use of tunnels, whereas in CIP and HAWAII host routes are used. Data packets within a HAWAII domain are encapsulated packets, thus introducing overhead to the network (which is not the case for CIP). In these protocols, the upstream path is preconfigured — that is, default routes have been established for the routing of outgoing packets regardless of their destination address. However, in HAWAII the default route is followed only when the traversed node does not maintain a routing entry for the packet's destination address. This enables an optimal routing in mobile-to-mobile communication when route optimization is used. In CIP, upstream packets always reach the gateway. MIP RR and HMIPv6 use standard IP routing for packets transmitted by MHs unless reverse tunneling has been requested. Mobile-to-mobile communication is performed through the GFA and the MAP, respectively, when route optimization is used.

As far as the handover functionality is concerned, we list in Table 5.3 the mechanisms used for handover management. In all protocols, handover is initiated by the MH except for the *Fast Handovers for MIPv6* scheme, where the OAR can also assume this role. In this protocol, L2 triggers are used for informing L3 of an impending handover, and therefore some degree of coupling between the two layers is assumed for assisting the handover procedure. Tunneling and forwarding techniques are used by all protocols but CIP in order to avoid packet loss while the host migrates to a new access point. Bicasting techniques can also be found in all protocols but MIP RR dealing with the same problem. The result of these techniques may be out-of-order for tunneling techniques and duplicated packets for bicasting techniques.

Finally, Table 5.4 presents a comparison of the protocols' paging functionality. Among the four protocols, only CIP uses data packets for performing implicit paging to MHs. Furthermore, HAWAII and HMIPv6 use IP multicast functionality for distributing paging requests within a PA. The configuration of their paging architecture consists of setting the multicast address of the routers belonging to a PA. Paging information for the hosts is located at the PFA and the PMAP entities of the respective domains, whereas selected nodes are augmented with paging functionality in CIP and HAWAII networks.

In MIP RR and HMIPv6, the PFA and the PMAP are responsible for initiating paging requests when data packets arrive for an idle host. HAWAII, on the other hand, defines an algorithm for dynamically balancing the paging load among the nodes of the network and that allows paging to be initiated by any node with paging information for the host. This certainly makes the network more robust against failures to paging-aware nodes at the expense of a greater processing load in the network. In CIP and HAWAII, the network becomes aware that the host has entered the idle mode in the absence of *route-update* and

TABLE 5.4   Paging

	CIP	HAWAII	MIPv4 RR	HMIPv6
Paging	Implicit/data packet is used for paging	Explicit, IP multicast is used for sending paging request to all BSs in a PA	Explicit, paging request sent by PFA to all FAs belonging to its subtree	Explicit, IP multicast is used for sending paging request by PMAP to all routers within a PA
Paging architecture configuration	Selected nodes configured by the administrator with paging functionality	Routers in a PA are configured with a multicast group address	PFAs are configured by the administrator	Routers in a PA are configured with an IPv6 multicast address obtained from the PAI
Paging information location	In selected nodes with paging functionality	In selected routers within PAs able to multicast paging requests	At the PFA of a PA	At the PMAP
Paging initiator	Data packets are used for paging; no explicit signaling message	Every router with paging information (see above)	PFA	PMAP
Entering idle mode	Let route entries time out	Let route entries time out	Home/Regional Registration with *Idle Mode Request* extension	BU with *Idle Mode Request* extension
Crossing a PA	*paging-update* ICMP messages sent to the gateway	*paging update* message sent to the gateway	Home/Regional Registration with *Idle Mode Request* extension	BU with *Idle Mode Request* extension
Paging entries update	paging-update, route-update, or data packets sent by MH	*refresh* aggregate messages	Home/Regional Registration with *Idle Mode Request* extension	BU with *Idle Mode Request* extension
Requirements from L2	No	Paging functionality in L2	No	No

refresh messages, respectively, whereas MIP RR and HMIPv6 use explicit messages to notify the network of this mode switch. All protocols use explicit signaling when a host enters a new PA or when refreshing its paging entries in the network. In addition, CIP allows data packets transmitted by MHs to be used as *refresh* messages for paging information. Last, among the four protocols, HAWAII requires the support of paging functionality from the underlying layer.

5.3.7   Discussion

So far, we have examined the operation of several micromobility protocols and have identified their design principles in performing *path update*, *paging*, and *handover*. All these principles try to address the same deficiencies of Mobile IP (i.e., reduce the signaling propagated to the network each time a host changes its point of attachment and enhance the handover performance). However, a micromobility protocol is not confined to simply addressing these two issues because additional requirements exist that should be met in order to prevent it from malfunctioning and performing poorly.

Future applications are becoming very demanding, and wireless users expect for their services quality similar to that perceived by wireline users. On top of that, the number of mobile users is continuously increasing, placing great challenges on such protocols. Therefore, on the one hand, micromobility protocols attempt to protect their operation from *overload conditions* in terms of resource consumption (both wired and wireless), processing overhead, and routing state maintenance and, on the other hand, to serve mobile users by mitigating the *disruption to their services*. Overload conditions are mainly caused by the great volume of *path update* control messages as well as by design choices regarding the number of nodes maintaining and processing routing information in the network. At the same time, support for fast and smooth handover is needed for ensuring minimal disruption in packet delivery to MHs while migrating to new access points. As a consequence, such issues should be taken into account by micro-mobility protocols.

Minimizing *path update* signaling is achieved by introducing a two-state operation to MHs, which means they do not have to transmit control traffic for each movement they perform across access points. Paging functionality is highly important and is taken into account by all four described protocols. However, paging should run independently of other mobility modules, albeit allowing for minimal interactions with them. A number of design issues affecting the scalability, robustness, and efficiency of a paging solution arise here. The most significant are (a) the places in the network to store location information (centralized or distributed architectures), which implies that these places are potentially the more loaded ones (processing and state maintenance); (b) how PAs are configured (statically or dynamically); and (c) whether paging is performed at L3 (i.e., independently of radio technology) or at L2. Dormant Mode Host Alerting (IP paging) is currently addressed at the IETF Seamoby working group, whose goal is to develop an architecture and set requirements to be met by such a protocol.

Although paging alleviates the problem of increasing signaling volume in the network, micromobility protocols should still take extra care in their *scalability* and *performance* properties because they might be jeopardized by the great number of served MHs and the expansion of the network over a larger geographical area. Issues in *path update* functionality (such as the number of nodes within the network that are mobility aware and consequently process and maintain routing information for MHs) are of great importance. As an example, the host-based forwarding schemes, although avoiding the tunncling overhead, face difficulties in scaling because for each MH forwarding entries are established in all nodes pertaining to the path to the gateway, as opposed to selected nodes in the case of tunnel-based schemes.

Network resources consumption is also affected by the tunneling overhead introduced by certain protocols. However, tunnel-based schemes have other benefits if we consider that only few nodes are mobility aware and, therefore, a gradual deployment and an easier upgrade (few nodes) of such schemes is facilitated. Moreover, the soft-state operation of a micromobility protocol also contributes to the resource consumption because it implies a great amount of control messages assigned with refreshing routing entries for hosts in a domain. Nevertheless, the soft-state nature of routing entries renders the protocol more robust and resilient to node or link failures, which is a desired feature. Single points of failure may also jeopardize the protocol's operation as, for example, single access gateways that hold location information for all hosts in a domain.

Apart from the protocol's design choices that affect scalability and efficiency, some additional requirements should be met for easing its deployment into any network architecture and for enabling its interworking with existing functionality. Therefore, it is essential that a micromobility protocol is topology independent. This could certainly provide freedom to network operators for deploying their network without architectural constraints. Moreover, the protocol's interoperation with existing network functionality (e.g., HAs, CHs, and MHs) is certainly required, and further, it should be ensured that such functionality experiences the minimal impact. Interworking with QoS schemes in an efficient way is also important for avoiding end-to-end QoS signaling. In particular, *path update* in conjunction with *handover control* are considered the functionalities that should tightly cooperate with QoS mechanisms because they deal with session continuation.

The primary concern of handover control should be to minimize the service disruption experienced by users (i.e., minimize the packet loss and the delay while communication is being restored on the new link). The main issue here is whether L3 handover is clearly separated from L2. This is important if we consider the sources of delay in an IP handover scheme assuming that hard handover is supported. First, a L2 handover is performed and the host regains connectivity on the new link. Second, the path-update procedure takes place, which certainly requires extra time to complete. These two factors may introduce significant delay to the handover procedure, which can be critical for real-time services.

Micromobility protocols have tried to make the path traversed by *path update* control messages as short as possible for minimizing the second factor of delay. Certain micromobility protocols have moved further to the separation of *path update* from *handover* signaling in order to enable the timely data delivery at the new access point before the actual path update takes place (by means of bicasting and tunneling techniques from OAR to NAR). Ideally, the L3 handover procedure completes before the actual L2 handover is performed. This is achieved only if the MH or the network is aware of an impending handover and therefore can initiate the L3 handover procedure before the L2 handover event (a *proactive* handover in contrast to a *reactive* handover). In this case, as soon as the L2 handover is performed, the host can start receiving packets along the already established IP path. The handover delay is reduced to the L2 handover delay and packet loss may be eliminated with the use of buffering techniques, for example. Likewise, if the IP handover procedure is not complete when the host migrates to the new access point, then the temporary forwarding of packets from the old to the new AR may be performed until the routing state on the new path is established.

The IETF Mobile IP working group has consolidated two handover schemes: *Low Latency Handovers in Mobile IPv4* and *Fast Handovers for Mobile IPv6*, which are based on the above principles. These protocols depend on obtaining timely information from L2 regarding the progress of a handover and accordingly establish tunnels from the OAR to the NAR for temporarily redirecting packets to the host's new point of attachment. *Handover signaling* concerns only the ARs (old and new), and *path update signaling* (MIP registration) follows the handover procedure. In particular, Low Latency Handovers define three handover methods. In the Pre-Registration method, L3 handover is initiated before L2 handover due to L2 triggers at the MH or the OAR. The host is able to register with its new FA and the HA through the old FA. It is hoped that the path is updated at the time that the L2 handover is complete and that packets already follow the new path. In the Post-Registration method, L2 triggers at the old or the new FA are used for the timely establishment of a tunnel between the concerned FAs so that service continuity is ensured before the actual path is updated. FAs communicate directly without the need for the MH to participate in the L3 handover procedure before the L2 handover. The combined method involves running Pre-Registration and Post-Registration methods in parallel to address the case (by means of tunneling from the old to the new FA) in which the L3 handover is not complete before the host's attachment on the new link. Likewise, Fast Handovers for MIPv6 establish a tunnel between the OAR and the NAR for the forwarding of packets until the MH establishes itself as an MIPv6 endpoint. Handover can be anticipated by means of any kind of trigger, in which case the tunnel is established before the attachment of the host on the new link. Otherwise, the tunnel establishment starts as soon as the host regains its connectivity on the new AR.

Apart from the need to support seamless handovers, it is also essential to establish at the NAR (in a timely manner) the MH's context, which is being maintained at the OAR before handover. This procedure is denoted as context transfer and is an active topic of investigation within the IETF Seamoby working group. Context transfer involves moving context from the OAR to the new one; which information may relate to Authentication, Authorization, and Accounting; Header Compression; and QoS services. By transferring configuration information for these services between the ARs, the MH refrains from participating from scratch in the reestablishment of these associations with possibly distant entities in the network. On top of that, the context transfer procedure uses the resources of the fixed infrastructure (AR-AR communication instead of MH-AR), which is highly important in the case of radio resource constraint networks. For the above reasons, context transfer appears to be an attractive procedure, which should coordinate with handover control and could possibly benefit as well from L2 triggers for its timely execution.

References

Campbell, A.T. et al., Cellular IP, Internet draft, draft-ietf-mobileip-cellularip-00.txt, Dec. 1999.

Campbell, A.T. et al., Cellular IP performance, Internet draft, draft-gomez-cellularip-perf-00.txt, Oct. 1999.

Campbell, A.T. et al., Design, implementation, and evaluation of Cellular IP, *IEEE Personal Communications Magazine*, 7(4), Aug. 2000.

Campbell, A.T. and Gomez-Castellanos, J., IP micromobility protocols, *ACM SIGMOBILE Mobile Computing and Communications Review*, 4, 4, Oct. 2000, pp. 45–53.

Campbell, A.T. et al., Comparison of IP micromobility protocols, *IEEE Wireless Communications Magazine*, 9, 1, Feb. 2002.

Deering, S. and Hinden, R., Internet Protocol, Version 6 (IPv6) specification, RFC 2460, Dec. 1998.

Eardley, P. et al., A framework for the evaluation of IP mobility protocols, PIMRC'00, Sep. 2000.

Eardley, P. et al., On the scalability of IP micromobility management protocols, MWCN'02, Sep. 2002.

Gustafsson, E. et al., Mobile IPv4 Regional Registration, Internet draft, draft-ietf-mobileip-reg-tunnel-07.txt, Oct. 2002.

Haverinen, H. and Malinen, J., Mobile IP Regional Paging, Internet draft, draft-haverinen-mobileip-reg-paging-00.txt, June 2000.

Hsieh, R. et al., Performance analysis on Hierarchical Mobile IPv6 with fast-handover over end-to-end TCP, Globecom 2002.

Johansson, F. and Johansson, T., AAA NAI for Mobile IPv4 Extension, Internet draft, draft-ietf-mobileip-aaa-nai-05.txt, March 2003.

Johnson, D. B. et al., Mobility support in IPv6, Internet draft, draft-ietf-mobileip-ipv6-21.txt, Feb. 2003.

Karagiannis, G. and Heijenk, G., Mobile IP, Ericsson State of the Art Report, July 1999.

Kempf, J., Dormant mode host alerting ("IP paging") problem statement, RFC 3132, June 2001.

Kempf, J., Problem description: reasons for performing context transfers between nodes in an IP access network, RFC 3374, Sep. 2002.

Kempf, J. et al., Requirements and functional architecture for an IP host alerting protocol, RFC 3154, Aug. 2001.

Koodli, R., Fast handovers for Mobile IPv6, Internet draft, draft-ietf-mobileip-fast-mipv6-06.txt, March 2003.

Malki, K.E. et al., Low latency handovers in Mobile IPv4, Internet draft, draft-ietf-mobileip-lowlatency-handoffs-v4-04.txt, June 2002.

Montenegro, G., ed., Reverse tunneling for Mobile IP, revised, RFC 3024, Jan. 2001.

Myles, A. and Skellern, D., Comparing four IP based mobile host protocols, *Computer Networks and ISDN Systems*, 26, 3, 1993, pp. 349-355.

Narten, T., Nordmark, E., and Simpson, W., Neighbor discovery for IP version 6 (IPv6), RFC 2461, Dec. 1998.

O'Neill, A. et al., Generalized IP handover, Internet draft, draft-oneill-craps-handoff-00.txt, Aug. 2000.

Perkins, C. E., Mobile IP joins forces with AAA, *IEEE Personal Communications Magazine*, 7(4), Aug. 2000.

Perkins, C. E., Mobile IP, *IEEE Communications Magazine*, May 2002.

Perkins, C. E., ed., IP mobility support for IPv4, RFC 3344, Aug. 2002.

Perkins, C. E. and Johnson, David B., Route optimization in Mobile IP, Internet draft, draft-ietf-mobileip-optim-11.txt, Sep. 2001.

Ramjee, R. et al., IP micromobility support using HAWAII, Internet draft, draft-ietf-mobileip-hawaii-01.txt, July 2000.

Ramjee, R. et al., Paging support for IP mobility, Internet draft, draft-ietf-mobileip-paging-hawaii-01.txt, July 2000.

Ramjee, R. et al., IP-based access network infrastructure for next-generation wireless data networks, *IEEE Personal Communications Magazine*, 7(4), Aug. 2000.

Ramjee, R. et al., HAWAII: a domain-based approach for supporting mobility in wide-area wireless networks, *IEEE/ACM Transactions on Networking*, 10(3), June 2002.

Reinbold, P. and Bonaventure, O., A comparison of IP mobility protocols, IEEE SCVT 2001 Proceedings.

Soliman, H. et al., Hierarchical Mobile IPv6 mobility management (HMIPv6), Internet draft, draft-ietf-mobileip-hmipv6-07.txt, Oct. 2002.

Valko, A. et al., Performance of Cellular IP access networks, PFHSN'99, 1999.

6

Quality of Service in Mobile IP Networks

Bongkyo Moon
King's College London

A.H. Aghvami
King's College London

A radio access network (RAN) typically provides the radio access (e.g., GSM, CDMA, or WCDMA) to mobile stations in a cellular network. In an Internet Protocol (IP)-based RAN, several wireless cells need to be interconnected at the IP level because mobility is basically a routing problem at the network layer. In addition, the Quality-of-Service (QoS) management should be tightly coupled with IP mobility management. As a result, all the network nodes in All-IP access networks must be integrated with Mobile IP[1] or IP micromobility protocols[2] in order to maintain seamless interoperability with the wired Internet.

Many real-time applications have been developed in the Internet, and the best-effort delivery model has become inadequate for these new applications. Transmission Control Protocol (TCP), which is being used widely in the current Internet, is not well suited for real-time applications. Instead, Real-time Transport Protocol (RTP) can usually be implemented on top of User Datagram Protocol (UDP), which is better adapted to real-time applications. However, this protocol mechanism is not enough to guarantee a specific QoS for a session between a sender and a receiver. Two different Internet QoS models have been proposed by the Internet Engineering Task Force (IETF): the Integrated Services (IntServ) model[3] and the Differentiated Services (DiffServ) model.[26] In the IntServ model, network nodes classify incoming

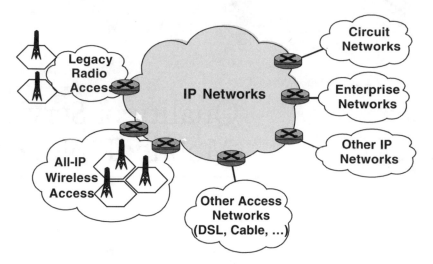

FIGURE 6.1 An example of an All-IP network.

packets, and network resources are explicitly identified and reserved. In the DiffServ model, instead of explicit reservation, traffic is differentiated into a set of classes for scalability, and network nodes provide priority-based treatment according to these classes. IPv6, which is a new version of the most successful protocol, IPv4, allows DiffServ-style QoS to be applied. However, some applications, such as streaming audio and video, would be much better served under the IntServ model because they have a relatively constant bandwidth requirement for a known period of time. Therefore, it is possible to consider DiffServ or the IntServ model in RANs in order to provide consistent end-to-end QoS behavior.

6.1   All-IP Networks

Recently there has been significantly increased interest in building future third-generation (3G) or fourth-generation (4G) networks based on IP technologies. The major motivation is to have a common service platform and transport for 3G or 4G networks based on IP. The trend in the wireless industry is to build an IP-based transport infrastructure for 3G wireless networks. This would help enable seamless delivery of IP services across wireline and wireless networks. Several standard organizations and industry forums, including the Third-Generation Partnership Project (3GPP), the Third-Generation Partnership Project 2 (3GPP2), and Mobile Wireless Internet Forum (MWIF), have initiated efforts on defining an All-IP based 3G network.[11,13] The main objective of the All-IP solution is to provide a flexible and scalable wireless network infrastructure for delivering data, voice, and multimedia services over 3G wireless networks.

The goal of the All-IP wireless access network is to build a wireless infrastructure to support current and future IP-based services in a globally common manner that allows ubiquitous service appearances to end users.[11,12] The All-IP network should also allow fast creation of new services, while taking advantage of existing and planned radio networks (see Figure 6.1). Toward achieving this goal, an All-IP architecture should be aligned with the existing Internet model to provide flexibility, scalability, and network-level reliability. This requires the use of IP transport throughout the network, both for wireline and wireless networks. All-IP wireless network architecture is designed to be independent of the access technology. Service creation and migration will be faster and less costly for wireless networks based on emerging access technologies. Moreover, the architecture should support a wide range of services, including real-time traffic, multimedia streams, and bulk data transfers. This is possible only if the All-IP architecture provides a means of supporting end-to-end QoS.[14]

On the other hand, the boundaries for IP-based RANs are typically the access points in base stations (BSs) for the radio transmission and reception and for the interfaces to the gateways (e.g., Mobile Switching Center [MSC] and SGSN/GGSN) that provide connections to the fixed network. The RAN

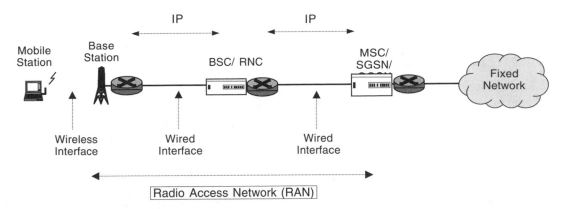

FIGURE 6.2 A typical RAN and its boundaries.

consists of a number of nodes, as shown in Figure 6.2. The Base Station Controller/Radio Network Controller (BSC/RNC) controls a number of BSs, including the radio channels and the connections to mobile stations (MSs). For a WCDMA RAN, the BSC/RNC provides soft handover combining and splitting between streams from different BSs belonging to the same MS.[15,16] Furthermore, the BSC/RNC is also responsible for the allocation of transport resources within the RAN. The transport is either between the BS and the BSC/RNC, between multiple BSC/RNCs, or between the BSC/RNC and the MSC/ SGSN (Mobile Services Switching Center/Serving GPRS Support Node).[15,16] The MSC provides support for circuit-switched services, and the SGSN/GGSN provide support for packet-switched services to MSs, including mobility management, access control, and call control, as well as interworking with external networks such as the Public Switched Telephone Network (PSTN) and the public Internet.

6.1.1 The 3GPP Network Architecture

In the 3GPP network architecture, UTRAN (Universal Mobile Telecommunications System [UMTS] Terrestrial Radio Access Network) transports both data and voice traffic at different priorities. IP is a good candidate to deliver the UMTS services for high-link utilization and supports data traffic more efficiently in the RAN. The UTRAN consists of a radio network subsystem (RNS) connected to the core network. Furthermore, an RNS consists of an RNC and one or more BSs. The BSs are connected to the RNC, which is responsible for control of the BSs, and to the radio link to the mobile station.[16] Each BS is integrated with an IP router. The BS router converts the UMTS packets into IP packets and forwards the packets to the proper outgoing interface into the IP network. The forwarding principle relies on a specific routing protocol, such as Cellular IP,[18] HAWAII,[19] and TeleMIP.[20] The large numbers of BSs are connected with a common transmission link of the tree topology. The transmission capacity of the link at the gateway is high because all the traffic from the BSs is aggregated at this point.

6.1.2 The 3GPP2 Network Architecture

The 3GPP2 network architecture takes advantage of 3G high data rates and Mobile IP to enhance the network architecture to provide IP capabilities. In this architecture, both the BSC and the base station transceiver (BTS) are contained in the IP-based RAN. The BSC can be an IP-based router containing some critical radio control functions (e.g., power control or soft handover frame selection). Typically, an RAN consists of a couple of BSCs and hundreds (up to tens of thousands) of BSs. The mobile terminal uses the Mobile IP–based protocol approach, which can provide mobility across different access networks such as wireless and wireline.[17] However, because it essentially uses address translation to provide mobility, it cannot do fast handover due to the latency of address updates from its home agent (HA). Cellular IP,[18] HAWAII,[19] and TeleMIP[20] propose some form of hierarchy with local/gateway routers, which can reduce latency by reducing updates from the remote HA. These schemes could be used to optimize Mobile IP application in 3GPP2. All the IP traffic from the BSs is aggregated at the core router, which interfaces with the core network.

6.1.3 The IP Mobility Protocols

6.1.3.1 Mobile IP

Mobile IP is the oldest and probably the most widely known mobility management proposal within IP. It was originally described in an IETF RFC2002.[1] Several published Internet drafts propose various improvements for Mobile IPv6. HAs and foreign agents (FAs) periodically broadcast their advertisements. A mobile node (MN) uses these advertisements to determine a new location it has moved to. If an MN needs to know the address of a potential agent immediately without waiting for the next advertisement, it can broadcast an Agent Solicitation message. If the node determines that it is attached to a foreign network (FN), it obtains a care-of address (CoA) from the FA or through another protocol such as Dynamic Host Configuration Protocol (DHCP) for a collocated CoA. The FA is responsible for unmarshalling the tunneled packets sent to it by the MN's HA and relaying them to the MN. It is also responsible for relaying packets from the MN to correspondent nodes (CNs) and to the HA. Once a node has obtained a new CoA, it registers this address with its HA. Registration with the HA is performed for the following reasons: (1) a node has moved to a new FN and therefore needs to register a new CoA, (2) a node needs to deregister an old CoA with its HA when it returns to the home network (HN), or (3) a node needs to reregister with the HA because its previous registration is about to expire.

In standard Mobile IP, when a CN sends packets to an HA for tunneling to the MN, the HA can assume that the CN is unaware of the MN's current CoA. However, route optimization extensions to Mobile IP would allow a CN to be informed of the MN's CoA so that it can send packets directly to the MN. This route optimization can thus solve the significant performance degradation problem due to triangle routing. That is, after the HA tunnels the packets from the CN, it sends an authenticated Binding Cache Update to the CN, which carries the CoA of the MN to which the HA is sending packets. Hence, this binding cache contains mappings from home addresses to the temporary CoAs. Once a CN updates its binding cache, it can tunnel packets directly to the MN's CoA, which solves the triangle routing problem.

In Mobile IPv6, many of the principles from Mobile IPv4 have been retained except that the FA is no longer needed. There are a number of improvements over the Mobile IPv4 protocol. Here we list the main differences:

- Triangular routing is no longer a problem. The packets are routed to the MN using the IPv6 routing header where the current CoA of the MN is the last intermediary address.
- The MN now uses its CoA as the source address of the packets that it sends. This is accomplished using a special "home address destination option" header.
- FAs are no longer needed. The MNs make use of Neighbor Discovery (RFC 2461) and Address Auto-configuration (RFC 2462) to operate without the support of an FA.
- Mobile IPv6 utilizes IP security (RFC 2401, RFC 2402, and RFC 2406) for all security requirements.
- Most packets are sent using the IPv6 routing header instead of using encapsulation. This means that fewer additional header bytes are sent.
- While the MN is away from the HN, the HA intercepts all packets destined for the MN and forwards them using IPv6-in-IPv6 encapsulation.

6.1.3.2 The IP Micromobility Protocol

Mobile IP has usually been optimized for macromobility and relatively slow-moving MNs. However, this requires frequent notification to the MN's HA during handover, causing much service disruption and high signaling overhead. The term *micromobility* is defined as the mobility within the same subnet. It is often used interchangeably with intrasubnet mobility. An intrasubnet handover occurs when an MN moves from one BS to another and both BSs are connected to the same router. Figure 6.3 shows intrasubnet and intersubnet handovers. Under micromobility, the IP address of the MN remains unchanged, and only the link-layer connection needs to be rerouted. Well-known IP micromobility

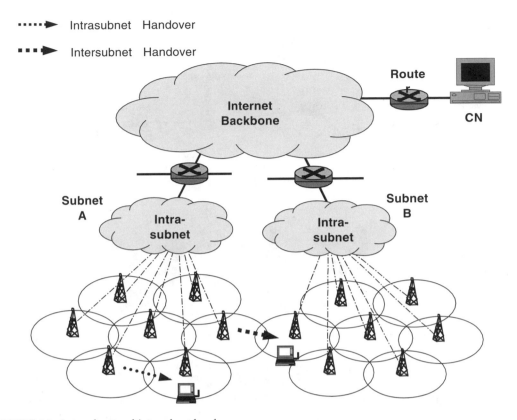

FIGURE 6.3   Intrasubnet and intersubnet handovers.

protocols[2,18–20] aim to handle local movement of MNs without interaction with the Mobile IP–enabled Internet. In other words, IP micromobility has the ability for an MN to move without notifying its HA. This has the benefit of reducing delay and packet loss during handover as long as mobile hosts (MHs) remain inside their local regions. This reduces the signaling load experienced by the core network in support of mobility by eliminating registration between MNs and possibly distant HAs. IP mobility protocols are discussed further in Chapter 5.

6.2   QoS Routing in a Radio Access Network

Typically, a RAN consists of potentially thousands of BSs and a significant number of BSCs/RNCs. Section 6.1 explained the architecture of an RAN (see Figure 6.2). The traffic volume (in terms of voice traffic) generated by these nodes can vary from a few up to 50 voice calls per BS and up to several thousand simultaneous calls per BSC/RNC site. Therefore, a router in the network has to handle many thousands of simultaneous flows.[15–17] Figure 6.4 shows the IP network configuration in the RAN. The transmission between BSs and the BSC/RNC is usually on leased lines. Because of the wide area coverage of the cellular network, the cost of transmission in this part is usually extremely expensive when compared with that of transmission in the backbone. Therefore, if the number of BSs is large, the cost for these leased lines becomes quite significant.

In an IP-based RAN, low-priority traffic can typically be transported between high-priority traffic if resources are sparse. A main goal of QoS routing is to maximize the resource efficiency. This means maximizing the number of QoS flow connections that are admitted into the network. That is also equivalent to minimizing the call-blocking ratio.[16] Thus, QoS routing and resource-reservation mechanisms must be used to provide resources to the real-time applications in the RAN. However, it is difficult

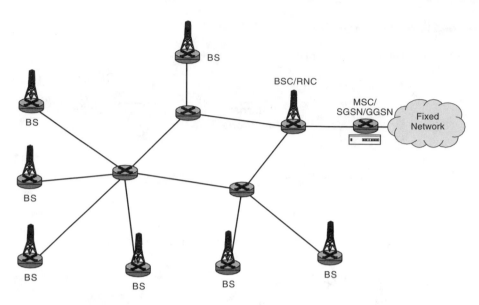

FIGURE 6.4 An example of an RAN based on an IP service bearer.

to guarantee the QoS of real-time traffic in an RAN because overload situations can often occur in case of high bandwidth variation or fluctuation.

On the other hand, the admission control is often considered a by-product of QoS routing and resource reservation. Admission control admits a flow only if there are sufficient resources to provide the requested QoS for the new flow; at the same time it ensures that the QoS level of the already admitted flows is not violated.[16] In both IntServ and DiffServ frameworks, admission control plays an essential role in protecting the network from overload. The blocking rate of QoS requests in an RAN typically depends on a number of parameters, such as the residual bandwidth in router interfaces in the RAN, the number of BSs in the RAN, and the number of ongoing connections in each BS. In addition, the handover dropping rate in a cell, which affects on-call admission control after handover events in an RAN, is based on the traffic load in each cell, the mobility information, and the call-duration statistics.

For a number of reasons, it is impossible to find an ideal QoS routing. Multiple QoS constraints often make the routing problem intractable. For example, finding a feasible path with two independent path constraints is NP complete. That is, all known algorithms for this path-finding problem require exponential runtime. It is unknown whether algorithms exist whose runtime is polynomial. Second, the Internet of the future is likely to carry both QoS traffic and best-effort traffic, which complicates the issue of performance optimization. Third, the network state changes dynamically due to transient load fluctuation and the variable number of connections, particularly when the wireless communication is involved.[45]

6.2.1 QoS Routing

QoS routing typically tries to find the least-cost path among all feasible paths that have sufficient residual resources to satisfy the QoS constraints of a connection. The routing protocol searches for an entry in the QoS routing table corresponding to the route request. Then it compares the available bandwidth of the shortest route with the destination and the requirements of the flow being processed. If the shortest path is selected and the available bandwidth on the path satisfies the bandwidth request, the flow is admitted. Otherwise, an admission of the flow is rejected.[50,51]

On the other hand, optimizing resource utilization and reducing the call-blocking rate are important for the QoS routing, which can be achieved by load balancing.[44–47] A balanced traffic distribution helps increase the call-admission ratio of future flow attempts. The probability of accepting a flow connection can be increased by multipath routing and path rerouting under resource contention. There are actually two kinds of multipath routing. First, routing messages are sent along multiple paths in parallel, and then the best path is selected and resources are reserved along the path. If the selected path does not meet the requirement, an alternative path is picked for resource reservation. Second, when there is no feasible path with sufficient resources, the QoS routing algorithm tries to find multiple paths whose combined resources satisfy the requirement. That is, a flow is split into these multiple paths and carried over the network.[44,45]

Rerouting helps balance the network traffic on the fly and improve the resource efficiency. It is especially useful when connections have different priorities. When no feasible path is available for a new connection, instead of rejecting the new connection the resources held by the existing connections with lower priority are preempted. Thus, some existing connections are rerouted in order to make room for the new connection. These approaches work well with network dynamics.[44,45]

Normally, the QoS routing is closely linked with resource reservation to provide the guaranteed service. To do so, the required resources (CPU time, buffer, bandwidth, etc.) must be reserved when a QoS connection is established. Whereas 3GPP architecture considers Resource Reservation Protocol (RSVP)/IntServ optionally to negotiate end-to-end QoS, 3GPP2 proposes that additional functional entities, such as a subscription QoS manager, monitor per-subscription usage of QoS. Both of them consider that the core network entity controls QoS and resource allocation based on DiffServ policy supported throughout the core network.

6.2.2 RSVP/IntServ-Coupled QoS Routing

Combining an RSVP with QoS-based routing allows fine control over the route and resources at the cost of additional RSVP state and setup time. RSVP may be used to trigger QoS-based routing calculations to meet the needs of a specific flow.[48,49] Thus, QoS routes are computed for each request, where requests explicitly express their resource requirements. That is, when a PATH message arrives, Tspec, which contains sender traffic characteristics, is passed to routing together with the destination address.

However, when QoS requests are more frequent, and the granularity of resource allocation is smaller, this approach has severe overhead in computing QoS routing paths for each request and in updating RSVP states to set up individual paths for each request.[52] Moreover, these QoS route computations and RSVP signaling attempts typically consume processing resources at the switches and introduce setup latency for each accepted connection. Minimizing this overhead is particularly important during periods of transient overload, such as bursty connection arrivals or rerouting of traffic after a mobility event in an RAN.

6.2.3 DiffServ-Coupled QoS Routing

QoS-based routing is typically used in conjunction with some form of resource-allocation mechanism. When DiffServ is coupled to the long-term traffic monitoring function, the different paths computed by QoS routing either are preestablished or change only infrequently.[52] In order to optimize various performance measures and thereby maximize network performance, QoS routing measures traffic patterns continuously and dynamically computes and shifts paths on which to route traffic aggregates. For example, a link may advertise its available bandwidth metric through the periodic update message, and multipath routing may establish more than one simultaneous route and fairly distribute traffic between these routes by using load balancing.[17,45,47] More specifically, when a connection with a QoS request arrives, the router typically computes a suitable route satisfying QoS requirements, using the most recent link-state information.[48,49] However, in the DiffServ-coupled QoS routing, it is difficult to provide explicit guarantees to individual flows because of the traffic aggregates and the network-wide traffic optimization in computing QoS paths.[52]

6.3 The Integrated Services Architecture

The IntServ architecture[3] recommends a set of extensions to Internet architecture in order to better support real-time applications. The IntServ architecture consists of two basic elements: the Integrated

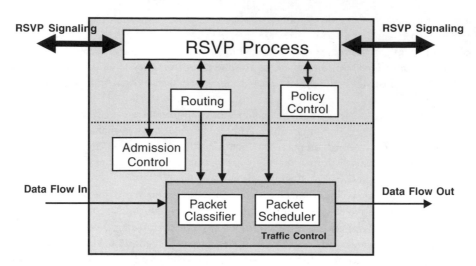

FIGURE 6.5 RSVP/IntServ router architecture.

Service model and the implementation reference model. The former defines two types of services for real-time applications: Controlled Load Service (CLS) and Guaranteed Service (GS). CLS is intended for adaptive real-time applications, which are highly sensitive to overload conditions in the network. GS is a quantitative service that provides guaranteed bandwidth and delay bounds and, as such, is intended for nonadaptive real-time applications with strict QoS requirements. Depending on the application's QoS requirements, typically one of two service treatments is invoked. For realization of the model, the latter defines several mechanisms such as scheduling, classification, admission control, and resource reservation. Figure 6.5 illustrates these functions in an Integrated Services router.

6.3.1   The RSVP Protocol

RSVP is an attempt to provide real-time service through the use of virtual circuits. It is a resource reservation setup protocol designed for the IntServ model. RSVP is not a routing protocol but a control protocol, which allows Internet real-time applications to reserve resources before they start transmitting data. That is, when an application invokes RSVP to request a specific end-to-end QoS for a data stream, RSVP selects a data path that depends on underlying routing protocols and then reserves resources along the path. Because RSVP is also receiver oriented, each receiver is responsible for reserving resources to guarantee requested QoS along its data path. Hence, the receiver sends a RESV message to reserve resources along all the nodes on the delivery path to the sender.

6.3.2   The RSVP/IntServ Router

RSVP is used to pass the QoS request originating in an end system to each router along the data path (see Figure 6.6). At each node along the path, the RSVP daemon passes a QoS request (flowspec) to an admission-control algorithm in order to allocate the node and link resources necessary to satisfy the QoS requested. Before a reservation can be established, the RSVP daemon must also consult policy control to ensure that the reservation is administratively permissible. Assuming that both admission control and policy control requests succeed, the daemon installs RSVP state in the local packet scheduler or other link-layer QoS mechanism to provide the QoS requested. The packet scheduler multiplexes packets from different reserved flows onto the outgoing links, together with best-effort packets for which there are no reservations. The RSVP daemon also installs the RSVP state in a packet classifier, which sorts the data packets forming the new flow into the appropriate scheduling classes. The state required to select packets for a particular QoS reservation is specified by the filter spec. Basically, an RSVP flow can be defined by the combination of the session (destination address, IP protocol ID, destination port) and the filter

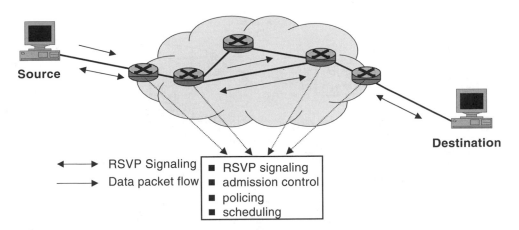

FIGURE 6.6 RSVP/IntServ framework.

(source address, source port) in IPv4 or the combination of the source address and the flow ID in IPv6. The IPv6 flow label allows routing to be based on the flow label as well as source and destination addresses. The RSVP daemon on the router identifies the flows from the information on the RSVP states (Reservation State and Path State); hence, RSVP includes a flow label in a PATH message of its IP/UDP packet.[9]

6.3.3 RSVP Messages

RSVP creates and maintains state by periodically sending control messages in both directions along the data paths for a session. RSVP messages are sent as IP datagrams and are captured and processed in each node (e.g., router or host) along the path to establish, modify, or refresh state.[3] The RSVP protocol defines seven types of messages; the fundamental ones are the PATH and RESV messages, which carry out the basic operation of RSVP.

A PATH message is initiated by a sender and travels addressed to the multicast or unicast destination address of the session to create a *path state* in each node. RSVP in each node queries routing in order to forward a PATH message along the exact path(s) that the corresponding data packets will traverse. Each PATH message contains information about the sender that initiated it, and the path state basically consists of a description of every sender to the given session.[3] RESV messages are periodically initiated by receivers to request reservations and travel against the direction of data flow to create *reservation states* in each node. Each RESV message is sent to the unicast address of the *previous* RSVP hop.[3]

The rest of the RSVP messages are used either to provide information about the QoS state or to explicitly delete the QoS states along the communication session path. PATH, PATH Tear, and RESV Confirm messages should be sent with the router alert option set. Figure 6.6 shows an example of RSVP signaling in IntServ networks.

6.3.4 RSVP and Routing

The RSVP and routing daemons shown in Figure 6.5 typically execute as background processes, and the dotted line represents the user/kernel boundary. Figure 6.7 shows the functionality of the RSVP daemon in the host and router. In an end system, the forwarding path is in the kernel, and RSVP and routing protocols will execute in user space. A routing module in an RSVP-enabled router typically sends its RSVP daemon a path change notification (PCN) message using the Routing Support for Resource Reservations (RSRR) interface, which is a specification for communication between RSVP and routing daemons.[23] RSVP resource reservation is accomplished by using this routing interface to allow RSVP to access to forwarding database entries whenever they change in the network.

In order to support real-time applications more effectively, the network routing protocol and RSVP need to cooperate to provide stable routes that deliver the requested QoS. For cooperation of routing

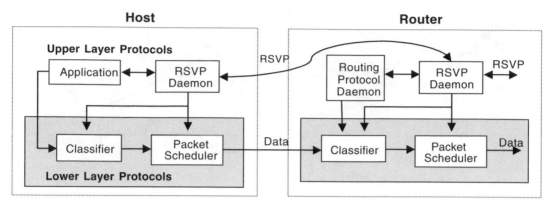

FIGURE 6.7 Functionality of the RSVP daemon in the host and router.

and RSVP, the routing protocol is responsible for identifying a route that has the possibility of meeting the requested QoS (e.g., QoS routing). The RSVP then verifies the availability of resources along the route and attempts to reserve resources for the user. If the RESV message fails, this process repeats with a new route for RSVP to try, or the reservation request is just rejected. Once reservation of a route is successful, the routing protocol maintains the route for the receiver.

6.4   The Differentiated Services Architecture

Differentiated Services (DiffServ, or DS)[27] is a policy-based approach for QoS support in the Internet, where traffic entering a particular network is classified into different classes, and thus classes are assigned to different behavior aggregate. DiffServ uses the DiffServ Code Point (DSCP) field in the IP packet header, which determines the service type of the data traffic, by specifying a per-hop behavior (PHB) for that packet.[41,42] A DS domain normally consists of one or more networks under the same administration; for example, an organization's intranet or an Internet service provider (ISP). The administration of the domain is responsible for ensuring that adequate resources are provisioned and reserved to support the service level agreements (SLAs) offered by the domain.[27] A DS domain has a well-defined boundary consisting of DS boundary nodes where traffic classification and conditioning functions are executed for service differentiation for individual or aggregated flows. That is, a meter measures the sending rate of a flow, and a marker sets the DSCP fields of packets in the flow at the boundaries of the network. When packets are forwarded within the core of the network, a dropper discards packets of different flows according to the PHB identified by the DSCP fields and the various dropping precedence policies.[29,33] Packets, which are marked into the same PHB class, experience similar forwarding behavior in the core nodes. PHBs are actually implemented by means of buffer management and packet-scheduling mechanisms in the core nodes. Figure 6.8 shows the DiffServ framework.

6.4.1   Packet Classification

A microflow is a single instance of an application-to-application flow of packets identified by flow ID. The flow ID attribute typically consists of five components (Source IP address, Destination IP address, Protocol ID, Source Port Number, Destination Port Number), and a DSCP byte can also be considered. Thus, a flow ID identifies a stream of IP datagrams sharing at least one common characteristic. The flow ID provides the necessary information for classifying the packets at a DiffServ boundary node. That is, a flow is identified by matching the fields in the packet headers with the flow ID attributes given in the Service Level Specification (SLS). A DiffServ traffic stream may consist of the set of active microflows that are selected by a particular classifier, such as a BA (Behavior Aggregate) classifier or an MF (Multi-Field) classifier. Packet classifiers select packets in a traffic stream based on the content of some portion

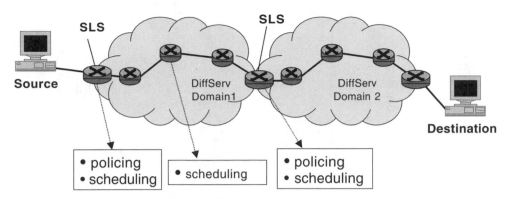

FIGURE 6.8 Differentiated Services framework.

of the packet header. The BA classifier classifies packets based on the DS code point only. The MF classifier selects packets based on the value of a combination of one or more header fields, such as the source address, destination address, DS field, protocol ID, source port and destination port numbers, and other information (such as the incoming interface[27]).

6.4.2 Service Level Agreement (SLA)

An SLA is a contract between a customer and an ISP that specifies the forwarding service.[27,33] Once hosts set up SLAs with their ISP, they can send DiffServ packets with a traffic rate on which they agree. When ISPs forward DiffServ packets into the Internet according to the DSCP marking, ISP networks queue and schedule DiffServ packets with their own queuing mechanisms or drop precedence policies. A DS region is a set of one or more contiguous DS domains. DS regions are capable of supporting differentiated services along paths that span the domains within the region. The DS domains in a DS region may support different PHB groups internally and different DSCP-to-PHB mappings. However, the peering DS domains must each establish a peering SLA that defines (either explicitly or implicitly) a traffic conditioning agreement (TCA).[27] ISPs thus set up SLAs between adjacent networks, and eventually end-to-end QoS can be provided for DiffServ traffic. That is, DiffServ is extended across a DS domain boundary by establishing an SLA between an upstream network and a downstream DS domain. The SLA may specify packet classification and re-marking rules and may also specify traffic profiles and actions to traffic streams that are in-profile or out-of-profile. The TCA between the domains is derived explicitly or implicitly from this SLA.[27] A software agent, called a bandwidth broker (BB),[31] actually automates the SLA negotiation among DiffServ domains. Upon negotiating an SLA for new incoming DiffServ traffic, BBs have to check if their network is able to support it without congestion.

6.4.3 Per-Hop Behaviors (PHBs)

A PHB is a description of the externally observable forwarding behavior of a DS node applied to a particular DS behavior aggregate. The PHB is the means by which a node allocates resources to behavior aggregates. PHBs may be specified in terms of their resource (e.g., buffer, bandwidth) priority relative to other PHBs. PHB groups will usually share a common constraint applying to each PHB within the group, such as a packet scheduling or buffer management policy.[27] PHBs are implemented in nodes by means of some buffer-management and packet-scheduling mechanisms. In general, a variety of implementation mechanisms may be suitable for implementing a particular PHB group. Furthermore, it is likely that more than one PHB group may be implemented on a node and utilized within a domain.[27] Several PHB group approaches have been introduced, but expedited forwarding (EF)[43] and assured forwarding (AF)[42] PHBs are currently considered to allow delay and bandwidth differentiation.

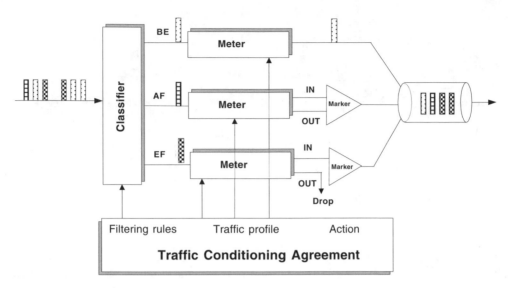

FIGURE 6.9   DiffServ edge router architecture.

6.4.3.1   Assured Forwarding (AF) PHBs

An AF PHB group[42] provides the means to offer different levels of forwarding assurances to IP packets. Four AF classes are defined, with three levels of drop precedence within each class. The level of forwarding assurance of an IP packet depends on the amount of resource assigned to the AF class, the current load on the class, and the drop precedence of packets in case of congestion. If congestion occurs rarely in its domain, supporting two levels of the drop ratio is satisfactory instead of implementing all four AF classes.[24] An AF PHB service flow, based on TCP traffic, would consume a fraction of the available bandwidth. A DiffServ node thus requires buffer-management mechanisms to control its effect on the TCP traffic under congestion, and packets belonging to different AF classes will be forwarded independently of one another. Furthermore, a DiffServ node must not reorder the AF packets belonging to the same microflow, regardless of their drop precedence.[27]

6.4.3.2   Packet Dropping Mechanisms

In a DiffServ network, an important element for AF service is a proper packet-dropping mechanism in overload conditions. Random Early Detection (RED)[27] is a router queue management scheme that can guarantee fair sharing of the available bandwidth among the TCP flows in congestion situations. In this scheme, when the queue length in a router exceeds a certain threshold, the packets to be dropped are selected randomly to prevent dropping packets of the same application data flow. In RIO (RED with In and Out) (Clark et al., 1997), the packets with in-profile and out-of-profile marking in a DSCP field share a common queue, but packets with out-of-profile marking are discarded earlier than packets with in-profile marking. This two-drop precedence policy makes RIO keep the dropping probability of in-profile packets low. In a network with a three-drop precedence policy,[43] the edge routers mark a packet as one of green, yellow, or red depending on the sending rate of the flow. Under overload conditions, the core routers thus drop the red packets first, the yellow packets second, and the green packets third. This dropping policy by three-color marking is expected to provide better performance. Figure 6.9 shows the architecture of a DiffServ edge router.

6.4.3.3   Expedited Forwarding (EF) PHB

The EF PHB[43] provides tools to build a low-loss, low-latency, low-jitter, assured-bandwidth, end-to-end service through DiffServ domains (RFC 2598). Thus, an EF PHB service flow, which has high priority and needs bandwidth and delay assurances, is based on UDP traffic and nonadaptive. An EF service thus requires admission control to prevent resource starvation of lower-priority traffic classes. The SLA (i.e.,

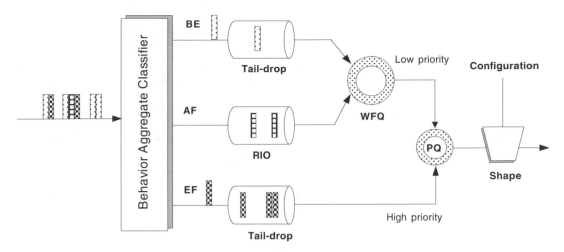

FIGURE 6.10 DiffServ Core Router Architecture.

SLS) specifies a peak bit rate, which a customer's applications will receive. However, it is the customer's responsibility not to exceed this rate because then their packets can be dropped. A DiffServ node in the forwarding path ensures that the particular aggregate has a minimum departure rate independently of the other traffic at the node. EF PHB implies a strict bit-rate control or a fast-forwarding treatment to the DiffServ boundary nodes and DiffServ internal nodes, respectively.[27]

EF PHB is implemented by different mechanisms, such as simple priority queuing (PQ), weighted round robin scheduling (WRR), and class-based queuing (CBQ). All these mechanisms implement the basic EF PHB properties, but different implementation choices result in different properties of the same service, such as, for instance, jitter as seen by individual microflows.[24] Figure 6.10 shows the architecture of a DiffServ core router.

6.4.4 Bandwidth Broker

A bandwidth broker (BB) is typically responsible for monitoring and managing the available bandwidth within a DiffServ domain.[30,31] When a user wants to transmit data to his or her DiffServ domain, the user sends a bandwidth request to the BB depending on the rules or some policies made with the ISP through an SLA contract. However, the signaling interface between a user and the border routers, which is needed to allow the user to send its QoS requirements to the BB of the domain, is still not defined. Various protocols can be used for the user "QoS request" within a DiffServ domain. They should be simple and scalable and should not maintain any state as well. Out-band signaling protocols,[31] such as such as CLI (Common Line Interface), Diameter, COPS, SNMP (Simple Network Management Protocol), RSVP, ICMP (Internet Control Message Protocol), and CORBA (Common Object Request Broker Agent), can be considered, and special IPv6 options/extension headers can be defined as in-band signaling for this purpose as well.

When data is transmitted across more than one DiffServ domain, bandwidth broker signaling is performed between adjacent BBs of neighboring DiffServ domains. The BBs negotiate the SLA between two domains through the requested QoS information, such as service type, desired data rate, maximum burst, and period of service request.[30,31] Figure 6.11 shows a resource-reservation mechanism by BBs between DiffServ domains.

6.5 Integrated Services over DiffServ Networks

Within the IntServ network, the processing of per-flow states in IntServ-capable routers leads to scalability problems when deployed to large networks. Meanwhile, DiffServ networks classify packets into one of a

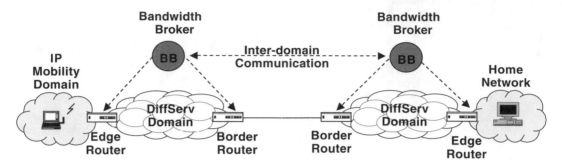

FIGURE 6.11 A resource management model for DiffServ networks.

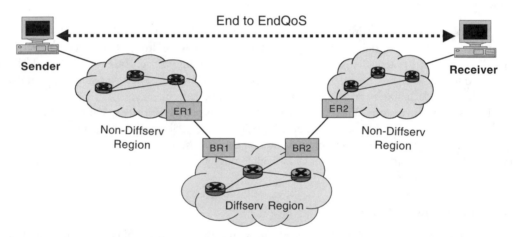

FIGURE 6.12 The reference network for the IntServ/RSVP over the DiffServ framework.

small number of classes based on the DSCP in the packet's IP header. DiffServ eliminates the need for processing per-flow states and therefore scales well to large networks. In a network with IntServ operation over DiffServ, the Integrated Services architecture is used to deliver end-to-end QoS to applications. RSVP signaling messages travel end-to-end between sending and receiving hosts to support RSVP reservations outside the DiffServ network region. From the perspective of IntServ, DiffServ regions of the network can be treated as virtual links connecting IntServ-capable routers or hosts. Therefore, this network includes some combination of IntServ-capable nodes and DiffServ regions. Individual routers may or may not participate in RSVP signaling regardless of where they reside in the network. This network includes DiffServ regions in the middle of a larger network supporting IntServ end to end. The DiffServ region contains a mesh of routers, at least some of which provide aggregate traffic control. The regions outside the DiffServ region (non-DiffServ regions) contain meshes of routers and attached hosts, at least some of which support the Integrated Service architecture. Figure 6.12 shows a reference network for the IntServ/RSVP over a DiffServ framework.

IntServ service requests specify an IntServ service type and a set of quantitative parameters known as a flowspec. These requests for IntServ services must be mapped onto the underlying capabilities of the DiffServ network region. In RFC 2998, this kind of mapping includes selecting an appropriate PHB (or set of PHBs) for the requested IntServ services, performing appropriate policing at the edges of the DiffServ region, exporting IntServ parameters from the DiffServ region, and finally performing admission control on the IntServ requests that takes into account the resource availability in the DiffServ region.[9,24]

6.5.1   RSVP Awareness in DiffServ Regions

The specific implementation of the IntServ operation over DiffServ depends on DiffServ resource management and on RSVP awareness in DiffServ regions.[24] Resource management in DiffServ can be achieved either statically (via human control) or dynamically (via protocols). A DiffServ region is considered RSVP aware if it has RSVP-enabled routers within its domain. Whether the routers at the borders of the domain would be only RSVP aware, or all the routers within DiffServ would be RSVP enabled, depends on the network administrator. The static SLA in the DiffServ region is negotiated manually with its customer (i.e., the IntServ region) in order to perform a number of actions conforming to the negotiated SLA to comply with service commitments. The dynamic SLAs are negotiated upon aggregated requests from the IntServ region in order to perform admission control and resource allocation within the DiffServ.

6.5.2   Service Mapping

If DiffServ is not used only as a transmission medium, the service mapping between IntServ and DiffServ is essential for IntServ operation over DiffServ networks. DiffServ service mapping on the IntServ request depends on appropriate selection of PHB, admission control, and policy control in the DiffServ. Typically, the GS can be mapped to EF PHB and the CLS can be mapped to AF PHB, but different priority levels of AF PHB depend on the tolerant latency of traffic flow. Thus, routers that are capable of handling both RSVP and DiffServ will perform the service mapping between IntServ and DiffServ. These routers will usually be located at the edges of the DiffServ region.[9,24]

6.6   RSVP/IntServ Extensions in IP Mobility Environments

Recently, as it becomes much easier to access the Internet from an MH, mobile users will be demanding the same real-time services available to fixed hosts. However, for a wireless network where hosts are likely to be mobile, the semantics of QoS are more difficult to achieve due to host mobility and the nature of the wireless medium. Mobility of hosts has a significant impact on the QoS parameters of a real-time application. It also introduces new QoS parameters at the connection and system levels. Currently, Mobile IP is based on the best-effort delivery model and does not take QoS into account. Furthermore, the RSVP model, which provides efficient resource reservation in the fixed endpoints, becomes invalid under host mobility. In this section, we investigate the problems of RSVP in wireless mobile environments. We then identify the advantages and drawbacks of the existing proposals under IP mobility.

6.6.1   Problems of RSVP under the Mobility Scenario

6.6.1.1   The RSVP Signaling Process after Handoff

For RSVP under wireless mobile networks, when an MN is a receiver and moves to another location, it must reestablish reservations along the new path from it to the source. In standard RSVP operation because each data source issues PATH messages periodically to automatically reserve resources along a new path after a routing change occurs, the MN must wait for a PATH message at its new location before it can send a RESV message back along the new path to the source for reservation of resources. In other words, whenever an MN performs a handover that incurs a path change, the new path to the MN from the CN would be discovered by RSVP only after the soft-state timers expire in network elements (such as routers). However, as a result, this mechanism is so slow that it may lead to large delays under a roaming environment with frequent host mobility.

Hence, in order to reserve resources along the new path with the support of both QoS and mobility, an RSVP signaling process must be invoked immediately instead of waiting for the next periodic RSVP state update associated with the RSVP soft-state mechanism. Simply reducing the soft-state timer may result in excessive signaling overhead when the host moves relatively slowly. This problem can be avoided by the mechanism that the CN triggers transmitting a PATH message after receiving a binding update message from the MN.

6.6.1.2 RSVP Signaling Delay and Overhead

For a guaranteed service, it is necessary for the network to offer tight delay bounds to applications. That is, it is required that network elements (e.g., routers) be able to estimate the maximum local queuing delay for a packet. Because of the long resource reservation delay during reestablishment of a flow after handover under RSVP, service disruptions could occur in providing real-time services. The longer the resource reservation delay due to RSVP signaling overhead in roaming environments, the greater the possibility for a larger number of packets being sent along the previous reservation path to the MN's old location; hence, there are a greater number of lost packets. Mechanisms are thus needed to minimize the resource reservation delay and the packet loss resulting from handovers under RSVP. Simply buffering to reduce packet loss would violate the semantics of guaranteed QoS for real-time services because it would introduce additional delay. Furthermore, the buffering requires extra storage and operation complexity to manage the buffers.

6.6.1.3 Service Disruption due to Mobility Events

In wireless networks, channel capacity, which is location dependent, will most likely cause an MH's negotiated QoS to be compromised when it moves into a highly populated cell. The effect of this movement may cause the QoS of other mobiles that are already in the cell to degrade. An MH must also renegotiate its desired QoS (using RSVP messages) to set up a new reservation path each time it moves to a new location, which incurs additional delay. There is also no guarantee that the same level of resources will be available under a new point of attachment to which an MN moves. This implies that there may be service disruption with the mobility of the host. This problem can be solved using information about how an MN is moving. In 3G cellular systems, the European Telecommunications Standards Institute (ETSI) has already defined this functionality on GSM phase 2+ under the name of SOLSA (Support of Localized Service Area). This new concept allows a cellular network operator to distinguish services offered to users or user groups depending on their location inside the network.

6.6.2 RSVP Extensions under Mobile IP

Several studies on RSVP supporting Mobile IP have been conducted. These protocols under macro-mobility are Mobile RSVP,[4] the multicast-based model,[7] the RSVP tunnel model with Mobile IP,[6] Mobile IP with Location Registers,[10] the Mobile IPv6 and RSVP integration model,[22] and the flow transparency-based model.[21]

6.6.2.1 Mobile RSVP

Talukdar et al.[4] proposed MRSVP (Mobile RSVP) as an extension to the conventional RSVP, for the provision of QoS guarantees to MNs independently of their movement throughout the access network. Figure 6.13 shows the reservation route of MRSVP when an MN is a receiver. In MRSVP, an MH can make advance resource reservations. To do this, it is necessary to specify the set of locations, called Mspec, that the MH may visit in the future. The MN thus establishes paths with sufficient resources to a possibly large set of attachment points that the MH may visit in the future. When the node arrives at a particular point of attachment, the path to that attachment becomes active, and the path to the previous attachment becomes passive, so that the data can still be delivered effectively. But this approach suffers from several drawbacks: Many resources are reserved, thus making them unavailable for other requests although they may never be used. It requires the BS to maintain a lot of state information regarding active and passive reservations during handovers. An MN originating a real-time flow has to wait until all the necessary resources in a possibly large set of attachments become available. Therefore, eventually the blocking rate of the flows originating for real-time applications may become very high. It is also difficult to determine accurately the possibly large set of attachment points of an MH.

6.6.2.2 Mobility Support Based on IP Multicast

Chen and Huang[7] proposed a new signaling protocol for MHs to reserve resources in Integrated Services Internet. Under this approach, the RSVP model is extended based on IP multicast to support MHs. RSVP

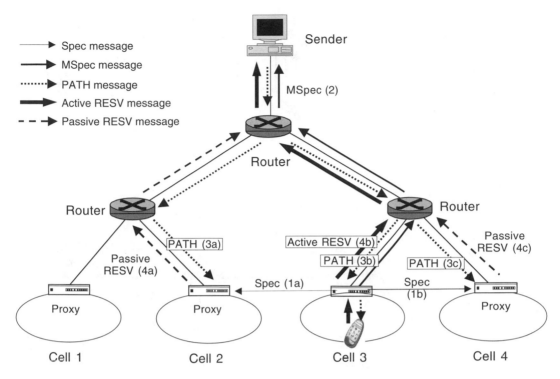

FIGURE 6.13 Reservation routes for MRSVP.

messages and actual IP datagrams are delivered to an MH using the IP multicast routing. The multicast tree, which is rooted at each source node, is modified dynamically every time an MH roams to a neighboring cell. Hence, the mobility of a host is modeled as a transition in multicast group membership. A new host joining a multicast group results in the establishment of a new branch in the multicast tree. A host leaving a multicast group may result in a branch in the multicast tree being "pruned." Once these new branches have formed, PATH messages from the sender are forwarded to mobile proxies along the multicast tree. Upon receiving the PATH messages, the Conventional Reservation message from the current mobile proxy and the Predictive Reservation messages from neighbor mobile proxies are propagated to the sender along the multicast tree. Figure 6.14 shows extended RSVP based on IP multicast when the MN is a receiver. When the mobile receiver is moving from one location to another, the flow of data packets can be switched to the new route as quickly as possible. Even though this method can minimize the service disruption due to rerouting the data path during handovers, it has disadvantages such as overload (to manage the multicast tree dynamically) and inefficiency (due to resource reservation in advance).

6.6.2.3 A Simple QoS Signaling Protocol

Terzis et al.[6] proposed a simple QoS signaling protocol by combining preprovisioned RSVP tunnels with Mobile IPv4. Figure 6.15 shows the RSVP tunnel model with Mobile IP when the MN is a receiver. When the MN moves to cell B, it informs its home agent of its new location. When the HA is informed about the MN's new location, it does two things: (1) It sets up a tunnel RSVP session between itself and the FA if one does not exist between them already, and (2) it encapsulates PATH messages from the sender and sends them through the tunnel towards the MN's new location. When the FA receives a RESV message from the visiting MN, it sends an RESV message for the corresponding tunnel session between itself and the HA. After sending the reservation request, the FA waits for a confirmation from the HA that the reservation over the tunnel was successful. This approach can be easily implemented with minimal changes to other components of the Internet architecture. However, when an MH roams far away from

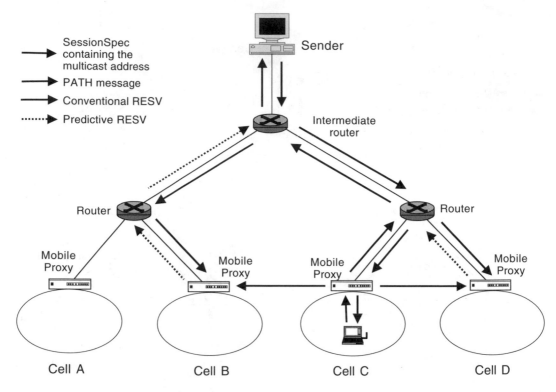

FIGURE 6.14 Extended RSVP based on IP multicast.

the HA, a triangle routing problem occurs, but there is no problem with using a shortest path mechanism, such as route optimization. In this protocol, there is also no solution for reducing service disruption due to frequent mobility of the host.

6.6.2.4   Mobile IP with Location Registers

Jain et al.[10] proposed a scheme, called MIP-LR (Mobile IP with Location Registers), which may be more suitable for 3G cellular systems. This scheme uses a set of databases, called a Location Register, to maintain the current CoA of the MH. When an MN moves from one subnet to another, it registers its current CoA with a database called a Home Location Register (HLR). When a CN has a packet to send, it first queries the HLR to obtain the MH's CoA and then sends packets directly to the MH. The CN caches the MN's CoA to avoid querying the HLR for every subsequent packets destined for the MN. MIP-LR not only eliminates the inefficiency of triangle routing in MIP, but also generally avoids tunneling and allows resource reservation using RSVP to provide QoS guarantees.

6.6.2.5   The Mobile IPv6 and RSVP Integration Model

Chiruvolu et al.[22] proposed a Mobile IPv6 and RSVP integration model. The main idea of Mobile IPv6 and RSVP interworking is to use RSVP to reserve resources along the direct path between the CN and the MN without going through their HAs because Mobile IPv6 provides route optimization. In this model, resources are initially reserved between the CN's and the MN's original location. Whenever the MN performs a handover that incurs a path change, a new RSVP signaling process is invoked immediately to reserve resources along the new path. Figure 6.16 shows the Mobile IPv6 and RSVP integration model when the MN is a receiver. When the MN performs a handover from subnet A to subnet B, it gets a new CoA and subsequently sends a Binding Update to the CN. The CN then sends a PATH message associated with the new flow from CN to MN. Upon receiving this PATH message, the MN replies with a RESV message immediately to reserve resources for the new flow. For each handover, the MN as receiver has to wait for a new PATH message from the CN, and only after getting the PATH message can it issue a

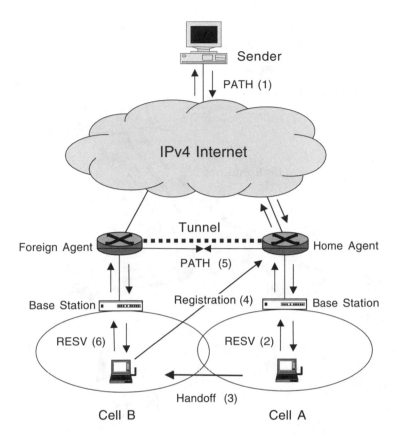

FIGURE 6.15 An RSVP tunnel with Mobile IP.

new RESV message to the CN. However, all these RSVP renegotiations are conducted end to end even though the path change may affect only a few routers within the whole path during a single handover. Hence, the long handover resource reservation delays and large signaling overhead caused by this end-to-end RSVP renegotiation process could lead to considerable service degradation in providing real-time services. Furthermore, during this period, there might not be enough resources in the newly added portion of the flow path between CN and MN.

6.6.2.6 Mobility Support Based on Flow Transparency

Shen et al.[21] proposed a method for solving the drawbacks in the existing IPv6 QoS with mobility support model, namely, long resource reservation delays and large signaling overhead. The node address changes because of node mobility, and, in turn, flow identity is different when the MN acts as the source or destination. Consequently, the same application data flow may be perceived as different flows at the network layer. Because each router needs to be processed based on flow, it is necessary to update the data in all intermediate routers along the flow path whenever a flow changes at the network layer. However, the routers that reside in the duplicated portion of the new and old flow path should be prevented from performing handover update, and only those routers that are in the newly added portion of the flow path need to be involved in the update process.

In this model, the RSVP session and flow identity at the network level should be consistently unique for the flow-handling mechanism (e.g., the packet classifier) in each router regardless of node mobility. Figure 6.17 shows the mobility support model based on flow transparency when the MN is a receiver. When the MN is acting as a receiver, instead of the CN, the crossover router (CR) issues a PATH message to the mobile receiver because the required information already exists in the CR from the previous RSVP message exchanges. In order to detect the route change of the receiver and to trigger Local Repair for the

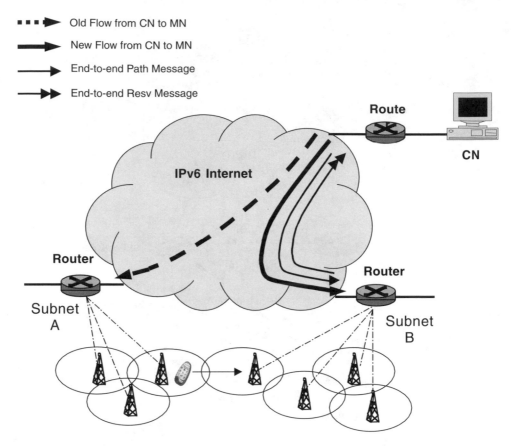

FIGURE 6.16 The Mobile IPv6 and RSVP integration model.

receiver, the receiver should be able to inform the CR of its handover information, which contains the flow destination and the MN's current address. This method automatically limits the handover RSVP renegotiation process within the newly added portion of the path between CN and MN. Therefore, handover resource reservation delays and signaling overhead can be minimized, which in turn minimizes the handover service degradation.

However, when the MN is acting as a sender, current Mobile IPv6 sets the source address (of packets originated from a MN) to the MN's CoA, and the MN's home address is moved into a Home Address Option in a Destination Options Header, which is not processed by intermediate routers during packet transmission. Merely changing the flow source to the MN's home address causes problems in router packet classification. Hence, implementation of this model requires some modification to the Mobile IPv6 specification. There is also no mechanism to guarantee the same level of resources at any new point of attachment to which the MH moves. This may cause service disruption with the mobility of the host.

6.6.3   RSVP Extensions under IP Micromobility

RSVP has been mainly considered with Mobile IP under macromobility, but network elements (e.g., the router) or intermediate nodes under micromobility need to be RSVP enabled to provide QoS guarantee for mobile users using real-time applications. Mahadevan et al.[25] proposed a resource-reservation scheme with class-based queuing (CBQ) in a microcellular network (intrasubnet) architecture. This scheme distinguishes between the two kinds of reservations: one between the sender and various BSs in neighboring cells in the wired network and another between the BS and the MN in the wireless region. In this scheme the RSVP protocol is used, modified to recognize passive and active reservations. In addition to making

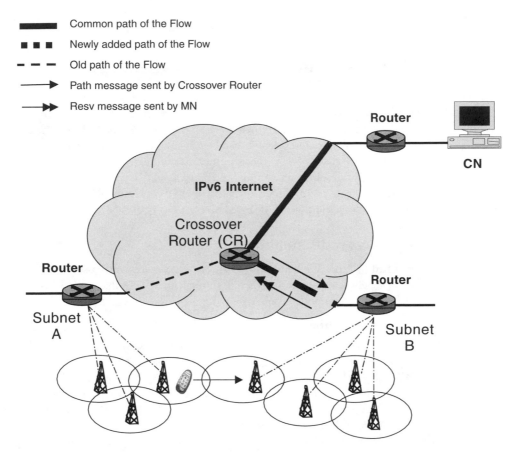

FIGURE 6.17 Mobility support based on flow transparency.

active reservation between the BS and the MN, the BS has to make passive reservations with all the BSs in the neighboring cells when the BS accepts the call. Once the MN then moves to the neighboring cell, both the MN and the BS know that a handoff occurred. At this point, PATH and RESV messages denoting an active reservation are exchanged between the old BS and the new BS and between the new BS and the MN. If the RSVP process has a reservation successfully along the path through the end-to-end QoS negotiation, the data sent from an application can use CBQ for packet scheduling and classification. CBQ makes forwarding decisions on the outgoing interface, which uses the particular link-layer medium.

An alternative approach involves changing the RSVP implementation at the network elements (NE) to which BSs are connected. In this scheme, whenever the MN moves into the neighboring cell, resource reservations are newly made over two-hop link between the NE and the MN. Once the BS detects the MN's handoff, it notifies the NE, and thus PATH and RESV messages are exchanged to make a new reservation between the NE and the MN. Therefore, the resource reservations delay over only a two-hop link will be very short, and guaranteed QoS in the newly modified portion of the path can be obtained easily. Admission-control and resource-allocation mechanisms can be used to help provide bandwidth required for an MN in the neighboring cells.

For more general approaches under micromobility, including various topologies (e.g., Cellular IP, HAWAII, the Hierarchical Mobile IP [HMIP]), a new path needs to be set up during a handover.[28] In setting up the new path, the remaining circuits must be reused because only a partial path is changed. For this kind of path reestablishment scheme, the new partial path between the crossover NE and the MN must be discovered and set up during handoff. Whereas the signaling overhead depends on the connection setup protocol, the number of hops required to set up the partial path during a handover depends on the crossover NE; end-to-end latency is a function of the route length. Hence, an efficient partial path discovery mechanism is required so that the RSVP resource reservation delay is reduced.

6.7 DiffServ Extensions in IP Mobility Environments

We focus on DiffServ for QoS support in RAN in this section. DiffServ will be deployed in the global wired Internet, such as IPv6. DiffServ is typically expected to provide a better QoS than existing services with a set of policies and rules that are predetermined between the user and the ISP through an SLA. However, DiffServ itself lacks a simple and scalable bandwidth-resource management scheme for a dynamic environment as an IP-based RAN. It is therefore necessary to extend the DiffServ model for admission control and on-demand resource reservation under IP mobility environments.

In this section, we first investigate the problems of DiffServ in IP mobility environments. We then present several examples of DiffServ solutions in IP-based access networks.

6.7.1 DiffServ Problems in IP Mobility Environments

DiffServ is typically expected to provide a better QoS than existing services with a set of policies and rules that are predetermined between the user and the ISP through an SLA. In this section, however, we investigate the problems of DiffServ under IP mobility environments.

6.7.1.1 Dynamic SLS Configuration

The SLS is a translation of an SLA into the appropriate information necessary for provisioning and allocating QoS resources within network devices at the edges of the domain. [31] An MH at its HN typically depends on the SLAs that have been negotiated between its HN and its CH's networks. When an MH visits several RANs managed by different ISPs, additional SLAs are required between the MH's visited network and the CH's network. Those SLAs may be negotiated between the BBs of the different networks. However, the admission control of BBs based on static SLAs is not suitable in highly dynamic mobile environments because the location and the QoS requirements of the users may change frequently. Therefore, for dynamically changing wireless networks, dynamic SLA renegotiations are essential. In other words, the current QoS requests of each MH need to be delivered to the BB more accurately using the signaling information. Dynamic SLA negotiation requires significant QoS signaling exchanges between the MH and the routers in DiffServ networks. Meanwhile, when the routing path is changed due to mobility in the DiffServ domain, the protocol flows necessary to reestablish QoS in a new subnet may be very time consuming for Mobile IP and other mobility protocols. In order to reestablish the specific service without having to perform the entire SLA negotiation between the MH and the CH, therefore, a QoS context such as SLS at the MH's HN has to be transferred to an MH's visited network. The BB in the visited network then configures the network according to the MH's SLS. Typically, a mechanism of transferring the MH's QoS context to the newly visited network could guarantee faster reestablishment of the MH's current service.[40]

6.7.1.2 Dynamic Routing for SLA Setup

Typically, the BB in a DiffServ domain should identify the path of the flow for the user's request and check whether there are enough available resources in the DiffServ routers along this path. Thus, interior routers or the BB should be able to be dynamically configured for admission control and resource scheduling in a DiffServ network. When transferring QoS context to the visited network, in addition to the MH's visited network or subnets, some additional routers in the path between the visited network and the CH's network might be involved. Therefore, when all the DiffServ routers on the data path of the user's request have enough available resources, the request could be accepted. In order to identify the data path satisfying a mobile user's QoS request for SLA setup, some scenarios could be considered from the QoS routing mechanisms for the fixed Internet. First, the BB might obtain the topology information of its domain by indirectly communicating with the domain's routing protocol (e.g., Mobile IP–enabled

Open Shortest Path First [OSPF], Cellular IP, or HAWAII). Second, a BB may calculate the QoS path by investigating routing tables from all routers of the domain using the signaling protocol. The third method is to find the QoS path by using a hop-by-hop route discovery mechanism, which depends on a specific routing protocol.

6.7.1.3 Transparent Flow Identification

In an IP-based RAN, the first and the border routers identify microflows associated with mobile users. The source or destination IP address fields in the header of IP packets, which an MH generates, might change when the MH roams among different subnets. More specifically, the original IP address in the IPv4 packet is changed when it is encapsulated by the IP tunneling mechanism. Moreover, the home address in the IPv6 packet is moved to its destination option header, and instead the CoA is put into the source address field. For multifield (MF) classification in an IP-based RAN, therefore, the first-hop router has to check the inner IPv4 header or IPv6 destination option header in order to identify the flow information of a mobile sender. It requires additional time or many modifications in router functions. In IP micromobility protocols, such as HAWAII or Cellular IP, the home address of an MH does not change within a domain even when the MH moves to a different network. Instead, the home address of an MH is replaced with the gateway node's address when packets are forwarded to the Mobile IP–enabled Internet. That is, the CoA of an MH known by a HA becomes the gateway node's address. Typically, when an MH moves from one DiffServ first-hop router to another within the same domain, no additional SLAs are required. For MF classification based on home IP address, however, the new first-hop router to which the MH connects must be able to identify the packets sent from the MH by its home address in order to provide the same service to the MH as the previous first-hop router. In addition, the new first-hop router must consult a BB in the same domain in order to get information about the service that the MH has received.

6.7.1.4 RAN Resource Provisioning

To provide an end-to-end QoS guarantee across an RAN, the air interface must be able to support the various kinds of traffic that an MH generates. Typically, the different types of traffic with different SLSs might be mapped into wireless QoS service classes in an RAN. The AF service class is well suited for an RAN because it gives assurances only on the delivered service. The service level of this class could be degraded in the case of dynamic resource availability in the RAN. The only difference between AF services in a fixed network and an RAN will be in the level of assurance. However, it may not be possible to allocate resources at every BS in support of an SLA that offers an EF service in an RAN because radio resources in heavily crowded areas may no longer be available after a handover. Instead, it is necessary to give a statistical guarantee on the handover success in an RAN. During a handover event, the resource-management scheme should be very fast and able to be invoked frequently and on demand. When a resource is reserved for AF and EF service classes at handover events an RAN, typically the wireless link becomes a main bottleneck. Therefore, the SLA in an RAN has to be negotiated based mainly on the available radio resource. It is also necessary to use some mechanism to limit usage of the radio resources, based on QoS service classes. This kind of mechanism would prevent network congestion or sudden degradation in the network performance. Moreover, priority can be given to a QoS service class based on the resource availability in the cell. If an MH that uses a high-priority service moves to a network in which sufficient resources are not available, the service of other MHs that use low-priority services in the same network may be downgraded.

6.7.2 DiffServ Proposals in IP Mobility Environments

Several studies have been conducted on DiffServ extensions under mobile environments. These schemes under IP mobility are the Internet Technologies Supporting Universal Mobile Operations (ITSUMO) architecture,[35] the MIR/HMIP/DiffServ architecture,[34] the SLS Transfer Mechanism for Mobile IP,[32] the Hierarchical QoS architecture,[26] and the Resource Management in DiffServ (RMD) framework.[36–38]

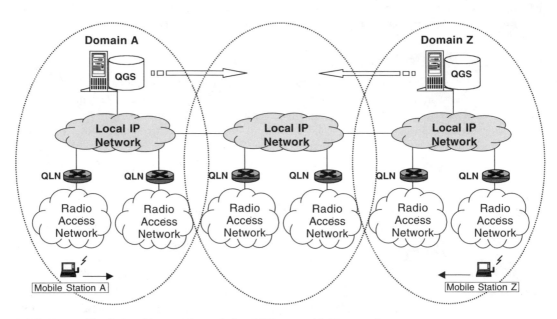

FIGURE 6.18   The QoS architecture for a wireless DiffServ-capable IP network.

6.7.2.1   The ITSUMO Architecture

Chen et al.[35] proposed a QoS architecture for wireless DiffServ-capable IP networks. This architecture shows that the DiffServ domain can be extended into the wireless mobile network by defining a set of QoS service classes and adding DiffServ-enabling network components: the QoS Global Server (QGS) and the QoS Local Node (QLN). Figure 6.18 shows the QoS architecture for a wireless DiffServ-capable IP network. QGS is added in the core network to provide QoS negotiation with the MN, whereas QLN is added near the RAN to act as a DS-domain ingress node that shapes and conditions traffic to and from the MN. The aim of this architecture and protocols is to provide an efficient and flexible way to support QoS in the mobile environment. The architecture separates the control and transport so that QGS handles the QoS signaling and QLN deals with the transporting of actual traffic. QLN provides the local information to the central QGS; thus, the QGS retains the global information of the domain. The merits of the architecture include flexibility, less QoS signaling message, integration with other protocols, ease of adjusting the reservation bandwidth, and provisioning of bandwidth in each local domain.

6.7.2.2   The MIR/HMIP/DiffServ Architecture

Le Grand et al.[34] proposed a Mobile IP Reservation Protocol (MIR), which is an extension of CLEP (Control Load Ethernet Protocol)[41] to mobile environments; it is intended to provide statistical QoS guarantees to MNs. MIR addresses the problem of bandwidth allocation and reservation in order to provide users of a shared medium with a guaranteed bandwidth. MIR is also integrated with HMIP and DiffServ in order to provide end-to-end QoS. In this architecture, there are two assumptions: Only the wireless cells need to run MIR, and MIR cannot be used all along the path for scalability reasons. In HMIP, an MN roams within a domain without updating the SLAs between the CN, the HA, and the visited domain. Figure 6.19 shows the visited domain with a GFA, and an HA with an HA, which can be described as follows: MIR operates on only the wireless part of the network between the FA and the MN. In addition, CLEP and HMIP are used within a domain, and DiffServ operates among domains. When an MN is roaming within a domain, the GFA with which it is registered is responsible for negotiating an SLA dynamically with the neighbor domains through a BB. When an MN is roaming within a visited network, it wishes to benefit from the same service it has in its HN.

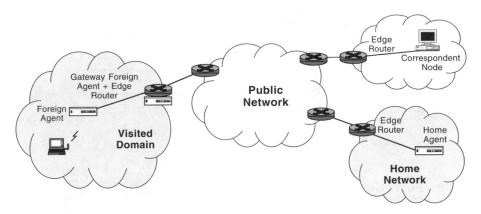

FIGURE 6.19 The visited domain with a GFA, and an HN with an HA.

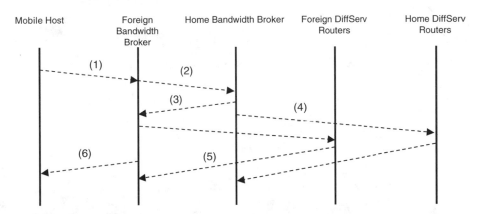

FIGURE 6.20 The SLS transfer to FN when an MH roams among different domains.

6.7.2.3 SLS Transfer Mechanism under Mobile IP

Stattenberger and Braun[32] proposed a QoS provisioning scheme for Mobile IP users using BBs. In this scheme, the SLS negotiation starts when an MN sends the desired SLS information to the BB in the home domain after registering at the HA in Mobile IP. The BB then sets up the routers in a way that best fits the current network topology according to the user's requirements. When an MN moves to a foreign domain, in order to transfer the MN's home SLS to the BB in the foreign domain, it signals the request message containing the MN's home IP address. The BB in the foreign domain then contacts the home domain's BB through the MN's home IP address to obtain the MN's SLS. The home BB then transmits the SLS to the foreign BB using the signaling message, and the foreign BB configures the routers in its network. The home BB then reconfigures the routers in the HN in order to release the resources occupied by the MN, and the foreign BB informs the MN about the success or failure of the SLS transfer. Failure can be caused by errors during the configuration or by unavailable bandwidth. Figure 6.20 shows the SLS transfer to the FN when an MH roams among different domains.

6.7.2.4 Hierarchical QoS Architecture

Garcia-Macias et al.[26] proposed a hierarchical QoS architecture for the DiffServ model on wireless local area networks (LANs). In Figure 6.21, each wireless cell is managed by an access route (AR) that forwards packets among MNs in a cell, and the AR connects it to an edge router. This architecture has two levels of management: intracell management and intercell management. The intracell QoS management is local to one cell and performed by the AR that manages fast-changing local situations. MNs inform the AR

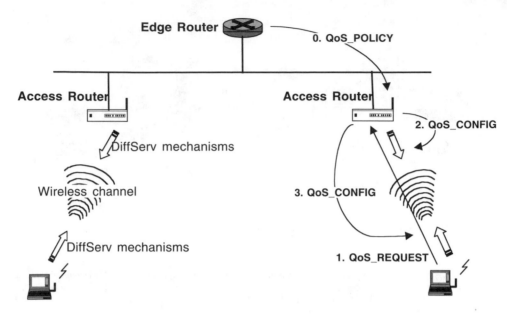

FIGURE 6.21 A hierarchical QoS architecture for wireless access networks on the DiffServ model.

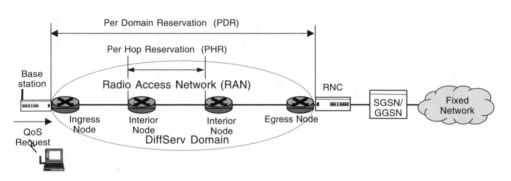

FIGURE 6.22 PDR and PHR protocols in the RAN.

of the required bandwidth, and the AR in turn configures their QoS mechanisms. The intercell QoS management concerns a set of wireless cells connected to an edge router. The edge router acts as a long-term global QoS manager for ARs managing cells. It sets policies, followed by ARs, such as admission control and resource reservation (QOS_POLICY). Bandwidth allocation follows the soft-state principle. The AR interprets requests for QoS allocation (QOS_REQUEST) and satisfies them if possible by appropriate configuration of DiffServ mechanisms (QOS_CONFIG). The QoS management module in the MN configures the output rate of the EF and AF/BE classes and fixes the proportion between the AF and BE classes.

6.7.2.5   The Resource Management in DiffServ (RMD) Framework

Heijenk et al.[36] and Westberg et al.[37,38] proposed the Resource Management in DiffServ (RMD) framework. This framework extends the DiffServ architecture with new reservation concepts and features of the IP-based RAN. It provides Per-Domain Reservation (PDR) and Per-Hop Reservation (PHR).[37] The PHR protocol extends DiffServ PHB with resource reservation and thus enables reservation of resources per PHB in each node within a DiffServ domain. The PDR protocol extends DiffServ per-domain behavior (PDB) with resource reservation, which describes the behavior experienced by a particular set of packets over a DiffServ domain. Figure 6.22 shows the PDR and PHR protocols in the RAN. RMD On Demand (RODA) PHR[38] is a unicast edge-to-edge protocol designed for a single DiffServ domain. It operates on a hop-by-hop basis on all nodes, both edge and interior, located in an edge-to-edge DiffServ domain. In the figure, the PDR protocol links the external "QoS request" and the PHR protocol. Once a QoS request arrives at the ingress node, the PDR protocol classifies it into an appropriate DSCP and calculates the associated resource unit for this QoS request (e.g., bandwidth parameter). The admission or rejection of "QoS requests" depends on the results of the PHR in the DiffServ domain.

6.8   Conclusions

In this chapter, we presented short overviews of the RSVP/IntServ and DiffServ modes and explained how to operate IntServ services over DiffServ networks. Next, we investigated the issues with existing RSVP in providing real-time services in wireless mobile networks and then identified several schemes for solving these issues. Finally, we described problems with the standard DiffServ model for IP-based access networks and then looked at schemes proposed for addressing these problems under IP mobility environments.

For additional information, the following Internet RFCs describe RSVP and RSVP features:

RFC 2205, Resource Reservation Protocol (RSVP), Version 1, Functional Specification
RFC 2209, Resource Reservation Protocol (RSVP), Version 1, Message Processing Rules
RFC 2210, The Use of RSVP with IETF Integrated Services
RFC 2211, Specification of the Controlled-Load Network Element Service
RFC 2215, General Characterization Parameters for Integrated Service Network Elements
RFC 2216, Network Element Service Specification Template
RFC 2747, RSVP Cryptographic Authentication
RFC 2961, RSVP Refresh Overhead Reduction Extensions
RFC 3209, RSVP-TE: Extensions to RSVP for LSP Tunnels

For more information, see the following Internet RFCs for descriptions of DiffServ features:

RFC 2474, Definition of the Differentiated Services Field (DS Field) in the IPv4 and IPv6 Headers
RFC 2475, An Architecture for Differentiated Services
RFC 2598, An Expedited Forwarding PHB
RFC 2597, Assured Forwarding PHB Group
RFC 2836, Per Hop Behavior Identification Codes
RFC 2983, Differentiated Services and Tunnels
RFC 3086, Definition of Differentiated Services Per Domain Behaviors and Rules for their Specification
RFC 3140, Per Hop Behavior Identification Codes
RFC 3246, An Expedited Forwarding PHB
RFC 3247, Supplemental Information for the New Definition of the EF PHB
RFC 3248, A Delay Bound alternative revision of RFC 2598
RFC 3260, New Terminology and Clarification for DiffServ
RFC 3289, Management Information Base for the Differentiated Services Architecture
RFC 3290, An Informal Management Model for DiffServ Routers

References

1. Perkins, C., Mobile IP specification, RFC 2002, Oct. 1996.
2. Campbell, Andrew, T., and Gomez-Castellanos, J., IP micromobility protocols, *ACM SIGMOBILE Mobile Computer and Communication Review (MC2R)*, 2001.
3. Braden, R., Zhang, L., Berson, S., Herzog, S., and Jamin, S., Resource ReSerVation Protocol (RSVP), Version 1 functional specification, RFC 2205.

4. Talukdar, A.K., Badrinath, B.R., and Acharya, A., MRSVP: a reservation protocol for an integrated services packet network with mobile hosts, Department of Computer Science Technical Report TR-337, Rutgers University.

5. Fodor, G., Person, F., and Williams, B., Proposal on new service parameters (wireless hints) in the controlled load integrated service, Internet Draft, draft-fodor-intserv-wireless-params-00.txt, July 2001.

6. Terzis, A., Srivastava, M., and Zhang, L., A simple QoS signaling protocol for mobile hosts in the integrated services Internet, InfoCom'99, pp.1011–1018, Vol. 3, 1999.

7. Chen, W.-T. and Huang, L.-C., RSVP mobility support: a signaling protocol for integrated services Internet with mobile hosts, InfoCom2000.

8. Fankhauser, G., Hadjiefthymiades, S., Nikaein, N., and Stacey, L., Interworking RSVP and Mobile IP Version 6 in wireless environments, 4th ACTS Mobile Communication Summit 1999, Sorrento, Italy, June 1999.

9. Bernet, Y., Ford, P., Yavatkar, R., Baker, F., Zhang, L., Speer, M., Braden, R., Davie, B., Wroclawski, J., and Flestaine, E., A framework for Integrated Services operation over DiffServ networks, RFC 2998, Nov. 2000.

10. Jain, R. and Burns, J., Bereschinsky, M., and Graff, C., Mobile IP with Location Registers (MIP-LR), Internet Draft, draft-jain-miplr-01.txt, July 2001.

11. Cisco Systems Inc., Cisco All-IP Mobile Wireless Network Reference Model Ver. 0.2, Apr. 18, 2000.

12. Yang, J. and Kriaras, I., Migration to All-IP based UMTS networks, 3G Mobile Communication Technologies Conference, pp. 19–23.

13. 3GPP working document, 1999, Architecture for an All IP network (feasibility study report), 3G TR23.922 version 0.1.4, Sep. 1999.

14. QoS in the All-IP Network, http://www.telecommagazine.com/issues/199910/tcs/qos.html.

15. Partain, D., Karagiannis, G., Wallentin, P., and Westberg, L., Resource reservation issues in cellular radio access networks, draft-westberg-rmd-cellular-issues-oo.txt, June 2001.

16. El Allali, H. and Heijenk, G., Resource management in IP-based radio access networks, *Proceedings CTIT Workshop on Mobile Communications*, Enschede, the Netherlands, ISBN 90-3651-546-7, Feb. 2001.

17. Kim, Y. and Dho, J.H., All IP standardization on IMT-2000 and beyond, *Telecommunications Review*, 11, 6, Nov. 2001, pp. 805–814.

18. Valko, A.G., Cellular IP: a new approach to Internet host mobility, *Computer Communication Review*, 29, 1, Jan. 1999, pp. 50–65.

19. Ramjee, R., La Porta, T., Thuel, S., Varadhan, K., and Wang, S.Y., HAWAII: a domain-based approach for supporting mobility in wide-area wireless networks, International Conference on Network Protocols, ICNP'99.

20. Das, S. et al., TeleMIP: telecommunication-enhanced Mobile IP architecture for fast intra domain mobility, *IEEE Personal Communications*, 7, 4, Aug. 2000, pp. 50–58.

21. Shen, Q., Lo, A., Seah, W., and Ko, C.C., On providing flow transparent mobility support for IPv6-based real-time services, Proc. MoMuC 2000, Tokyo, Oct. 2000.

22. Chiruvolu, G., Agrawal, A., and Vandenhoute, M., Mobility and QoS support for IPv6 based real-time wireless Internet traffic, 1999 IEEE International Conference on Communications, 1, 1999, pp. 334–338.

23. Zappala, D. and Kann, J., RSRR: a Routing Interface for RSVP, IETF Internet Draft, draft-ietf-rsvp-routing-02.txt, July 1998.

24. Rexhepi, A.V., Interoperability of Integrated Services and Differentiated Services architectures, Open Report, Revision A, Oct. 2000.

25. Mahadevan, I. and Sivalingam, K.M., An architecture and experimental results for quality of service in mobile networks using RSVP and CBQ, ACM/Baltzer Wireless Networks, July 1999.

26. Garcia-Macias, J.A., Rousseau, F., Berger-Sabbatel, G., Toumi, L. and Duda, A., Quality of service and mobility for the wireless Internet, Proceedings of the First Workshop on Wireless Mobile Internet 2001, Rome, Italy.

27. Blake, S., Black, D., Carlson, M., Davies, E., Wang, Z. and Weiss, W., An architecture for Differentiated Services, IETF RFC 2475, Dec. 1998.
28. Moon, B., and Aghvami, A.H., Reliable RSVP path reservation for multimedia communications under IP micromobility scenario, *IEEE Wireless Communications*, Oct. 2002.
29. Nichols, K., Jacobson, V., and Zhang, L., A two-bit Differentiated Services architecture for the Internet, Internet Draft, Nov. 1997.
30. Terzis, A., Ogawa, J., Tsui, S., Wang, L., and Zhang, L., A two-tier resource management model for the Internet, IEEE Global Internet'99, Dec. 1999.
31. Internet2 Qbone BB Advisory Council, A discussion of bandwidth broker requirements for Internet2 Qbone deployment, Aug. 1999.
32. Stattenberger, G. and Braun, T., QoS provisioning for Mobile IP users: applications and services in wireless networks, Hermes Science Publications, 2001.
33. Kilkki, K., Differentiated Services for the Internet, Macmillan Technical Publishing U.S.A., 1999.
34. Le Grand, G., Othman, J. Ben, and Horlait, E., Providing quality of service in mobile environments with MIR, ICON 2000, Sep. 2000, Singapore.
35. Chen, J.-C., McAuley, A., Caro, A., Baba, S., Ohba, Y., and Ramanathan, P., QoS architecture based on Differentiated Services for next generation wireless IP networks, draft-itsumo-wireless-diffserv-00.txt, July 2000.
36. Heijenk, G., Karagiannis, G., Rexhepi, A.V., and Westberg, L., DiffServ resource management in IP-based radio access networks, WPMC'01, 2001, pp.1469–1474, Aalborg, Denmark,.
37. Westberg, L., Jacobsson, M., Karagiannis, G., Oosthoek, S., Partain, D., Rexhepi, V., Szabo, R., and Wallentin, P., Resource Management in DiffServ (RMD) Framework, draft-westberg-rmd-framework-01.txt, Feb. 2002.
38. Westberg, L., Jacobsson, M., Karagiannis, G., De Kogel, M., Oosthoek, S., Partain, D., Rexhepi, V., and Wallentin, P., Resource Management in DiffServ On DemAnd (RODA) PHR, draft-westberg-rmd-od-phr-01.txt, Feb. 2002.
39. Cheng, Y. and Zhuang, W., DiffServ resource allocation for fast handoff in wireless mobile Internet, *IEEE Communications*, May 2002.
40. Kempf, J., Problem description: reasons for performing context transfers between nodes in an IP access network, RFC 3374, Sep. 2002.
41. Horlait, E. and Bouyer, M., CLEP (Controlled Load Ethernet Protocol): bandwidth management and reservation protocol for shared media, draft-horlait-clep-00.txt, July 1999.
42. Jacobson, V., Nichols, K., and Poduri, K., An expedited forwarding PHB, RFC 2598, 1999.
43. Heinanen, J., Baker, F., Weiss, W., and Wroclawski, J., Assured Forwarding PHB group, RFC 2597, 1999.
44. Tran, T.H. and Do, V.T., An analytical model for analysis of multipath routing scheme, Proc. SPECTS, San Diego, CA, July 2002.
45. Chen, S. and Nahrstedt, K., An overview of quality-of-service routing for the next generation high-speed networks: problems and solutions, *IEEE Network*, Nov./Dec. 1998.
46. Nelakuditi, S. and Zhang, Z.-L., On selection of paths for multipath routing, IWQoS'01, Karlsruhe, Germany, June 2001.
47. Su, X. and De Veciana, G., Dynamic multi-path routing: asymptotic approximation and simulations, 2001 ACM SIGMETRICS, Cambridge, Mass., 2001.
48. Retvari, G., Issues on QoS based routing in the Integrated Services Internet, EUNICE summer school, 13–15, Sep. 2000.
49. Aukia, P. and Qy, N., Quality of service based routing, Proceedings of the HUT Internetworking Seminar, May 1996.
50. Guerin, R., Orda, A., and Williams, D., QoS routing mechanisms and OSPF extensions, draft-qos-routing-ospf-00.txt, Nov. 1996.
51. Crawley, E., Nair, R., Rajagopalan, B., and Sandick, H., A Framework for QoS-based Routing in the Internet, RFC 2386, Aug. 1998.
52. Apostolopoulos, G., Guerin, R., Kamat, S., and Tripathi, S., Quality of service based routing: a performance perspective, Proceedings of SIGCOMM'98, Vancouver, British Columbia, Sep. 1998.

7

Multicast in Mobile IP Networks

Hong-Yon Lach
Motorola, Inc.

Christophe Janneteau
Motorola, Inc.

Imed Romdhani
Motorola, Inc.

Alexandru Petrescu
Motorola, Inc.

7.1 Multicast Concepts in the Internet

7.1.1 History and Motivation of IP Multicast

Multicasting is a good way to save bandwidth. Rather than sending data to a single receiver (unicasting) or to all the receivers on a given network (broadcasting), multicasting aims to deliver the data to a set of selected receivers. In IP multicast,[23] a single data packet is sent by the source. The network duplicates the packet as required until a copy of the packet reaches each one of the intended receivers. Thus, IP multicast avoids processing overhead associated with replication at the source and the bandwidth overhead due to sending duplicated packets on the same link. Sending multicast data is slightly different from sending unicast data. In fact, the traditional *Any Source Multicast (ASM)* model (RFC 1112)[23] defines a special IP destination address called *multicast address* to identify the group of interested receivers. Senders (multicast sources) send to the multicast address without prior knowledge of the multicast receivers.

The ASM model does not require senders to a group to be members of the group. To receive multicast traffic, an interested receiver needs a mechanism to join and to eventually leave the multicast group. The receiver notifies its local router that it is interested in a particular multicast group address; the receiver accomplishes this task by using a membership protocol like *Internet Group Management Protocol (IGMP)* for IPv4 hosts[9] or *Multicast Listener Discovery Protocol (MLD)* for IPv6 hosts.[22] To build a distribution tree from the senders to all the receivers, multicast capable routers need a multicast routing protocol to handle the duplication of multicast traffic and conveying of multicast packets across the built tree. Thus, we distinguish two components to handle multicast. The host part defines the IP multicast addressing, the membership protocols, and the network part that is typically concerned with multicast tree construction. The two previous components are radically different for the ASM model and the *Source Specific Model (SSM)*.[15,22] The SSM model has recently been proposed as an alternative to the currently deployed ASM model. The multicast address range of the SSM model is different from that of the ASM. In addition, SSM defines a new terminology for joining and leaving multicast groups and allows receivers to explicitly specify the source address. In fact, the pair (S, G), where S is a specific multicast source and G is a

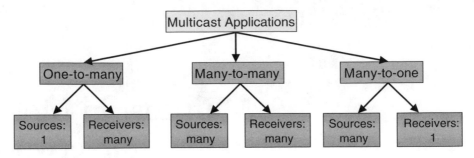

FIGURE 7.1 Classification of multicast applications.

TABLE 7.1 Sample of Multicast Applications and Related Software

Application	Software
Shared text editing	NTE (Network Text Editor)
Shared whiteboard	Wb (Whiteboard), wbd, etc.
Bourse	No free software
Software distribution	File servers, mirroring, etc.

multicast address from the SSM range, identifies a unique *channel*. This implies that receivers have to subscribe to this channel when they want to receive multicast data from the source S and unsubscribe where they leave the group. Despite these differences, the two models support both one-to-many and many-to-many multicast communication. The most serious criticism of the SSM is that it does not support shared trees, which may be useful for supporting many-to-many applications.

To deploy and experiment with the IP multicast in the Internet, the *Internet Engineering Task Force (IETF) Mbone Deployment working group* is the appropriate forum for coordinating the deployment, engineering, and operation of multicast routing protocols and procedures in the global Internet. Since 1992, the Mbone, which is a virtual overlay network that connects several multicast networks through IP tunnels and that allows multicast packets to travel through routers that are set up to handle only unicast traffic, provides an opportunity for network research groups throughout the world to experiment with IP multicast conferencing applications and IP multicast routing protocols. The experimentations can cover both IPv4 and IPv6 multicasting.

7.1.2 Multicast Applications

Within the context of multicast, various types of multicast communication can be differentiated, depending on the number of senders and receivers. Three general categories of multicast applications exist (see Figure 7.1):

- **One-to-many**: A single host sending to two or more receivers. Scheduled audio/video distribution, push media (news headlines, weather updates, sports scores, etc.), file distribution, and caching are the best samples of this kind of multicast application.
- **Many-to-many**: Any number of hosts sending to the same multicast group address, as well as receiving from it. For example, multimedia conferencing, shared distributed databases, distributed parallel processing, shared document editing, distance learning, chat groups, and multiplayer games are multicast applications in which each member may receive data from multiple senders while the member also sends data to all of them.
- **Many-to-one**: Any number of receivers sending data back to a (source) sender via unicast or multicast.

Table 7.1 describes some types of applications and the related public and free software.

TABLE 7.2 Multicast Applications Used on the MBone

Application	Version	Solaris	SunOS	Irix	Linux	FreeBSD	Windows	IPv6	Source
RAT	V3.0.35	Yes	Yes	Yes	Yes	Yes	Yes		Yes
RAT	V4.2.20	Yes			Yes	Yes	Yes	Yes	Yes
VIC	V2.8	Yes	Yes	Yes	Yes	Yes	Yes	Yes	Yes
WB	V1.6.x	Yes		Yes		Yes			
WBD	V1.x	Yes	Yes						Yes
NTE	V2.3	Yes	Yes		Yes	Yes	Yes	Yes	Yes
SDR	V3.0	Yes	Yes		Yes	Yes	Yes	Yes	Yes
SPAR	V1.2	Yes	Yes	Yes	Yes	Yes	Yes	Yes	Yes
ReLaTe	V2.1	Yes		Yes			Yes	Yes	Yes
MMCR-cli	V1.1	Yes		Yes	Yes	Yes	Yes		Yes
MMCR-srv	V1.1	Yes							
UTG-cli	V1.3	Yes		Yes	Yes	Yes	Yes		Yes
UTG-srv	V1.3	Yes							
TAG	V1.4	Yes	Yes	Yes	Yes	Yes	Yes	Yes	
RTP-QM	V1.0.x								Yes

Source: Modified from www.mice.cs.ucl.ac.uk.

1111 1111	00PT	Scope	Reserved	Plen	NetworkPrefix	Group Identifier

FIGURE 7.2 IPv6 multicast address format.

Table 7.2 highlights some multicast applications that are currently used on the Mbone conferencing (for more information, see http://www.mice.cs.ucl.ac.uk/multimedia/software/main.html). For each multicast application, we outline the current version and the different operating systems on which the application may be run.

7.1.3 Multicast Addressing and Routing

7.1.3.1 IPv6 Multicast Addressing

As we stated above, a multicast address is a topological identifier used to indicate a set of hosts to deliver IP datagrams to. The multicast address can be allocated from a specific range of addresses dedicated for sending to groups, and it has a specific scope, which limits the flooding of multicast packets. In this section, we detail the scoping and allocation of IPv6 multicast addresses. Different mechanisms have been suggested by IETF to describe how multicast addresses should be allocated. These mechanisms are required to avoid confusion and reduce the probability of IP multicast address collision in both the link-layer and the IP layer. IETF has defined guidelines that explain how to assign permanent IPv6 multicast addresses, allocate dynamic IPv6 multicast addresses, and define permanent IPv6 multicast group identifiers. These guidelines include a detailed description of the currently defined multicast address formats for IPv6; similar guideline exists for IPv4 networks. Currently, the IPv6 multicast address architecture[21] has been revised, and new extensions have been introduced that simplify the dynamic allocation of multicast addresses and embed IPv6 unicast prefixes in multicast addresses. By delegating multicast addresses at the same time as unicast prefixes, network operators will be able to identify their multicast addresses without needing to run an interdomain allocation protocol.[6] Figure 7.2 illustrates the current proposal for IPv6 multicast address formatting.

The first 8 bits are set to 1, and they identify the address as being a multicast address. The second field contains a P bit that indicates whether the multicast address is assigned based on the network prefix, where the T bit indicates a permanently assigned multicast if it is assigned to 1 and a transient multicast address otherwise. A permanent multicast address is independent of the scope value, whereas the transient one is meaningful only within a given scope. The value of the scope field (4 bits) is used to limit the

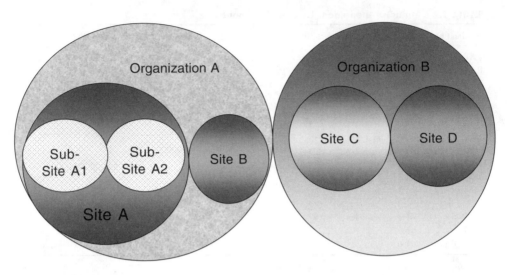

FIGURE 7.3 Multicast scopes.

scope of the multicast group and control the distribution of the multicast traffic. In addition to the reserved and unassigned scopes, we distinguish four scopes: node-local, link-local, site-local (IPv6 site-local addresses might be deprecated in the future), and organization-local scope (see Figure 7.3). The node-local scope is the minimal enclosing scope and hence is not further divisible. For instance, an organization may have multiple sites. Each site might have its own site-local scope zone, each of which would be an instance of the site-local scope. A reserved field (8 bits) follows the scope field. Then, the *Plen* field shows the number of bits in the network prefix field. The *network prefix* field identifies the network prefix of the unicast subnet owning the multicast address, and it should be initialized according to the multicast address allocation guides described in references 5 and 31. The network prefix portion of the multicast address will be at most 64 bits. Finally, the group identifier (32 bits) identifies the multicast group, either permanent (the global identifier of a particular multicast service) or transient, within the given scope. Thus, the total length of an IPv6 multicast address is 128 bits, whereas an IPv4 multicast address is 32-bits long.

7.1.3.2 Multicast Membership Protocol

To join or to leave a multicast group at any time, multicast receivers require a membership protocol. IETF has defined the MLD[22] protocol to be used by IPv6 routers to discover the presence of interested receivers of a given multicast group. Similarly, IETF has proposed IGMP[9] be used by IPv4 hosts and routers. IGMP and MLD have similar functionalities that we will explain by choosing the MLD protocol.

The current version of MLD (MLDv2) overcomes the limits of the previous one (MLDv1) and offers the opportunity for receivers to filter specific multicast sources. The MLDv2 specifies two kinds of messages. The first message is the *Multicast Listener Query*. This message is sent periodically by multicast routers to all the local hosts to learn for each of the multicast addresses whether they have listeners on that link. Source-specific queries add the capability for a multicast router to discover which multicast sources have listeners among the neighboring nodes. The multicast listener queries are used to build and refresh the multicast state of routers on attached links. In fact, a multicast router implementing MLDv2 keeps state per multicast address and per attached link. The multicast address state consists of a filter mode that may be *include* or *exclude*, a list of sources, and various timers. For each attached link, a multicast router records the listening state for that link, which consists of a set of records that indicate the IPv6 multicast address, the multicast address timer, the filter mode, and the source records. Each source record contains the IPv6 source address and the associated timer. As a response to the multicast listener query, multicast receivers respond to these queries by sending a *Multicast Listener Report (MLR)*. The MLRs are the second kind of message specified by MLD, and they are sent with an IPv6 link-local

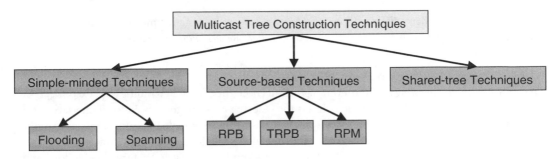

FIGURE 7.4 Multicast tree construction techniques.

source address, which allows other neighbors to delay their report messages. The multicast receiver may express interest in listening or not listening to traffic from particular sources by sending specific reports to explicitly notify the multicast router about the changes. At the same time, each multicast receiver maintains and computes multicast listening state for each of its interfaces. This state contains the IPv6 multicast address, the filter mode, and the source list. On the basis of the membership information, multicast routers start to join the multicast delivery tree of the specified multicast group by using the appropriate multicast routing protocol.

7.1.4 Multicast Routing Algorithms and Tree Construction

As mentioned in the previous section, multicast receivers are allowed the freedom of joining and leaving the multicast session at will. On the other hand, the multicast routing protocols are responsible for the construction of the multicast delivery tree. For this purpose, multicast routers need a form of multicast routing algorithms to distribute the membership information and to build the entire multicast delivery tree to forward multicast packets. The techniques of creating multicast trees can be classified into three different classes according to how the multicast routes span between multicast sources and multicast receivers. As shown in Figure 7.4, the three classes are *simple-minded techniques, source-based techniques,* and *shared-tree techniques*.[13] We discuss the strengths and weakness of each class in the following subsections.

7.1.4.1 Simple-Minded Techniques

The simple-minded techniques are based on the concept of flooding. They are simple, but they tend to waste the network bandwidth. We can classify these techniques in two subclasses: flooding and spanning.

7.1.4.1.1 Flooding

Flooding is the simplest technique. To forward multicast packets, a multicast router does not need to maintain a routing table. However, the multicast router has to determine whether it has seen a received multicast packet before. If this is not the case, the packet is flooded to all interfaces except the one on which it arrived. Otherwise, the packet is discarded. Consequently, this technique leads to a significant duplication of packets, and it needs that multicast router to maintain the track of the most recently seen packets. Thus, this technique requires a large amount of computational resources, and it uses all available paths across the internetwork instead of a restricted number.

7.1.4.1.2 Spanning

The spanning-tree technique is a more effective solution than flooding. Compared with the previous technique, this one selects a subset of the internetwork paths, which form a spanning tree. Only one active path connects two multicast routers, and multicast packets are replicated only when the tree branches diverge. This technique seems to be well suited for several cases, but it is difficult to compute a spanning tree.

7.1.4.2 Source-Based Techniques

The source-based techniques differ from the spanning tree in the way and the number of the multicast delivery trees it needs to build. These techniques build only one spanning tree for each potential multicast source. Thus, we can have different spanning trees rooted at each active multicast source. The main routing algorithms of such techniques are the so-called *Reverse-Path Forwarding (RPF)* algorithms, described below.

7.1.4.2.1 Reverse-Path Broadcasting (RPB)

The *reverse-path broadcasting (RPB)* routing algorithm checks first whether a multicast packet has arrived from the shortest path back to the source. If it is the case, the packet is forwarded to all interfaces of the multicast router except the incoming interface. Otherwise, it is discarded. To perform this operation, it is clear that every multicast router needs to maintain a global database for its entire routing domain. By using such a database, the network is used efficiently, and multicast packets are forwarded following the shortest path from the source to the interested destination. The main limit of the RPB algorithm is that it does not take into consideration multicast group membership when building the delivery tree for the source.

7.1.4.2.2 Truncated Reverse-Path Broadcasting (TRPB)

The *truncated reverse-path-broadcasting algorithm (TRPB)* uses membership information to build the multicast delivery tree. Thus, no multicast packets are sent to leaf subnetworks if there is no interested receiver. In fact, a multicast router truncates the delivery tree if a leaf subnetwork has no group members.

7.1.4.2.3 Reverse-Path Multicasting (RPM)

The *reverse-path multicasting algorithm (RPM)* is designed to overcome the limitations of the two above algorithms. This algorithm enables the creation of a multicast delivery tree that spans only a subnetwork with group members as well as routes to subnetworks along the shortest paths. By using the prune technique, a leaf multicast router can send a prune message with a time-to-live (TTL) equal to 1 to the upstream router to notify it that the subnetwork connected to it has no group members. Consequently, the prune message helps upstream routers build their prune states. A multicast router prunes a branch only when all the downstream routers have sent prune messages.

The main disadvantage of this algorithm is that multicast packets are periodically flooded across every router onto every leaf subnetwork until the *Prune State* is constructed.

7.1.4.3 Shared-Tree Techniques

Compared with the source-based techniques, the shared-tree techniques construct a single multicast delivery tree, not a source-based tree for each pair (multicast source, multicast group). Thus, all the multicast members have to explicitly join the shared tree. Then, the multicast traffic is sent in the same tree regardless of the sending multicast source.

No single multicast routing protocol incorporates all the tree construction algorithms described in this section. These routing protocols will be described next.

7.1.5 IP Multicast Routing Protocols

Multicast receivers are allowed the freedom of joining and leaving the multicast session at will, whereas the multicast routing protocols are responsible for the construction of the multicast delivery tree. For this purpose, multicast routers need a routing algorithm to distribute the membership information and to build the necessary multicast tree to forward multicast packets. In the next sections, we detail some intradomain and interdomain multicast routing protocols.

7.1.5.1 Intradomain Multicast Routing Protocols

Intradomain multicast routing protocols are required to build a multicast delivery tree within a given domain. Each administrative domain can decide which is the appropriate multicast routing protocol to be used.

7.1.5.1.1 Distance-Vector Multicast Routing Protocol (DVMRP)

The *Distance-Vector Multicast Routing Protocol (DVMRP)*[24] was the first protocol to be deployed in the Mbone. The Mbone is a virtual overlay network that connects several multicast network through IP tunnels. The DVMRP is based on the concepts of the *Routing Information Protocol (RIP)*, but it uses its own unicast routing protocol. DVMRP constructs two separated routing tables for unicast and multicast routings. The first version of this protocol specifies the construction of a source-based delivery tree using the RPF algorithm. However, DVMRPv1 is obsolete and was never largely deployed. DVMRPv2, the second version of DVMRP, is the current implementation used throughout the Mbone. The last version of DVMRP (DVMPR v3) defined by the IETF Inter-Domain Multicast Routing (IDMR) working group uses the RPM algorithm to construct the delivery tree and implements a mechanism to quickly graft back a previously pruned branch. A graft message cascades hop by hop back to the source until it reaches the nearest live branch point. In DVMRP, a router maintains for each interface a routing metric and a threshold. The metric is used for unicast routing as a distance while exchanging the distance vectors. A threshold value is used to limit the scope of the flood. When a router receives a packet, it propagates it only if the inbound interface is the one attached to the shortest path to the source (RPF) and if the lifetime (TTL [IPv4]/*Hop Limit* [IPv6]) of the received packet is greater than the threshold of the interface.

The DVMRP concept is simple, and the DVRMP router requires a modest processing capacity. However, DVMRP requires its own underlying unicast routing protocol. In addition, it needs to periodically flood multicast traffic to rebuild its trees, which affects its scalability when thousands of multicast groups share the same network infrastructure.

7.1.5.1.2 *Multicast Extensions to Open Shortest Path First (MOSPF)*

The *Multicast Extensions to Open Shortest Path First (MOSPF)*[16] protocol uses link-state databases to build multicast source-based trees and get an overview of the whole subnetwork according to the shortest path. When a subnet (or area) includes more than one router, one router must be selected as the *MOSPF-designated router* and another router as a *backup designated router*. Both the designated and the backup routers have responsibility for listening and monitoring host membership reports. The designated router performs the transmission of host membership queries solely. To build a multicast tree, the MOSPF-designated router first sends a *Link-State Advertisement (LSA)* message to all other routers. Once a router receives the LSA message, it adds the group membership information to its link-state database.

MOSPF is currently at "proposed standard" status. However, most members of the IETF IDMR working group doubt that MOSPF will scale to any large degree and therefore agree that MOSPF cannot be a standard for IP multicasting. With MOSPF, we get a faster convergence than DVMRP. In addition, this protocol adapts rapidly to changes in group membership or the availability of network resources. Moreover, MOSPF does not require the flooding packets to determine either group membership or network topology. As the tree building in MOSPF is based on the *Dijkstra algorithm*, the processing cost of this protocol to compute the shortest-path tree can become a limiting factor as the network gets larger. The other disadvantage is that MOSPF does not allow manipulating tunnels. In addition, it is conceived for use within *Autonomous Systems* only and so is not well suited for handling sparse-mode topologies.

7.1.5.1.3 *Protocol-Independent Multicast (PIM)*

The *Protocol-Independent Multicast (PIM)* is a multicast routing protocol that does not rely on any unicast routing protocol. It is thus capable of using any unicast protocol to perform multicast routing. In addition, PIM supports both sparse-mode and dense-mode topologies. If a network has many multicast members and there is enough bandwidth, dense-mode routing is used. However, if bandwidth is limited and multicast members are thinly distributed, sparse mode is used. In other words, there are two types of PIM protocol: *PIM-Dense Mode* and *PIM-Sparse Mode*.

7.1.5.2 Protocol-Independent Multicast Dense Mode (PIM-DM)

The *Protocol-Independent Multicast Dense Mode (PIM-DM)*[4] is similar to DVMRP in the way it floods multicasts out of all interfaces except the source interface. PIM-DM uses the RPM algorithm and prune messages. PIM-DM differs from PIM-SM in two essential ways. First, no periodic joins are transmitted, only explicitly triggered prunes and grafts. Second, there is no *Rendezvous Point (RP)*. In PIM-DM, when a source starts sending, multicast packets are flooded to all nodes of the network. However, PIM-DM uses RPF to prevent looping of multicast packets while flooding. As Figure 7.5 illustrates, when the multicast routers R2 and R3 receive multicast data from S that is addressed to a multicast group G and

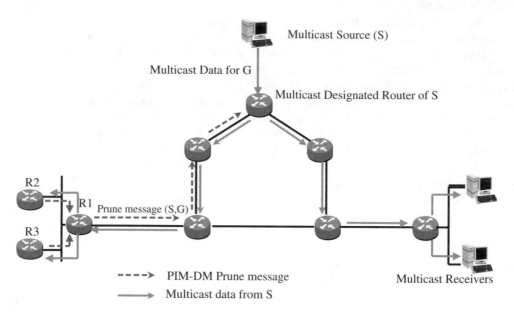

FIGURE 7.5 PIM-DM protocol signalling.

they have any interested receivers for this group, they instantiate a prune state associated to the (S, G) pair and they send a *PIM-DM Prune* message toward the designated router of S. The *Prune* state has a finite lifetime. When a new multicast receiver appears in a pruned area, the pruned branch is turned back into a forwarding branch. On the other hand, the problem of duplicated packets is eliminated when sending an *Assert* (S,G), which includes the distance between the sender and the source.

Actually, a new Internet draft[4] has been published by the *IETF PIM working group*, and this specification proposes to extend the PIM-DM protocol efficiency over large IP multicast–enabled networks. PIM-DM is simple and flexible because it does not depend on a specific topology discovery protocol often used by a unicast routing protocol. In addition, there are no RPs, and thus the risk of a single point of failure is eliminated. Unfortunately, the simplification offered by PIM-DM generates more overhead by causing flooding and pruning to occur on some links. Moreover, the flooding mechanism, especially in initial source transmission, makes PIM-DM not scalable for a wide coverage.

7.1.5.3 Protocol-Independent Multicast Sparse Mode (PIM-SM)

The *Protocol-Independent Multicast Sparse Mode (PIM-SM)*[7,12] is used for geographic distributions where flooding protocols cannot be deployed. Currently, much of the effort in IETF toward a working multicast protocol is focused on PIM-SM. This protocol is based on use of RPs, where receivers meet sources. In PIM-SM, a *Designated Router (DR)* is necessary for each multiaccess local area network (LAN). The DR is selected according to the highest IP address, and it has no responsibility for transmitting host membership query messages. The DR election process was recently updated in the latest PIM-SM specification of the IETF PIM working group.[7] In this specification, the DR election process is based on the highest priority when every PIM router has enabled the appropriate optional feature (e.g., priority). Otherwise, that is, if only one of the PIM neighbors does not exhibit that optional feature, the DR election process remains based on the highest IP address (as specified in PIM-SMv1). The PIM-SM protocol offers a switching mechanism to the source delivery tree technique (when needed), in order to optimize the propagation time and the concentration of traffic. In fact, a DR can be removed from the RP-tree by constructing a shortest-path tree to the source. Once it switches, it sends a prune message to the RP. Unlike PIM-DM, PIM-SM requires that multicast members explicitly join a delivery tree. The RPs are used by senders to announce their existence and by receivers to learn about new senders of a group. In PIM-SM, when a receiver wishes to join a given group G, it sends a report message to the DR. The DR creates a multicast forwarding entry for (*,G) (any source, G) pair and transmits a unicast *PIM-Join*

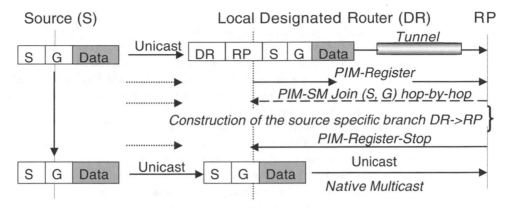

FIGURE 7.6 Multicast source registration on the PIM-SM protocol.

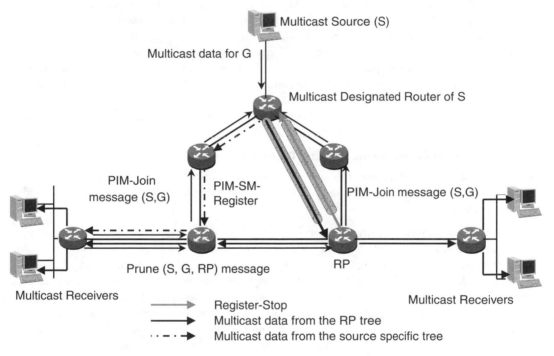

FIGURE 7.7 The PIM-SM protocol.

message to the RP. The RP does not acknowledge these join messages. The join messages are resent periodically as long as the receiver remains in the group. When all receivers on an LAN leave the group, the DR will send a *PIM (*,G) Prune* message to the RP for that multicast group. When the source starts sending data addressed to a multicast group, first the source's DR takes the data packets, unicast-encapsulates them to obtain *PIM-SM Register* packets, and sends them directly to the RP (see Figure 7.6). The RP receives these encapsulated data packets, decapsulates them, and forwards them to the receivers through the shared tree (see Figure 7.7).

PIM-SM is well suited for both intradomain and interdomain routing. This protocol offers the possibility for a DR to switch a group from the RP to a shortest-path tree once a session is started. More generally, the use of PIM-SM can reduce the traffic on Wide-Area Network (WAN) links. However, the dependence of PIM-SM on RP can lead to bottlenecks at the RP router. Because PIM is independent of any unicast routing protocol, it must be interoperable with unicast-dependent multicast protocols. That is not obvious to obtain today, as we must consider all features of each unicast-dependent multicast protocol.

7.1.5.3.1 Core-Based Tree Protocol (CBT)

The *Core-Based Tree Protocol (CBT)*[2,3] was designed in order to overcome problems related to dense-mode protocols, which are not well adapted to wide networks such as the Internet. The CBT protocol is suited for intra- and interdomain routing. This protocol builds and maintains a shared delivery tree for a multicast group. In fact, CBT constructs a single delivery tree per group, which is shared by all the group's senders and receivers. CBT can have a single router or a set of routers, which act as the *core* of a multicast delivery tree. This protocol may use a separate multicast routing table or that of underlying unicast routing. To create a CBT tree, when a host is interested in joining a multicast group, it first multicasts (locally) a membership report. Once a local CBT-aware router receives the report message, it will generate and send a *Join request* message to the core router. This join message is then acknowledged (*Join ack*) either by the core router or by an intermediate router that is on the path between the sending router and the core. Both the *Join request* and *Join ack* messages are sent hop by hop.

A new version of CBT (CBT v2) has been specified.[3] The most significant changes compared with version 1 are, first, the introduction of new LAN mechanisms, including the incorporation of a HELLO protocol. Second, each group-shared tree has only one active core router. The primary advantage of the shared-tree approach is that it typically offers more favorable scaling characteristics than all other multicast algorithms. More generally, because CBT does not incur source-specific state, it is particularly suited to many sender applications. This protocol makes more efficient use of network bandwidth; it does not require the multicast packets to be periodically forwarded to all routers in the Internet. Furthermore, the CBT protocol is relatively simple compared with most other multicast routing protocols, which can lead to enhanced performance. The weakness of the CBT is that multicast traffic is concentrated at the core of the tree. The core router maintains connectivity among all group members and can thus constitute a single point of failure.

7.1.5.4 Interdomain Multicast Routing Protocols

Interdomain multicast routing protocols attempt to construct a multicast tree among domains. Here we limit the scope of our discussion to the MSDP and BGMP protocols, which are the most popular interdomain multicast routing protocols.[18]

7.1.5.4.1 Border Gateway Multicast Protocol

The *Border Gateway Multicast Protocol (BGMP)*[14] is an interdomain multicast routing protocol. BGMP is associated with another protocol, *Multicast Address Set-Claim (MASC)*, which forms the basis of hierarchical address allocation architecture. BGMP runs on domain border routers and constructs a bidirectional shared tree that connects individual multicast trees built in each domain. In addition, BGMP optionally allows receiver domains to build source-specific, interdomain, distribution branches. The need to construct such branches arises when the shortest path from the current domain to the source domain does not coincide with the bidirectional shared tree from the domain. In other words, BGMP supports both SSM and ASM multicast models. Building on concepts from PIM-SM and CBT, BGMP requires that each multicast group be associated with a single root. However, in BGMP, the root is an entire domain rather than a single router. BGMP maintains group-prefix state in response to messages from BGMP peers and notification from the multicast routing protocol used for tree construction within a domain (*Multicast Interior Gateway Protocol, M-IGP*). In addition, BGMP routers build group-specific bidirectional forwarding state as they process the BGMP join. They optionally build source-specific unidirectional forwarding state.

7.1.5.4.2 Multicast Source Discovery Protocol (MSDP)

One of the initial proposals for interdomain multicast routing protocol consisted of using PIM-SM and PIM-DM. The PIM-DM is used as the intradomain routing protocol and the PIM-SM constructs shared tree that connects multiple source-specific trees maintained at every PIM-DM domain. This solution lacks scalability and consumes a large amount of control overhead due to RPs advertisement and the maintenance of soft-state entries in PIM-SM. The *Multicast Source Discovery Protocol (MSDP)* was

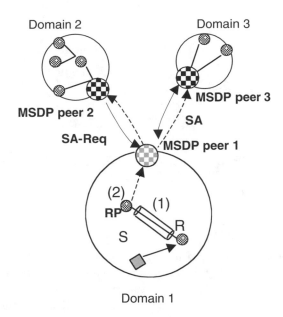

FIGURE 7.8 The MSDP protocol.

subsequently proposed to overcome these limitations. Actually, the MSDP protocol is the short-term IETF standard proposal. MSDP is applicable to shared-tree protocols such as PIM-SM and CBT, as well as other protocols that keep active source information at the borders. In the case of PIM-SM domains, each domain relies on its own RPs. As Figure 7.8 shows, MSDP works on the following way: When a source S goes active for a given group G in a PIM-SM domain (domain 1), the first-hop multicast router (R) informs immediately the RP of the group G by using a *PIM-SM Register* message. The RP informs its *MSDP Speaker* (typically a boundary RP of the PIM-SM domain, MSDP peer 1) of the existence of the active source in its local domain. This is accomplished via the *MSDP Source Active Message (SA)*. The *MSDP Speaker* informs the other MSDP peers in domain 2 and 3 (i.e., MSDP peer 2 and MSDP peer 3) by establishing Transmission Control Protocol (TCP) connections with them. The SA message contains the IP address of the source, the multicast group address, and the IP address of the RP that has sent the SA message and, optionally, the initial encapsulated multicast data from the source.

Each MSDP peer caches the SA messages in order to reduce the join latency. In fact, instead of awaiting the arrival of periodic SA messages, the MSDP peer scans its cache and can send immediately a PIM-SM (S, G) join to the connected domain where the source S is announced active. Thus, the SA messages are used to enable domains to discover multicast sources in other domains for internal multicast groups and to reduce the problem of locating active sources. The originating RP continue to send the SAs periodically as long as there is data being sent by the source S. MSDP peers 2 and 3 can request a list of active sources for a given multicast group by sending an *SA Request* message (SA-Req). In response, MSDP peer 1 responds to them by using an *SA Response* message.

Using MSDP, no shared tree is built across domains. Therefore, each domain can depend solely on its own RP. The main advantage of MSDP is that it is easy to implement. Robustness is achieved using RPF among MSDP peers before the forwarding data. However, the maintenance of interdomain state for every source that is not located in the same domain as the receivers is an important issue that can hardly be ignored.[18]

7.1.6 Multicast Transport Protocols

This section gives a brief overview of several higher-level protocols used with IP multicast, such as RTP, RTCP, RSTP, and RSVP (see Figure 7.9).

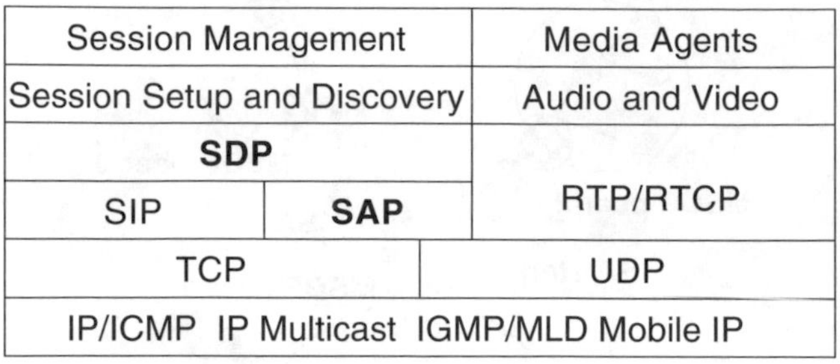

Session Management	Media Agents
Session Setup and Discovery	Audio and Video
SDP	RTP/RTCP
SIP	**SAP**	
TCP	UDP
IP/ICMP IP Multicast IGMP/MLD Mobile IP	

FIGURE 7.9 Stack of higher-layer protocols.

7.1.6.1 Real-Time Protocol (RTP)

The *Real-Time Protocol (RTP)* provides end-to-end delivery service that differs from the functionalities of TCP. The RTP services vary from the identification of the type of the payload to the sequence numbering and the time stamping. Thus, a receiver can choose which payload to accept or to reject and can change the coding to accommodate its bandwidth requirements. Using the time stamping, the receiver can reconstruct the timing originally produced by the source. The delivery in RTP is monitored by the RTCP protocol.

7.1.6.2 Real-Time Control Protocol (RTCP)

In an RTP session, every participant periodically transmits RTCP control packets to all other receivers. The main functionalities of the *Real-Time Control Protocol (RTCP)* are to provide information to applications, such as the identity of the RTP source; convey minimal session control information; and control the transmission interval.

7.1.6.3 Real-Time Streaming Protocol (RTSP)

The *Real-Time Streaming Protocol (RTSP)* is considered more of a framework than a protocol. RTSP was initially designed to work on top of the RTP protocol to both control and deliver real-time content. Moreover, RTSP is intended to control multiple data delivery sessions and provide a means for choosing delivery channels, such as User Datagram Protocol (UDP), TCP, IP multicast, and delivery mechanisms based on the RTP protocol described in the previous section.

7.1.6.4 Resource Reservation Protocol (RSVP)

The *Resource Reservation Protocol (RSVP)* is an Internet control protocol (like IGMP and ICMP), and it is used for *Quality of Service (QoS)* resources reservation and allocation. In the IP multicast context, a multicast receiver starts by sending its membership report and, after that, sends its resource-reservation message to reserve resources along the delivery path of the multicast group. The multicast routers between the source and the receivers should maintain QoS reservation states that are requested in only one direction (i.e., from receivers to the multicast source).

7.1.7 Multicast Session Protocols

7.1.7.1 Session Announcement Protocol (SAP)

The *Session Announcement Protocol (SAP)* is a protocol designed by the IETF (RFC 2974). This protocol defines the behavior of a session announcer and listeners. The session announcer periodically announces available multicast sessions by sending announcement packets to a well-known multicast addresses and ports without prior knowledge of the existing of multicast members. Interested multicast members listen on these addresses and ports to learn of all the sessions being announced. Once announced, a session

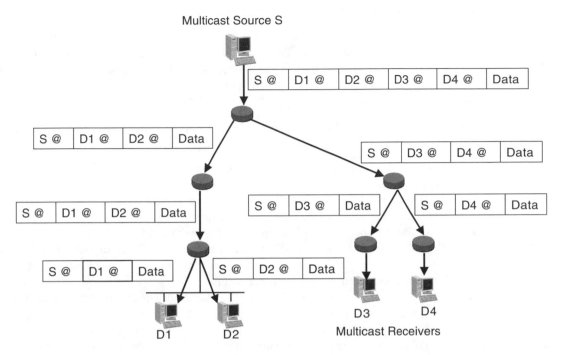

FIGURE 7.10 Explicit multicast.

may be deleted or modified by specific announcement packets from the announcer. The session announcement packet contains a session description that is to be conveyed to multicast members.

7.1.7.2 Session Description Protocol (SDP)

The *Session Description Protocol (SDP)* (RFC2327) defines the syntax to be used to describe a multimedia session with sufficient information for the participants in that session. The session description may be sent using any number of existing protocols, such as SAP or RTSP. A session description includes the following information:

- Session name and purpose
- Time the session is active
- The media comprising the session
- Information to receive those media (addresses, ports, format, and so on)
- Information about the bandwidth to be used by the multicast application
- Contact information for the manager responsible for the session

7.1.8 Explicit Multicast

The *Explicit Multicast (Xcast)*[19] is a new scheme to support IP multicast when groups are of limited size. This scheme requires that the source keeps track of the destinations and creates a packet header that contains a list of destination addresses. Consequently, there is no need for multicast group membership and multicast routing protocols. From the source to the destinations, the list of destination addresses is updated. In fact, the intermediate routers forward within their areas the packets with relevant IP addresses and remove the other addresses. This procedure is done hop by hop until all the receivers are reached (see Figure 7.10).

This approach is not scalable. However, it may be practical for a small and closed/preestablished multicast group.

7.2 Multicast for Mobile Nodes

This section introduces the basic concepts of the IP mobility. Because the global features of the IPv4 mobility[10] and the IPv6 mobility are similar, we illustrate these features by taking the IPv6 protocol as an example. Then, we describe how to support IP multicast in the Mobile IP environment (IPv4 and IPv6).

7.2.1 Terminology

Before describing IP multicast in the Mobile IPv6 environment, we start by defining the following Mobile IPv6 technical terms:

- **Mobile node (MN):** A host that can be moved by a nomadic user from one IP subnet to another without reconfiguration.
- **Home address (HoA):** A unicast routable address assigned to an MN within the mobile node's home link. The HoA is the permanent address of the MN.
- **Care-of address (CoA):** A unicast routable address associated with an MN while visiting a foreign IP subnet. The MN can form this address by using stateless (RFC 2462) or stateful address configuration (RFC 2461) including the mechanism of *Duplication Address Detection (DAD)*. The MN repeats this operation whenever the handover occurs.
- **Handover:** A change in an MN's point of attachment to the Internet such that it is no longer connected to the same IP subnet as it was previously. The handover can be detected by different facilities including the *Neighbor Discovery Protocol* (RFC 2461), signal strength, or signal quality information. The handover induces an IP address change and acquisition of a new topologically correct CoA when the MN moves from one IP subnet to another. The movement detection and the handover itself can be initiated by the MN or by the network. If the MN makes quality measurements of signal strengths and decides to switch to the best one with the help of the network layer, the handover is called *Mobile Initiated Handover*. In the second scheme, called *Network Initiated Handover*, the network makes the decision on the handover. Depending on the prediction and the timing of the handover, a small or large amount of data can be lost by the MN during this time. Once the handover is achieved and a new CoA is obtained, the MN registers its CoA with its *Home Agent*.
- **Home agent (HA):** A router on an MN's home IP subnet. While the mobile node is away from its home network, the HA is responsible for forwarding data addressed to the HoA of the MN. The HA must be able to maintain a data structure called *"Binding Cache"* containing an entry per MN to associate the permanent HoA with the transient CoA.
- **Correspondent node (CN):** A network host that has an active session with the MN.
- **Binding Update (BU):** The association of the HoA of a MN with its current CoA is called a *binding*. The binding has a fixed lifetime, and the MN updates and refreshes the binding when the lifetime expires thanks to *Binding Update* messages sent to the HA.

7.2.2 Background on IPv6 Mobility

To support IPv6 mobility, IETF has proposed the Mobile IPv6 protocol.[11] The main goal of Mobile IPv6 is to allow an MN to continue communicating with its CN while moving. The MN can move from its home link to a foreign one without changing its identity (IP *Home Address* HoA). Furthermore, the movement of the MN is transparent to higher-layer protocols and applications. A Mobile IPv6 node is identified by its IPv6 HoA (the same is true for Mobile IPv4). When the MN is away from its home link, it gets one or more CoAs from an *Access Router (AR)* located on the visited subnet. To maintain the transport and higher-level communications when moving, the MN maintains its HoA and registers one of the CoAs with its HA. The registered address is called the *primary CoA,* and the association between this address and the HoA is called a *binding*. To perform the registration of the primary CoA, the MN sends a packet containing a *binding update destination option,* which is an extension to the IPv6 destination option. The MN's HA uses proxy *Neighbor Discovery* to intercept any packet addressed to the MN's HoA

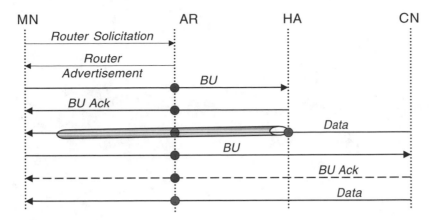

FIGURE 7.11 Mobile IPv6 signaling.

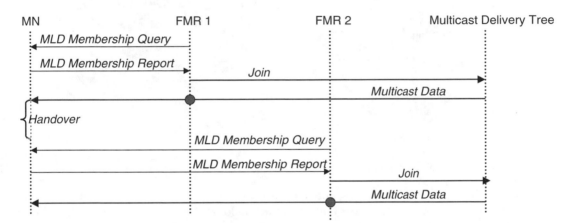

FIGURE 7.12 Messaging for remote subscription.

on its home link. Then, the HA encapsulates and sends the packet through an IPv6 tunnel to the primary CoA on the MN's visited link. As illustrated in Figure 7.11, when the MN receives a tunneled packet, it sends a BU to notify the CN of its primary CoA. The correspondent node caches and dynamically updates this binding and can request a new BU before the expiration of the existent one by sending a *Binding Update Request* to the MN. Similarly, the MN's HA learns and caches the binding whenever the MN's CoA changes.

7.2.3 Mobile Multicast with Remote Subscription

The remote subscription (RS) approach proposes that the MN joins the multicast group via a local multicast router on the foreign link being visited (see Figure 7.12).

In Figure 7.13, an MN visiting a foreign network (FN) (FN1) joins a multicast group G through the local multicast system — that is, via the local multicast router of the foreign network 1 (FMR1). In this approach, the MN does nothing specific; it just sends its MLD report messages onto the visited IPv6 subnet and performs any multicast related tasks through the local multicast system as any fixed host in FN1 would do. A *MLDv2 Report* message is sent with an IPv6 link-local source address of the MN. Upon receiving the MLD report, FMR1 will join the multicast group, which will result in the creation of a new branch in the multicast tree (see Figure 7.14).

Once the local multicast router has joined the multicast group, the multicast traffic from the source is forwarded down the multicast tree to FMR1, which forwards it to any node in FN1 that is interested in this traffic, including the MN.

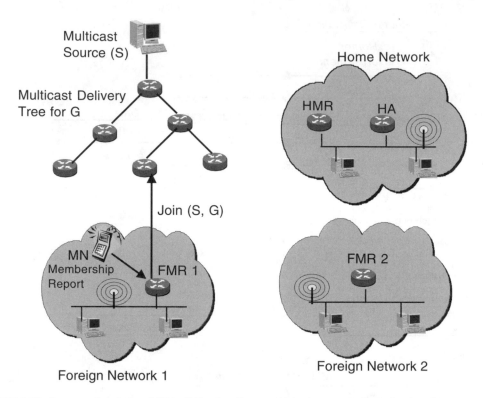

FIGURE 7.13   Remote subscription: MN in FN1 subscribes to multicast group G via the local multicast router.

When MN moves to a new IPv6 foreign network (FN2), it joins the multicast group through the new local multicast router FMR2 as any fixed host in FN2 would do (see Figure 7.15). Again FMR2 will join the multicast group, which will result in the creation of a new branch to FMR2 in the multicast tree. Before leaving FN1, the MN could leave the multicast group explicitly by issuing an *MLD Done* message. In any case, if the MN is the only receiver for this multicast group in FN1, the multicast branch to FMR1 will be pruned automatically after a short time because FMR1 will not receive anymore MLD report messages for this group.

Once the new multicast branch is established, the multicast traffic is delivered by FMR2 to the MN as usual (see Figure 7.16).

The above description focused on the example of a mobile multicast receiver. However, it is worth mentioning that the remote subscription approach is not applicable to handle movement of mobile *Source-Specific Multicast* sources. This limitation is described in the following section.

7.2.3.1   Advantages and Limitations

This approach eases optimal routing because the multicast delivery tree will be reconstructed to route multicast packets from the source to the receivers on the shortest possible path. The reconstruction of (a branch of) the multicast tree at each movement of the MN introduces additional latency due to the resubscription to the group and propagation of the subscription up to the tree. As such, the remote subscription may be highly vulnerable to frequent handovers that will introduce regular and possibly long interruptions of the multicast session. This is especially true in case the MN moves to a new location topologically far from the multicast tree, as the length of the new multicast branch to be created will be larger. This delay introduced by the resubscription to the multicast group is to be balanced with the delay of the Mobile IPv6 registration to the HA for the home subscription approach. Similarly, network resource consumption is also optimized because multiple MNs, in the same foreign link and subscribed to the

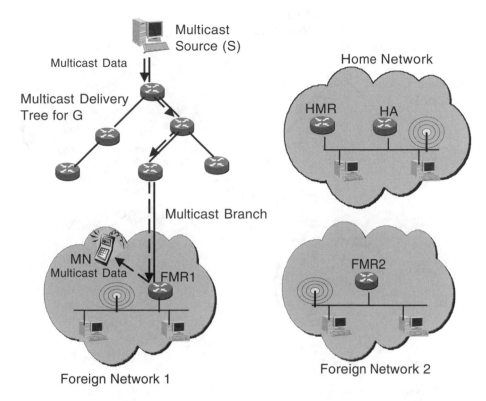

FIGURE 7.14 Remote subscription: MN in FN1 receives multicast traffic from the local multicast router.

same group, will not require duplication of packets as for the home subscription approach (described in the next section).

Another significant advantage of the remote subscription approach is its independence with respect to any IP unicast mobility protocol. In fact, the MN is not required to run Mobile IPv6, or any other IP unicast mobility protocol, to maintain its ongoing IP multicast sessions while moving among IPv6 subnets. In addition, remote subscription natively supports per-multicast flow handover. An MN equipped with several active network interfaces can join various multicast groups though different interfaces and move each multicast session from one interface to another independently from the other sessions. It just has to join this particular multicast group through the new network interface.

Last but not least, as mentioned earlier, the remote subscription approach is not suitable for a mobile SSM source. *Source-Specific Multicast (SSM)* refers to the ability of a receiver of a multicast group G to choose the source(s) of this group it wants to receive traffic from. In fact, the remote subscription approach requires the multicast source to use a topologically correct address — that is, a CoA — in the source field of the multicast packets it sends in order to pass the RPF check at each multicast router in the tree. As a consequence, a mobile SSM source should change the source address in the packets it sends when moving from one IPv6 subnet to another, each time using a topologically correct address (the new CoA). If this is not a problem from the point of view of the source, this has a major impact on the whole multicast delivery tree and receivers because the source IPv6 address of the multicast packet is used to identify the multicast source in traditional IPv6 multicast. As a consequence, each movement of an SSM source would result in a change of identity of the multicast source. This results in a loss of the multicast session for receivers because they are not aware of the change of IPv6 address of the source. In addition, if multicast receivers become aware of the change and want to pursue the session, they have to subscribe to this new source (i.e., the same source with the new topologically correct address, the new CoA). Consequently, a reconstruction of the multicast tree occurs. In addition, in the case of PIM-SM and CBT

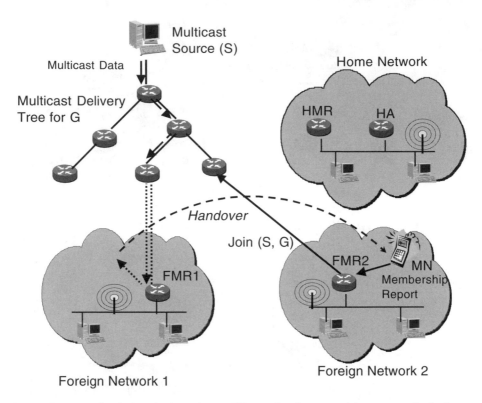

FIGURE 7.15 Remote subscription: MN moving to FN2 resubscribes to multicast group G via the new local multicast router.

protocols, the receiving core or RP of the multicast group will interpret the source as a new sender and will trigger the reconstruction of a new branch between the source and the core/RP.

7.2.4   Mobile Multicast with Home Subscription

The home subscription approach described in reference 15 proposes that an MN subscribe to a multicast group through its home network, via its Mobile IPv6 bidirectional tunnel to its HA. The MN tunnels its multicast group membership control packets to its HA, and the HA forwards multicast packets down the tunnel to the MN. This approach requires the HA either to be a multicast router or to proxy multicast group membership requests between the MN and the multicast router in the home network.

As illustrated in Figure 7.17, an MN moving into FN1 will first send a *BU* to its HA and establish its Mobile IPv6 bidirectional tunnel. The MN, away from home, may then subscribe to a multicast group as if it were at home by sending its MLD report messages through the bidirectional tunnel to its HA. If the HA is itself a multicast router, it will then join the multicast group (e.g., with *PIM_REGISTER* messages in case of PIM-SM), which will result in the creation of a new branch in the multicast tree for this group. Of course, if the multicast delivery tree is already covering the home network (because another host already had registered to the group), the HA does not need to explicitly join the tree again.

Once the HA has successfully joined the multicast group (see Figure 7.18), IPv6 packets from the multicast source are routed down the tree to the HA, which will, in turn, forward these packets to the MN through the Mobile IPv6 bidirectional tunnel. Note that if the HA functionality is co-located with the edge multicast router of the home network, this router may have to duplicate incoming multicast packets to serve both local receivers in the home networks and MNs away from home that have subscribed to the same multicast group. Similarly, the HA will have to duplicate received multicast packets to forward a copy to each MN that has subscribed to this group.

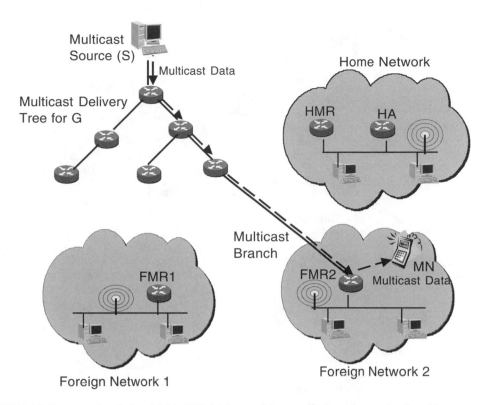

FIGURE 7.16   Remote subscription: MN in FN2 receives multicast traffic from the new local multicast router.

When the MN moves to a new IPv6 subnet FN2 (see Figure 7.19), it will send a new Mobile IPv6 *BU* message to its HA to inform it about its new location. As a result, the Mobile IPv6 bidirectional tunnel will be updated with the new CoA of the MN as a termination point, and incoming multicast traffic will continue to be delivered to the MN through this updated tunnel. It is worth mentioning that the overall handover delay (link-layer handover plus Mobile IPv6 movement detection and signaling) will result in a loss of multicast packets to be received by the MN.

A similar scenario is applicable for a mobile multicast source. Because a multicast source does not have to be a member of the multicast group it sends a packet to, the source will not have to send any MLD messages. The source has to send its multicast packets only to its home multicast system through the Mobile IPv6 bidirectional tunnel. The MN will always use its HoA as the source address of its multicast packets regardless of its current location. It is worth mentioning that the HA, receiving the multicast packet through the Mobile IPv6 tunnel, must undertake its own multicast RPF check on the multicast packet using the Mobile IPv6 *Binding Cache* instead of the unicast routing state before forwarding the packet to the multicast tree.

The above description uses an example in which the HA is a multicast router. However, there may be situations where this assumption is not true. In this case, the HA must implement a so-called *proxy MLD* functionality encompassing the following features:

- First, the HA must maintain a *multicast subscription table* for each Mobile IPv6 tunnel (i.e., for each MN) it handles.
- Second, the HA must maintain a global synthesis of these multicast subscriptions in order to join all the multicast groups that attached MNs would like to join.

7.2.4.1   Advantages and Limitations

The main advantage of the home subscription approach is its simplicity and the transparency of handover of MNs to multicast operation (because the MN does not need to resubscribe to the multicast group

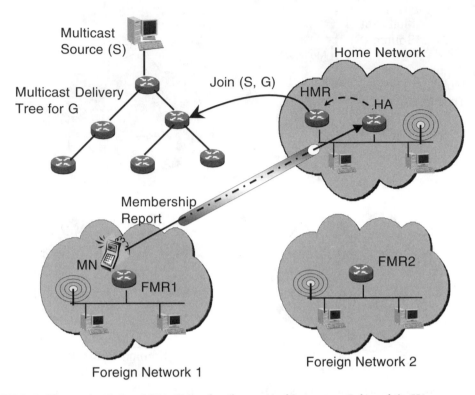

FIGURE 7.17 Home subscription: MN in FN1 subscribes to a multicast group G through its HA.

whenever it moves from one network to another). In particular, mobility of a multicast source will be completely hidden (except for possible interruptions of the session due to the handover delay) from the routers in the multicast tree and receivers because the HoA is always used as IPv6 source address of multicast packets regardless of the location of the MN. In addition to providing true transparency of node mobility to the multicast tree, this approach is quite simple to implement.

On the other hand, the home subscription approach introduces a nonoptimal routing of the multicast traffic (from mobile sources or to mobile receivers) due to the forwarding of the packets through the Mobile IPv6 tunnel. This worsens if the MN is far from its home network.

Another issue resides in the fact that the HA represents a single point of failure. If the HA crashes, all multicast sessions of MNs attached to this HA will be interrupted. Similarly, scalability may be a concern when a large number of multicast members expect to receive traffic from the same HA, overloading it with duplication of incoming multicast packets to be forwarded in each tunnel. In addition, the bidirectional tunnel breaks the multicast nature of the IP multicast traffic conveyed, which results in a nonoptimal use of the network resources. This is an important issue when a large number of MNs, registered to the same multicast group, receive the same data through different bidirectional unicast tunnels to the same HA. This is even more critical if many of these receivers are in the same FN. This is known as the *multicast avalanche* problem.[17]

Unlike remote subscription, home subscription is not very well suited to handle per-flow handover of multicast sessions. Per-flow handover is related to the ability of an MN equipped with multiple network interfaces to redirect a multicast session seamlessly from one interface to another while maintaining other ongoing multicast sessions on the same interface(s). This is because, as currently specified, Mobile IPv6 does not support this feature for unicast traffic. In fact, an MN is allowed to register one and only one primary CoA to its HA (per MN's HoA), whereas it may actually acquire several CoAs (one per interface). Several extensions to Mobile IPv6 have been proposed to solve this problem, but none of them has been integrated in the protocol specification yet.

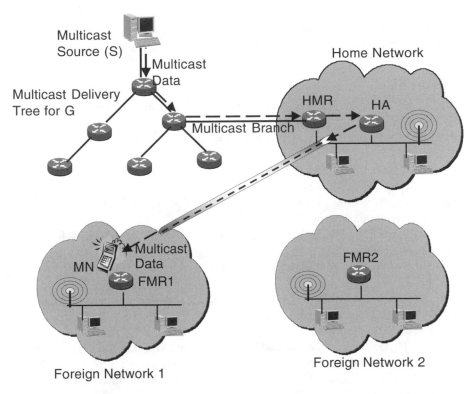

FIGURE 7.18 Home subscription: MN in FN1 receives multicast traffic through the tunnel to its HA.

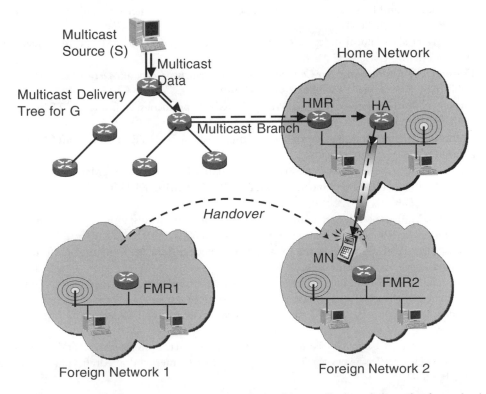

FIGURE 7.19 Home subscription: MN moving to FN2 receives multicast traffic through the updated tunnel to its HA.

Both the home and remote subscriptions constitute the basic IETF approaches that are largely used to develop new solutions that attempt to overcome the limits of these approaches and enhance the support of mobile multicasting. This research area is very active and attracts more and more researchers in the world.

References

1. Conta, A. and Deering, S., Generic packet tunneling in IPv6 specification, IETF Standards Track, RFC 2473, Dec. 1998.
2. Ballardie, A., Core Based Trees (CBT) multicast routing architecture, IETF Standards Track, RFC 2201, Sep. 1997.
3. Ballardie, A., Core Based Trees (CBT version 2) multicast routing protocol specification, IETF Standards Track, RFC 2189, Sep. 1997.
4. Adams, A. and Nicholas, J., Protocol Independent Multicast–Dense Mode (PIM-DM): Protocol Specification (revised), Internet draft, draft-ietf-pim-dm-new-v2-01.txt, Feb. 15, 2002.
5. Haberman, B. and Thaler, D., Unicast-prefix-based IPv6 multicast addresses, IETF Standards Track, RFC 3306, Aug. 2002.
6. Haberman, B., Allocation guidelines for IPv6 multicast addresses, IETF Standards Track, RFC 3307, Aug. 2002.
7. Fenner, B., Handley, M., Holbrook, H., and Kouvelas, I., Protocol Independent Multicast—Sparse Mode (PIM-SM): Protocol Specification (revised), IETF Internet draft, daft-ietf-pim-sm-v2-new-05.txt, 1 Mar. 2002.
8. Fenner, B., He, H., Haberman, B., and Sandick, H., IGMP/MLD-based multicast forwarding (IGMP/MLD proxying), IETF Internet draft, draft-ietf-magma-igmp-proxy-01.txt, Nov. 2002.
9. Cain, B., Deering, S., Fenner, B., Kouvelas, I., and Thyagarajan, A., Internet Group Management Protocol, Version 3, RFC 3376, Obsoletes: 2236, Oct. 2002.
10. Perkins, C., IPv4 mobility support, IETF Standards Track, RFC 2002, Oct. 1996.
11. Johnson, D.B. and Perkins, C., Mobility support in IPv6, IETF Internet draft, draft-ietf-mobileip-ipv6-20.txt, Jan. 20, 2003.
12. Estrin, D. et al., Protocol Independent Multicast-Sparse Mode (PIM-SM): protocol specification, IETF Standards Track, RFC 2362, June 1998.
13. Kosiur, D., *IP Multicasting: The Complete Guide to Interactive Corporate Networks*, Wiley Computer Publishing, New York, 1998, ISBN 0-471-24359-0.
14. Thaler, D., Border Gateway Multicast Protocol (BGMP): protocol specification, IETF Internet draft, draft-ietf-bgmp-spec-03.txt, June 2002.
15. Holbrook, H. and Cain, B., Source-specific multicast for IP, IETF Internet draft, draft-ietf-ssm-arch-00.txt, Nov. 21, 2001.
16. Moy, J., Multicast extensions to OSPF, IETF Standards Track, RFC 1584, Mar. 1994.
17. Lee, J., Multicast avalanche avoidance in Mobile IP (MAAMIP), IETF Internet draft, draft-lee-maa-mip-00.txt, Dec. 27, 2001.
18. Ramalho, M., Intra- and inter-domain multicast routing protocols: a survey and taxonomy, *IEEE Communications Surveys & Tutorials*, First Quarter 2000, 3, 1, 2000.
19. Ooms, D. et al., Explicit Multicast (Xcast) Basic Specification, IETF Internet draft, draft-ooms-xcast-basic-spec-03.txt, June 2002.
20. Wittmann, R. and Zitterbart, M., *Multicast Communication Protocols and Applications*, Morgan Kaufmann Publishers, San Francisco, 2001, ISBN 1-55860-645-9.
21. Hinden, R. and Deering, S., IP Version 6 addressing architecture, IETF Standards Track, RFC 2373, July 1998.
22. Vida, R., Costa, L., Zara, R., Fdida, S., Deering, S., Fenner, B., Kouvelas, I., and Haberman, B., Multicast Listener Discovery Version 2 (MLDv2) for IPv6, Internet draft, draft-vida-mld-v2-05.txt, Oct. 16, 2002.

23. Deering, S., Host extensions for IP multicasting, IETF Standards Track, RFC 1112, Aug. 1989.
24. Pusateri, T., Distance Vector Multicast Routing Protocol, IETF Internet draft, draft-ietf-idmr-dvmrp-v3-10.txt, Aug. 2000.

8

Secure Mobility in Wireless IP Networks

Pekka Nikander
Ericsson Research Nomadiclab

Jari Arkko
Ericsson Research Nomadiclab

8.1 Introduction

In this chapter, we will look at the mobility-related security aspects of Identity Protocol (IP) networks. The main part of the chapter consists of two detailed case studies. The first (Section 8.3) delves into the security aspects of Mobile IPv6 route optimization. The second (Section 8.4) introduces the Host Identity Protocol (HIP), which is an architectural approach to providing better security. These case studies are preceded by two introductory sections. The first discusses the roots of the mobility-related security issues in IP networks in a fairly general level, and the second discusses the security details. The chapter closes with a discussion of the lessons learned.

8.1.1 Background

The TCP/IP protocol suite was originally designed in the late 1970s and early 1980s, with singly homed stationary hosts in mind. As a consequence, it was decided to use IP addresses to identify hosts and to use IP addresses in upper-layer protocols (Transmission Control Protocol [TCP] and User Datagram Protocol [UDP]) to identify the endpoints of communication.[1,2] In those early days, the IP addresses served beautifully as identifiers for the hosts because hosts were rarely, if ever, moved.

The decision to use IP addresses directly in upper-layer protocols has had a number of long-term consequences. First, it has served as a weak form of security in the Internet. That is, when you send a packet to an IP address, the packet is either delivered to that particular IP address or lost. This property is a direct consequence of the very structure of the Internet: The primary purpose of all the routers and links forming the Internet is to deliver packets to their given destination IP addresses. Because the upper-layer protocols use the IP addresses to directly name connections, and because it is impossible to change those IP addresses as long as the connection is in use, it is fairly hard to hijack existing connections.

However, mobility support may change this situation, opening new threats that must be dealt with. (Note that it is not impossible to hijack existing connections even without mobility support; however, only attackers on the path between the communicating parties can do it easily. A more resourceful attacker may, for instance, be able to play tricks with the routing infrastructure, to fully or partially hijack connections even when it is not on the path. However, those kinds of attacks go well beyond the scope of this chapter.)

Second, the decision has inadvertently made both mobility and multihoming difficult. As we will discuss at greater length shortly, the current routing structure of the Internet requires that IP address allocation be location based. That is, a topological location in the Internet determines the range of IP addresses available at that particular location. Consequently, if a host changes its location or is available through several topological paths at the same time, it is assigned several different IP addresses. Because the upper-layer connections are bound to the unchangeable IP addresses, a change of IP address necessarily breaks existing connections.

As was discussed in Chapter 5, in order to solve the mobility problem it becomes necessary to be able to change the active IP address independently of the existing upper-layer connections. That, in turn, creates a security problem: how to be sure that it is really the communicating peer that has moved to a new IP address and not an attacker who just claims that the peer has moved. Although this appears to be a fairly simple and innocent problem, it is not. As we will see in Section 8.2, the potential attacks and their consequences are serious and would threaten the integrity and availability of all Internet communications.

The purpose of this chapter is to provide a good understanding of the range of security issues related to IP mobility. To do so, in the rest of this section we first discuss the nature of IP networking, including IP-based mobility and multihoming. Although this discussion repeats some material from Chapter 5, we are looking at the issue from a different point of view here, focusing on security. Hence, we urge even the impatient reader to consider reading this brief introduction for it outlines the background for what follows. Otherwise, it may be hard to fully comprehend some of the details.

After the introductory discussion, we will be ready to dive into the details of the security problems in Section 8.2. Once we are well armed with an understanding of the problems, we look at two different kinds of solutions: In Section 8.3 we discuss the security aspects of Mobile IPv6 route optimization,[3] and in Section 8.4 we introduce the HIP[4,5] and show how, by changing the architecture, it is possible to transform the nature of the security problems so that they can be solved trivially. Finally, in Section 8.5, we discuss the lessons learned.

8.1.2 Terminology

Before considering the nature of the IP infrastructure and its implications for mobility and multihoming, we will say a word about terminology. In the rest of this chapter we will mostly follow the terminology defined in Mobile IPv6.[3] A *mobile node (MN)* is a mobile device hosting applications — that is, an end host that can change its point of attachment. For the purposes of this chapter, a *mobile host* is synonymous with an MN. A *correspondent node (CN)* or *correspondent host* is a peer of a mobile host — that is, a host that communicates with the mobile host. A correspondent host may itself be mobile, or it may be stationary. A *home agent (HA)* is a stationary node that provides mobility-related services to one or more MNs. These services include, but are not limited to, packet forwarding. A *forwarding agent (FA)* is another type of stationary node that provides services to MNs. It is used in HIP to implement the HA functionality, but the term itself is slightly more general, as we will see.

In Mobile IP, a *home address (HoA)* is a permanent address assigned to a mobile host. An HA listens at the HoA whenever the mobile host is away from its home network. A *care-of address (CoA)*, on the other hand, is a temporary address assigned to the mobile host at its current location.

Some other abbreviations are listed in Table 8.1.

8.1.3 On The Nature of the Existing IP Infrastructure

To understand the roots of the security problems that arise with IP mobility, it is necessary to understand the design constraints behind the current IP infrastructure. In particular, it is necessary to understand

TABLE 8.1

Abbreviation	Explanation
AAA	Authentication, Authorization, and Accounting; an architecture and set of protocols defined by the Internet Engineering Task Force (IETF) AAA WG.
DHCP	Dynamic Host Configuration Protocol; a protocol used to assign an IP address and other information to hosts by a server.
ESP	Encapsulated Security Payload; an optional IP header used to protect the integrity, confidentiality, and authenticity of IP packets.
GPRS	GSM Packet Radio System; an extension to the GSM wireless phone network to carry packet data.
HI	Host Identifier; a host identifier proposed by the HIP architecture.
HIP	Host Identity Payload; a proposal by Robert Moskowitz and others to add a new layer above the IP layer but below the transport layer.
HIT	Host Identity Tag; a 128-bit representation of a host identifier, generated by taking a cryptographic hash of the HI.
IPSEC	IP Security; a suite of protocols used to secure IP packets (see also AH and ESP).
IRTF	Internet Research Task Force; the research branch of the IETF.
MAP	Mobile Anchor Point; a network node used in LMM. *LMM agent* in an alternative name for MAP.
NAT	Network Address Translation; a practice of the network changing IP addresses on the fly.
NSRG	Name Space Research Group; an IRTF research group.
REA	Readdressing packet; used in HIP to implement mobility.
RR	Return Routability; a security mechanism used in Mobile IPv6.
RSIP	Realm Specific IP; an alternative for NAT.
SCTP	Stream Control Transport Protocol; a new transport layer protocol, somewhat similar to TCP and UDP.
TLI	Transport Layer Identifier.
WLAN	Wireless Local Area Network.

how the routing infrastructure is constrained by the necessity of keeping the IP address structure hierarchical and how the routers must trust each other.

As we will see, Mobile IP and HIP take fairly different approaches to solving the security problems. They both live within the restrictions of the current IP infrastructure; neither requires any changes to the IP protocols themselves, nor to existing routers. The levels of changes that they require at the end hosts are fairly different, however. One of the original design goals in Mobile IP was to make mobility possible without changing too much.[6] In particular, Mobile IP was designed to provide mobility support even when the correspondent host was not at all aware of Mobile IP. This was especially important for IPv4, with its large installed base. The same design goals were more or less implicitly inherited by Mobile IPv6. This must be contrasted to the approach taken in HIP, where the protocol stack is changed in all endpoints, both in mobile hosts and in correspondent hosts.

To understand the security aspects, it is important to understand the Mobile IPv6 and HIP views regarding the base IP protocol and infrastructure. The most important assumptions are as follows:

1. In both Mobile IPv6 and HIP, the IP addresses available to a host are determined by the current topological location of the host, and therefore the host must change its IP address as its moves.
2. In Mobile IPv6 route optimization, the routing infrastructure is assumed to be secure and well functioning, delivering packets to their intended destinations as identified by the destination address.
3. In both Mobile IPv6 and HIP, it is expected that the upper-layer protocols (TCP, UDP, etc.) will name their connections using stable identifiers that are independent of the current, location-dependent IP addresses.

Although these assumptions may appear trivial, let us explore them a little further. Firstly, in current operational practice, IP addresses are distributed in a *hierarchical* manner. This limits the number of routing table entries that each single router needs to handle. An important implication is that the topology determines what (globally routable) IP addresses are available at a given location. That is, a node cannot freely decide what IP address to use; rather, it must rely on the IP address(es) received from a Dynamic Host Configuration Protocol (DHCP) server or, in IPv6, formed on the basis of routing prefixes served by the local routers. In other words, from this point of view IP addresses are just what their name says: *addresses*, or locators (i.e., names or labels of locations).

Second, in the current Internet structure, the routers collectively maintain a distributed database of the network topology and forward each packet to the location determined by the destination address carried in the packet. To maintain the topology information, the routers *must* trust each other, at least to an extent.[7] The routers learn the topology information from other routers, and they have no option but to trust their neighbor routers regarding distant topology. (For a different approach, see reference 10.) At the borders of administrative domains, *policy rules* are used to limit the amount of untrusted routing table information received from peer domains. Although this is mostly used to weed out administrative mistakes, it also helps with security. The aim is to maintain a reasonably accurate idea of the network topology even if someone feeds faulty information to the routing system. This system is far from perfect, but it works reasonably well, and most of the current Internet communications infrastructure relies on it to provide baseline integrity.

In the current Mobile IPv6 design, it is explicitly assumed that the routers and the policy rules are configured in a reasonable way and that the resulting routing infrastructure is trustworthy enough. That is, it is assumed that the routing system maintains an accurate idea of the network topology and that it is therefore able to route packets to their destination locations, if possible. If this assumption is broken, the Internet is broken in the sense that packets go to wrong locations. Under such circumstances it does not matter how hard the mechanisms above try to make sure that, for example, problems packets are not delivered to wrong addresses because of Mobile IP security. HIP, on the other hand, takes a different approach. Although it relies on the routing system, it uses public-key cryptography to establish the security baseline. Thus, it is more resilient to attackers that are able to infiltrate the routing infrastructure and alter routing information.

Finally, these IP-layer assumptions must be contrasted with the assumptions made by the upper layers. Whereas the routing system assumes that IP addresses are used as names of locations, the upper layers in the TCP/IP stack assume that IP addresses can be used as names of hosts. That is, the basic structure of TCP, UDP, Steam Control Transmission Protocol (SCTP),[8] and Datagram Congestion Control Protocol (DCCP)[9] assumes that an IP address statically names a host — that is, a device capable of hosting processes — and that the address is stable at least as long as there is any ongoing communication between the two hosts. In the face of mobility, this is clearly incompatible with the routing system assumptions.

8.1.4 Mobility

With this basic understanding of the limitations of the current IP routing system and the assumptions made by the upper-layer protocols, we can now proceed to look at the situation from a mobility point of view. As discussed earlier in Chapter 5, there are two fundamental approaches to solving the IP mobility problem:[6] packet forwarding and bindings updates. *Packet forwarding* suffers from an intrinsic performance penalty: The hosts cannot use optimal routes. *Binding updates (BUs)* provide for optimal routing but suffer from a number of security problems.[6] To start with, unsecured BUs would allow various man-in-the-middle, masquerade, and denial-of-service attacks (see Section 8.2). Thus, one must somehow secure BUs, which, in turn, requires creating some kind of *authorization* relationship between the parties involved. That is, the CN must know, for sure, that the MN is authorized to change the bindings.

In the general case, the parties involved in mobility do not know each other and do not trust each other. Therefore, such an authorization relationship cannot be based on any existing relationship between them. In theory, it would be possible to create a global authorization infrastructure that would allow any two parties to create mobility-related authorization relationships. However, such an infrastructure does not exist, and creating one would be extremely difficult in practice.

8.1.5 Multihoming

It is now time to consider the relationship between end-host mobility and end-host *multihoming*. Multihoming refers to a situation where an end host has several parallel communication paths that it can use.

A mobile host uses several IP addresses, one after another. A multihomed host, on the other hand, also uses several IP addresses but all at the same time. Thus, in a sense, mobility and multihoming can be considered semantic duals of each other. Furthermore, many mobility-related security problems also occur in multihoming, at least if one wants to allow a multihomed host to use its parallel IP addresses in an interchangeable way.

Today, most multihomed end hosts are servers, but, for example, a future mobile terminal with both General Packet Radio Service (GPRS) and wireless local area network (WLAN) connections would also be a multihomed end host. Sometimes certain kinds of multihoming are called multipath or multiaccess; in this chapter we do not make any distinction between these and just use the term *multihoming*.

8.1.6   Identifiers and Locators

To really understand why the current Internet architecture makes end-host mobility unnecessarily hard, we have to restate the facts about the structure of the TCP/IP stack. Today, there are exactly two name spaces relevant to mobility. First, we have network layer *addresses*. As stated earlier, these addresses are determined by the network topology. Second, we have so-called *Transport Layer Identifiers* (*TLIs*; e.g., ports in TCP and UDP). Of these, the network-layer addresses usually have a global scope (or are translated into global scope addresses, if NAT is used), and the TLIs are unique within the scope of a single IP address or an IP address pair. Because we have only these two name spaces, the communicating processes must be named with (pairs of) <IP address, TLI> 2-tuples, effectively binding the connection names to topological locations in the network. This situation makes it difficult to give unique names to processes that are mobile, for example, hosted on an MN. That is, because the IP addresses must change due to mobility, the names of the end points must also change.

Using the terminology introduced by J. Noel Chiappa,[11] we can say that in the current architecture an IP address functions in the dual role of being simultaneously an *end-point identifier (EID)* and a *locator*. That is, for the upper-layer protocols, an IP address identifies the node that is hosting the peer end point. On the other hand, at the IP layer, an IP address names a location and acts as a routing destination name.[12]

8.2   Security Problems

As we already briefly mentioned, a number of security problems are associated with IP-based mobility and multihoming. These problems stem from the need to assign several IP addresses to a single host and from the desire to use these addresses in an interchangeable way. That is, in an ideal mobility and multihoming solution, a correspondent host can use any of the mobile or multihomed host's addresses and change to a different address on the fly as needed. The problem is how to make sure that the IP addresses really belong to the right host and that any changes in the address set are properly authorized.

In this section, we take a detailed look at the security problems. First, we briefly introduce the two basic kinds of attacks: address stealing and address flooding. Next, we discuss what constraints exist on how an attacker can attack a particular target at a particular time and location and how mobility changes these constraints. After these preliminaries, we discuss the various identified attacks and threats in detail.

8.2.1   Address Stealing and Flooding

While the basic security problem can be described as an authorization problem, the consequences of various attacks can be described as resulting in *address stealing* or *address flooding*. In an address-stealing attack, a malicious node claims to "own" an address that some other node is currently using, with the intention of launching a masquerade, man-in-the-middle, or denial-of-service (DoS) attack against the real "owner" of the given address. In an address-flooding attack, a malicious node (or group of nodes) causes a large number of other nodes to believe that the malicious node has become (better) available at a target address (although in reality it most probably is not at the target address at all), causing the other nodes to flood the target address with unwanted traffic.

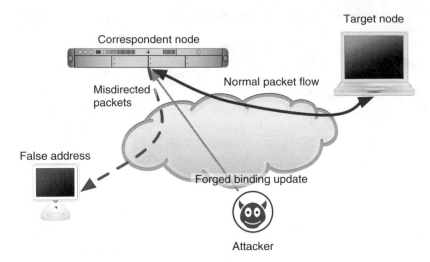

FIGURE 8.1 Address stealing attack.

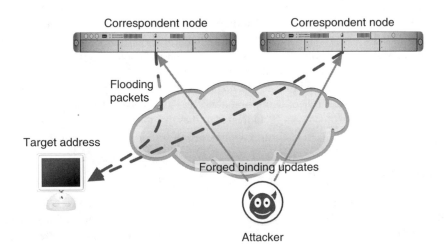

FIGURE 8.2 Flooding attack.

It is easy to see how an attacker can launch the first attack. It simply sends a BU to the CN(s) of the target node, claiming that the target node has moved to a new address. Unless the recipient of this BU is able to securely verify that the sender of the update is indeed the target node, all the future traffic destined to the target node is directed elsewhere, resulting in a masquerade, DoS, or man-in-the-middle situation, depending on the other actions and the actual location of the attacker. The basic form of attack is illustrated in Figure 8.1.

A flooding attack results from failing to check that the sender of an address update is indeed reachable at the claimed new address. If a recipient of a BU blindly starts to send messages to the new address, the messages may be delivered to an innocent third party that now receives excess traffic. Although one node sending bogus packets may not be that bad, hundreds or thousands of nodes sending extra traffic at the same time are likely to fill the communication link causing a DoS. See Figure 8.2 for an illustration of the basic form of these kinds of attacks.

Although these security problems can be solved to an extent with the so-called Return Routability (RR) procedure (see Section 8.3.2.1), it is hard to completely solve the problems within the current architecture. There is no way of checking that a node claiming to be at a given address is actually the node who is indeed located at the address, topology wise. That is, when Bob is communicating with someone who Bob thinks is Alice, at an address A, Bob cannot be sure that the someone is actually Alice and not an attacker node at some topologically intermediate point between Alice and Bob, or at a point close to Alice or Bob, and able to hear all the messages that Bob sends to Alice. Furthermore, if Bob wants to talk to Alice, supposedly at A, and ends up talking to Carol at some intermediate point, Carol can now tell Bob that she has moved to a new address D. Bob has no other option but to believe what Carol claims and start diverting traffic destined to A to the new destination D.

Within the scope of the current architecture, the only really secure solution would be to provide an authorization infrastructure providing credentials that bind addresses to cryptographic keys, thereby creating the possibility of binding nodes and addresses in a stronger sense. However, such an infrastructure does not exist, and it is very unlikely that anything resembling it could be created any time soon.[13]

8.2.2 Dimensions of Danger

On the basis of the discussion of Section 8.1, it should now be clear that the dangers in IP mobility lie in creation (or deletion) of bindings between endpoint identifiers and locators. In Mobile IP, both the end-point identifiers and locators are IP addresses, whereas in HIP the end-point identifiers and locators live in different name spaces. In Mobile IPv6 route optimization terms, the danger is in the possibility of unauthorized creation of *binding cache entries (BCEs)*. In HIP, the binding between host identifiers and IP addresses always exists, and it does not have any specific name.

To better understand the attacks, we have to be able to consider their effects and limitations. The effects differ depending on the *target of the attack*, *the timing of the attack*, and *the location of the attacker*. Correspondingly, the architecture imposes limitations on an attacker's range of targets and the timing and location of attacks.

8.2.2.1 Target

Let us first consider the range of targets and ways they can be attacked. Basically, the target of an attack could be any node or (sub)network in the Internet, stationary or mobile. The basic differences in the types of attacks lie in the nature of the attack goals: Does the attacker aim to *divert* (steal) the traffic destined for or sent by the target node, or does it aim to cause DoS to the target node or network?

Whether the target is an MN or a stationary one is unimportant because the actual target node may not play an active role in the attack scenario at all. As an example, consider a case where an attacker targets Alice, at the address A, by contacting a large number of other nodes, claiming itself to be Alice, and diverting traffic at these other nodes to the real Alice. Alice herself need not be involved at all before she starts to receive the deluge of bogus traffic. Note that Alice does not need to be mobile at all; she may well be stationary or she may not exist at all if the attacker is targeting a certain subnetwork instead of a single node.

To better understand the range of potential targets, let us consider the nature of IP addresses. As stated before, Mobile IP uses the same class of identifiers, IP addresses, as HoAs, CoAs, and stationary node addresses. Thus, it is impossible to distinguish an MN's HoA from a stationary node's address. Consequently, an attacker can take advantage of this by taking any IP address and using it in a context where normally only an HoA or a CoA would appear. This means that certain attacks that otherwise would only concern mobiles are, in fact, threats to all IP nodes.

In fact, the role of an MN appears to be the safest one because in that role the node does not need to maintain state about the whereabouts of remote nodes. Conversely, the role of a CN appears to be the weakest point. It is much easier for an attacker to fool a CN into believing that an MN is somewhere where it is not than to fool an MN into believing something similar. On the other hand, because it is possible to attack any node by using its peers as attack weapons, all nodes are equally vulnerable in some sense. Furthermore, an MN often plays the role of being a CN, too because it may talk to other MNs. Thus, in practice, all nodes and networks in the Internet are roughly equally vulnerable to attacks related to IP mobility, regardless of whether they implement any particular IP mobility mechanism themselves.

8.2.2.2 Timing

An important aspect of understanding IP mobility–related dangers is timing. In a stationary IP network, an attacker must be between the communicating nodes at the same time that the nodes communicate. With IP-level mobility, a new danger arises. In Mobile IPv6, for example, an attacker could attach itself between the HA and a CN for a while, create a BCE at the CN, leave that position, and then continue to send bogus BUs to the CN. This would cause the CN to send packets destined to the MN to an incorrect address as long as the BCE remained valid. The opposite would also be possible: An attacker could also launch an attack by first creating a BCE and then letting it *expire* at a carefully selected time. If a large number of active BCEs carrying large amounts of traffic expired at the same time, the result might be an overload at the HA or home network.

8.2.2.3 Location

In a static IPv4 Internet, an attacker can receive packets destined for a given address only if it is able to attach itself on the path between the sender and the recipient. Alternatively, if the attacker is able to control a node on the path, or change the path by attacking the routing system, it may also be able to launch an attack. However, such attacks are fairly hard to perform.

Although it is relatively hard to eavesdrop on packets (unless the attacker happens to be at a suitable location), an attacker can easily send spoofed IP packets from almost anywhere in the Internet. Hence, if an IP mobility solution allowed sending unprotected BUs, an attacker could create bogus bindings on any CN from anywhere in the Internet simply by sending a fraudulent BU to the CN.

In summary, by introducing the binding structure at the CNs, any BU-based IP mobility protocol introduces the dangers of time and space shifting. Without proper protection, the mere existence of such a protocol in any larger number of nodes would allow an attacker to act from anywhere in the Internet, possibly well before the time of the actual attack. In contrast, in the static Internet the attacker must be present at the time of the attack, and it must be positioned in a suitable way, or the attack would not be possible in the first place. Moreover, it must be understood that the potential targets of some of the attacks are all nodes and (sub)networks in the Internet and not just the mobile or CNs that implement the mobility mechanism.

8.2.3 Attacks and Threats

In this section, we describe a representative set of known attacks against BU-based IP mobility protocols, mainly Mobile IPv6 route optimization. Although many of these attacks are also possible in HIP (see Section 8.4.4), this section looks at the situation mostly from a Mobile IPv6 perspective. In Mobile IPv6 terms, one goal of the attacker might be to corrupt a CN's binding cache and to cause packets to be delivered to the wrong address. This can compromise the secrecy and integrity of communication, cause DoS at the communicating parties, or cause DoS at an address that receives unwanted packets. The attacker might also exploit features of the security protocol carrying BUs to exhaust the resources of the MN, the HA, or the CNs. The aim of this section is to describe a number of known attacks, thereby allowing the reader to better understand the rational behind the security mechanisms described later in Section 8.3 and Section 8.4.

It is essential to understand that some of the described threats are more serious than others. Some can be mitigated but not removed, some may represent acceptable risk, and some may be considered too expensive to prevent.

We consider only active attackers. The rationale behind this is that in order to corrupt the binding structure, the attacker must sooner or later send one or more messages. Thus, it makes little sense to consider attackers that only observe messages but do not send any. In fact, some active attacks are easier for the average attacker to launch than any passive one would be. That is because in many active attacks the attacker can initiate the BU protocol at any time, whereas most passive attacks require the attacker to wait for suitable messages to be sent by the target nodes.

8.2.3.1 Attacks against Address "Owners," a.k.a. Address "Stealing"

The most obvious danger in Mobile IPv6 is address "stealing" — that is, an attacker illegitimately claiming to be a given node at a given address and then trying to "steal" traffic destined to that address.[14] There are several variants of this attack. We first describe the basic variant, followed by a description of how the situation is affected if the target is a stationary node, and continuing with more complicated issues related to confidentiality, integrity, and DoS aspects.

8.2.3.1.1 Basic Address Stealing

If BUs were not authenticated at all, an attacker could fabricate and send spoofed BUs from anywhere in the Internet. All potential CNs would be vulnerable to this attack. As explained in Section 8.2.2.1, there is no way of telling which addresses belong to MNs that could send BUs and which addresses belong to stationary nodes.

Consider a node A sending IP packets to another node B. The attacker could redirect the packets to an arbitrary address C by sending a BU to A. The HoA in the BU would be B and the CoA would be C. After receiving this BU, A would send all packets intended for node B to address C (See Figure 8.1 and Figure 8.2).

The attacker might select the CoA to be either its own current address (or another address in its local network) or any other IP address. Ingress filtering at the attacker's local network does not prevent the spoofing of BUs but forces the attacker either to choose a CoA from inside its own network or to use the *alternate care-of address* suboption.[3]

8.2.3.1.2 Stealing Addresses of Stationary Nodes

To perform the basic address-stealing attack, the attacker needs to know or guess the IP addresses of both the source of the packets to be diverted (A in the example above) and the destination of the packets (B). This means that it is difficult to redirect *all* packets to or from a specific node because the attacker would need to know all the IP addresses of the nodes that the target node is communicating with. On the other hand, if the IP address of the target node is known, stealing *some* of the traffic is fairly easy.

Given this, nodes with well-known addresses, such as servers, are most vulnerable because their addresses are most easily available. Furthermore, nodes that are a part of the network infrastructure, such as Domain Name System (DNS) servers, are particularly interesting targets for attackers and particularly easy to identify. In contrast, nodes that frequently change their address or use randomly assigned addresses are relatively safe. However, if they register their address into DynDNS, they become more exposed. Similarly, nodes on publicly accessible networks such as public-access WLANs risk revealing their addresses. IPv6 addressing privacy features mitigate these risks to an extent, but it should be noted that addresses cannot be completely recycled while there are still open connections that use those addresses.

Thus, it is *not* the MNs that are most vulnerable to address stealing attacks; it is the well-known static servers. Furthermore, such servers often run old or heavily optimized operating systems and may not have any mobility-related code at all. Thus, the security design definitely cannot be based on the idea that MNs might somehow be able to detect whether someone has stolen their address (in which case they could reset the state at the CN).

8.2.3.1.3 Attacks against Confidentiality and Integrity

By spoofing BUs, an attacker could redirect all packets exchanged by two IP nodes to itself. By sending a spoofed BU to A, it could capture the data intended for B. That is, it could pretend to be B and hijack A's connections with B or establish new spoofed connections. The attacker could also send spoofed BUs to both A and B and thereby insert itself into the middle of all connections between them (the man-in-the-middle attack). Consequently, the attacker would subsequently be able to see and modify the packets exchanged by A and B, as illustrated in Figure 8.3.

Strong end-to-end encryption and integrity protection, such as IPSec or SSL/TLS, can prevent all attacks against data confidentiality and integrity. When the data is cryptographically protected, spoofing

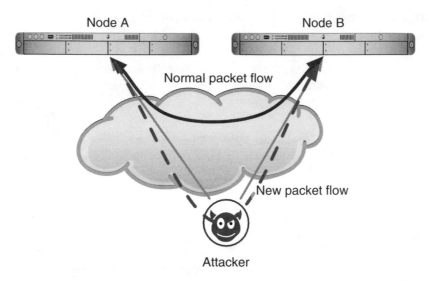

FIGURE 8.3   The man-in-the-middle attack.

of BUs could result in DoS (see below) but not in disclosure or corruption of sensitive data beyond revealing the existence of the traffic flows. Two fixed nodes could also protect the communication between themselves by refusing to accept BUs from each other. Ingress filtering, on the other hand, does not help, because the attacker is using its own address as the CoA and is not spoofing source-IP addresses.

8.2.3.1.4   *Basic Denial-of-Service Attacks*

By sending spoofed BUs, the attacker could redirect all packets sent between two IP nodes to a random or nonexistent address. In this way, it might be able to stop or disrupt communication between the nodes. This attack is serious because any Internet node could be targeted, including fixed nodes critical to the infrastructure (e.g., DNS servers).

8.2.3.2   Attacks against Other Nodes and Networks (Flooding)

Although the most obvious attack appears to be address stealing, it turns out that address flooding is much nastier in many respects. If there were no protection, an attacker could redirect traffic to any IP address by sending spoofed BUs. This could then be used to bomb that IP address with huge amounts of traffic. The attacker could also target a (sub)network instead of a node by redirecting data to one or more IP addresses within the network. There are two main variations of flooding: basic flooding and return-to-home flooding. Here we consider only basic flooding, for it is easier to understand. Return-to-home flooding is described in reference 13.

8.2.3.2.1   *Basic Flooding*

In the simplest flooding attack, an attacker would arrange a heavy data stream from some node A to another node B and then redirect this data stream to the target address C. However, A would soon stop sending the data because it would not be receiving acknowledgments from B.

A slightly more sophisticated attacker would itself act as B (see Figure 8.4). It would first subscribe to a data stream (e.g., a video stream) and then redirect this stream to the target address C. By doing so the attacker receives the initial packets of the data stream; therefore, it can easily spoof any acknowledgments and thus cause A to continue sending data. It would even be able to accelerate the data rate by simulating a fatter pipe.[15]

During the Mobile IPv6 design process, the effectiveness of this attack was debated. It was mistakenly assumed that the target node C would send a TCP Reset to the source of the unwanted data stream, which would then stop sending. In reality, all practical TCP/IP implementations fail to send the TCP Reset. The target node drops the unwanted packets at the IP layer because it does not have a Mobile IPv6

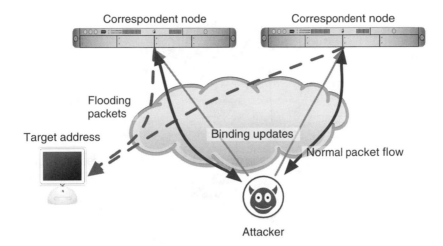

FIGURE 8.4 The basic flooding with attacker acting as node B.

Binding Update List entry corresponding to the routing header on the incoming packet. Thus, the flooding data is never processed at the TCP layer of the target node and no Reset is sent. This means that the attack using TCP streams is more effective than was originally believed.

This attack is serious because the target can be any node or network, not only a mobile one. What makes it particularly serious compared with the other attacks is that the target itself cannot do anything to prevent the attack. For example, it does not help if the target network stops using route optimization. The damage is the worst if these techniques are used to amplify the effect of other distributed denial-of-service (DDoS) attacks. Ingress filtering in the attacker's local network prevents the spoofing of source addresses, but the attack would still be possible by setting the *alternate care-of address* suboption to the target address.

So far we have considered the attacks that would be available if there were no security mechanisms used. Now we turn our attention to the potential problems within the security mechanisms themselves.

8.2.3.3 Attacks against BU-Based IP Mobility Protocols

Security protocols that successfully protect the secrecy and integrity of data can sometimes make the participants more vulnerable to DoS attacks. In fact, the stronger the initial authentication mechanism is, the easier it may be for an attacker to use the protocol features to exhaust the resources of the mobile or CNs. Alternatively, an attacker may be able to use a security protocol as a means of protecting itself, making it hard to track the real attacker. In this section, we describe basic variants of both types of attacks.

8.2.3.3.1 Inducing Unnecessary BUs

When an MN receives an IP packet from a new CN via the HA, it might make sense to automatically initiate the BU protocol. If so, an attacker could exploit this by sending the MN a spoofed IP packet (e.g., ping or TCP SYN packet) that appears to come from a new CN. Because the packet arrives via the HA, the MN would automatically start the BU protocol with the CN, thereby expending resources unnecessarily.

In a real attack, the attacker would induce the MN to initiate the BU protocol with a large number of CNs at the same time. If the spoofed addresses given as CN addresses are real addresses of existing IP nodes, then many cases of the BU protocol might even complete successfully, creating entries in the Binding Update List that are correct but useless. In this way, the attacker could induce the mobile host to execute the protocol unnecessarily and to create unnecessary state, both of which will drain resources.

A CN (i.e., any IP node) can also be attacked in a similar way. The attacker sends spoofed IP packets to a large number of MNs with the target node's address as the source address. These mobiles will initiate the BU protocol with the target node. Again, most of the protocol executions would complete successfully. By inducing a large number of unnecessary BUs, the attacker is able to consume the target resources.

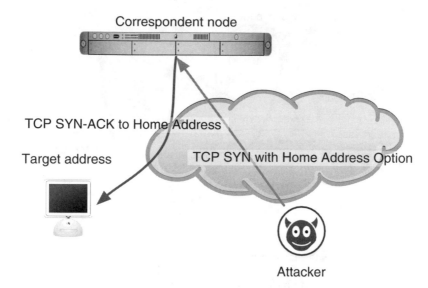

Correspondent node

TCP SYN-ACK to Home Address

Target address   TCP SYN with Home Address Option

Attacker

FIGURE 8.5   Using triangle routing for reflecting TCP SYN-ACKs.

This attack is possible against any BU-based IP mobility protocol. The more resources the protocol consumes, the more serious the attack. Hence, strong cryptographic authentication protocols are more vulnerable to this attack than are weak or unprotected protocols. Ingress filtering helps a little because it makes it harder to forge the source address of the spoofed packets, but it does not completely eliminate this threat.

8.2.3.3.2   *Reflection and Amplification*

Attackers sometimes try to hide the source of a packet flooding attack by reflecting the traffic off other nodes. That is, instead of sending the flood of packets directly to the target, the attacker sends the packets to other nodes, tricking them into sending the same number of packets, or more, to the target. Such reflection can hide the attacker's address even when ingress filtering prevents source-address spoofing. Reflection is particularly dangerous if the packets can be reflected multiple times, if they can be sent into a looping path, or if the nodes can be tricked into sending many more packets than they receive from the attacker, because such features can be used to amplify the traffic by a significant factor. When designing protocols, one should avoid creating services that can be used for reflection and amplification.

Mobile IP's triangular routing creates opportunities for reflection: a CN receives packets (e.g., TCP SYN) from the MN and replies to the HoA given by the MN in the Home Address Option (HAO). This is illustrated in Figure 8.5. The mobile host might not really be mobile, and the HoA could actually be the target address. The target would see only the packets sent by the CN and not the attacker's address (even if ingress filtering prevents the attacker from spoofing its source address).

A badly designed IP mobility protocol could also be used for reflection: The correspondent would respond to a data packet by initiating the protocol, which usually involves sending a packet to the HoA. In that case, the reflection attack by the MN can be discouraged by requiring it to include its CoA to the messages it sends to the CN.

These types of reflection and amplification can be avoided by ensuring that the CN *responds only to the same address from which it received a packet and only with a single packet of the same size.* These principles have been applied to Mobile IPv6 route optimization security design.

8.2.4   Summary of Vulnerabilities

In this section, we have covered the threats and attacks potentially created by adding IP-layer mobility to the current stationary IP network. We have seen how adding any such mechanism is a concern not only for the participating nodes but for any nodes in the Internet. We have also seen how a badly designed mobility protocol would allow attackers to benefit from time and location shifting, thereby being able to better cover their tracks and to attack a much larger number of nodes than they otherwise could. To give a concrete understanding of the potential problems, we described a representative set of known threats, including various forms of address stealing, address flooding, and attacks against the BU protocol itself.

With this understanding of the potential vulnerabilities in hand, we can now proceed to consider solutions. In this chapter we examine only two examples: Mobile IPv6 route optimization and the HIP. We have chosen these cases because they illustrate two fairly different approaches. Furthermore, both of them are interesting alone; they represent unconventional security protocols, in one sense or another.

Both of the protocols can be described as infrastructureless or by the fact that they do not require any external security infrastructure in order to provide the necessary level of security for IP-based mobility and multiaddress multihoming. Mobile IPv6 attempts to act within the limitations of the current TCP/IP stack structure, whereas HIP proposes an architectural change to the structure of the TCP/IP protocol suite.

8.3   Case 1: Mobile IPv6 Route Optimization

In this section, we will take a look at Mobile IPv6 route optimization and especially at its security aspects. We start by rehashing the basic Mobile IPv6 route optimization design. After that, we discuss *Return Routability (RR)*, the solution adopted in the Mobile IPv6 design.

Mobile IP is based on the idea of providing mobility support on the top of the existing IP infrastructure, *without requiring* any modifications to the routers, the applications, or the stationary end hosts. However, in Mobile IPv6 (as opposed to Mobile IP in IPv4) the stationary end hosts are supposed to (though not absolutely required to) provide additional support for mobility — that is, to support *route optimization*. In this respect, Mobile IPv6 and HIP are similar because they both require changes to both the MN and the CN.

As a security goal, the Mobile IPv6 design aimed to be "as secure as the (non-mobile) IPv4 Internet" was at the time of the design, during 2001 to 2002. In particular, that means that there is little protection against attackers that are able to attach themselves between a CN and an HA. The rationale is simple: In the 2001 Internet, if a node was able to attach itself to the communication path between two arbitrary nodes, it was able to disrupt, modify, and eavesdrop on all the traffic between the two nodes, unless cryptographic protection was used. Even when cryptography *was* used, the attacker was still able to selectively block communication by simply dropping the packets. The attacker in control of a router between the two nodes could also mount a flooding attack by redirecting the data flows between the two nodes (or, more practically, an equivalent flow of bogus data) to a third party.

8.3.1   Basic Design

The Mobile IP design aims to solve two problems at the same time. First, it allows transport-layer connections (TCP connections, UDP-based transactions) to continue even if the underlying hosts move and change their IP addresses. Second, it allows a node to be reached through a static IP address, an *HoA*. The latter design choice can also be stated in other words: Mobile IPv6 aims to preserve the *identifier* nature of home IP addresses. That is, Mobile IPv6 takes the view that IP addresses can be used as natural identifiers of nodes, as they have been used since the beginning of the Internet.

The basic idea in Mobile IP is to allow an HA to work as a stationary proxy for a MN. Whenever the MN is away from its *home network,* the HA intercepts packets destined for the node and forwards the packets by tunneling them to the node current address, the CoA. The transport layer (TCP, UDP) uses the HoA as a stationary identifier for the MN. Figure 8.6 illustrates this basic arrangement.

The basic solution requires tunneling through the HA, thereby leading to longer paths and degraded performance. This tunneling is sometimes called *triangular routing*.

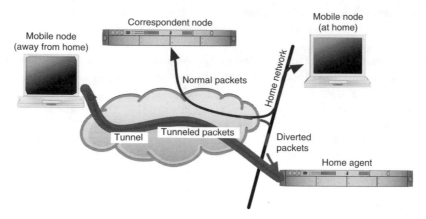

FIGURE 8.6   The basic Mobile IP arrangement.

To alleviate the performance penalty, Mobile IPv6 includes a mode of operation that allows the MN and its peer, a CN, to converse directly, bypassing the HA completely after the initial setup phase. This mode of operation is called *route optimization (RO)*. When RO is used, the MN sends its current CoA to the CN using BU messages. The CN stores the binding between the HoA and the CoA in its *binding cache*.

Whenever RO is used, the CN effectively functions in two roles. First, it is the source of the packets it sends, as usual. Second, it acts as *the first router* for the packets, effectively performing *source routing*. That is, when the CN is sending out packets, it consults its binding cache and *reroutes* the packets if necessary. A BCE contains the HoA and CoA of the MN and records the fact that packets destined to the HoA should now be sent to the destination address. Thus, it represents a local routing exception.

The packets leaving the CN are *source routed* to the CoA. Each packet includes a routing header that contains the HoA of the MN. Thus, logically, the packet is first routed to the CoA and then *virtually* from the CoA to the HoA. In practice, of course, the packet is consumed by the MN at the CoA, and the header just allows the MN to select a socket associated with the HoA instead of one associated with the CoA. However, the mechanism resembles source routing because there is routing state involved at the CN and a routing header is used.

8.3.2   The Solution Selected for Mobile IPv6

The current Mobile IPv6 route optimization security has been carefully designed to prevent or mitigate the threats that were discussed in Section 8.2. The goal has been to produce a design whose security is close to that of a static IPv4-based Internet and whose cost in terms of packets, delay, and processing is not excessive. As discussed, the result relies heavily on the assumption of an uncorrupted routing infrastructure and builds on the idea of checking that an alleged MN is indeed reachable through both its HoA and its CoA. Furthermore, the lifetime of the state created at the CNs is deliberately restricted to a few minutes in order to limit the opportunities for time shifting.

8.3.2.1   Return Routability

Return Routability (RR) is the name of the basic security mechanism deployed in Mobile IPv6 route optimization. Basically, it means that a node verifies that there is a node that is able to respond to packets sent to a given address. The check yields false positives only if the routing infrastructure is compromised or if there is an attacker between the verifier and the address to be verified. With these exceptions, it is assumed that a successful reply indicates that there is indeed a node at the given address and that the node is willing to reply to the probes sent to it.

The basic return routability mechanism consists of two checks: an *HoA check* and a *CoA check*. The purpose of the HoA check is to verify that the mobile host is reachable at the HoA. Correspondingly, the CoA check verifies that the mobile host is able to receive packets sent to the CoA. When the procedures

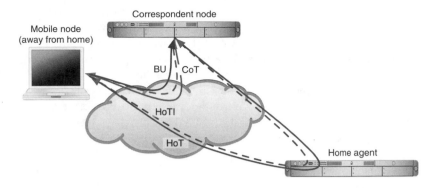

FIGURE 8.7 Return Routability.

are combined in a BU message (see below), the CN gains assurance that the mobile host is able to receive traffic at both addresses.

The Return Routability packet flow is depicted in Figure 8.7. First the MN sends two packets to the CN: a *Home_Test_Init (HoTI)* packet is sent through the HA, and a *Care-of_Test_Init (CoTI)* is sent directly. The CN replies to both of these independently by sending a *Home_Test (HoT)* in response to the HoTI and a *Care-of_Test (CoT)* in response to the CoTI. Finally, once the MN has received both the HoT and CoT packets, it sends a BU to the CN.

It must be noted that the real return routability checks are the message pairs < HoT, BU > and < CoT, BU >, not the < HoTI, HoT > nor the < CoTI, CoT > pairs. The init packets are needed only to *trigger* the test packets, and the BU acts as a combined routability response to both of the tests.

Two main rationales lie behind this design:

- Avoidance of reflection and amplification (see Section 8.2.3.3.2)
- Avoidance of state exhaustion DoS attacks (see Section 8.3.2.2)

The reason for sending two init packets instead of one is the avoidance of amplification. The CN is replying to packets that come out of the blue. It does not know anything about the MN, and therefore it just suddenly receives an IP packet from some arbitrarily looking IP address. In a way, this is similar to a server receiving a TCP SYN from a previously unknown client. If the CN would send two packets in response to an initial trigger, that would create a DoS amplification effect, as discussed in Section 8.2.3.3.2.

Reflection avoidance is directly related. If the CN were to reply to another address than the source address of the packet, that would create a reflection effect. Thus, the only safe approach is to reply to the received packet with just one packet and to send the reply to the source address of the received packet. Hence, two initial triggers are needed instead of just one.

Let us now consider the two return routability tests separately.

8.3.2.1.1 HoA Check

The HoA check consists of a HoT packet and a subsequent BU packet. It is triggered by the arrival of an HoTI. A CN replies to an HoTI by sending an HoT to the source address of the HoTI. The source address is assumed to be the HoA of a MN, and therefore the HoT is assumed to be tunneled by the HA to the MN. The HoT contains a cryptographically generated token, the *home keygen token,* which is formed by calculating a hash function over the concatenation of a secret key K_{cn} known only by the CN, the source address of the HoTI packet, and a nonce. An index to the nonce is also included in the HoT packet, allowing the CN to more easily find the appropriate nonce:

$$\text{home keygen token} = \text{hash}(K_{cn}|\text{source address}|\text{nonce}|0) \tag{8.1}$$

The token allows the CN to make sure that the subsequently received BU is created by a node that has seen the HoT packet; see Section 8.3.2.2.

In most cases the HoT packet is forwarded over two different segments of the Internet. It first traverses from the CN to the HA. On this trip, it is not protected and any eavesdropper on the path can learn its contents. The HA then forwards the packet to the MN. This path is taken inside the IPSec Encapsulated Security Payload (ESP) protected tunnel, making it impossible for outsiders to learn the contents of the packet.

At first it may sound unnecessary to protect the packet between the HA and the MN because it traveled unprotected between the CN and the HA. If all links in the Internet were equally insecure, the situation would indeed be so — that protection would be unnecessary. However, in most practical settings the network is likely to be more secure near the HA than near the MN. If the CN is a big server, all the links on the path between it and the HA are likely to be fairly secure. On the other hand, the MN is probably using wireless access technology, making it sometimes trivial to eavesdrop on its access link. Thus, it may be relatively easy to eavesdrop on packets that arrive at the MN, and protecting the path from the HA to the MN is likely to provide real security benefits even when the path from the CN to the HA remains unprotected.

8.3.2.1.2   CoA Check

From the CN's point of view, the CoA check is very similar to the HoA check. The only difference is that now the source address of the received CoTI packet is assumed to be the CoA of the MN. Furthermore, the token is created in a slightly different manner in order to make it impossible to use home tokens for care-of tokens, or vice versa:

$$\text{care-of keygen token} = \text{hash}(K_{cn}|\text{source address}|\text{nonce}|1) \qquad (8.2)$$

The CoT traverses only one leg, directly from the CN to the MN. It remains unprotected all along the way, making it vulnerable to eavesdroppers near the CN, on the path from the CN to the MN, or near the MN.

8.3.2.1.3   Forming the First BU

When the MN has received both the HoT and CoT messages, it creates a binding key K_b by taking a hash function over the concatenation of the tokens received:

$$K_b = \text{hash}(\text{home token} | \text{care-of token}) \qquad (8.3)$$

This key is used to protect the first and subsequent BUs, as long as the key remains valid.

Note that the key K_b is available to anyone who is able to receive both the CoT and the HoT messages. However, they normally arrive through different routes through the network, and the HoT is transmitted over an encrypted tunnel from the HA to the MN.

8.3.2.2   Creating State Safely

The CN may remain *stateless* until it receives the first BU. That is, it does not need to record receiving and replying to the HoTI and CoTI messages. The HoTI/ HoT and CoTI/CoT exchanges take place in parallel but independently of each other. Thus, the CN can respond to each message immediately, and it does not need to remember doing that. This helps in potential DoS situations: no memory needs to be reserved when processing HoTI and CoTI messages. Furthermore, HoTI and CoTI processing is designed to be lightweight and can be rate limited if necessary.

When receiving an initial BU, the CN goes through a rather complicated procedure. The purpose of this procedure is to ensure that there is indeed an MN that has recently received an HoT and a CoT that were sent to the claimed home and care-of addresses, respectively, and to make sure that the CN does not unnecessarily expend CPU time or other resources while performing this check.

Because the CN does not necessarily have state when the BU arrives, the BU itself must contain enough information so that relevant state can be created. The BU contains the following pieces of information:

- *Source address.* The source address must be equal to the source address used in the CoTI message.
- *HoA.* This must be the same address that was used as the source address for the HoTI message and as the destination address for the HoT message.

- Two *nonce indices*. These are copied over from the HoT and CoT messages, and together with the other information, they allow the CN to recreate the tokens sent in the HoT and CoT messages and used for creating K_b. Without them the CN might need to try the 2–3 latest nonces, leading to unnecessary resource consumption.
- *Message Authentication Code (MAC)*. The BU is authenticated by computing a MAC function over the CoA, the CN's address, and the BU message itself. The MAC is keyed with the key K_b.

Given the addresses, the nonce indices (and thereby the nonces), and the key K_{cn}, the CN can recreate the home and care-of tokens at the cost of a few memory lookups and computation of one MAC and one hash function.

Once the CN has recreated the tokens, it hashes the tokens together, giving the key K_b. This key is then used to verify the MAC that protects integrity and origin of the actual BU. If the BU is authentic, K_b is cached together with the binding. Note that the same K_b may be used for a while, until either the MN moves (and needs to get a new CoA token), the care-of token expires, or the home token expires.

Note that because the CN may remain stateless until it receives a valid BU, the MN is solely responsible for retransmissions. That is, the MN should keep sending the HoTI/CoTI messages until it receives a HoT/CoT, respectively. Similarly, it may need to send the BU a few times in cases where it is lost in transit.

8.3.2.3 Quick Expiration of the BCEs

A BCE, along with the key K_b, represents the return routability state of the network *at the time* the HoT and CoT messages were sent. Now, it is possible that a specific attacker is able to eavesdrop on an HoT message at some point in time but not later. If the HoT had an infinite or a long lifetime, that would allow the attacker to perform a *time-shifting* attack (see Section 8.2.2.2). That is, in the current IPv4 architecture, an attacker in the path between the CN and the HA is able to perform attacks only as long as the attacker is able to eavesdrop (and possibly disrupt) communications on that particular path. A long-lived HoT would enable sending valid BUs for a long time, allowing an attacker to continue its attack even after the attacker is no longer able to eavesdrop on the path.

To limit the seriousness of this and other similar time-shifting threats, the validity of the tokens is limited to a few minutes. This effectively limits the validity of the key K_b and the lifetime of the resulting BUs and BCEs.

Although short life times are necessary for the reasons given, they are clearly detrimental for efficiency and robustness. That is, a HoTI/HoT message pair must be exchanged through the HA every few minutes. These messages are unnecessary from a purely functional point of view and thus represent overhead. Even worse, they make the HA a single point of failure. That is, if the HoTI/HoT messages were not needed, connections from an MN to other nodes could continue even when the HA fails, but the current design forces the bindings to expire after a few minutes.

This concludes our brief walk-through of the selected security design. The cornerstones of the design were the use of the return routability idea in the HoT, CoT, and BU messages; the ability to remain stateless until a valid BU is received; and the limiting of the binding lifetimes to a few minutes.

8.4 Case 2: HIP

In this section, we describe the experimental HIP and discuss its security aspects. In particular, we consider how HIP provides security for end-host mobility and multihoming and the properties of the HIP-based mobility and multihoming solution. First, we describe the basic architectural ideas behind HIP. Next, we outline the basic design ideas. Finally, we discuss how HIP solves the problems related to IP mobility and multiaddress multihoming, and we discuss the security solution.

8.4.1 Introduction

The HIP originated as an idea for addressing a number of architectural and security problems in the Internet architecture. However, it was clear from the beginning that HIP would make IP-based mobility

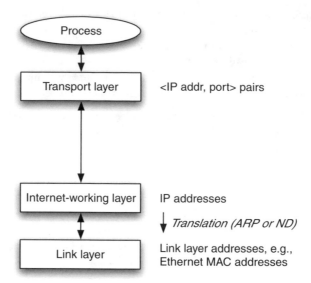

FIGURE 8.8   The current internetworking architecture.

much easier. On the other hand, the details of how exactly to implement mobility with HIP, and how to integrate multiaddress multihoming with the mobility features, were established only later.

The design of HIP is radically different from that of Mobile IPv6. In Mobile IPv6, there are two types of IP addresses: HoAs and CoAs. In HIP there is only one type of IP address, which corresponds most closely to Mobile IPv6 CoAs. The function of HoAs is taken by identifiers from a new name space: host identifiers. That is, the HIP architecture introduces a new name space and uses names from that name space to identify hosts.

In practice, an HIP host identifier is a cryptographic public key. Each host is assumed to generate one (or more) public key pair(s). The public key of such a pair is used to identify the host. For the purposes of this discussion, it is safe to think of each host as being uniquely identified by a single key pair. However, there are reasons (such as anonymity) to allow a node to represent itself as a set of endpoints and to allow the endpoints to move among hosts.[4]

The fact that the nodes generate their key pairs themselves may lead to a suspicion that there is a possibility of key collision — that is, two nodes generating the same key pair. Fortunately, the chance of such a collision is extremely small.

Let us consider RSA-based public keys as an example. According to the Prime Number Theorem, the density of primes near a given number n is about $1/\log n$. Thus, using 1024-bit RSA keys, and correspondingly primes of about 512 bits, the density of primes around 2^{512} is larger than $1/512$, meaning that there are more than $2^{(512-9)}$ primes of that size. According to the birthday paradox, one can expect about one collision in a population of approximately $\sqrt{2n}$. Because the number of primes is about 2^{503}, the size of a population required to produce an almost sure collision is about $\sqrt{2 \cdot 2^{503}} = \sqrt{2^{504}} = 2^{252}$. Hence, as long as good random number generators are used, the probability of a key collision is negligible and can be safely ignored.

8.4.2   Basic Design

In introducing the new name space, HIP changes the structure of the TCP/IP stack. In particular, it introduces a new protocol layer between the IP layer and the upper layers.

Figure 8.8 shows the current structure of the TCP/IP stack, without HIP. In it, processes are bound to transport-layer sockets, and the sockets are identified by IP addresses and ports. As a result, this structure binds processes to IP addresses and therefore to specific topological locations.

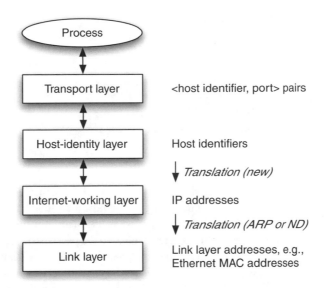

FIGURE 8.9 The HIP architecture.

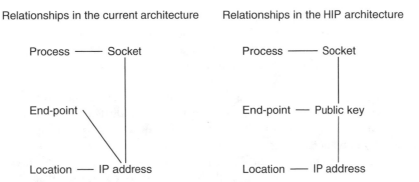

FIGURE 8.10 Relationships among processes, end-points, and locations.

The new structure is shown in Figure 8.9. In the HIP architecture, transport-layer sockets are no longer named with IP addresses but with host identifiers. The new Host Identity layer translates the host identifiers into IP addresses. This is achieved by *binding* a host identifier to one or more IP addresses. Typically, this binding is simultaneously a dynamic relationship, enabling mobility, and a one-to-many relationship, enabling multihoming.

The HIP architecture[4] does not define exactly how bindings from host identifiers to IP addresses are created. However, a viable possibility for creating the initial bindings is to store both host identifiers and some of the corresponding IP addresses into the DNS repository. (Note that we are *not* proposing that DNS would be used to track rapidly changing addresses. See the discussion in Section 8.4.3.1.) As a result, a connecting host receives both a host identifier *and* one or more addresses in a reply to a DNS query. If it receives a host identifier, it can use the new architecture; if no host identifiers are received, it falls back to the old architecture. In the new architecture, the HIP protocol is used to check that there is indeed a remote endpoint at the given address and that the remote endpoint actually matches the host identifier received from the DNS. (See Section 8.4.4.2 for the security details.)

The HIP architecture changes the relationships among the various entities and identifiers, as illustrated in Figure 8.10. In the current architecture, IP addresses are used to name both hosts (endpoints) and topological locations. In the HIP architecture, these functions have been separated, and the hosts (endpoints)

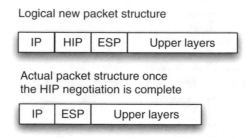

FIGURE 8.11 The packet structures.

are named by host identifiers. Furthermore, the binding between a host identifier and the IP address(es) is made dynamic. Because of the cryptographic nature of the host identifiers, it is fairly easy to secure the signaling messages needed to update this binding, as we will explain in Section 8.4.4.

At the logical level, the HIP architecture also requires changes to the packet structure. That is, each packet must logically include the host identifiers of the sender and recipient. However, whenever IPSec is used, the IPSec security associations can be used as a shortcut for host identifiers, resulting in packets that are syntactically similar to those used today. This is illustrated in Figure 8.11 and further discussed in Section 8.4.4.2.

8.4.3 Mobility and Multihoming

It should be obvious by now that basic mobility and multihoming becomes trivial in the HIP architecture. That is, all that is needed to support mobility is to make sure that the binding between a host identifier and a (set of) IP address(es) is dynamic. Respectively, all that is needed to support multihoming is to make the binding a one-to-many binding. In practice, both things are achieved with the HIP version of BUs (see Section 8.4.4.2).

To be more specific, in the HIP architecture the host identifiers are used to identify the communication endpoints, and they have no permanent relationship with locations or IP addresses. IP addresses, on the other hand, are used to identify only the topological locations, not hosts. As a result, the *actual* IP addresses used in a packet do not matter as much as they do in the current architecture because the endpoints are not identified with them. All that is required is that hosts be able to determine the current addresses of their active peers. Furthermore, if the integrity of the packets is protected with, say, IPSec ESP, the recipient is always able to verify that a received packet was sent by the alleged peer no matter what the source and destination addresses are. Thus, by binding IPSec security associations to host identifiers instead of IP addresses, the destination address becomes pure routing information, and the source address becomes almost obsolete.[4] (Note that it has always been easy to spoof source addresses. Using the source address as a destination address in a reply makes sense only if there is no better address to use. In HIP, for example, the hosts know their peer's current addresses, and therefore it is more reliable to use an address from the set of known addresses than one received in an incoming packet.) Only during connection setup, when the hosts have not authenticated each other, does the source address play any substantial role. Once the peer hosts have secure bindings between the HIs and the addresses, the source address is no longer needed by the hosts. Its only function becomes to carry information about the topological path the packet has taken.[16]

Going beyond the basic mobility and multihoming achieved with the separation of the identifier and location natures of IP address, additional issues must be addressed. In particular, to fully support mobility and end-host multihoming, the base architecture must include *FAs*.

8.4.3.1 Forwarding Agents

In HIP, basic mobility support requires that the mobile host sends signaling messages (BUs) to its correspondent hosts. These inform the CNs of changes in the address(es) that they can use to reach the mobile host. Thus, in the basic case, explicit packet forwarding is not needed because the hosts are able

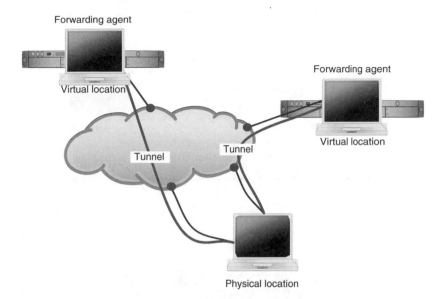

FIGURE 8.12 The virtual interface model.

to send packets directly to each other. However, this leaves two problems unaddressed. First, there must be a mechanism that allows a mobile host to be contacted independent of its current location. Second, if two hosts move at the same time, it is possible that the BUs cross each other and never reach their intended destinations. Introducing packet FAs allows us to solve these two problems.

In HIP, an FA is a network node that forwards all packets sent to a given IP address (virtual address) to another IP address (real address).

As discussed Section 8.1.5, a multihomed host can be considered present at several locations at the same time. In functional terms, that means that the host is able to receive packets sent to several different IP addresses. On the other hand, if we have an FA, the end host is also able to receive packets sent to the forwarding address. Thus, in a sense the FA can be considered to represent a *virtual address* of the host, and the host can be considered to be *virtually present* at the location of the FA.

Consider the situation from the point of view of a correspondent host. It sees only a number of addresses, and it knows that the multihomed host is reachable through all these addresses. From the correspondent host's point of view, there is no difference whether packets sent to a given address are forwarded or directly received by the host. All that matters is that the packets eventually reach the right destination. This situation is illustrated in Figure 8.12.

8.4.4 Security Design

The introduction of the host identifiers opens up new security vulnerabilities. In the original TCP/IP architecture, a host's identity is implicitly but weakly authenticated by the routing infrastructure, as we described earlier in Section 8.1.3. In the HIP architecture, there is no implicit binding between the host identifiers and the routing infrastructure. Thus, the implicit authentication does not exist anymore.

Looking at the situation more closely, it becomes clear that the new vulnerabilities are actually caused not by the new host identifiers, but by the dynamic binding between the host identifiers and IP addresses. They are clearly visible in Mobile IPv6, as was discussed in Section 8.2.3.

Fortunately, introducing host identifiers that are themselves cryptographic public keys makes it easy to address the vulnerabilities.

8.4.4.1 Host Identifiers

The cryptographic nature of the host identifiers is the security cornerstone of the HIP architecture. The host is supposed to keep its private key(s) secret and not to disclose it (them) to anybody. (Note that a

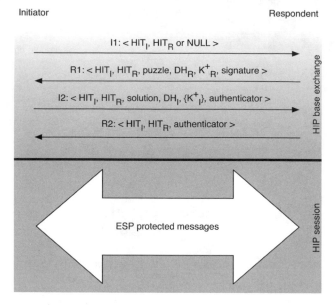

FIGURE 8.13 A typical HIP session.

single *user* typically would want to have several different public keys, and hence several host identifiers, in a single computer.)

The use of the public key as the name makes it possible to directly check that a party is actually entitled to use the name. A simple public-key authentication protocol, such as the one included in the HIP base exchange, is sufficient for that. In contrast to solutions where names and cryptographic keys are separate, the key-oriented naming used in HIP requires no external infrastructure to authenticate identity. In other words, no explicit Public Key Infrastructure (PKI) is needed. Because the identity is represented by the public key itself, and because any proper public key authentication protocol can be used to check that a party indeed possesses the private key corresponding to a public key, a proper authentication protocol suffices to verify that the peer is indeed entitled to the name. Nobody's certification is needed.

This property of being able to verify the identity of any party without explicit external infrastructure is an important aspect of the HIP architecture. It allows the architecture to scale naturally, free of extra administrative overhead.

8.4.4.2 HIP Base Exchange

The HIP base exchange is the host-to-host authentication protocol in the HIP architecture. The details of the current protocol proposal are available as an Internet draft[5] and explained briefly in this section. The protocol takes place between two parties, called the *initiator* and the *respondent*. The initiator sends the first message and thereby must know the respondent (or at least its IP address). The respondent, on the other hand, might not know anything about the initiator before the first message arrives.

The HIP base exchange is a four-way handshake protocol, consisting of messages I1, R1, I2, and R2 (see Figure 8.13). The first message, I1 (sent by the initiator), is a very simple trigger message that requests the respondent to send an R1. At this point, the respondent does not yet create any state, and the amount of processing required is minimal. It would be unsafe to create state or perform any larger amount of processing because the I1 packets essentially may come from any host in the Internet, and expending resources on them would create opportunities for various DoS attacks.

The R1 packet contains a cryptographic puzzle. In addition to the puzzle, R1 contains the public key of the respondent, as well as a precomputed signature over the security critical contents of the packet. This allows the initiator to verify that it received R1 from the right respondent.

To continue the HIP handshake, the initiator has to use some CPU time to solve the puzzle. Only once it has solved the puzzle can it continue and send I2. The respondent, in turn, is able to easily verify that

I2 is indeed a reply to a recently sent R1 and that the initiator has successfully solved the puzzle. In this way, the respondent can be sure that there is indeed an initiator (because it received R1) and that the initiator indeed wants to open an HIP connection (because it was willing to consume resources to solve the puzzle).

In addition to the puzzle, the public key, and the signature, the R1 packet acts as the first packet of an authenticated Diffie-Hellman exchange. That is, it contains the respondent's public Diffie-Hellman key, allowing the initiator to start to compute the session key while waiting for a response to I2. I2, in turn, contains the initiator's public Diffie-Hellman key, as well as the initiator's public key and signature. For privacy reasons, the public key is encrypted with the Diffie-Hellman session key, thereby protecting the initiator's identity from third parties.

Once the respondent has successfully performed the initial verification of I2, as described above, it proceeds to process the Diffie-Hellman components. It uses the public Diffie-Hellman key in the packet to create the session key, decrypts the initiator's public-identity key with the session key, and finally verifies that the signature matches. This allows the respondent to create the HIP connection in one step and still be sure that there is indeed an initiator, identified with the given public-identity key, and willing to open the HIP connection.

The base exchange is finished with R2, which contains the respondent's signature. It allows the initiator to authenticate the Diffie-Hellman session key with the respondent's identity. After exchanging these initial messages, both communicating hosts know that there is indeed an entity at the other host that possesses the private key corresponding to its host identifier. Additionally, the exchange creates a pair of IPSec ESP security associations, one in each direction. The hosts use the ESP security associations to protect the integrity of the packets flowing between them; optionally, ESP can also be used to encrypt the packets.

In addition to using ESP to protect the network layer integrity of the payload traffic, the created security associations are used to secure the signaling messages exchanged between the hosts. For example, once the initial messages have been exchanged and the security associations are in place, the hosts inform their peers of their current IP addresses. In effect, this shares information about the current multihoming situation of the hosts. Each host has complete freedom to select which IP addresses to announce to the peer.

To the peer, it is immaterial whether the announced IP addresses are real or virtual. All it needs to know is that the announcing host is indeed *reachable* through the claimed IP addresses. The reachability needs to be checked, or otherwise the mechanism could be used to launch a number of DoS attacks, as we discussed in Section 8.2.3.

Once the initial multihoming situation is established and verified, the hosts can communicate in a secure and resilient way. As the connectivity statuses of the hosts change, they may signal the changes in the situation as needed. That is, if an end host loses connectivity on an interface, acquires a new interface, or moves an interface from one location to another, it typically wants to signal the change to its peers. For this purpose, the HIP protocol includes a kind of BU message called readdressing packet (REA). Naturally, all REA packets must be secured, and any new addresses must be checked for reachability.

8.4.4.3 Privacy

Using public keys as primary identifiers is clearly a potential source of privacy problems. If each user had just a single public key and that key were used repeatedly by the user, the very nature of public key cryptography would lead to a situation where it is fairly easy to link together all the transactions made by the user. In the case of the HIP architecture, the situation is not that bad. Because a single computer can host several end points and therefore have several host identifiers, it is easy for a user to have several public keys instead of just one. One public key can be used as a more permanent identifier, allowing others to contact the user, whereas other keys can be completely temporary and periodically replaced with new ones. A temporary host identifier needs to be valid only as long as active connections are associated with it.

Furthermore, in the HIP architecture the actual public keys are present in the network only during the HIP base exchange (see Figure 8.11 and Figure 8.13). Thus, packets other than the very first ones used to establish communication with a new host do not include the public key. A user who does not want to be identified by the operator or third parties in the vicinity of his or her real location is able to remain locally anonymous by tunneling the first few packets through an FA and only then telling the peer about the current address. In that way, the public keys and signatures do not openly cross the network at the current location at all, thereby allowing users to remain anonymous to all but their peers.

Nevertheless, the HIP architecture does not directly address the location privacy problem. Because the architecture assumes that hosts inform their peers about their current addresses, the peers see the locations and movements. On the other hand, a host that wants to keep its movements hidden from a specific peer can do so by using a semipermanent FA, at some cost in performance.

8.4.4.4 Security Summary

From the discussion so far, it should be clear that addressing the mobility and multihoming-related security issues is much easier with cryptographic host identifiers. As an added benefit, their use can enhance privacy.

We are using public keys as the names for hosts; therefore, identity authentication is trivial. All that is required is a good authentication and key agreement protocol; no certificates or external entities are needed. Basic multihoming and mobility introduces the requirement of checking each new IP address for reachability, to make sure that the host currently really is reachable at the IP addresses where it claims to be. Note, however, that the reachability problem is a slightly easier one to solve than the so-called address ownership problem, which we briefly discussed in Section 8.2.3.1. There is no need to verify that the host somehow "owns" the address, or that the address is somehow a legitimate name for the host. Instead, the host is named by the public key, and it just happens to be reachable at the address. If some other host were reachable at the same address at the same time, even that would not be a security problem.[14]

In summary, the use of public keys as host identifiers leads to a natural security solution for end-host mobility and multihoming. In the next section, we briefly consider the lessons learned from both the HIP design and the Mobile IPv6 route optimization security design.

8.5 Lessons Learned

In this section, we briefly sum up the more general lessons to be learned from Mobile IPv6 route optimization and HIP security design. In particular, we discuss how important it is to understand the environment where a particular mechanism is used and the dangers present in that environment. Mobility, with its potential ability to track people's physical locations, threatens privacy, and therefore we consider the issue of privacy separately.

8.5.1 Know Your Enemy

First, consideration of the security aspects of IP mobility has once again shown how important it is to analyze the environment in which a given mechanism functions. The intuition that IP mobility would be a danger only to mobile hosts and their correspondent peers has turn out to be completely wrong. Instead, IP mobility mechanisms appear to be of greatest danger to old hosts that understand nothing about mobility, due to the new potential DoS attacks enabled by the ability to redirect existing data streams to new IP addresses.

Historically, the Mobile IPv6 design proceeded for a number of years, until late 2000, without anybody paying any real attention to the security aspects. The designers relied on IPSec, assuming that just using it would adequately address the security concerns. In late 2000 and early 2001 came the realization that IPSec would not solve the problems, partly due to the lack of a suitable global security infrastructure and the difficulty of creating one (see Section 8.2.1), and partly because IPSec alone does not protect from flooding attacks. As a result, the standardization of Mobile IPv6 was delayed by approximately 2 years, the time it took to design the proper security solutions.

HIP, on the other hand, has its roots deeply in security. One motivation for developing HIP has been to simplify IPSec key management and to provide a simpler alternative to the standard IPSec key management protocol, IKE. Hence, HIP has had basic security mechanisms available from the very beginning. However, even the HIP design lacked proper return routability tests until fairly recently. The reason for this was that there was no general understanding of the potential DoS problems caused by IP mobility until the Mobile IPv6 route optimization security design took place. Consequently, HIP adopted a proper return routability procedure only after one was designed for Mobile IPv6. On the other hand, due to the basic nature of HIP, return routability in HIP is simpler than in Mobile IPv6, with fewer messages and fewer potential vulnerabilities.

8.5.2 Semantics vs. Security

Although understanding the environment is crucial for good security design, it is equally important to properly understand the semantics of the mechanism to be secured. In particular, it is extremely important to understand how state is created at the nodes in the network and the potential consequences if any piece of this state gets compromised.

In the case of Mobile IPv6, the state is represented in the binding cache. As explained in Section 8.3.1, the state can be understood as a set of local routing exceptions, causing source routing to be performed on those addresses that have been entered into the binding cache. In the case of HIP, on the other hand, the state is the binding between a host identifier and a dynamic set of IP addresses. To be more specific, HIP also includes a piece of state for each individual address, recording whether the return routability test has been performed on that particular address, along with other operational data.

The consequences of this state getting compromised are more or less similar in both Mobile IPv6 and HIP: Some packets get delivered to the wrong location. That, in turn, may compromise confidentiality and integrity or cause DoS. However, because HIP additionally protects the actual data flow using cryptographic means, the danger of a confidentiality or integrity breach is smaller. With Mobile IPv6, the same can be achieved with IPSec, at the cost of extra signaling. Both seem to be fairly similar with respect to the potentiality of DoS attacks.

8.5.3 Privacy

In addition to security problems, the main concern in this chapter, mobility brings forth privacy problems. In particular, mobile communication devices potentially allow their users to be tracked as they move around. In Mobile IP, users are always implicitly identified with a stable HoA. Both the HA and the CoAs are always carried around in route-optimized Mobile IP packets, allowing anyone who has access to those packets to keep track of the current location of the user. In HIP, the host identifiers are transmitted on the wire only during the initial four-way handshake and in the readdressing packet, making it harder to keep track of what packets belong to a particular user and, therefore, where a particular user currently is. Furthermore, there is ongoing work that will allow hiding the host identity even in the base exchange and subsequent signaling packets, thereby allowing full identity privacy in the presence of eavesdropping third parties.

In both Mobile IP and HIP, the real privacy danger lies in unnecessarily including potentially dangerous information in packets sent on the network. There is usually a good reason for including such information, for example, increased reliability, security, or even basic connection management. Consequently, unless privacy is explicitly considered during the protocol design, the problems will not even be identified.

Given that in the near future people are likely to carry several mobile IP devices on their persons, it is becoming extremely important to make sure that the base IP protocols do not unnecessarily cause privacy problems. On the other hand, crime investigation and national security mandate that it must be possible to somehow track users, if needed. A balance must be found between the privacy needs of individual citizens and the law enforcement needs mandated by society. Only future user reactions will show what is the appropriate level of privacy and how the various alternatives fulfill the conflicting needs.

In general, perhaps the most important lesson to learn is that it pays to consider security and privacy issues very early in the protocol design process.

Acknowledgments

We want to thank Tuomas Aura, Catharina Candolin, Tom Hendersson, Juha Heinänen, Miika Komu, Yki Kortesniemi, Gabriel Montenegro, Glenn Morrow, Martti Mäntylä, Erik Nordmark, Alexandru Petrescu, Jarno Rajahalme, Teemu Rinta-Aho, Göran Schultz, Tim Shepard, Vesa Torvinen, and Zoltan Turanyi for their constructive comments on various versions of this chapter and the preceding papers on which this chapter is based. Finally, we want to express our special thanks to Ken Rimey, who kindly read the final form of the chapter, corrected our language, and provided many constructive comments on the structure. All the remaining mistakes are ours.

References

1. Saltzer, J.H., On the naming and binding of network destinations, in *Local Computer Networks*, Ravasio, P. et al., North-Holland, Amsterdam, 1982. Also available as RFC 1498, University of Southern California, Marina Del Rey, Aug. 1993.
2. Saltzer, J.H., Reed, D., and Clark, D., End-to-end arguments in system design, *ACM Transactions on Computer Systems*, 2, 227, Nov. 1984.
3. Johnson, D., Perkins, C., and Arkko, J., *Mobility support in IPv6*, forthcoming, IETF, 2004.
4. Moskowitz, R. and Nikander, P., *Host Identity Payload Architecture*, work in progress, http://www.ietf.org/internet-drafts/draft-moskowitz-hip-arch-03.txt, Nov. 2003.
5. Moskowitz, R. and Nikander, P., *Host Identity Payload and Protocol*, work in progress, http://www.ietf.org/internet-drafts/draft-moskowitz-hip-07.txt, Nov. 2003.
6. Bhagwat, P., Perkins, C., and Tripathi, S.K., Network layer mobility: an architecture and survey, *IEEE Personal Communications*, 3, 54, June 1996.
7. Smith, R. et al., Securing distance-vector routing protocols, in *Proc. Symp. on Network and Distributed System Security*, Internet Society, Feb. 1997.
8. Stewart, R. et al., Stream Control Transmission Protocol, RFC 2960, IETF, Oct. 2000.
9. Kohler, E. et al., *Datagram Congestion Control Protocol (DCCP)*, work in progress, http://www.ietf.org/internet-drafts/draft-ietf-dccp-spec-04.txt, June 2003.
10. Perlman, R., *Network Layer Protocols with Byzantine Robustness*, Ph.D. thesis, MIT, Boston, Aug. 1988.
11. Chiappa, J.N., *Endpoints and Endpoint Names: A Proposed Enhancement to the Internet Architecture*, unpublished, http://users.exis.net/~jnc/tech/endpoints.txt.
12. Lear, E., ed., *What's in a name: report from the name space research group*, work in progress, http://www.ietf.org/internet-drafts/draft-irtf-nsrg-report-09.txt, Mar. 2003.
13. Nikander, P. et al., *Mobile IP version 6 route optimization security design background*, work in progress, http://www.ietf.org/internet-drafts/draft-nikander-mobileip-v6-ro-sec-01.txt, June 2003.
14. Nikander, P., Denial-of-service, address ownership, and early authentication in the IPv6 world, in *Security Protocols, 9th International Workshop*, LNCS 2467, Springer, 2002, 12.
15. Savage, S. et al., TCP congestion control with a misbehaving receiver, *Computer Communication Review*, 29, 71, Oct. 1999.
16. Candolin, C. and Nikander, P., IPv6 source addresses considered harmful, in *Proc. NordSec 2001, Sixth Nordic Workshop on Secure IT Systems*, Nielson, H.R., Technical University of Denmark, Nov. 2001, 54.

9

Security Issues in Wireless IP Networks

Subir Das
Telcordia Technologies, Inc.

Farooq Anjum
Telcordia Technologies, Inc.

Yoshihiro Ohba
Toshiba America Research, Inc.

Apostolis K. Salkintzis
Motorola, Inc.

9.1   Introduction

Wireless access networks are rapidly becoming a part of the ubiquitous computing environment whether based on IEEE 802.11 LANS in enterprise networks and public hotspots (e.g., in airports, hotels, and the home) or second-, third-, or fourth-generation (2G/3G/4G) wide area networks (e.g., General Packet Radio Service [GPRS], Wideband CDMA [WCDMA], Code Division Multiple Access 2000 [CDMA2000], etc.), as well as wireless ad hoc networks, often deployed in infrastructureless environments. Also, the widespread availability of miniature wireless devices such as personal digital assistants (PDAs), cellular phones, Pocket PCs, small fixtures on buildings, and sensors is one step toward making the vision of wireless nirvana a reality. Wireless nirvana would make it possible for any wireless device to seamlessly communicate with any other wireless device regardless of the time and place. But we are still a long way off from this goal.

Although current advancements in technology are making possible our drive to this goal of wireless nirvana, many serious obstacles still remain. One such serious barrier is related to security. The basic characteristics of the underlying radio communication medium provide serious exposure to attacks against wireless networks. In addition, when considering ad hoc wireless networks, one cannot depend on traditional infrastructure found in enterprise environments. In this chapter, we describe the security issues in Wireless Local Area Network (WLAN) and 3G networks. In particular, we concentrate on the vulnerabilities and attacks in existing standards. We also describe the emerging standards and protocols that are necessary to secure these wireless networks.

9.2   Security Attributes

Security, in general, is defined by two important parameters: (1) trustworthiness of a system, and (2) weakness of a system. A system is called trustworthy if it can be assured that the system will perform as expected.[1] Related to this are the concepts of dependability and survivability.[2] Survivability is the capability of a system to fulfill its mission in a timely manner, whereas dependability is the ability to deliver service that can justifiably be trusted. The authors in reference 2 conclude that all these concepts are essentially equivalent in their goals and address similar threats, although some minor differences remain. For example, survivability and trustability have the threats explicitly listed in the definitions; dependability does not. Thus, a trustworthy system must address threats from hostile attacks, environmental disruptions, and human and operator errors. The attributes related to the trustworthiness of a system are as follows:[1–3]

- *Authentication:* Authentication is the ability to positively verify the identity of a user, device, or other entity in a computer system. This is generally a prerequisite to allowing access to resources in a system. Without authentication, an adversary can gain unauthorized access to the various resources and information of a system.
- *Availability:* Availability is related to the assurance that information and communications services will be ready for use when expected.
- *Confidentiality:* Confidentiality is the assurance that information will be kept secret, with access limited to appropriate persons.
- *Integrity:* Integrity is the assurance that information will not be accidentally or maliciously altered or destroyed.
- *Nonrepudiation:* Nonrepudiation is the method by which the sender of data is provided with proof of delivery and the recipient is assured of the sender's identity, so that neither can later deny having processed the data.

Weakness of a system, on the other hand, considers various flaws that might be present in the system and that make it possible for intruders to access the system. The attributes related to the weaknesses of a system are as follows:[1,2]

- *Threat:* Threat is the means through which the ability or intent of a threat agent to adversely affect an automated system, facility, or operation can be manifested. This is a potential violation of security and reflects the capabilities of the intruder. Threats depend on the types of attackers; for example, nation states can be expected to possess capabilities for mounting any type of attacks. In fact, all methods or things used to exploit a weakness in a system, operation, or a facility constitute threat agents.
- *Vulnerability:* Vulnerability is any hardware, firmware, or software flaw that leaves an information system open for potential exploitation. The exploitation can be of various types, such as gaining unauthorized access to information or disrupting critical processing.
- *Attack:* An attack is an attempt to bypass the security controls on a computer. The attack may alter, release, or deny data. Whether an attack will succeed depends on the vulnerability of the computer system and the effectiveness of existing countermeasures. Attacks can further be divided into two main categories:
  - *Passive attacks:* Passive attacks are those attacks whereby the attacker only listens to the packets passing by without modifying or tampering with the packets. These attacks mainly target the confidentiality attribute of the system.
  - *Active attacks:* Active attacks are those attacks in which the attacker takes action that involves modification of packets in addition to spoofing of the packets.

9.3   Security Issues in WLANs

The use of wireless links makes WLANs vulnerable to attacks ranging from passive eavesdropping to active interference. An attacker just needs to be within radio range of an access point (AP) in order to intercept network traffic. Easy access to wireless infrastructure networks is coupled with easy deployment. It is easy to simply purchase an AP and connect it to the corporate network without authorization, thus creating a rogue AP. War drivers have published results indicating that a majority of APs (enterprise or public hotspots) are put into service with only minimal modifications to their default configuration and often do not have security features activated. These APs are easily accessible by anyone and can be used as a free ride to the Internet. Also, intruders may be able to gain access to enterprise networks and inflict damage to sensitive information. The Service Set Identifier (SSID) is meant to differentiate networks from one another. Initially, APs are set to a default depending on the manufacturer. Usually, these default SSIDs are so well known that not changing it makes your network much easier to detect. Another common mistake is setting the SSID to something meaningful, such as the AP's location or department, or setting it to something easily guessable. By default the AP broadcasts the SSID every few seconds in what are known as *beacon frames*. Although this makes it easy for authorized users to find the correct network, it also makes it easy for unauthorized users to find the network name. This feature is what allows most wireless network detection software to find networks without having the SSID up front. There are, however, a limited number of vendors (Cisco Aironet 350, etc.) who support this setting whereby SSID broadcasting on the AP can be turned off. APs broadcast their SSID with every second packet, reducing the authorization time it takes for a client to connect to the AP. By turning off SSID broadcasting, the clients must send out a probe packet on startup requesting the SSID they are configured for. Only when the AP receives the probe packet from the client will it send out a broadcast. This type of configuration drastically reduces the chance of an attacker identifying your AP's SSID during a random war drive. In addition to disabling broadcasting, it is also recommended that you select a random, not easy-to-guess SSID for your clients and APs. The use of a password generation program is recommended.

IEEE 802.11 networks that are still using only Wired Equivalent Privacy (WEP) are essentially not secure at all. Unauthenticated intruders can easily associate with APs and gain access to network resources. Intruders can also recover Wired Equivalent Privacy (WEP) keys and gain access to protected data. RC4 (Ron's Code #4, a stream cipher) combines the 40-bit WEP key with a 24-bit Initialization Vector (IV) to encrypt the data. The 40-bit keys are inadequate for any system, and in the case of WEP, it has been shown that keys of any length will not protect the system.[14]

9.3.1 Vulnerabilities and Attacks in WLANs

In the following subsections we describe the vulnerabilities and attacks that are possible in currently deployed 802.11b-based WLAN networks.

9.3.1.1 Vulnerabilities

In this section, we highlight the various vulnerabilities and attacks that a WLAN system is susceptible to. Note that although the lists of vulnerabilities and attacks considered here are by no means exhaustive, an attempt has been made to make the lists representative. The weaknesses in infrastructure WLAN include the following:

- The use of wireless links makes infrastructure networks vulnerable to attacks ranging from passive eavesdropping to active interference. An attacker simply needs to be within radio range of an AP to intercept network traffic.
- Medium Access Control (MAC) protocols are susceptible to attacks. Because they all follow predefined procedures to avoid (or recover from) collisions, a misbehaving node can easily change the MAC protocol behavior that could resemble a Denial-of-Service (DoS) attack.
- Easy access to wireless infrastructure networks is coupled with straightforward deployment. Simply purchasing an AP and connecting it to the corporate network without authorization, thus creating a rogue AP, requires little effort.
- War drivers have published results indicating that a majority of APs are put into service with only minimal modifications to their default configuration and often do not have security features activated. These APs allow access to the network by anyone.
- WLANs operate in unlicensed frequency bands.

- Infrastructure wireless networks usually are connected to a back-end network that may contain sensitive data and run mission-critical operations. Intruders may be able to gain access to this back-end network and inflict damage if they are able to compromise the wireless portion of the network.
- There is a centralized point of failure at the access node. By compromising the AP, the intruder can disrupt communications.
- APs are initially set to a default depending on the manufacturer; these default SSIDs are so well known that not changing them makes the network much easier to detect.
- By default, the AP broadcasts the SSID every few seconds, which makes it easy for unauthorized users to find the network name (see our earlier discussion).
- 802.11 networks that are still using only WEP are essentially not secure at all. Unauthenticated intruders can easily associate with APs and gain access to network resources. Intruders can also recover WEP keys and gain access to protected data.
- WEP-specific vulnerabilities include the following:[12]
    - Vendor-specific built-in functions in wireless cards may also increase the chance of WEP IV collision. As noted in reference 5, a common wireless card from Lucent resets the IV to 0 each time a card is initialized and increments the IV by 1 with each packet. This means that two hosts that inserted their cards at roughly the same time will provide an abundance of IV collisions for an attacker. Adding to this problem, changing the IV with each packet is defined as optional by the Institute of Electrical and Electronics Engineers (IEEE) 802.11 standard, thereby increasing the chance of IV collision.
    - WEP also suffers from lack of integrity protection for the data. WEP uses CRC-32 to provide integrity guarantees on the data. CRC-32 and other linear block codes are inadequate for providing cryptographic integrity. Message modification is possible (i.e., the message can be altered) while retaining the same Cyclic Redundancy Check (CRC).
    - Another problem with WEP is key management. When WEP is enabled according to the 802.11b standard, the network operator has to visit each wireless device to be used and type in the proper WEP key. Although this can initially be rolled into new client setup, if the key is compromised for any reason the key has to be changed. This is not a big deal with only a few users, but what if it is an entire university campus or hundreds of corporate users on a network? In those scenarios, changing the WEP key quickly becomes a logistical nightmare.
    - Authentication is another completely weak area in WEP. The authentication protocol in WEP is based on a challenge-response paradigm. But the way that the protocol is designed ensures that an intruder is able to get the information needed for authentication by capturing the two packets containing the challenge and the response.

9.3.1.2 Attacks and Resulting Compromise

Infrastructure networks are subject to both passive and active attacks. The currently known active attacks are described in the following subsections.

9.3.1.2.1 *Eavesdropping*

Eavesdropping is easy in the radio environment because radio signals travel in air. So, when one sends a message over the wireless medium, everyone equipped with a suitable transceiver in the range of the transmission can potentially decode the message and obtain sensitive information, such as bank access codes and passwords. The sender or the intended receiver has no means of detecting if the transmission has been eavesdropped. Even messages encrypted with WEP security protocol can be decrypted within a reasonable amount of time using easily available hacking tools. These passive intruders put businesses at risk of exposing sensitive information to corporate espionage. Though the physical transmission of packets on WLANs uses spread spectrum, the hopping sequence (in Frequency Hopping Spread Spectrum [FHSS]) and the 11-chip Barker code (in Direct Sequence Spread Spectrum [DSSS]) are the same for all WLAN users, unlike in 2G/3G CDMA systems. Hence, the physical layer does not provide any useful privacy in WLAN systems.

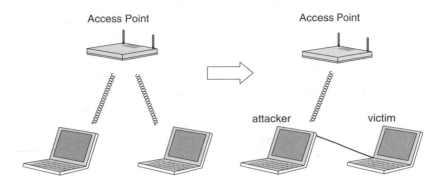

FIGURE 9.1 Man-in-the-middle attack.

The ease of eavesdropping in the wireless LAN environment justifies costly procedures to guarantee the confidentiality of the network traffic. In all WLAN standards, this is addressed by some kind of link-level ciphering done by MAC entities, but the safety gained with these algorithms in many cases is not good enough for the most demanding applications.

9.3.1.2.2 *Virtual Private Network Attacks*
The attacker combines a rogue AP with a Virtual Private Network (VPN) server of the same type as the one on the target network. When a user tries to connect to the real server, the spoofed server sends a reply back, leading the user to connect to the fake server. Usually the credentials of the user are asked and thus stolen. Afterwards, the fake AP goes offline and the unknowing user connects to the real network. The attacker can then use legal credentials to access the network at a later time when the real user goes offline.

9.3.1.2.3 *Man-in-the-Middle Attacks*
In this type of attack, the attacker places his or her machine in the line of communication between the valid wireless station and its AP. Thus, all the frames exchanged between the victim station and its AP pass through the attacker's machine transparently without any knowledge of either the victim station or the AP. After successfully launching the Man-in-the-Middle (MITM) attack, the attacker may steal information from the frames or may change the content of the data inside the frame. MITM is also possible by placing a rogue AP within the range of an authorized AP. Figure 9.1 shows a typical example of the MITM attack. Here the attacker makes the victim's connection to the AP and goes through itself.

Here are the steps involved in launching an MITM attack in an IEEE 802.11 WLAN:

1. The attacker sniffs the traffic flowing in the WLAN, selects a victim machine, and collects the MAC addresses of the AP and the victim machine. It also gathers WLAN-related information such as the SSID and the frequency channel through which the victim is connected to the AP.
2. The attacker sends a deauthentication management frame to the victim machine with a spoofed source address of the AP (i.e., it acts as a rogue AP).
3. Because 802.11 does not mandate a wireless station to authenticate the management frames, the wireless station (victim) will immediately disconnect from the AP it is connected to.
4. The victim station will retry to connect to the AP again.
5. The attacker will now configure its own machine as an AP with the same SSID as the original AP. It will use a different frequency channel. It will also try to boost its signal strength so that the victim station gets connected to its machine instead of the original AP. One way to do that is to place its machine nearer the victim machine than the original AP.
6. The victim machine will connect to the attacker's machine thinking it as a valid AP. The attacker will then associate itself with the original AP and transparently route traffic between the victim and the original AP.

9.3.1.2.4 DoS

DoS attacks are also easily applied to wireless networks, where legitimate traffic cannot reach clients or the AP because illegitimate traffic overwhelms the frequencies. DoS attacks are possible at various layers; we will describe a few such attacks, namely, physical-layer, MAC-layer, and network-layer attacks:

- *Jamming attacks:* Because WLANs operate either in unlicensed frequency bands (2.4-GHz ISM and 5-GHz UNII), they experience interference from other systems such as microwave ovens, cordless phones, and baby monitors. Excessive deployment of one or combination of these systems can result in high interference for WLAN users, rendering the communication medium useless. Malicious users may even consider using radios that operate in those frequencies without resorting to the specific devices mentioned above.
- *MAC-layer attacks:* Nodes in a WLAN follow the Carrier Sense Multiple Access/Collision Avoidance (CSMA/CA) specification to fairly share the wireless bandwidth. It is easy for a malicious node to not follow the standard defined procedures for sensing and back off. It may choose to use a nonstandard back-off policy, causing legitimate users to receive either little or no throughput. Note that some commercial products such as some Cisco APs allow the user to change the standard CSMA/CA mechanism. To ensure that the legitimate users in a system suffer, an intruder can use this approach along with rogue APs.
- *Dynamic Host Configuration Protocol (DHCP) server attacks:* A malicious node can send large number of discover/request messages to a DHCP server by simply changing its waveLAN card and thus exhaust a pool of Internet Protocol (IP) addresses for a given subnet. On the other hand, when legitimate users query the DHCP server for an IP address, they cannot get one. This results in DoS for legitimate users. Although this attack is not specific to WLANs (it can happen to wireline networks as well), it is easy to launch this type of attack in wireless environments.

9.3.1.2.5 Impersonation (Identity Theft)

The theft of an authorized user's identity poses one the greatest threats. SSIDs that act as crude passwords and MAC addresses that act as personal identification numbers are often used to verify that clients are authorized to connect with an AP. However, existing encryption standards are not foolproof and allow knowledgeable intruders to pick up approved SSIDs and MAC addresses to connect to a WLAN as an authorized user with the ability to steal bandwidth, corrupt or download files, and wreak havoc on the entire network. This attack therefore also requires spoofing of MAC addresses in addition to obtaining the SSID information. MAC address spoofing is very easy with some of the currently available software.

9.3.1.2.6 Brute-Force Attacks

In this type of attack, the attacker attempts to learn either the WEP key or a password by using brute force. APs often use a single key that is shared among all clients. A successful brute-force dictionary attack would give the intruder access to the network once the password has been found. Most APs use well-known Simple Network Management Protocol (SNMP) community strings (passwords) for management, and those that provide a Hypertext Transfer Protocol (HTTP) interface are accessible by anybody who happens to know the IP address of the device.

9.3.1.2.7 Dictionary Attacks

In case of WEP-enabled WLANs, a table or dictionary can be maintained to obtain access to confidential information. The small space of possible IVs allows an attacker to build a decryption table. Once the attacker learns the plaintext for some packet, he or she can compute the RC4 keystream generated by the IV used. This key stream can be used to decrypt all other packets that use the same IV. Over time, perhaps using the techniques above, the attacker can build up a table of IVs and corresponding key streams. This table requires a fairly small amount of storage (about 15 GB); once it is built, the attacker can decrypt every packet that is sent over the wireless link. Because the IV is only 24 bits long, there are only so many permutations of the IV for RC4 to pick from. Mathematically, there are only 16,777,216 (2^{24}) possible values for the IV. Although that may seem like a lot, it takes so many packets to transmit useful data that 16 million packets can go by in hours on a heavily used network. So, eventually the RC4 mechanism starts picking the same IVs over and over. By passively listening to the encrypted traffic and picking the repeating IVs out of the data stream, an intruder can begin to infer the data that is protected. For example, the attacker can build a dictionary of IV and key streams so that the message can be decrypted in real time.

9.3.1.2.8 *Partial Information-Based Attacks*

When the attacker has partial knowledge of the information that was encrypted using WEP, the chances of launching successful attacks greatly improves. We consider three special cases:[4]

9.3.1.2.8.1 Active Attacks to Inject Traffic — This attack is a direct consequence of the problems with WEP. Suppose an attacker knows the exact plaintext for one encrypted message. He or she can use this knowledge to construct correctly encrypted packets. The procedure involves constructing a new message, calculating the CRC-32, and performing bit flips on the original encrypted message to change the plaintext to the new message. The basic property is that $(RC4(X) \oplus X) \oplus Y = RC4(Y)$, where $\oplus$ represents the bitwise Exclusive OR operation. This packet can now be sent to the AP or mobile station (MS), and it will be accepted as a valid packet.

A slight modification to this attack makes it much more insidious. Even without complete knowledge of the packet, it is possible to flip selected bits in a message and successfully adjust the encrypted CRC (as described in the previous section) to obtain a correctly encrypted version of a modified packet. If the attacker has partial knowledge of the contents of a packet, he or she can intercept it and perform selective modification on it. For example, it is possible to alter commands that are sent to the shell over a Telnet session or via interactions with a file server.

9.3.1.2.8.2 Active Attacks from Both Ends — The previous attack can be extended further to decrypt arbitrary traffic. In this case, the attacker makes a guess not about the contents but rather about the headers of a packet. This information is usually quite easy to obtain or guess; all that is necessary to guess is the destination IP address. Armed with this knowledge, the attacker can flip appropriate bits to transform the destination IP address to send the packet to a machine he or she controls, somewhere in the Internet, and transmit it using a rogue MS. Most wireless installations have Internet connectivity; the packet will be successfully decrypted by the AP and forwarded unencrypted through appropriate gateways and routers to the attacker's machine, revealing the plaintext. If a guess can be made about the Transmission Control Protocol (TCP) headers of the packet, it may even be possible to change the destination port on the packet to be port 80, which will allow the packet to be forwarded through most firewalls.

9.3.1.2.8.3 Known Plaintext Attacks — Once an attacker obtains a collection of cipher texts, he or she can easily mount attacks to get the plaintext. In this way, the attacker can take advantage of the fact that IP traffic contains many known plaintext strings.[18] This enables recovery of key stream of length N for a given IV. Without a keyed Message Integrity Check (MIC), it also allows forging packets of size N. Or the attack can be based on partially known plaintext, in which only a portion of the plaintext is known (e.g., the IP header). It is possible to recover M octets of the key stream (M < N). With repeated attempts, attackers can obtain the whole key stream.

9.4 Emerging WLAN Security Schemes

In light of several security issues related to WEP and IEEE 802.11 security, two new task groups, IEEE 802.1X and 802.11i, have been formed to come up with better authentication and encryption algorithms for WLANs. Whereas IEEE 802.1X provides an architecture for port-based network access control and an Extensible Authentication Protocol (EAP) framework to authenticate the WLAN to the wireless client, 802.11i proposes WEP key management with stronger security algorithms, namely TKIP (Temporal Key Integrity Protocol) and CCMP (Counter Mode CBC [Cipher Block Chaining] MAC [Message Authentication Code] Protocol). TKIP is supposed to be a short-term fix for WEP, whereas CCMP is envisioned to be the long-term WLAN security solution. In the following subsections, we first describe the WEP security briefly, followed by a discussion of the IEEE 802.1X and 802.11i security architecture, protocols, and security algorithms.

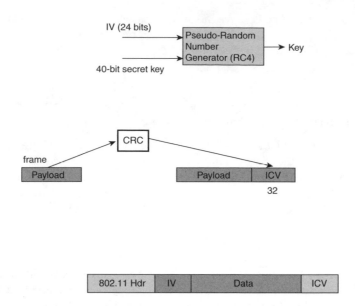

FIGURE 9.2   WEP operation and frame format.

9.4.1   Wired Equivalent Privacy (WEP)

Figure 9.2 shows the WEP operation. The WEP algorithm is used to protect wireless communication from eavesdropping. A secondary function of WEP is to prevent unauthorized access to a wireless network. WEP relies on a secret key that is shared between an MS and an AP. The secret key is used to encrypt packets before they are transmitted, and an integrity check is used to ensure that packets are not modified in transit. The standard does not discuss how the shared key is established. In practice, most installations use a single key that is shared between all MSs and APs. The WEP key is typically 64 bits long and is generated using a 24-bit IV and a 40-bit secret key using a Pseudo Random Number Generator (RC4). A 32-bit CRC is generated using the user data. The CRC is called the Integrity Check Value (ICV). User data along with the ICV is encrypted using a bitwise exclusive operation with the key. The actual 802.11 frame consists of the physical/MAC 802.11 header, the IV (so the receiver can decrypt the packet after combining it with the secret key), and the body. WEP also provides authentication functions. It allows three different types of authentication:

- *Open authentication:* Open authentication provides no security whatsoever. Anyone who requests access is granted it.
- *Key based:* Shared-key authentication implements the challenge-response protocol. Even though the response is encrypted, the challenge is sent in cleartext. By listening to this information, attackers can get the information needed for authentication.
- *MAC-address based:* WEP also allows authentication based on MAC-layer filtering. The first problem with this approach is that MAC addresses are passed in the clear even when encryption is turned on, thereby making it easy to spoof a legitimate user by using a standard network sniffer program and reconfiguring the wireless card. The second problem is that wireless cards can easily have their MAC address changed via a simple software reconfiguration.

9.4.2   The 802.1X Framework

IEEE 802.1X[6] is designed to work over a point-to-point IEEE 802 LAN segment in order to provide port-based access control for the LAN. A port is a single point of attachment to the LAN infrastructure, which can be formed on each physical port of an IEEE 802 bridge or on an association between a station and

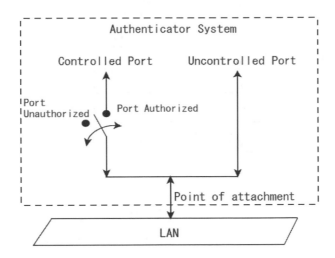

FIGURE 9.3 Uncontrolled and controlled ports.

an AP in an IEEE 802.11 WLAN.[7] An important functionality of IEEE 802.1X is to carry EAP messages at the IEEE MAC layer in order for LAN switches or APs to authenticate the attached client devices. The encapsulation is known as *EAP over LANs*, or *EAPOL*. IEEE 802 devices that act as an EAP peer and an EAP authenticator are referred to as *supplicant* and *authenticator*, respectively. In a nutshell, 802.1X provides an architecture for port-based network access control and an EAP framework to authenticate the 802.11-based wireless LAN network to the wireless client.

An authenticator system has two types of ports: uncontrolled ports and controlled ports, as illustrated in Figure 9.3. Uncontrolled ports allow the uncontrolled exchange of Protocol Data Units (PDUs) between the system and the other systems on the LAN, regardless of the authorization state of the supplicants, whereas controlled ports allow the exchange of PDUs only when the supplicants are authorized for the use of the ports. IEEE 802.1X initial packets are allowed only on uncontrolled ports because supplicants are not authorized for the use of controlled ports. The format of IEEE 802.1X packets when carried over IEEE 802 Ethernet and the list of packet types are shown in Figure 9.4.

EAP is the heart of the 802.1X framework and can be used by a variety of authentication types, known as EAP methods. A family of such methods have been developed specifically for wireless networks based on public-key certificates and the Transport Layer Security (TLS) protocol, including EAP-TLS, EAP-TTLS, and PEAP, among others. Before we describe these protocols in detail, let us look briefly at the message exchanges that typically occur in an 802.1X framework (Figure 9.5):

- The supplicant (MS) starts the 802.1X exchange with an EAPOL-Start message.
- The normal EAP exchange begins. The authenticator (AP) issues an EAP-Request/Identify frame.
- The supplicant replies with an EAP-Response/Identity frame, which is passed on to the Remote Authentication Dial-In User Service (RADIUS) server as a Radius-Access-request packet.
- The RADIUS server replies with a Radius-Access-Challenge packet, which is passed on to the supplicant as an EAP-Request of the appropriate authentication type containing any relevant challenge information.
- The supplicant gathers the reply from the user and sends an EAP-Response in return. The response is translated by the authenticator into a Radius-Access-Request with the response to the challenge as a data field.
- The RADIUS server grants access with a Radius-Accept packet, so the authenticator issues an EAP-Success frame. The port is authorized and the user can begin accessing the network.
- When the supplicant is finished accessing the network, it sends an EAPOL-Logoff message to put the port back to unauthorized state.

Octet Number

PAE Ethernet Type (0x88-8E)	1-2
Protocol Version	3
Packet Type	4
Packet Body Length	5
Packet Body	6-N

(a) IEEE 802.1X Packet Format

Packet Type	Value
EAP Packet	0x00
EAPOL-Start	0x01
EAPOL-Logoff	0x02
EAPOL-Key	0x03
EAPOL-Encapsulated-ASF-Alert	0x04

(b) IEEE 802.1X Packet Types

FIGURE 9.4 802.1X packet formats and packet types.

9.4.2.1 EAP

EAP[8] is an authentication protocol that was originally defined for Point-to-Point Protocol (PPP)[32] but has been revised to fulfill the increasing demand for the protocol to run on top of any Layer-2 technologies and even on top of any higher layer.

Before EAP was designed, only a few standard authentication protocols were supported in PPP, including Challenge Handshake Protocol (CHAP).[9] Although CHAP supports several authentication algorithms, such as MD5,[10] SHA-1,[11] and MS-CHAP,[12] the protocol is limited to supporting challenge-response type authentication. The main purpose of EAP is to support not only challenge-response type authentication with one or more multiple message exchange round-trips, but also a variety of authentication algorithms in a single framework protocol. Authentication algorithms supported by EAP are referred to as Request/Response type authentication methods. More than 30 authentication methods already exist, including MD5-Challenge,[8] One-Time Password (OTP),[13] and TLS.[14]

Figure 9.6 shows the relationship among entities involved in EAP. EAP runs between an authenticator and a peer on a link, where the authenticator is the entity that initiates authentication methods and the peer is the entity that responds to the authenticator. If the authenticator itself is not able to authenticate the peer, it may use an authentication service to communicate with a back-end authentication server that is responsible for authenticating the peer. In this case, the authentication service may use an Authentication, Authorization, and Accounting (AAA, pronounced "triple A") protocol such as RADIUS[15] or Diameter[16] to carry authentication information (i.e., EAP messages) between the authenticator and the back-end authentication server.

The EAP message format is shown in Figure 9.7. The Code field contains a message type; currently the following four values are defined: 1 (Request), 2 (Response), 3 (Success), and 4 (Failure). The Identifier

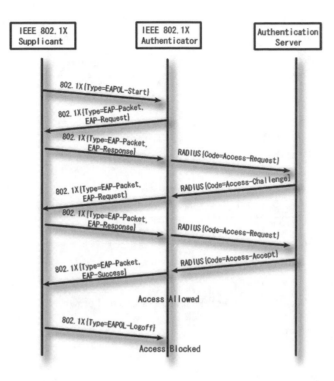

FIGURE 9.5 A typical EAPOL exchange.

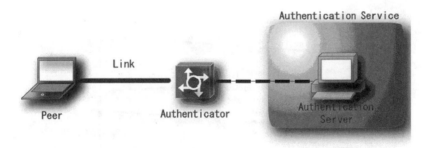

FIGURE 9.6 EAP authentication model.

field is used for matching Responses with Requests. The Length field has a length of 2 octets and indicates the length of the EAP packet, including the Code, Identifier, Length, and Data fields. The Data field contains zero or more octets of data that are specific to the message type indicated by the Code field. For Request and Response messages, the Data field consists of a 1-octet Type field and a variable-length Type-Data, as shown in Figure 9.7.

A basic messaging sequence in EAP is illustrated in Figure 9.8. The authenticator initiates EAP by sending a Request message. A trigger for the authenticator to send the first Request message is provided outside the EAP protocol (e.g., via a link-layer indication). The first Request/Response exchange is typically for Type 1 (Identity) that is used for carrying the identity of the peer. The identity exchange can be omitted if the identity can be known by using some other method. The subsequent message exchange is for running an authentication method, which is MD5-Challenge in this example. When the EAP conversation completes successfully (or fails for some reason), a Success (or Failure) message is returned to the peer.

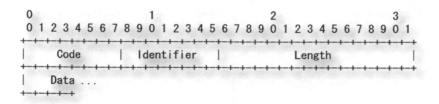

(a) EAP Message Format

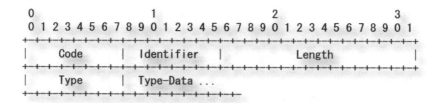

(b) EAP Request/Response Message Format

FIGURE 9.7 EAP message format.

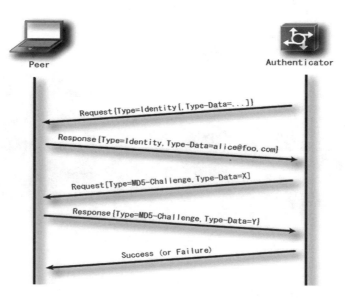

FIGURE 9.8 A sample EAP message sequence.

9.4.2.2 EAP TLS

There is an EAP authentication method that provides not only authentication but also key derivation. EAP-TLS[14] is the first EAP authentication method that supports key derivation functionality. EAP TLS carries encapsulated TLS packets in TLS record format[17] in the Type-Data field of EAP Requests and Responses. EAP TLS relies on TLS for both authentication and key derivation. In EAP TLS, mutual authentication can be performed based on the exchange of certificates between the authenticator and the peer. PPP encryption keys (or any type of cipher keys to protect subsequent data traffic in general) are derived from the TLS master secret that is securely established in the TLS handshake between the authenticator and the peer.

9.4.2.3 EAP TTLS and PEAP

Although EAP TLS provides stronger authentication based on public keys (such as certificates) than other authentication methods based on symmetric keys (such as passwords), there are some environments, such as roaming among multiple Internet Service Providers (ISPs), where certificate management (i.e., issuing a certificate and revoking a certificate) becomes a problem. In such environments, password-based authentication methods are still in demand. Similar to EAP TLS, EAP TTLS (Tunneled TLS)[18] also carries TLS packets for key deriving keys. The difference is that EAP TTLS allows nesting of EAP messages in which inner EAP messages are carried in TLS packets that are carried in outer EAP messages (i.e., EAP TTLS messages). A secure TLS communication channel between the peer and a TLS server provides encryption or integrity protection of inner EAP messages, and the inner EAP authentication method can be of any type. The TLS handshake performed in EAP TTLS does not require the peer to present a certificate to the network, which makes this authentication method suitable for roaming usage. Note that the TLS handshake still requires the TLS server to present the server certificate to the peer in order to prevent an MITM attack on the handshake.

Protected EAP (PEAP)[19] is very similar to EAP TTLS. A major difference is that in PEAP, the EAP Success/Failure message is carried over a protected TLS channel instead of being sent in cleartext.

9.4.2.4 RADIUS

RADIUS[15] is an integral part of the 802.1X authentication framework. It is a client/server type protocol that provides AAA functionality. RADIUS uses UDP as its transport because of its simplicity.

In RADIUS, a Network Access Server (NAS) operates as a client of RADIUS. The client is responsible for passing user information to designated RADIUS servers and then acting on the response that is returned. RADIUS servers are responsible for receiving user connection requests, authenticating the user, and then returning all configuration information necessary for the client to deliver service to the user. To support user devices roaming among multiple administrative domains, RADIUS also allows an additional protocol entity called a *proxy*, which forwards the request to a remote RADIUS server, receives the reply from the remote server, and sends that reply to the client, possibly with changes to reflect local administrative policy.

RADIUS messages exchanged between the client and the RADIUS server are authenticated through the use of a shared secret, which is never sent over the network. Information carried in RADIUS messages is not encrypted, with the assumption that any user passwords are sent encrypted between the client and the RADIUS server, to eliminate the possibility of password spoofing by attackers on the signaling path.

The RADIUS server can support a variety of methods to authenticate a user. When it is provided with the user name and original password given by the user, it can support PPP PAP or CHAP, EAP, UNIX login, and other authentication mechanisms.

All transactions consist of variable-length Attribute-Length-Value 3-tuples. New attribute values can be added without disturbing existing implementations of the protocol.

Figure 9.9 shows the RADIUS message format and list of message types. The Code field includes a message type. The Identifier field is used for matching requests and replies. The RADIUS server can detect a duplicate request if it has the same client source IP address and source UDP port and Identifier within a short span of time. The Length field indicates the length of the packet, including the Code, Identifier, Length, Authenticator, and Attribute fields. The Authenticator field is mainly used to authenticate the reply from the RADIUS server. The Authenticator fields in an Access-Request and an Access-Response packet are referred to as a Request Authenticator and a Response Authenticator, respectively. A Request Authenticator is a random value. A Response Authenticator is a one-way MD5 hash calculated over a stream of octets consisting of the RADIUS packet, beginning with the Code field, including the Identifier, the Length, the Request and Authenticator field from the Access-Request packet, and the response Attributes, followed by the secret shared between the RADIUS client and the RADIUS server.

Access-Request packets are sent from a RADIUS client to a RADIUS server to convey authentication information. Access-Challenge packets are sent from a RADIUS server to a RADIUS client to convey authentication information. Access-Request and Access-Challenge packets also serve as a response to

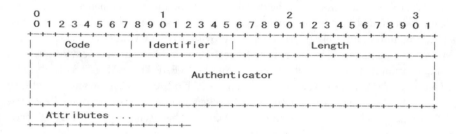

(a) RADIUS Message Format

```
1        Access-Request
2        Access-Accept
3        Access-Reject
4        Accounting-Request
5        Accounting-Response
11       Access-Challenge
12       Status-Server (experimental)
13       Status-Client (experimental)
255      Reserved
```

(b) RADIUS Code Values defined in RFC 2865

FIGURE 9.9 RADIUS message format and values.

outstanding Access-Challenge and Access-Request messages, respectively. When authentication completes successfully (or fails), an Access-Accept (or an Access-Reject) is sent from a RADIUS server to the RADIUS client. An Access-Accept packet may contain additional authorization information as well. Basic attributes that are allowed in each message type are listed in Figure 9.9. Vendor-specific attributes are defined to allow vendors to define their own attributes. Additional attributes are described in RADIUS accounting[29] and RADIUS extensions[30] specifications.

9.4.2.5 Weakness in the 802.1X Framework

Although the 802.1X framework reduces the vulnerabilities in WLAN networks by providing stronger user authentication, authorization, and data security, it has still some weaknesses. Although most of the above-mentioned attacks are irrelevant in the 802.1X framework, the MITM attack is possible.

9.4.2.5.1 MITM

EAP TTLS, PEAP, and other authentication methods that are classified as compound authentication methods address an issue of vulnerability to a new class of MITM attack.[22] A compound authentication method is an authentication method in which two or more authentication methods are performed in a sequenced or a nested manner (or a combination of both sequencing and nesting). A compound authentication method may be used for several reasons — for example, because a weak authentication method such as MS-CHAP is widely deployed and needs to be protected with another authentication method, or because a single entity is identified with multiple identities that are authenticated using different methods.

A sample MITM attack is explained in Figure 9.10. Here a tunneling authentication method, such as EAP-TTLS or PEAP, is performed to establish a secure tunnel between a peer and the back-end authentication server to secure a subsequent authentication method used for authenticating the peer. Because the subsequent authentication method is used for authenticating the peer, the tunneling method does not require the peer to be authenticated, which allows an attacker to reside between the legitimate pair of peer and authenticator. Then the attacker can request that the legitimate peer use the tunneled method

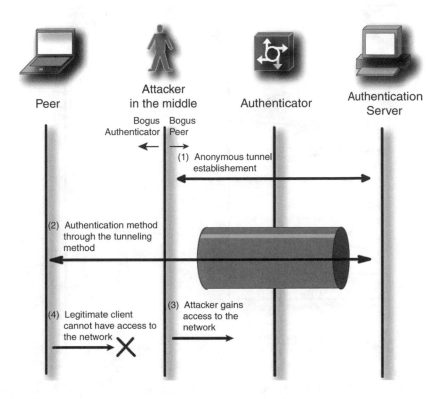

FIGURE 9.10 Man-in-the-middle attack in the compound authentication method.

but with or without use of a tunneling method between the legitimate peer and the attacker; that way, it makes both the legitimate authenticator and the back-end authentication server communicate with the legitimate client while running the method inside the tunneling method. As a consequence, the attacker is able to gain access to the network without using his or her credentials.

A solution for this class of MITM attack is to cryptographically bind the authentication methods of a compound authentication method. Two types of solutions are considered: the compound MAC (Message Authentication Code) method and the compound key method. In the compound MAC method, a MAC is used for protecting the authentication messaging as well as subsequent data traffic from a set of sequentially applied component MACs, where each component MAC is calculated regarding the corresponding component authentication method. The compound key method is based on deriving a key used for protecting the authentication messaging as well as subsequent data traffic from the component keys, where each component key is derived from the corresponding component authentication method.

9.4.3   IEEE 802.11i

When the IEEE 802 LAN operates over wireless media, security is one of the biggest concerns, especially because of the shared access nature of 802.11. WEP,[7] the encryption mechanism used in 802.11, is known to be insecure[4] and, therefore, 802.11i[23] is proposing new security algorithms.

802.11i defines the concept of the *Robust Security Network Association (RSNA)*. RSNA provides two basic subsystems, one based on the RC4 algorithm called the TKIP and the other is based on the Advanced Encryption Standard (AES) algorithm. The latter consists of the Counter CBC-MAC Protocol (CCMP) algorithm. Both of these subsystems rely on IEEE 802.1X to transport EAP messages, to provide authentication services, and to deliver cipher keys (derived from the authentication services). The cipher keys are used for protecting 802.11 MAC frames exchanged among stations and an AP or among stations.

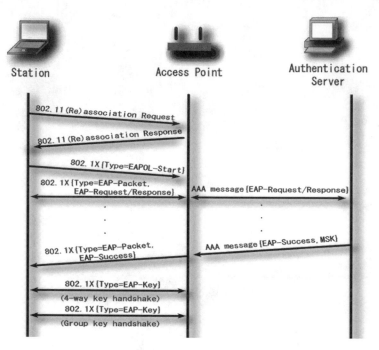

FIGURE 9.11 A sample sequence for 802.11i.

9.4.3.1   TKIP

As the predecessor of TKIP, WEP was full of security holes that were fixed by TKIP. TKIP was written to replace WEP so that they both run on the same MAC processor that is capable of using RC4 encryption. TKIP uses 128 bits for encryption and 64 bits for authentication. By use of a cryptographic message integrity check (MIC), forged messages are avoided. Another difference between the two algorithms is their use of IVs. WEP allows the reuse of IVs, which enables an attacker to decrypt data by monitoring and to record the traffic over a period of time. This can also be done because RC4 contains many weak keys that can be exploited. TKIP, on the other hand, specifically performs various functions to avoid this same problem: It never uses the same IV value more than once for any particular session key and uses incrementing IV values, regularly generating a new random session key before the IV counter overflows and a mixing of the IV value and the session key to derive the packet's RC4 key. By refreshing the keys often, the system is less vulnerable to key reuse. Furthermore, built into the TKIP system are preventive countermeasures that activate when a possible attack is in progress. These include MIC failure detection and key refreshing. Finally, TKIP also enforces the use of an EAP-based algorithm to authenticate.

9.4.3.2   CCMP

CCMP is an AES-based algorithm designed to replace the RC4-based ciphers in the long run. First, a new cipher is required, which means this new system cannot be implemented immediately with existing hardware because it is not possible to force current host processors to perform computations. This algorithm uses the CTR mode to provide encryption of data and uses CBC-MAC to provide data integrity; two separate passes, though, are needed to ensure both these attributes.

A sample sequence for the 802.11i protocol exchange is illustrated in Figure 9.11. Before IEEE 802.1X authentication is performed, the station associates with the AP. Once an association is created, the EAP message exchange starts. EAP messages are exchanged between the station and the back-end server through the AP. The AP acts as a pass-through device of EAP messages, where EAP messages are encapsulated in IEEE 802.1X packets and AAA messages (e.g., RADIUS or Diameter messages), when it communicates with the device and the authentication server, respectively. As a result of successful EAP authentication, a Master Session Key (MSK) needs to be established between the station and the back-end server, which, in turn, is used for establishing 802.11i cipher keys between the station and the AP. This means that a specific class of EAP authentication methods that are capable of deriving keys as well as authenticating a peer must be used for IEEE 802.11i. When the EAP authentication is successful, IEEE 802.1X EAPOL-Key messages are exchanged between the station and the AP in order to make sure that the two entities share the same MSK and to derive unicast and multicast cipher keys from the MSK. Note that IEEE 802.1X EAPOL-Key messages are also used in the other mode of IEEE 802.11i, which is based on using a preshared secret that is preconfigured between the station and the AP instead of using EAP.

9.5   3G Networks

Third-generation (3G) cellular systems have better security features. This is clearly because 3G cellular systems have evolved from the first- and second-generation (2G) cellular systems. For example, the 2G cellular systems have significantly enhanced the security along with using digital radio, unlike Advanced Mobile Phone Service (AMPS). Along the same lines, 3G cellular systems have further enhanced the security of 2G systems by making them quite robust to unauthorized use and eavesdropping. In spite of many security enhancements, there are still some security risks, albeit small, in 3G systems.

9.5.1   Vulnerabilities and Attacks in 3G Systems

In the following subsections, we describe the vulnerabilities and attacks that are possible in current 3G systems.

9.5.1.1   Vulnerabilities

The 3G-system security is built to enhance that of 2G.[24] In spite of the many security improvements, there are still some potential issues with 3G security:[25–27]

- The International Mobile Subscriber Identity (IMSI) is the permanent identifier of the MS. For security purposes, the network derives a Temporary Mobile Subscriber Identity (TMSI) from the IMSI and asks the mobile to use the TMSI. However, the IMSI is sent in cleartext to the Visitor Location Register/Serving Node (VLR/SN) when allocating the TMSI to the user. This is needed because no Authentication and Key Agreement (AKA) procedures can be performed before the device identity is known.
- The transmission of the IMEI (International Mobile Equipment Identity) is not protected. Though the IMEI is not a security feature in itself, the fact that the IMEI is sent to the network only after the authentication of the SN reduces the vulnerability of the system.
- A user can be enticed to camp on a false base station (BS). Once the user camps on the radio channels of a false BS, the user is out of reach of the paging signals of the SN.
- Hijacking outgoing/incoming calls in networks with disabled encryption is possible. The intruder poses as an MITM and drops the user once the call is set up.
- The architecture does not protect against compromised authentication vectors that have not yet been used to authenticate the Universal Subscriber Identity Module (USIM). Thus, the network is vulnerable to attacks using compromised authentication vectors that have been intercepted between generation in the authentication center and usage in the network.
- The links between the Home Environment (HE) and the SN are assumed to be trusted links.

Some attacks that can be launched taking advantage of the vulnerabilities above are discussed in references 25, 26, and 27.

9.5.1.2   Attacks

We now present some of the attacks possible and discuss how current 3G security features address them.

9.5.1.2.1   *Impersonation of the Network*

There are at least two ways to launch an impersonation of the network attack. These attacks are typically aimed at eavesdropping on user data or devising a way around data confidentiality measures. The first way to launch an impersonation of the network attack is by suppressing encryption between the user and the attacker. In this attack, the attacker impersonates a false BS and, while communicating with the user, attempts to trick the user into suppressing communication encryption. 3G security can thwart this attack if the device's security configuration is set appropriately. The device can be set to require encrypted communications, in which case the attack will fail. However, it is incumbent upon the device user to ensure that this setting is properly configured. Therefore, an intelligent attacker could lie in wait, attempting to suppress encryption with all devices it encounters until it finds a suitably misconfigured device.

The second way to launch an impersonation of the network attack is by forcing the use of a compromised cipher key. In this case, the attacker impersonates a BS and must possess a compromised authentication vector. With this information, the attacker can force the user to use the compromised vector and its associated key, because the user has no control over the cipher key to be used in communication. In most cases, 3G security will prevent this type of attack. As discussed earlier, a sequence number in the challenge generated by the USIM is used to verify freshness of the cipher key. Therefore, a replay attack would be thwarted. However, if the attacker were able to capture the key after it was generated by the authentication center but before it was used in the network, then the key would be compromised but still fresh. In this case, the attack would be successful. Therefore, it is feasible that an impersonation attack could be successfully launched, and the attacker would capture all communications.

9.5.1.2.2 Impersonation of the User

There are at least three ways to launch an impersonation-of-the-user attack. In this type of attack, the attacker attempts to impersonate the valid user for at least some part of a communication sequence in order to gain some illicit advantage. In the first type of attack, the attacker uses a compromised authentication vector to pose as a legitimate user. As has been previously demonstrated, the use of a sequence number by the USIM during the authentication response prevents a successful replay attack. However, if the vector is stolen after it has been created by the authentication center but before it has been used, then the attack will be successful. The second type of impersonation attack involves a simple replay attack. In this case, the attacker eavesdrops on an authentication sequence between a valid user and the network. The attacker then attempts to authenticate itself with the network by replaying the authentication sequence. In this case, 3G security will thwart the attack because of the sequence number used by the USIM in the authentication response. The network will realize that the sequence number has been previously used and will simply ignore the authentication sequence.

The third type of impersonation attack involves channel hijacking. Channel hijacking can be achieved in various ways, but the basic concept is that the attacker fools the network into setting up an outgoing call channel for the user and then takes over the channel and fraudulently makes calls in the name of the user. If encryption on the user device is enabled, then 3G security will thwart this type of attack. The data integrity protection for signaling messages that 3G network provides will prevent this type of attack from occurring. However, if the user has disabled encryption, then the attack may be successful because the network will be unable to determine that the user communication is, in fact, coming from the attacker.

9.5.1.2.3 Eavesdropping

Eavesdropping is difficult even when network level encryption is disabled. This is because the air link performs mandatory scrambling of user data as part of the radio signal processing. Successful eavesdropping is possible, for example, in the CDMA2000 3G system if the combination of a 42-bit-long code mask, the long code, and digital logic are somehow solved by the eavesdropper. If network-level encryption is enabled on top of the physical-layer scrambling, eavesdropping becomes exceedingly difficult.

9.5.1.2.4 DoS Attacks

There are many ways in which one could attempt a DoS attack. Some of these attempts can be thwarted by the current 3G security schemes, whereas the rest cannot be. We break them up into two groups: those that are accomplished at the network layer and those that are accomplished at the physical layer.

9.5.1.2.4.1 DoS Attacks at the Network Layer — The first method of attack is through user deregistration request spoofing. In this attack, the attacker spoofs a deregistration message from a mobile device to the network, causing the network to terminate its session with the device. 3G security addresses this attack by providing entity authentication and data integrity assurances by encrypting all signaling communication, as discussed in Section 9.5.2.

The second method of launching a DoS attack is through location update request spoofing. This is similar to the previous attack except that instead of a deregistration request being sent, a location update request is sent and the user can no longer reach the network because the network believes the user is in an incorrect location. Again, 3G security addresses this attack by providing entity authentication and data integrity assurances by encrypting all signaling communication.

The third DoS attack involves "camping" on a false BS. In this attack, the attacker impersonates a BS and the user associates itself with this false BS, thereby losing connectivity with the real 3G network. 3G security does not address this attack, and in fact users are vulnerable to this kind of attack. In this case, the attacker would be able to capture several encrypted messages. The attacker's ability to cause harm is dependent on the underlying strength of the key and encryption algorithms that are used.

The fourth DoS attack involves "camping" on a false BS or MS. In this attack, the BS/MS acts as a repeater in relaying messages between the real network and the user. The false station can permit some messages to pass while dropping others. Because of the data confidentiality and integrity assurances provided by 3G, however, the false station will not be able to alter or snoop messages. Rather, certain messages may be dropped. In this case, 3G security does not address this type of attack.

9.5.1.2.4.2 DoS Attacks at the Physical Layer — Power control is an important part of 3G systems operation.[28] Appropriately increasing or decreasing the transmitted signal power leads to the advantages of improved Radio Frequency (RF) signal quality, improved battery life, reduced interference, and system capacity gains. Power control can be used as a tool to launch DoS attacks. For example, a node can artificially ask the BS to keep increasing the transmitted signals power although the received signal has a high Signal to Interference Noise Ratio (SINR). Because power is a key resource in CDMA systems, by stealing an inappropriate amount of power, mobiles can effectively reduce the total available power in the system, thereby leading to reduced throughputs and high error rates and even blocking new calls from being admitted into the system. Similarly, a node may choose to neglect the power-down messages from the BS and continue to increase its transmit power on the reverse link. Because CDMA systems attempt to equalize the powers of all receiving signals to counter the well-known near-far problem, a single malicious node can cause the transmit power of all other legitimate mobiles' transmit power to go up, resulting in a considerable reduction of their battery life.

Evolution of CDMA2000 (i.e., 1x-EV-DO and 1x-EV-DV) and UMTS (i.e., High Speed Downlink Packet Access [HSDPA]) use fast channel feedback procedures on the reverse link. Typically, such channel quality estimates are used by the BS to do channel-sensitive scheduling, like Proportional Fair. Although channel state–dependent schedulers minimize system resources needed to meet certain target throughput requirements, they also suffer from vulnerabilities of opportunistic communications. A mobile that is located in a poor RF location can continue to report that its channel condition (e.g., on the DRC channel in 1x-EV-DO) is very good and hog most of the resources of the BS. This will result in reduced or even a DoS to other MSs. Autonomous data rate selection on the reverse link of 1x-EV-DO poses problems similar to that of abusing the power control algorithm. A group of malicious stations can improperly choose to transmit at high data rates on the reverse link, causing the noise floor to rise and hence increasing the transmission power of other benign users.

9.5.1.2.5 *Identity Caching Attacks*

There are at least two ways to launch an identity caching attack. The first method is a passive identity caching attack. In this attack, an attacker monitors network communication in the hopes that the network may request that the user send its identity to the BS. 3G security thwarts this attack in several ways. First, all signaling information is encrypted; therefore, the attacker will likely be unable to capture the identity in plaintext form. Second, 3G systems make use of temporary identities that are associated with permanent identities. So, even if the attacker is able to cache the user's identity, it will most likely be the temporary identity only. Permanent identities are rarely used, so attempting to cache it is highly inefficient.

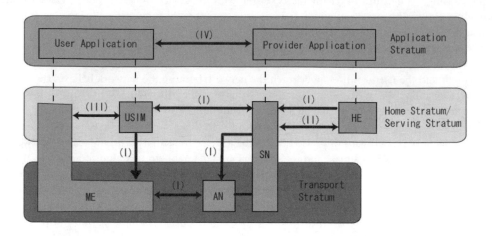

FIGURE 9.12 3G security architecture and the 3G security model.

The second identity caching attack is an active attack. In this attack, an attacker with a false BS requests that the user send its permanent identity in plaintext. 3G security addresses this type of attack through the use of a group encryption key. With group encryption, the attacker will be able to narrow the user's identity to only a certain group of users but will be unable to explicitly identify the user.

9.5.1.2.6 *Attacks in All-IP Radio Networks*

Asynchronous Transfer Mode (ATM)-based transport was used in the earlier version of UTRAN (Release 99) and CDMA2000 RAN. However, the newer specifications (Release 5) are moving to an IP-based radio access network (RAN), or more generally, to All-IP networks (i.e., both access and core networks). In general, it is fair to say that IP-based transport would be more vulnerable than an ATM RAN from a security standpoint, especially because multiple service providers are likely to share a high-bandwidth UTRAN to maximize the advantages of IP transport. Similarly, the move toward IP-based core and radio access networks in CDMA2000 systems also needs to address the issue of maintaining high security. In the absence of appropriate security measures, malicious users can now access the user information through the back-haul as opposed to the air link (which is already secure to some extent). Moreover, network elements in an RAN can freely access user traffic belonging to other RANs if they wanted to. Additionally, IP-based RAN or All-IP networks open themselves to all the attacks prevalent in an IP network. Hence, using IPSec-based VPN tunnels among different network elements such as BSs (Node B) and BS Controllers (RNC) would greatly improve the confidentiality and integrity of user traffic.

9.5.2 3G Security Architecture

The 3G-security architecture[31,32] consists of five security feature groups, with each meeting certain threats and accomplishing certain security objectives. The five groups are network domain security, user domain security, application domain security, visibility and configurability of security, and network access security (see Figure 9.12).

9.5.2.1 Network Domain Security

Network domain security (NDS) enables nodes in the provider domain to securely exchange signaling data and protect against attacks on the wired network. NDS is the set of security features that enable nodes that are within a service domain to protect against wireline attacks. Specifically, a 3G network provides protection for both IP (for packet-switched part RANs) and MAP (Mobile Application Part, used for circuit-switched part of RANs) protocols.

For the IP protocol, NDS supports IPSec. This is known as NDS/IP. NDS/IP provides services for data integrity, data origin authentication, antireplay protection, confidentiality, and limited protection against traffic flow analysis. Using IPSec in the Encapsulating Security Payload (ESP) mode provides all these services. Among different security gateways (SEGs), key management and distribution is handled by the Internet Key Exchange (IKE) protocol. IKE is used to negotiate, establish, and manage security associations among different parties that need to establish secure connections. For the MAP protocol, NDS supports MAPsec. This is known as NDS/MAP. NDS/MAP provides services for data integrity, data origin authentication, antireplay protection, and confidentiality. These services are provided by using MAPsec in one of three different modes. Mode 0 offers no protection; Mode 1 offers integrity and authenticity protection; Mode 2 offers integrity, authenticity, confidentiality, and antireplay protection.

9.5.2.2   User Domain Security

User domain security is a set of security features that secure access to MSs, and it includes user-to-USIM authentication and USIM-terminal links.

- *User-to-USIM authentication:* This feature ensures that access to the USIM is restricted until the USIM has authenticated the user. The user that provides a shared secret (such as a Personal Identification Number) to the USIM completes the authentication process. Once the USIM verifies the user input, the authenticity of the user can be established and access to USIM is allowed.
- *USIM-terminal link:* This ensures that access to a terminal is restricted to authorized USIMs. The USIM should store a shared secret, which can be provided to the terminal for authentication. Failing to provide the shared secret by the USIM will result in denied access to the terminal.

9.5.2.3   Application Domain Security

Application domain security contains the set of features that enables applications in the user and in the provider domain to securely exchange messages. The security feature group includes, but is not limited to, secure messaging between the USIM and the network. Secure messaging between the USIM and the network enables messages to be securely transferred over the network to applications on the USIM, according to the level of security chosen by the network operator or the application provider. Operators and application providers are also given the capability to create applications on the USIM with the use of USIM Application Toolkit.

9.5.2.4   Visibility and Configurability of Security

Visibility and configurability of security, though not affecting the underlying security implementation of the 3G network, are nevertheless important user-level security features. Visibility and configurability of security enables the user to determine whether a security feature is in operation and whether the use and provision of services should depend on the security feature.

9.5.2.5   Network Access Security

This security feature group contains the features that provide users with secure access to 3G services and protects against attacks on the radio access link. Its primary concern is to protect the radio access link from any attacks. There are four security features in this group: (1) user identity confidentiality, (2) entity authentication, (3) confidentiality, and (4) data integrity. These security features are further discussed below.

9.5.2.5.1   User Identity Confidentiality

This security feature seeks to ensure that the permanent user identity (the IMSI) of a user is not compromised. Three related aspects are that an attacker cannot learn the IMSI of a user by eavesdropping on the radio access link, the location of the user cannot be determined by eavesdropper, and an eavesdropper cannot determine whether different services are delivered to the same user as required by user untraceability. To achieve these objectives, users are identified in the visited serving network (SN) by a TMSI and a packet (P-TMSI). The Visited Location Register (VLR) associates users' IMSI to their TMSI,

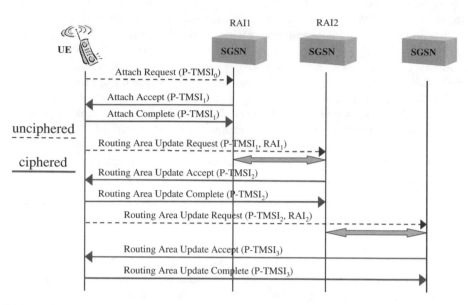

FIGURE 9.13 Using P-TMSI as a temporary user identity.

whereas the Serving GPRS Support Node (SGSN) associates users' IMSI to their P-TMSI. The temporary identities should change during communication of long periods to ensure user untraceability.

The main mechanism for providing user identity confidentiality is by identifying a user with a temporary identity or with an encrypted permanent identity. Below we explain how the temporary identity is used. For the purposes of illustration we consider a UMTS network as a typical example of a 3G system.

9.5.2.5.1.1 Identification with Temporary Identities — Assume that a User Equipment (UE) needs to attach to the packet-switched (PS) domain of a UMTS network. For this purpose, it should send an *Attach Request* message, which should embed some kind of UE identity in order to make it possible for the network to perform identification and authentication procedures. The problem at this point is that the *Attach Request* message cannot be transmitted ciphered because some ciphering parameters (e.g., a ciphering key) need to be first negotiated between the UE and the network (the SGSN in particular). For this reason, the messages transmitted by the UE before and during the ciphering negotiation must be transmitted unciphered. This fact makes the system vulnerable to malicious monitoring, and specific mechanisms should be devised to provide an acceptable level of protection against sensitive data, such as the user identity.

To accomplish this goal, the network allocates a temporary identity to the UE. As noted earlier, this is referred to as P-TMSI for the PS domain. Note that for the purposes of the following discussion we will focus only on the PS domain. You should keep in mind, though, that similar procedures apply for the CS domain as well. P-TMSI is unique within a specific routing area and is allocated either explicitly (with the P-TMSI Reallocation procedure; see reference 33) or implicitly in the context of some mobility management procedures.

P-TMSI is used as illustrated in Figure 9.13. Whenever a P-TMSI is allocated to a UE, this value is stored in the USIM. Subsequently, when the UE sends an *Attach Request* it identifies itself with the P-TMSI value stored in USIM. If no valid P-TMSI is stored in USIM (e.g., at the very first *Attach Request*), the UE uses its permanent identity (the IMSI) instead. In the example shown in Figure 9.13, the used P-TMSI value is denoted as $P\text{-}TMSI_0$. The $P\text{-}TMSI_0$ combined with the old RAI (which is also included in the *Attach Request* message) points to a unique Mobility Management (MM) context in the network, which contains the UE's permanent identity (the IMSI).

When the network (i.e., an SGSN in this example) responds to the Attach Request with an *Attach Accept* message, it allocates a new P-TMSI value to the UE, say, $P\text{-}TMSI_1$. Note that the *Attach Accept* is typically transmitted in ciphered mode; hence, P-TMSI$_1$ is protected against malicious monitoring. In this way, the new P-TMSI value allocated to the UE is considered to be known only to the UE and the network.

The next time the UE will need to perform a mobility management procedure, it will use P-TMSI$_1$ to identify itself. For example, as shown in Figure 9.13, a subsequent *Routing Area Update Request* will include P-TMSI$_1$. (In fact, the *Routing Area Update Request* does not explicitly contain the P-TMSI value, but this value can be derived by other means, which are not important in the current discussion.) Note that this request message is transmitted unciphered. In response, the SGSN will allocate a new P-TMSI value, P-TMSI$_2$, to the UE within the *Routing Area Update Accept* message. In this way, the network allocates a new P-TMSI to the UE *every time its previous one is transmitted unciphered*.

With the above mechanism, the UE uses P-TMSI to identify itself instead of its IMSI. It should be noted that P-TMSI is meaningful only to the serving network, which contains an association between the allocated P-TMSIs and the IMSIs. Typically, P-TMSI provides no information to a malicious user about the UE's permanent identity.

9.5.2.5.1.2 P-TMSI Signature — Suppose now that an unauthorized user sends a *Routing Area Update Request* using a P-TMSI value that is arbitrarily chosen. If this P-TMSI value is allocated to an authorized user, the network may erroneously believe that the *Routing Area Update Request* comes from the legitimate user. Therefore, using only P-TMSI, one can provide identity confidentiality but may leave "open holes" in the system for malicious attacks. To work around this security problem, the network will need to initiate an authentication procedure before processing the *Routing Area Update Request*. However, this takes up valuable resources (e.g., additional messages sent to the UE) and adds considerable delays.

To avoid the frequent authentication procedures, the *P-TMSI Signature* may be used. P-TMSI is an additional parameter with a 3-octet length and is used as follows. Each time the SGSN allocates a new P-TMSI value to a UE, it also allocates a *P-TMSI Signature* that is associated with that P-TMSI. On the UE side, *the P-TMSI Signature* is stored in the USIM together with P-TMSI, whereas on the network side, both the *P-TMSI Signature* and the *P-TMSI* currently allocated to a UE are stored in the UE's MM context.

Every time a UE sends a message that includes a P-TMSI value, it should also include the associated *P-TMSI Signature* value. Hence, the SGSN receives a [*P-TMSI, P-TMSI Signature*] pair. The SGSN checks whether the received P-TMSI has been allocated, and if so, it checks the validity of this P-TMSI by means of the received *P-TMSI Signature*. If the received *P-TMSI Signature* is correct, the received P-TMSI is considered valid; otherwise, it is considered invalid. This way, the probability an unauthorized user to choose a valid [*P-TMSI, P-TMSI Signature*] pair becomes negligible.

9.5.2.5.2 Entity Authentication

The entity authentication aims at providing the following:

- *User authentication*, which ensures that the serving network confirms the identity of the user
- *Authentication mechanism agreement*, which ensures that the user and the serving network can securely negotiate the mechanism for authentication and key agreement that they will use subsequently
- *Network authentication*, which ensures that the user confirms that he or she is connected to a serving network authorized by the user's HE to provide 3G services; this includes the guarantee that this authorization is recent

Before we discuss in detail how entity authentication is provided, we must become familiar with the authentication vectors — that is, a special set of parameters used to authenticate an entity.

9.5.2.5.2.1 Authentication Vectors — The information used to authenticate the user *and the network* is composed of a number of so-called 3G (or UMTS) authentication vectors (see Figure 9.14). Each vector contains five authentication parameters: (1) a random challenge, RAND (128 bits long); (2) an expected user response, XRES (variable length); (3) a cipher key, CK (128 bits long); (4) an integrity key, IK (128 bits long); and (5) a network authentication token, AUTN. Note that a GSM/GPRS authentication vector contains only the first three parameters.

Authentication Vectors				
1- RAND	XRES	CK	IK	AUTN
2- RAND	XRES	CK	IK	AUTN
....				
N- RAND	XRES	CK	IK	AUTN

FIGURE 9.14 UMTS authentication vectors.

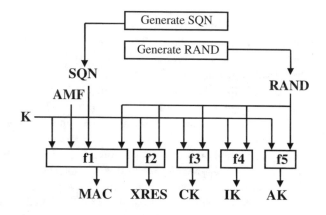

$$\text{AUTN} := \text{SQN} \oplus \text{AK} \parallel \text{AMF} \parallel \text{MAC}$$

FIGURE 9.15 The calculation of an authentication vector.

The purpose of the RAND is to provide the MS with a random number to be used to calculate the authentication response RES and both the CK and IK. The same calculation is performed on the network side. The XRES is the response expected by the user after an authentication challenge. The actual user response RES is compared with the XRES to determine whether the mobile, or more correctly, the USIM card, is authentic. The CK is used for the ciphering of information, both at the network and at the mobile. Likewise, the integrity key is another key used at both the network and the mobile to provide data integrity. Finally, the authentication token AUTN is used to provide the mobile with a means to authenticate the network. Through the AUTN parameter, the mobile authenticates the HE, which supplied the authentication vectors.

Figure 9.15 indicates how the various authentication parameters are calculated. Each time the network needs to generate a new authentication vector, fresh RAND and Sequence Number (SQN) values are generated. As shown in Figure 9.15, the RAND and the long-term secret key K are used as inputs to the authentication functions f2, f3, and f4, which output XRES, CK, and IK, respectively. The long-term secret key K is generated when a new subscription is created. This key is never transmitted and is securely stored at the MS (in the USIM) and at the Authentication Center (AuC) in the network. The authentication functions f1 to f5 shown in Figure 9.15 can be operator specific. They are executed by the USIM when a mobile is challenged with a RAND value. On the network side they are executed in the AuC when new authentication vectors need to be calculated.

As illustrated in Figure 9.15, the AUTN is the concatenation (denoted as ||) of three individual parameters: (1) the parameter derived by the exclusive OR between the SQN and the anonymity key, AK; (2) the authentication management field, AMF; and (3) the message authentication code, MAC. The

SQN is a 6-octet sequence number generated before a new authentication vector is calculated. A method to generate SQN is specified in appendix C of reference 31. The AK is an anonymity key used to conceal the SQN because the latter may expose the identity and location of the user. The concealment of the SQN is to protect against passive attacks only. If no concealment is needed, then f5 ≡ 0, AK = 0. The AMF is a parameter that may be used to indicate the algorithm and the key used to generate a particular authentication vector. By means of AMF, a mechanism is provided to support the use of multiple authentication and key agreement algorithms. This mechanism is useful, for example, for disaster-recovery purposes. The MAC parameter will be explained in Section 9.5.2.5.2.3.

9.5.2.5.2.2 Acquisition of Authentication Data — Assume that an SGSN needs to authenticate a UE; also assume that the SGSN maintains no authentication vectors for that UE. These authentication vectors may be acquired (1) from the UE's HE (typically, an HLR) or (2) from another SGSN, which previously served the UE and has already acquired the necessary authentication information. The second method is typically more efficient because, first, it minimizes long-distance accesses to the HE (in case of roaming) and, second, because it prevents overloading the HE and the corresponding bottlenecks. In addition, it may be considered faster when long-distance circuits are involved between the SGSN and the HE.

When an SGSN needs to authenticate a UE and the authentication vectors cannot be retrieved from another SGSN, it sends a *Send Authentication Info* (IMSI) command to the appropriate HLR, which contains subscription information for that particular UE. Before sending the command, the SGSN needs to know the IMSI of the UE. This IMSI value is used as a parameter to the *Send Authentication Info* command and is used for deriving the SS7 address of the appropriate HLR. For roaming UEs, the HLR and the SGSN belong to different 3G networks.

After receiving the command *Send Authentication Info* (IMSI), the HLR returns to the requesting SGSN a number of authentication vectors, typically from 1 to 5. The HLR either produces these vectors at the moment it receives the *Send Authentication Info* command (by sending another command to the AuC) or it simply retrieves these vectors from its storage, if these are already available (e.g., from a previous operation). Evidently, because an SGSN always retrieves authentication information from the user's HE, the authentication procedure does not depend on the SN where the UE is located.

Every time the SGSN needs to authenticate the UE, it uses an authentication vector that *has not been used before*. If all the available authentication vectors have already been used, the SGSN will request a fresh set of authentication vectors from the HLR. Also, when the SGSN transfers the authentication vectors to another SGSN (e.g., during an inter-SGSN routing area update), it sends only the authentication vectors marked as not used. Note that in GSM an authentication vector can be used more than once (the exact number is specified by the operator); however, in UMTS it is mandatory that each vector be used only once.

9.5.2.5.2.3 The UMTS Authentication Procedure — When the SGSN needs to authenticate a specific UE and already maintains a list of usable authentication vectors, it chooses an authentication vector from that list and sends to the UE an *Authentication and Ciphering Request* message that includes the RAND and AUTN values of the chosen authentication vector. Optionally, when a mobile is an old GSM mobile that does not support the UMTS authentication procedure, the network may not send the AUTN. In such cases, the mobile infers that the network perform a GSM authentication challenge (as opposed to the UMTS authentication challenge), and it will use the GSM-specific algorithms[34] to calculate the authentication response and the ciphering key (in GSM, this key is denoted as Kc).

Upon receiving the RAND and the AUTN, the USIM is expected to perform the calculation illustrated in Figure 9.16. Using the RAND and K, and by applying the authentication functions f2, f3, f4, the RES, the CK, and the IK are derived. Also, the expected MAC, XMAC, is calculated, as well as the SQN. Subsequently, USIM verifies (1) that the derived XMAC is equal to the received MAC, and (2) that the SQN is in the allowed range. In such case, the mobile considers the network as authentic and trusted for safe provision of 3G mobile services. Therefore, it responds to the authentication challenge with an authentication response message that includes the derived RES value. However, if XMAC is not equal to MAC, or if the SQN is not in the allowable range, the network authentication fails and the mobile responds with an authentication failure message.

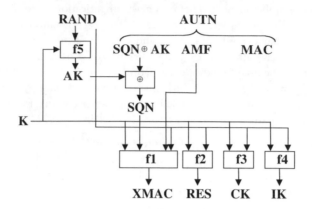

Verify MAC = XMAC

Verify that SQN is in the correct range

FIGURE 9.16 The calculation performed by the USIM after being challenged by the network.

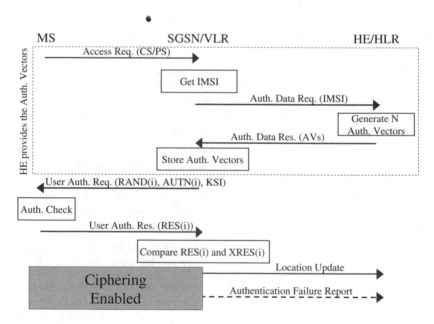

FIGURE 9.17 Typical steps during authentication and key agreement.

When the network receives the authentication response from the mobile, it simply checks whether the RES matches with the XRES. If they do, the mobile is authenticated. However, when a mismatch between RES and XRES is identified, the USIM is not considered authentic and the network responds with an authentication reject message.

The mechanism for providing authentication (of both user and network) and key agreement is depicted in Figure 9.17. This figure effectively illustrates the sequence of messages and events that take place in the context of this mechanism.

In this example, we assume that the mobile tries to attach to the PS domain of a serving network. Therefore, the mobile station transmits some kind of access request message, which is routed to the SGSN. (Note that the terms MS and User Equipment [UE] are used interchangeably.) In a similar manner, when the MS requests access to the CS domain, the access request is routed to the VLR. Here the access request can be considered a generic message corresponding to any kind of access request message, such as *Attach Request, Routing Area Update Request,* and so forth. The SGSN finds the IMSI of the MS, which either is included in the access request or is derived from the P-TMSI, as discussed previously. If the SGSN needs to authenticate the UE and has no authentication vectors for that UE, it must fetch these vectors from the HLR. As already mentioned, the HLR usually responds with one to five authentication vectors, which are stored at the SGSN. Subsequently, the SGSN uses an authentication vector, say with index i, to challenge the mobile. Note that the transmitted authentication request message, besides RAND(i) and AUTN(i), contains the Key Set Identifier (KSI), also referred to as CKSN.

As illustrated in Figure 9.17, if the mobile and network authentication is successful (i.e., RES(i) is equal to XRES(i)), the ciphering mode is enabled and the SGSN updates the HLR with the new location of the mobile. If, however, the authentication fails, the SGSN sends an Authentication Failure Report back to the HLR to report about this event.

The UMTS network access security mechanism contains another enhancement as compared with the equivalent GSM security mechanism. In particular, the USIM contains a mechanism that limits the amount of data that is protected by a given set of ciphering and integrity keys or that limits the time period in which the same set of CKs can be used. In other words, the USIM keeps track of the time period a given key set is being used, and upon expiry, it may initiate itself an authentication procedure. This procedure follows the same principles discussed already but could be triggered by the MS.

9.5.2.5.3 *Confidentiality*
The confidentiality of user data and signaling is provided by means of ciphering. The following security features are provided for establishing confidentiality on the network access link:

- *Cipher key agreement:* The UE and the SN agree on a cipher key that they will use subsequently.
- *Cipher algorithm agreement:* The UE and the SN *securely* negotiate the cipher algorithm that they will use subsequently.
- *Confidentiality of user data:* User data cannot be overheard on the radio access interface.
- *Confidentiality of signaling data:* Signaling data cannot be overheard on the radio access interface.

As already seen, the cipher key agreement is realized during the UE and network authentication procedure and results in the establishment of a common CK at both the network and the UE (stored in the USIM). Also, the cipher algorithm agreement is realized during the security mode negotiation (see reference 31). In the following, we will discuss how confidentiality of user data and signaling is provided by means of ciphering.

Figure 9.18 illustrates the processing performed for ciphering (at the transmitting side) a plaintext block (i.e., a clear block of user data or signaling) and the processing performed for deciphering (at the receiving side) a ciphertext block (i.e., a ciphered block of user data or signaling). This processing takes place at the Radio Link Control (RLC) layer, and therefore, the plaintext block is effectively an RLC block. As shown in Figure 9.18, the ciphering of the plaintext block is performed by XORing the plaintext block with a keystream block. The latter is derived at the output of the UMTS Encryption Algorithm (UEA), which is generally denoted as function f8. The parameters applied to the input of f8 are CK, COUNT-C, BEARER, DIRECTION, and LENGTH. At the receiving side, the same keystream block, used to cipher a plaintext block, is derived by applying exactly the same inputs to function f8. Finally, the original plaintext block is derived by XORing again the received ciphertext block with the keystream block. Next we briefly explain the input parameters to the ciphering function f8. More detailed information can be found in reference 31.

The CK used to cipher a plaintext block can be either the CK corresponding to PS domain (CK_{ps}) or the CK corresponding to the CS domain (CK_{cs}). Because the RLC layer at the mobile does not know whether CS or PS is the destination of a message, it cannot apply the CK on a per-message basis. The general rule that governs the selection of the CK is that the mobile will apply the CK of the CN domain for which the most recent security mode negotiation took place. Hence, if the mobile is attached only to the CS domain, it will use CK_{cs}. If it subsequently attaches to the PS domain and a security mode

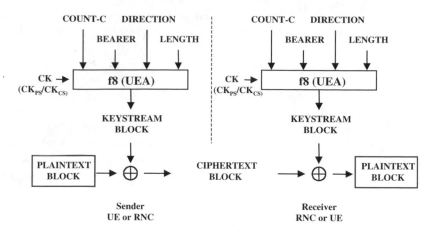

UEA0: No encryption
UEA1: Kasumi

FIGURE 9.18 Ciphering/deciphering method.

negotiation takes place with the PS domain, then the mobile will use CK_{ps} for all subsequent uplink messages (going to either CS or PS domain).

COUNT-C is the ciphering sequence number and is 32 bits long. The update of COUNT-C depends on the transmission mode and the logical RLC channel used. For example, in RLC Acknowledged Mode, one part of the COUNT-C (12 bits) contains the RLC sequence number that is available in each RLC PDU and the other part (20 bits) contains the RLC Hyper-Frame Number.[31] Therefore, a different COUNT-C value is applied to every plaintext block.

BEARER is the radio bearer identifier and is 5 bits long. There is one BEARER parameter per radio bearer associated with the same user and multiplexed on a single 10-ms physical layer frame. BEARER is input to the ciphering function f8 to avoid an identical set of input parameter values being used for a different key stream.

DIRECTION is the direction identifier and is 1 bit long. The value of the DIRECTION is 0 for messages on the uplink and 1 for messages on the downlink.

LENGTH is the length indicator and is 16 bits long. The length indicator determines the length of the required keystream block. LENGTH affects only the length of the keystream block but not the actual bits in it.

Up to now, only two UEAs have been defined: UEA0, which corresponds to no encryption, and UEA1, which applies the so-called Kasumi algorithm.

9.5.2.5.4 *Data Integrity*
The following security features are provided with respect to integrity of data on the network access link:

- *Integrity algorithm agreement:* The UE and the SN can securely negotiate the integrity algorithm that they will use subsequently.
- *Integrity key agreement:* The UE and the SN agree on an IK that they may use subsequently.
- *Data integrity and origin authentication of signaling data:* The receiving entity (UE or CN) is able to verify that signaling data has not been modified in an unauthorized way because it was sent by the sending entity (SN or UE) and that the data origin of the signaling data received is indeed the one claimed.

We have already seen that IK agreement is realized during the UE and network authentication procedure and results in the establishment of a common IK at both the network and the UE (stored in the USIM). Also, the integrity algorithm agreement is realized during the security mode negotiation (see

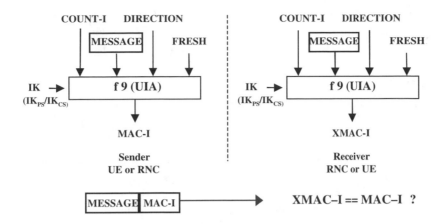

UIA1: Kasumi

FIGURE 9.19 Data integrity check and origin authentication method.

Reference 31). Next, we discuss how integrity protection of signaling data is provided. It must be noted that, after a Radio Resource Control (RRC) connection establishment and after the security mode setup, all dedicated control signaling messages (e.g., for mobility/session management, call control, radio resource control, etc.) are integrity protected.

Figure 9.19 illustrates the method used to provide integrity protection and origin authentication. This method is realized in the RRC layer, as opposed to ciphering, which is realized in the RLC layer. The transmitting side transmits along with a message a kind of electronic signature, called *Message Authentication Code* (denoted as MAC-I). The receiving side processes the message and derives the *expected MAC* value, XMAC, which is compared with the MAC sent by the originator. If the originator is the expected one (i.e., the one using the Integrity Key [IK], FRESH, and UIA previously negotiated), then, under normal conditions, the comparison of the expected and received MAC values will be successful and the message will pass the integrity check and will be accepted.

The input parameters to the UMTS Integrity Algorithm (UIA), generally called integrity function f9, must be synchronized for the integrity protection to work correctly. The IK applied for the integrity protection can be either the one negotiated with the CS domain (IK_{cs}) or the one negotiated with the PS domain (IK_{ps}). Because the RRC layer at the mobile does not know whether CS or PS is the destination of a message, it cannot apply the IK on a per-message basis. The general rule that governs the selection of the IK is the one mentioned for the CK in the previous section. That is, the mobile will apply the IK of the CN domain for which the most recent security mode negotiation took place.

The MESSAGE applied to the input of f9 is the original RRC message, which has to be integrity protected.

COUNT-I is the integrity sequence number and is 32 bits long. One part of COUNT-I (4 bits) contains the RRC sequence number that is available in each RRC PDU, and the other part (28 bits) contains the RRC Hyper-Frame Number.[31] Therefore, a different COUNT-I value is applied in every RRC message.

DIRECTION is the direction identifier and is 1 bit long. Its value is 0 for messages on the uplink direction and 1 for messages on the downlink direction. It is used to avoid computing message authentication codes for uplink and downlink messages with an identical set of input parameters to the integrity function f9.

FRESH is a random 32-bits parameter selected by the network during the security mode activation. There is one FRESH parameter value per user. As already seen, at connection setup the RAN generates a random FRESH value and sends it to the user in the RRC security mode command. This FRESH value is subsequently used by both the network and the user throughout the duration of a single connection.

The input parameter FRESH protects the network against replay of signaling messages by the user. In other words, this mechanism assures the network that the user is not replaying any old MAC-Is.

Up to this writing, only one UIA has been defined: UEA1, which is also called the *Kasumi* algorithm.

9.6 Summary

We presented several security issues in WLAN and 3G wireless networks. As we all know, wireless networks are in general more vulnerable to various attacks than are fixed wireline networks. However, it is evident from the discussion that 3G systems are far more secured than WLANs. This is because 3G security builds on top of GSM security (which has been deployed for more than a decade), fixed some security holes, and introduced new features. On the other hand, WLANs are still in their infancy. But due to the cost-effectiveness and ease of deployment, the WLAN is becoming the de facto wireless data network and is proliferating rapidly. Researchers and standards organizations are trying to repair the security holes that are present in the current standard. Although the IEEE 802.1X and 802.11i standards provide better authentication and data security and can eliminate most of the attacks, it will be always a challenging task because attackers are becoming smarter and devices are getting cheaper day by day. The scenario is little bit different in the case of 3G networks. Not only is it costly to install a 3G BS and launch attacks, but 3G systems are inherently secured. The enhanced authentication protocols have strengthened security by providing mutual authentication between the user and the serving network, as well as guaranteeing the freshness of cipher/integrity keys to the user. Encryption has been enhanced both in terms of longer encryption keys and stronger ciphering algorithms.[35] Encryption now terminates in the radio network controller instead of the BS, thus providing greater coverage of protected data. Core network security has also increased because signaling between network nodes is protected.

Acknowledgments

The authors would like to acknowledge Arun Roy, Kamesh Medepalli, and Latha Kant for their help in preparing this chapter.

References

1. Schneider, F. ed., *Trust in Cyberspace*, National Academy Press, 1999.
2. Avizienis, A., Laprie, J., and Randell, B., Fundamental concepts of dependability, UCLA CSD Report 010028, Aug. 2001.
3. Shirey, R., Internet security glossary, RFC 2828, Informational, May 2000.
4. Borisov, N., Goldberg I., and Wagner, D., Intercepting mobile communications: the insecurity of 802.11, Proceedings of International Conference on Mobile Computing and Networking, July 2001, pp. 180–189.
5. Blunk, L. and Vollbrecht, J., PPP Extensible Authentication Protocol, RFC 2284, http://www.ietf.org/rfc/rfc2284.txt.
6. IEEE Standard for Local and Metropolitan Area Networks, Port-based network access control, IEEE Std. 802.1X-2001.
7. IEEE Standard for Local and Metropolitan Area Networks, Wireless LAN Medium Access Control (MAC) and Physical Layer (PHY) Specifications, ANSI/IEEE Std. 802.11, 1999 Edition, 1999.
8. Blunk, L. and Vollbrecht, J.J., PPP Extensible Authentication Protocol (EAP), RFC 2284, March 1998.
9. Simpson, W., PPP Challenge Handshake Authentication Protocol (CHAP), RFC 1996, Aug. 1996.
10. Rivest, R., The MD5 Message-Digest algorithm, RFC 1321, Apr. 1992.
11. Eastlake, D. and Jones, P., US Secure Hash Algorithm 1 (SHA1), RFC 3174, Sep. 2001.
12. Zorn, G. and Cobb, S., Microsoft PPP CHAP (MS-CHAP) extension, RFC 2433, Informational, Oct. 1998.

13. Haller, N. and Metz, C., A one-time password system, RFC 1938, May 1996.

14. Aboba, B. and Simon, D., PPP EAP TLS Authentication Protocol, RFC 2716, Oct. 1999.

15. Rigney, C., Willens, S., Rubens, A., and Simpson, W., Remote Authentication Dial In User Service (RADIUS), RFC 2865, June 2000.

16. Calhoun, P., Loughney, J., Guttman, E., Zorn, G., and Arkko, J., Diameter Base Protocol, RFC 3588, Sept. 2003.

17. Dierks, T. and Allen, C., The TLS Protocol Version 1.0, RFC 2246, Jan. 1999.

18. Funk, P. and Blake-Wilson, S., EAP Tunneled TLS Authentication Protocol (EAP-TTLS), Internet draft, work in progress.

19. Palekar, A., Simon, D., Zorn, G., Saloway, J., Zhou, H., and Josefsson, S., Protected EAP Protocol (PEAP), Version 2, Internet draft, work in progress.

20. Rigney, C., RADIUS accounting, RFC 2866, June 2000.

21. Rigney, C., Willats, W., and Calhoun, P., RADIUS extensions, RFC 2869, June 2000.

22. Puthenkulam, J., Lortz, V., Palekar, A., and Simon, D., The compound authentication binding problem, Internet draft, work in progress.

23. IEEE Standard for Local and Metropolitan Area Networks, Wireless Medium Access Control (MAC) and physical layer (PHY) specifications: Medium Access Control (MAC) security enhancements, IEEE Std 802.11i/D4.0, May 2003.

24. 3GPP Technical Specification, 3GPP TS 33.120, Security Principles and Objectives, http://www.3gpp.org/ftp/Specs/2003-06/R1999/33_series/33120-300.zip.

25. 3G Security Principles, http://choices.cs.uiuc.edu/MobilSec/posted_docs/3G_Security_Overview.ppt.

26. 3GPP Technical Specification, 3GPP TS 21.133, Security Threats and Requirements, http://www.3gpp.org/ftp/Specs/2003-06/R1999/21_series/21133-320.zip.

27. 3G TR 33.900, A Guide to 3rd Generation Security, ftp://ftp.3gpp.org/TSG_SA/WG3_Security/_Specs/33900-120.pdf.

28. 3GPP2 C.S0002-B, Physical Layer Standard for CDMA2000 Spread Spectrum Systems, Release B, Apr. 19, 2002.

29. Bender, P., Black, P., Grob, M., Padouani, R., Viterbi, A., and Sindhurshayana, A., CDMA/HDR: A bandwidth efficient high speed data service for nomadic users, *IEEE Communications Magazine*, July 2000.

30. Jalali, A., Padouani, R., and Pankaj, R., Data throughput of CDMA-HDR: A high efficiency, high data rate personal communication wireless system, In *Proceedings of the IEEE Semiannual Vechicular Technology Conference*, VTC 2000-Spring, Tokyo, Japan, May 2000.

31. 3GPP Technical Specification, 3GPP TS 33.102, Security Architecture, http://www.3gpp.org/ftp/Specs/2003-06/R1999/33_series/33102-3d0.zip.

32. Walker, M., On the Security of 3GPP networks, http://www.esat.kuleuven.ac.be/cosic/eurocrypt2000/mike_walker.pdf.

33. 3GPP Technical Specification, 3G TS 23.060, General Packet Radio Service (GPRS); Service Description; Stage 2, http://www.3gpp.org/ftp/Specs/2003-06/R1999/23_series/23060-3f0.zip.

34. 3GPP Technical Specification, 3GPP TS 33.105, Cryptographic Algorithm Requirements, http://www.3gpp.org/ftp/Specs/2003-06/R1999/33_series/33105-380.zip.

35. Goodman, D.J., *Wireless Personal Communications Systems*, Addison-Wesley Wireless Communications Series, Prentice Hall, New York, 1997.

10

Header Compression Schemes for Wireless Internet Access

Frank H.P. Fitzek
Acticom GmbH

Stefan Hendrata
Acticom GmbH

Patrick Seeling
Arizona State University

Martin Reisslein
Arizona State University

10.1 Introduction

The wireless market is widely believed to be the most important future market for data services. Despite the efforts that were made to increase usability and abilities of all different wireless devices, the wireless protocol domain is still challenging.

As new services and protocols emerge for wired networks, the need to incorporate those services and protocols in wireless communication systems arises. Existing wireless networks of the second generation (also known as 2G) are mostly circuit switched and have been developed and optimized for voice transmission. Wireless networks of the third generation (3G) have to support a broad range of application scenarios. 3G networks and terminals such as smart phones, personal digital assistants (PDAs), and laptops are marketed with new services, most importantly multimedia.

Internet Protocol (IP)-based multimedia applications, however, require more bandwidth than do traditional voice services. Multimedia applications comprise several different application scenarios, including audio, video, and gaming.[5] The bandwidth needs of these applications are higher than offered in 2G systems. In addition, multimedia applications require more stringent network Quality of Service (QoS). One of the key issues for QoS is the suitability for real-time applications. Therefore, delay and jitter are key considerations within the QoS domain. Multimedia applications often use Real-Time Transport Protocol (RTP), User Datagram Protocol (UDP), and IP as protocols. Each of the protocol

Table 10.1   Theoretical Upper-Bound Savings
(in Terms of Bandwidth) for Voice Traffic

Codec	Mean Bit Rate (kbps)	IPv4 $\overline{S}$ (%)	IPv6 $\overline{S}$ (%)
LPC	5.6	74	81
GSM	13.2	55	65
ITU-T G.711	60.0	21	29

layers adds header overhead. Thus, the bandwidth requirements derive from the application (i.e., the payload) and the protocol overhead. Compression of the multimedia payload is mostly achieved on the application level (e.g., with current voice and video compression schemes). Attempting additional compression on the payload thus yields no benefits. Significant compression of the packet traffic can be achieved by reducing the amount of overhead information. For Linear Predictive Coding (LPC)-coded voice, for instance, the IP overhead is 81%, as detailed in Table 10.1. In general, for multimedia services, header compression achieves a dramatic savings in bandwidth. Given the high license fees of 3G bands and the migration of IP-based services into the wireless format, it is necessary to reduce the header overhead of IP-based traffic. IP is the underlying network layer protocol used for most multimedia application scenarios. Focusing on this protocol domain thus promises the highest gains. IP header compression mechanisms have always been an important part of saving bandwidth over bandwidth-limited links. Many header compression schemes exist already, but most of them are not suitable for the wireless environment. Wireless links typically have a very high and variable bit error probability (BEP) due to shadow- and multipath fading and mobility. With a reduction of the required bandwidth, the latency and Packet Error Probability (PEP) can be improved. This is because the probalility that a given packet is affected by link errors is reduced for smaller packets. For multimedia services in wireless environments, Robust Header Compression (ROHC) was introduced. ROHC was standardized by the Internet Engineering Task Force (IETF) in RFC 3095[6] and will be an integral part of the Third-Generation Partnership Project-Universal Mobile Telecommunications System (3GPP-UMTS) specification.[7] This compression scheme was designed to operate in error-prone environments by providing error detection and correction mechanisms in combination with robustness for IP-based data streams. A connection-oriented approach removing packet inter- and intradependencies yields a significant reduction of the IP header and other headers.

In this chapter, we present header compression schemes for the wireless domain. We start by outlining the motivations for and theoretical upper limits of header compression in general. Next, we describe various header compression schemes for the wired Internet. We examine the drawbacks and shortcomings of these compression schemes when applied in the wireless domain. Next, we introduce the ROHC scheme that has been developed for wireless channels. We study the performance of ROHC for audio and video traffic and discuss the ROHC deployment in 3G wireless systems.

10.1.1   Motivation for Header Compression

The motivation for IP header compression is based on the facts that (1) the multimedia payload is typically compressed at the application layer, (2) the headers occupy a large portion of the packet for some services, and (3) the headers have significant redundancy. In Figure 10.1, the combined header for a real-time multimedia stream with IPv4, resulting in 40 bytes protocol overhead, is illustrated. The protocol headers include the 20-byte IPv4 header, the 8-byte UDP[8] header, and the 12-byte RTP[9] header. With IPv6, there is a total of 60 bytes of overhead. In case of Global System Mobile (GSM)-coded audio transmission, the payload is only 33 bytes (13.2 kbps × 20 ms) long; the header of IPv4 accounts for 55% of the packet. For IPv6, this ratio is even larger. There are some redundancies among the various headers (IP, UDP, and RTP) of a given packet, but typically there are even larger redundancies among contiguous packets of a given IP flow. Thus, there are two types of header redundancies:

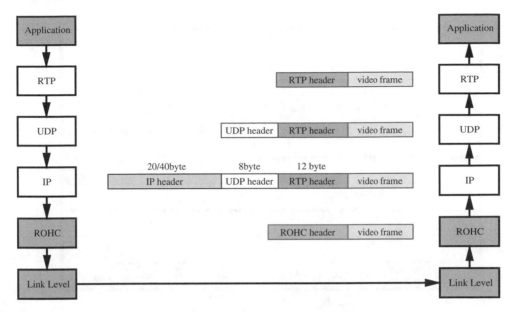

FIGURE 10.1 Header structure and protocol stack with relevant layers (ROHC, IP, UDP, and RTP).

- *Intrapacket:* The headers for the different protocols within a single packet carry identical or deducible information.
- *Interpacket:* The headers among consecutive packets have only marginal (incremental) differences.

The structure of typical multimedia packets is illustrated in Figure 10.2 for IPv4 and Figure 10.3 for IPv6. The specific header fields for the RTP/UDP/IP headers are marked with respect to their dynamics. The header fields are classified into two groups: *nonchanging* and *changing*. The nonchanging group consists of static, static-known, and inferred header fields. Figures 10.2 and 10.3 illustrate that a large portion of the header fields are static or static known (20.5 bytes in IPv4 and 44.5 bytes in IPv6). The compression of these fields can be achieved with low to moderate complexity. In fact, these fields could be completely omitted after the first successful transmission. The Segment Length (2 bytes) and Packet Length (2 bytes) fields (as well as the Header Checksum in IPv4 with 2 bytes) are referred to as *inferred*. These entries can be inferred from other header fields and are also relatively easily compressed. The changing group consists of not-classified-*a-priori*, rarely changing, static or semistatic changing, and alternating changing header fields. These header fields are more difficult to compress, and it depends on the applied header compression scheme as to how the compression is achieved.

10.1.2 Potential Savings of Header Compression

To get a basic idea of the possible savings of header compression, we compute an upper bound for voice communication and for audio streaming. For this upper-bound calculation, we assume that with header compression the overhead due to IP, UDP, and RTP is zero. The upper bound on the savings, denoted by Si, for packet i is

$$S_i = 1 - \frac{Packet(i)}{Header + Packet(i)} = \frac{Header}{Header + Packet(i)}$$

where *Packet(i)* denotes the size of the payload data and *Header* denotes the size of the (uncompressed) IP, UDP, and RTP headers. Clearly, the potential savings depend only on the mean packet length. The packet length depends on the service type used. The mean saving

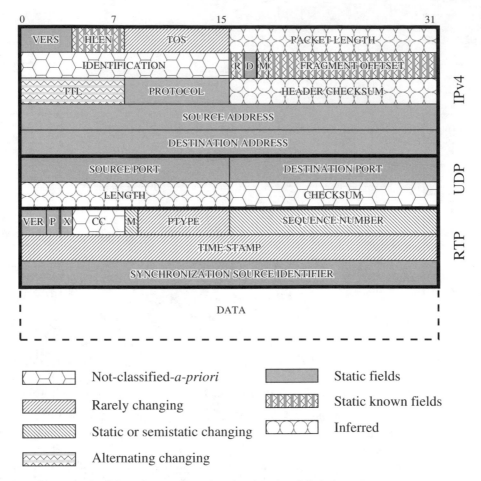

FIGURE 10.2 Header fields for RTP/UDP/IP packets (version 4) and their dynamics.

$$\bar{S} = \frac{1}{N}\sum_{i=1}^{N}S_i$$

gives the portion of bandwidth that a wireless network provider can potentially save for a session with N packets. The potential savings for voice service are given in Table 10.1. We consider the LPC with 5.6 kbps, the GSM codec with 13.2 kbps, and a codec following the ITU-T standard G.711 with 60.0 kbps. Note that LPC gives acceptable quality for voice communications; the GSM and ITU-T G.711 codecs provide higher quality suitable for audio streaming. For the calculation it is assumed that a packet is generated every 20 ms. Therefore, S_i is the same for all IP packets (and thus equal to $\bar{S}$).

The potential savings for IPv6 are larger than for IPv4 because of the IP header length (IPv4 with 40 bytes and IPv6 with 60 bytes). The second indication from Table 10.1 is that smaller IP packets correspond with larger savings $\bar{S}$. This is due to the larger ratio of the header compared with the smaller packets. Overall, we observe there is a large potential for bandwidth savings. The savings are most significant for low-bit-rate streams, which tend to be common in wireless networks.

In contrast to voice streams with fixed frame and packet sizes, the frame sizes of a video stream vary over time.[10,11] The size of the video frames depends on the content of the video sequence and the applied encoding scheme and settings. We encoded the generally accepted video reference streams *container*, *bridge*, *carphone*, *claire*, *foreman*, *grandma*, *highway*, *mother*, *news*, *salesman*, and *silent* (as given in Table 10.2) in the Quarter Common Interchange Format (QCIF) with 176×144 pixel (see also

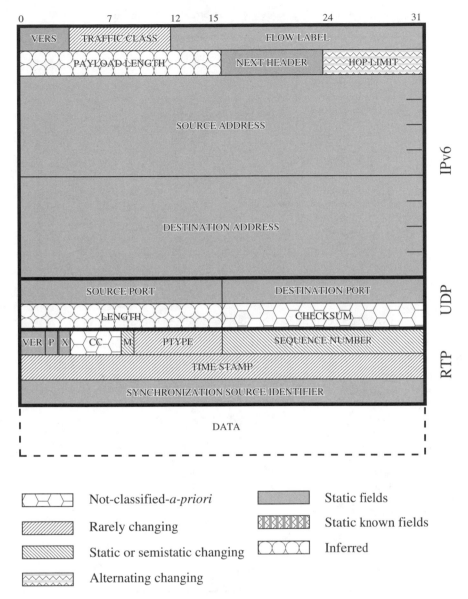

FIGURE 10.3 Header fields for RTP/UDP/IP packets (version 6) and their dynamics.

Chapter 11, Section 11.2 for a more detailed explanation) with the H.26L encoder[12,13] using the IBBPBB-PBBPBB Group of Picture (GoP) structure. To evaluate the upper bound on the mean savings $\overline{S}$, we assume that each video frame is carried in one packet. The measured frame sizes give the upper bounds for the mean potential savings $\overline{S}$ for H.26L encoded video streams given in Table 10.3. Here we assumed that no layered coding is used, which would further increase the header compression gain. This is because with layered coding, each video layer is transmitted in its own RTP session and has its own headers.

10.2 Header Compression Schemes

Several schemes for compressing the network and transport-layer protocol headers emerged in the 1990s. These schemes all have in common that their respective focus is only on a certain combination of protocols. All are mainly based on the same compressor/decompressor concept. In Figure 10.4 the general

TABLE 10.2 Transmitted Video Streams

Sequence	Quantization Scale	Frames
container	10	300
container	20	300
container	30	300
container	40	300
container	51	300
bridge (close)	30	2001
carphone	30	382
claire	30	494
foreman	30	400
grandma	30	870
highway	30	2001
mother and daughter	30	961
news	30	300
salesman	30	449
silent	30	300

TABLE 10.3 Upper Bound on Total Bandwidth Savings $\bar{S}$ for H.26L Encoded Video

Video Sequence	Quant. Scale	Mean Bit Rate (kbps)	IPv4 $\bar{S}$ (%)	IPv6 $\bar{S}$ (%)
container	10	855.0	1.1	1.6
container	20	213.0	4.3	6.3
container	30	65.8	12.7	17.9
container	40	24.1	28.5	37.5
container	51	9.1	51.0	61.0
bridge close	30	69.9	10.3	14.6
carphone	30	135.4	6.6	9.6
claire	30	44.3	17.8	24.5
foreman	30	121.9	7.28	10.5
grandma	30	56.4	14.5	20.3
highway	30	57.2	12.3	17.3
mother and daughter	30	66.4	12.6	17.8
news	30	100.9	8.7	12.5
salesman	30	81.5	10.5	15.0
silent	30	101.7	8.6	12.4

Note: The video sequences in the QCIF format were encoded with different quantization scales (smaller quantization scales give higher encoded video quality).

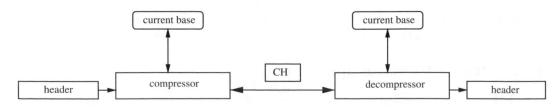

FIGURE 10.4 General concept of header compression: A current base header is maintained at the compressor and the decompressor. Header redundancy with respect to the base header is removed to obtain the compressed header (CH).

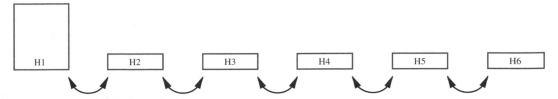

FIGURE 10.5 Jacobson:[1] Delta coding with respect to immediately preceding header.

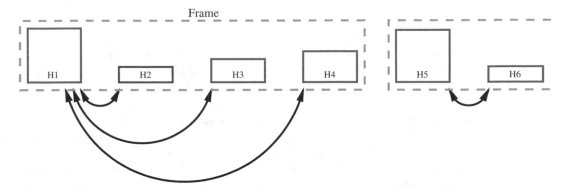

FIGURE 10.6 Perkins and Mutka:[2] Delta coding with respect to first (uncompressed) header in frame.

concept of header compression is illustrated. On the sender side, the compressor removes redundancy from the incoming packet with respect to a reference (base) header. This reference is also known (and maintained) at the receiver and allows the receiver to decompress the incoming compressed packet headers. In the following sections, we present the most common of these header compression schemes. Refer to the referenced literature for more details on these schemes.

10.2.1 Compressed Transport Control Protocol (CTCP, VJHC)

The first proposed IP header compression scheme, Compressed Transport Control Protocol (CTCP), also known as Van Jacobson Header Compression (VJHC), for the Internet was introduced by Van Jacobson in 1990[1] as RFC 1144 and focuses on the TCP protocol. VJHC processes TCP and IP headers together and compresses the 40-byte TCP+IP header to a 4-byte compressed header. A second benefit from the combined processing is the reduced complexity of the used algorithms. VJHC is based on delta coding, as illustrated in Figure 10.5. The differences between two packet headers are referred to as the *delta*. Instead of transmitting the entire header, VJHC transmits only the delta. This approach achieves high compression. On the downside, it introduces vulnerability. If only one delta-coded header is corrupted, all the following packets are erroneous. To recover from these errors and reestablish the current base header, VJHC sends all TCP retransmissions uncompressed. Thus, VJHC does not require any signaling between compressor and decompressor. The disadvantage is the sensitivity to error-prone links as investigated in references 2 and 14 to 17.

10.2.2 Refinements of CTCP

In reference 2, robustness at the cost of less compression was introduced by Perkins and Mutka. As illustrated in Figure 10.6, the delta coding for the adjacent packets has been replaced by a reference frame. Several consecutive packets are aggregated to a frame. The first packet of a frame is sent uncompressed, and the following packets use delta coding with respect to the first (uncompressed) packet in the frame. Clearly, the differences to packets at the end of a frame are larger than for those at the beginning. The compression gain is thus limited (and lower than for VJHC). The advantage of this approach is the usage of shorter delta coding ranges. Corrupted packets do not necessarily lead to the loss of synchronization. This is a clear advantage over VJHC. An optimization for the header compression of Perkins was introduced by Calveras et al.[15] and Calveras and Paradells.[16] This optimization minimizes the overhead by adapting the frame length as a function of the channel state and updating the base with compressed headers. The header compression schemes introduced by Perkins and Mutka[2] and Calveras et al.[15] obtain base updates by sending both compressed and uncompressed headers. Whenever one of these base updates is lost due to transmission errors, the synchronization between compressor and decompressor is lost and performance degrades.

Rossi et al.[18,19] proposed the following modification to the base update procedure. Whenever the compressor decides to update the current base, it sends the new base to the decompressor following the standard delta coding mechanism (i.e., the compressor sends the new base in a compressed header taking the current base as the reference base). In addition, a *request flag* is set in the compressed base. The decompressor receives the new base, decompresses it using the current base, and temporarily stores it as the new proposed base. The new base is at this time only a proposed base; the decompressor will start to use it only upon specific indication by the compressor. When the decompressor receives a new base, a TCP ACK including the sequence number of the packet carrying the new base is transmitted to the compressor. When the compressor receives the ACK with the sequence number of the packet carrying the new base, it knows that the proposed base has been properly received by the decompressor. Therefore, it can safely start to use the new base for compression. This change of the used base is communicated to the decompressor by a *bistable flag*. The compressor learns from the TCP ACKs that the decompressor has started to use the new base. So, a minimum of two consecutive TCP ACKs is needed to change the base used for compression. The header base update is performed as frequently as possible to minimize the delta field's resource occupancy. Because the header base update involves two adjacent ACKs, this is used as the rate of the header base change. Further details and comparisons with other proposed algorithms can be found in references 18 and 19.

10.2.3 IP Header Compression (IPHC)

IP Header Compression (IPHC)[17] provides a number of extensions to VJHC. The most important extensions are support for UDP, IPv6, and additional TCP features. With the explicit support of UDP come additional features, such as multicast. Nevertheless, support for RTP is still not given, which makes the scheme unsuitable for many multimedia applications. Similar to VJHC, IPHC relies on the change of header fields as well as on the derivation of header field contents. The encoding also uses the delta scheme, transmitting only the changes in the header fields. The error-correction schemes of VJHC are used for TCP packets. For non-TCP packets, no differential encoding between sender and receiver is used. Thus, the compression for UDP-based streams is worse than for TCP-based streams, but the context is not affected by packet losses. Generally, context update packets are sent periodically to maintain the state at both ends. For the application of differential coding over lossy links, two new algorithms are introduced:

- *TWICE:*[20] The decompressor computes a checksum to determine if its context information has been updated properly. If this fails, a lost header-compressed packet is assumed. The decompressor recomputes the checksum by skipping the packet and assuming that the lost and current packet had the same delta values. If the second checksum computation is successful, the decompressor adjusts the base header (context) by two deltas.
- *Header Request:* When the decompressor is unable to repair the context after a loss, it requests a complete header from the compressor. This is possible only on bidirectional links because the decompressor must communicate with its compressor. The decompressor sends a context state message that includes all compressed packet streams in need of a context update.

The performance of IPHC for packet voice over the wired internet and the interactions between the IPHC and the interleaving of source coder symbols are studied in reference 21. We also note that a header compression scheme specifically for IPv6-based communication with mobile wireless clients has been developed.[22] This IPv6 header compression scheme uses (1) address reassignment (translation), which maps the 16-byte IPv6 address to a shorter address; (2) LZW coding[23,24] for the initial header; and (3) incremental comparator encoding for subsequent headers.

10.2.4 Compressed Real-Time Protocol (CRTP)

The Compressed Real-Time Protocol (CRTP) scheme presented in RFC 2508[25] compresses the 40-byte header of IP to 4 bytes if the UDP checksum is enabled or to 2 bytes if it is not. This is possible by compressing the RTP/UDP/IP headers together, similar to the VJHC approach. With the characteristics of the RTP protocol, the changes for the RTP header fields become partially predictable. In addition, changes in some fields are constant over long periods of time. Thus, the expected change in these fields can be implied without even transmitting the differences. These implied fields are also referred to as *first-order* changes. They are stored with the general context for each specific connection. The differences within fields that have to be compressed are referred to as *second-order* differences. An example for these are video frame skips. Video frames are generally transmitted every 40 ms. In case a frame cannot be encoded (e.g., due to lack of processing power or because of a slower play-out ratio), the implied time no longer is accurate. Therefore, the new first order is set to the second order and the connection context is updated. CRTP cannot use a repair mechanism as VJHC does because UDP/RTP are unidirectional protocols without retransmissions. CRTP uses a signaling message from decompressor to compressor to impart that the context is out of synchronization. For lossy links and long round-trip delays, CRTP does not perform well. After a single lost packet, several sequential packets are lost within the round-trip time. Thus, CRTP is not suitable for cellular links, where the header compression currently is envisioned to be implemented in the wireless terminal and the radio network controller, resulting in significant round-trip times (see Section 10.3.13). A performance evaluation for CRTP is given in reference 26.

Robust Checksum-based Compression (ROCCO)[27] is a refinement of CRTP. ROCCO includes a checksum over the original (uncompressed) header in the compressed header. The checksum facilitates local recovery of the synchronization. In addition, ROCCO incorporates compression profiles (tailored for specific applications, e.g., audio or video streaming) and has a code with hints on the change of header fields in the compressed header. These mechanisms improve the header compression performance, especially for highly error-prone links and long round-trip times.[27,28] Similarly, Enhanced Compressed RTP (ECRTP)[29] is a refinement of CRTP. ECRTP uses local retransmissions to more efficiently recover from wireless link errors.

10.3 Header Compression Schemes for Wireless Channels

10.3.1 Drawback of Existing Schemes on Wireless Links

The majority of the header compression schemes were designed for wired links. Wired links are characterized by low error rates. Wireless links, on the other hand, are characterized by higher and bursty transmission error rates. In addition, wireless services — such as those for 3G — often require real-time protocol support. Thus, schemes designed with different goals have several shortcomings:

- *VJHC and its refinements:*[1,2,18,19] These schemes were designed for usage with TCP. They are therefore unsuitable for multimedia applications running on UDP. The delta coding scheme makes these protocols vulnerable to link errors. As noted above, the packets from the instant an error occurs to the end of the TCP timeout window are retransmitted uncompressed. This considerably reduces the achieved performance in wireless environments.
- *IPHC, CRTP:*[17,25] Neither IPHC nor CRTP offer the efficiency and robustness needed for wireless links.[30] The local recovery mechanisms of ROCCO are very helpful in ensuring efficiency and robustness; these basic ideas are incorporated in the design of ROHC.

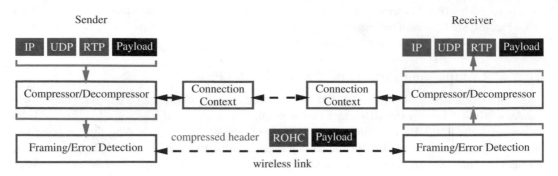

FIGURE 10.7 Transmission of compressed packets between the ROHC compressor and the ROHC decompressor.

10.3.2 Robust Header Compression (ROHC)

ROHC is envisioned as an extensible framework for robust and efficient header compression over highly error-prone links with long round-trip times. This design is motivated by the large bit error rates (typically on the order of 10^{-4} to 10^{-2}) and long round-trip times (typically 100–200 msec) of cellular networks. The design of ROHC is based on the experiences from the header compression schemes reviewed earlier. In particular, ROHC incorporates elements from ROCCO and Adaptive Header Compression (ACE),[31] which may be viewed as a preliminary form of ROHC.

ROHC in its original specification (as in RFC 3095) is a header compression scheme with profiles for three protocol suites: RTP/UDP/IP, UDP/IP, and ESP/IP. The IP protocol header can be version 4 or 6. In case any other protocol suite is used, ROHC does not perform compression by using the uncompressed profile. Note that other profiles are in development to support more protocol suites, including IP only and TCP/IP (see, for instance, references 32–37). Also note that there have been research efforts toward a general, protocol-independent header compression framework.[38] This framework combines (1) a high-level header description language for flexible specification of the header properties of a specific protocol, and (2) a code generator tool for automatic generation of the corresponding header compression code. As shown in Figure 10.7, ROHC is located in the standard protocol stack between the IP-based network layer and the link layer. The need for saving bandwidth is limited to the wireless link. So the compression should work only between two wireless nodes, whereas for the rest of the Internet this operation remains transparent.

In the simplest configuration, the wireless sender contains the compressor and the decompressor is located in the wireless receiver (see Section 10.3.13 for the actual location of the ROHC compressor and decompressor in 3G systems). ROHC controls the interaction between these two nodes in order to achieve two main goals:

1. The network providers desire a significant bandwidth-saving obtainable by reducing the headers to a shorter ROHC header.
2. Despite the compression, it is necessary to ensure a QoS acceptable for the customers of the network providers.

The compressor can sacrifice the bandwidth saving in order to keep the decompressor synchronized even if errors occur on the link. ROHC can thus sacrifice compression efficiency for error-correction capability and therefore does not always work at the peak of its compression abilities. Different levels of compression, called *states*, are used within ROHC. This state-based approach is a new robust solution against the perils of the wireless link. In this section, we give a basic overview of how ROHC achieves robust header compression. For details on the algorithms and schemes, refer to the original specification in RFC 3095.

10.3.3 Context and States

In general, data packets that are transferred over a wireless link are not independent from each other but share certain common parameters, such as equal source and destination addresses. Moreover, they usually can be grouped together logically (e.g., data packets that constitute an audio stream and data packets that make up the accompanying video stream). Thus, it makes sense to use a stream-oriented approach in ROHC to compress packet headers. Each stream or flow is identified by its parameters that are common to all packets belonging to this stream. The compressor and decompressor maintain a context for each stream, which is identified by the same context identifier (CID) on both sides. A context, being a set of state data, contains, for example, the static and dynamic header fields that define a stream.

The ROHC compressor and decompressor can each be regarded as a state machine with three states. Compressor and decompressor start at the lowest state, which is defined as *no context established*. In this state, compressor and decompressor have no agreement on compressing or decompressing a certain stream. Thus, the compressor needs to send an ROHC packet containing all the stream and packet information (static and dynamic) to establish the context. This packet is the largest ROHC header that the compressor sends. In the second state, the static part of the context is regarded as established between compressor and decompressor, whereas the dynamic part is not. In this state, the compressor sends ROHC headers containing information on the dynamic part of a context. These headers are smaller than those sent in the first state but slightly larger than the headers used in the third state. In the third and final state, the static as well as the dynamic part of a context are established and the compressor needs to send only minimal information to advance the regular sequence of compressed header fields. Fallbacks to lower states occur when the compressor detects a change or irregularity in the static or dynamic part (i.e., pattern) of a stream or when the decompressor detects an error in the static or dynamic part of a context.

The compressor strives to operate as long as possible in the third state under the constraint of being confident that the decompressor has enough and up-to-date information to decompress the headers correctly. Otherwise, it must transit to a lower state to prevent context inconsistency and to avoid context error propagation.

10.3.4 Compression of Header Fields

The compression of the static part of headers for a stream is similar to the other header compression schemes. Header fields that do not change need to be transmitted only at context establishment and remain constant afterward. More sophisticated algorithms are needed for compression of the dynamic part. With ROHC, the values of the dynamic header fields are derived as linear functions from the sequence number of each packet, which in general increases by one for each new packet. These functions for the dynamic header fields are expected to change, and the compressor must therefore be able to effectively communicate these changes to the decompressor.

For compressing and decompressing dynamic header fields, either directly or by the use of a function, ROHC uses two basic algorithms:

- *Self-describing variable length values:* This algorithm reduces the number of bits needed to transmit an absolute value. Small values are described with fewer bits than large values.
- *Windowed Least Significant Bits (W-LSB) encoding:* Dynamically changing values usually have their characteristic dynamic behavior, e.g., always incrementing by one for sequence numbers. Using this knowledge of the dynamic behavior, a window is constructed around a reference value, which is a previous, correctly transmitted value. Depending on the distance of the new value from the reference value and the relative position of the window regarding the reference value, the number of bits to transmit the compressed new value is determined. These bits thus describe the advancement of a value from a reference value, and their number is small for header field values that do not randomly change but continuously increase or decrease by usually small differences. The

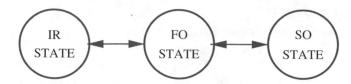

FIGURE 10.8 Compressor states.

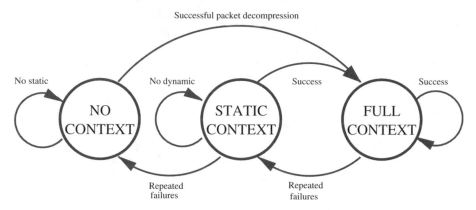

FIGURE 10.9 Decompressor states for all modes.

advantage of this algorithm is the consideration of the dynamic behavior of the value when defining the position of the window, resulting in a minimization of the average number of bits needed to be transmitted in order to describe the sequence of values for a header field in a stream.

The W-LSB compression algorithm in combination with an elaborate protection scheme for sensible data in ROHC-compressed headers contribute to the robustness of ROHC.

10.3.4.1 Compressor States

The three compressor states illustrated in Figure 10.8 are the following:

- Initialization and refresh state (IR)
- First-order state (FO)
- Second-order state (SO)

In the IR state, no context is available for compression. Thus, the compressor and decompressor have to transit to a higher state as soon as possible for effective compression. When confident of its success to establish a context, the compressor can change to the SO state immediately. In the SO state, only the transmission of a sequence number is necessary, and the values of all other header fields are derived from it. These SO state ROHC headers are the smallest ones and generally are 1 byte in size. If an irregularity in a stream occurs, the compressor falls back to the FO state. Depending on the irregularity, different ROHC headers with sizes of 2, 3, or more bytes are used in this state. If the stream returns again to a regular behavior (pattern), the compressor transits up to the SO state.

10.3.4.2 Decompressor States

The three decompressor state names depicted in Figure 10.9 refer to the grade of context completeness. In the No Context (NC) state, the decompressor lacks the static and dynamic part of a context. Consequently, it can only decompress IR packets (i.e., packets sent with uncompressed header fields in the IR compressor state). In the Static Context (SC) state, the decompressor lacks only the dynamic part (fully or partially) and therefore needs packets that contain information on dynamic header fields in order to complete the context again. The decompressor usually works in the Full Context (FC) state, which is

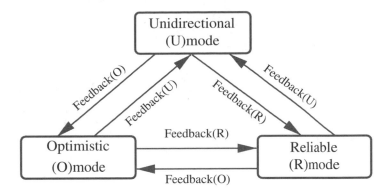

FIGURE 10.10 Mode transitions.

reached after the entire context has been established. In case of repeated failures in decompression attempts, the decompressor always transits to the SC state first. Then it often is sufficient to correctly decompress an FO packet to recover to the FC state. Otherwise, receiving further errors leads to the transition to the NC state.

10.3.5   Modes and State Transitions

To offer the ability to run over different types of links, ROHC operates in one of three modes: Unidirectional, Bidirectional Optimistic, and Bidirectional Reliable mode. Similar to the states, ROHC starts at the most basic mode (Unidirectional) but can then transit to the other modes if the link is bidirectional. Contrary to the states, modes are not directly related to the level of compression. The modes differ from each other in the amount of coupling between compressor and decompressor by the use of feedback packets sent by the decompressor to the compressor. For example, the Unidirectional mode does not make use of feedback packets at all, whereas the Bidirectional Reliable mode tightly couples compressor and decompressor by requiring a feedback packet for each update of the context. Mode transitions can be initiated by the decompressor at any time for an established context. To do so, the decompressor inserts a mode transition request in a feedback packet, indicating the desired mode, as Figure 10.10 illustrates.

10.3.5.1   Unidirectional Mode (U-Mode)

This mode is designed for links without a return channel. There is no way for the compressor to be certain whether the decompressor has received the correct context information and is thus decompressing correctly. It can only optimistically assume that the decompressor has received the context data correctly by repeatedly sending the same information. However, the decompressor must have a chance to update and correct its context in case of context errors. The compressor therefore periodically times out and falls back to the FO and IR states, as illustrated in Figure 10.11.

Typically, the timeout period for fallbacks to the IR state is longer than the timeout period for fallbacks to the FO state. The decompressor uses the periodically sent FO and IR packets to verify and possibly correct its context. The compressor also falls back to the FO state whenever the pattern of header field evolution changes. Whenever the compressor is in the IR or FO state, it sends multiple packets with the same lower level of compression until it is confident that the decompressor has established the flow context. The compressor then optimistically transits upward to the higher compression FO or SO state. This adds robustness against single packet errors. The U-Mode is the least robust and least efficient mode of the three ROHC modes.

10.3.5.2   Bidirectional Optimistic Mode (O-mode)

As an extension to the Unidirectional mode, the Bidirectional Optimistic mode uses feedback packets that are sent from the decompressor to the compressor in order to accelerate state transitions at the

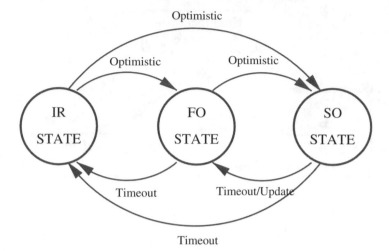

FIGURE 10.11 Compressor state transitions for Unidirectional mode: After repeatedly sending a packet with lower (IR or FO state) compression, the compressor optimistically transitions to the FO or SO state. Updates in the header field pattern and a periodic timeout return the compressor to the FO state. A longer period timeout takes the compressor to the IR state.

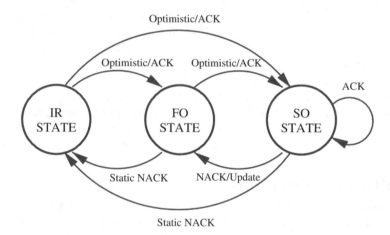

FIGURE 10.12 Compressor state transitions for Bidirectional Optimistic mode: After receiving an (optional) ACK from the decompressor or repeated transmissions (and optimistic assumption of established context), the compressor transits upward. The compressor returns to the FO state to update the header field development pattern or upon request (NACK) from the decompressor. With a Static-NACK, the decompressor requests a static context update.

compressor and to avoid the periodic fallbacks to the FO and IR states. As shown in Figure 10.12, context update acknowledgments (ACKs) are used to notify the compressor that the decompressor has successfully received context information. (These ACKs from the decompressor are optional; thus, the compressor may still need to use the optimistic upward transitions.) In case of a context error, a context recovery request (NACK) is sent to the compressor, causing a retransmission of context information to update and repair the context at the decompressor. With a Static-NACK, the decompressor forces the compressor back to the IR state, thus reestablishing the static context. With these context updates on request, the compressor can achieve a higher compression efficiency compared to the Unidirectional mode.

Because of the mostly weak protection (3-bit CRCs) of context-updating data sent in the Bidirectional Optimistic mode, there is still a not-to-be-neglected probability of context damage that can result in a

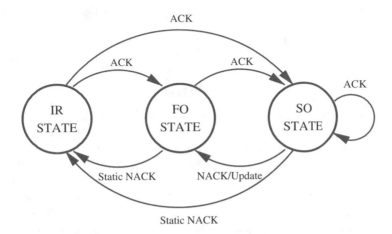

FIGURE 10.13 Compressor state transitions for Bidirectional Reliable modes: The compressor transits upward only after receiving ACKs from the decompressor. Updates of the context or decompressor requests (NACK, Static-NACK) cause downward transitions.

sequence of incorrectly transmitted packets. For applications that prefer a more reliable transmission with a lower probability of incorrect packet transmission, the Bidirectional Reliable mode was conceived.

10.3.5.3 Bidirectional Reliable Mode (R-mode)

To achieve a lower probability of incorrect packets, a more powerful error correction (7-bit CRCs) is used for context-updating information in the Bidirectional Reliable mode. In addition, the compressor transits to the FO or SO state only after receiving an ACK from the decompressor, as illustrated in Figure 10.13. The compressor transits downward to update the context or upon decompressor request (NACK, Static-NACK). With this mode, the behavior of the compressor and decompressor are thus even closer coupled than with the optimistic mode. The rare NACKs provide a quick context recovery in case of errors. Therefore, the compressor always knows in which state the decompressor is and when to make a state transition.

10.3.6 Timer-Based Compression of RTP Timestamps

A special, more efficient compression method for the RTP timestamp header field can be applied under certain conditions. If the application hands over RTP packets in real time to the ROHC compressor, the time difference between the handover of two RTP packets is proportional to the difference of the RTP timestamp header field values in these two packets. Provided that the transmission channel has a low delay jitter, the decompressor can then use the time difference between the reception of two compressed packets to estimate the new RTP timestamp value based on the previous value. With timer-based compression, the compressor needs to send even fewer bits for compression of the RTP timestamp than with the standard W-LSB encoding based method. These bits are used to refine the estimation of the decompressor. The number of bits needed for refinement depends on the amount of delay jitter on the transmission channel that has to be determined.

10.3.7 List Compression

Some data in headers of data packets that belong to a stream contain a variable number of items, such as the Contributing Source (CSRC)[9] list in the RTP header. They also rarely change within a stream. In order to compress these lists, an elaborate scheme was developed in ROHC called "list compression." With list compression, the compressor can refer to items if it is confident that the decompressor has received them correctly in previous packets using a short item identifier. The list is maintained and updated between compressor and decompressor by list operations such as insertions, removals, and so forth that describe changes to the list.

In addition to CSRC lists, the list compression scheme is used to compress extension headers, such as GRE headers,[39,40] authentication headers, minimal encapsulation headers, and IPv6 extension headers.

10.3.8 More Profiles: TCP, IP-Only, UDP-Lite

As already mentioned, ROHC is defined in RFC 3095 as a framework and is thus extensible with new profiles. In addition to already specified profiles for compressing IP/UDP/RTP, IP/UDP, and IP/ESP headers, profiles for compressing the IP/TCP and IP headers are only in development as of the time of this writing. Contrary to previous TCP compresssion schemes, the ROHC TCP profile is capable of referring to already established TCP compression contexts when establishing a new context resulting in a reduced overhead for new compression contexts. This improves compression efficiency for sequences of so-called "short-lived TCP transfers," which often occur with HTTP transfers used by World Wide Web (WWW) browsers. The IP-only profile is intended for header combinations that contain an IP header but are not covered by the ROHC profiles specified so far. The compression of the IP header can at least reduce the size of the headers to some degree. As a modification to the already existing profiles for compression of the IP/UDP and IP/UDP/RTP headers, two new profiles are proposed that specify compression for IP/UDP-Lite and IP/UDP-Lite/RTP headers. With the standard UDP protocol, the UDP payload is either entirely or not at all protected by a CRC checksum. UDP-Lite[41] offers flexibility in the amount of payload data that is CRC protected. This is advantageous for several audio and video codecs that prefer the delivery of erroneous packets (and apply error concealment[42] at the decoder level). This approach is attractive for real-time streaming, because it allows the decoder to make the best out of the delivered data.

10.3.9 ROHC over Wireless Ethernet Media

The specification for ROHC does not describe the interoperation with the underlying link-layer in detail. Only a few requirements that have to be met by the link-layer are mentioned (e.g., no packet reordering or duplication on the channel between compressor and decompressor). Additional drafts and RFCs specify how ROHC operates on top of certain link-layer protocols (e.g., PPP[43]). A large group of link-layer technologies are formed by the Ethernet-based network technologies. Among them, the wireless local area variants, such as IEEE 802.11 (WLAN) or Bluetooth, exhibit similar bit-error characteristics. Consequently, there are efforts under way to standardize the operation of ROHC on top of wireless Ethernet-like media. An exemplary application that can benefit from the use of ROHC is voiceover IP telephony in a WLAN environment.

10.3.10 Signaling Compression

Another significant outcome of the ROHC working group (along with RFC 3095) was RFC 3320,[44] specifying Signaling Compression (SigComp). SigComp uses an entirely different approach to compression than ROHC. With SigComp, a Universal Decompressor Virtual Machine (UDVM) executes code that is sent by the compressor to decompress packets. This allows for greater flexibility in utilizing different compression strategies and algorithms. SigComp is envisioned for the compression of Session Initiation Protocol (SIP) (RFC 3261[45]) packets, with typical sizes of a few hundred bytes. SIP, in turn, is expected to play an important role in the development toward an All-IP structure in 2.5G and 3G networks.

10.3.11 ROHC Summary

ROHC is an efficient compression protocol that is especially suitable for transmission of real-time multimedia data over wireless links that have high error probabilities. Sophisticated compression and encoding methods and elaborate schemes that provide robustness against transmission errors make ROHC superior to the previously described header compression protocols. Although a highly increased

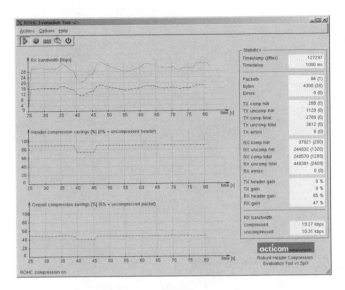

FIGURE 10.14 NetMeter tool for the evaluation of ROHC compression.[3]

complexity of the compression algorithms in ROHC results in an increased demand for computing power, the advances in microelectronics during the past few years make it possible to cost-effectively provide this computing power in small, mobile devices. Ongoing research efforts primarily focus on the tuning of the parameters of ROHC (see for example, reference 46) the performance evaluation of ROHC, and its adaptation to specific wireless systems (e.g., cellular networks or WLANs).

10.3.12 Evaluation of ROHC Performance for Multimedia Services

In this section, we give an overview of performance evaluations of ROHC for wireless multimedia. We consider voice and video traffic and present the achieved compression and audio/video quality over wireless links.

For the voice-traffic evaluations, voice files were coded with the GSM encoder and transmitted over an error-prone network; see reference 47 for details. At the receiver side, the data was decoded and compared with the original data. To obtain the network traffic and compression metrics during the transmission, the NetMeter tool[3] (see Figure 10.14) was used. Different objective quality measures were implemented to measure the voice quality. These measures estimate the speech intelligibility and have high correlations to subjective listening tests. Whereas listening tests require only the distorted voice files, objective measures compare the undistorted with the distorted files. The distorted file is not synchronized to the original file. A synchronization algorithm[47] to allow for computation of the objective measures was developed. This algorithm allows for a frame-wise computation of the objective quality measures.

Three different voice samples were used to investigate the ROHC performance. All measurements yielded a compression gain of approximately 85% for the header and about 46% for the entire packet. From the network provider's view, this allows for nearly a doubling of the capacity in terms of the total number of supported customers. This is a very interesting and promising result, given the costs of the 3G frequency bands. As already mentioned at the beginning of this chapter, first compression schemes offered a good compression gain but were vulnerable in presence of the wireless link. Therefore, the achieved compression gain has to be evaluated in conjunction with the achieved voice quality. Figure 10.15 gives the the voice quality (segmental signal to noise ratio [sSNR] in dB) as a function of the bit error probability on the wireless link for the transmission with ROHC and without header compression. (The bit errors are uniformly distributed in this experiment, which is a reasonable model for the dedicated channels in UMTS; evaluations for the common UMTS channels that are reasonably modeled by bursty bit errors are ongoing.)

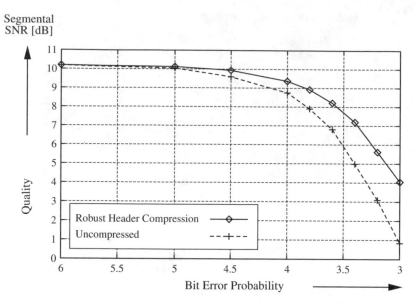

FIGURE 10.15 Objective voice quality (segmental SNR) as a function of bit-error probability on a wireless link for transmission with ROHC and without header compression: ROHC improves the voice quality for moderate to large bit-error probabilities while using roughly 46% less bandwidth than transmission without header compression.

We observe that for both approaches the voice quality decreases as the bit-error probability increases, as is to be expected. We additionally observe that with increasing bit-error probability, ROHC achieves higher voice quality than the transmission without header compression. This indicates that ROHC efficiently compresses the headers and thus the packets, making the small packets less vulnerable to wireless link errors. At the same time, ROHC is robust largely by avoiding incorrect header decompression due to the packets that do suffer from wireless link errors. Overall, the results indicate that ROHC has the potential to (almost) double the number of supported voice users while providing the individual users with improved voice quality.

For the ROHC evaluations for video traffic, the publicly available video sequences in Table 10.2[10] were encoded with the H.26L encoder.[12,13] These sequences were encoded with different quantization scales. The quantization scales range from 1 to 51, with 1 giving the best quality and least compression. The sequence *container* was encoded with different quantization scale settings to evaluate the effects of the transmission with and without ROHC on various video quality levels. The other sequences were encoded at an intermediate quality level (quantization scale of 30) to evaluate the effect of different video content. The result of these encodings are bitstreams with different characteristics (the different mean bit rates, frame size variabilities, etc.). These different mean bit rates and the packetization of each video frame into one IP packet (giving different mean ratios of IP headers to packet payload) result in different upper bounds on the compression gain; see Table 10.3. To evaluate the ROHC performance, these sequences were transmitted over the same error-prone link as the audio data. The received video sequences was decoded and their objective quality in terms of the Peak Signal to Noise Ratio (PSNR) was determined with the VideoMeter tool,[4] which is illustrated in Figure 10.16. The results are presented in Table 10.4. The achieved header compression is close to the theoretical limit, indicating the effectiveness of the ROHC scheme.

Interestingly, we observe that the sequence *container* has a relatively low PSNR quality for the highest quality encoding setting. This results from the large sizes of the individual video frames as only low compression is achieved for high-quality encoding. The subsequent packetization results in large IP packets. The larger a packet, the more likely the packet suffers from bit errors. Thus, more high-quality packets are damaged or lost during the transmission. This results in a significant loss in quality due to the bit errors (in conjunction with the error propagation in the predictively encoded P and B frames).

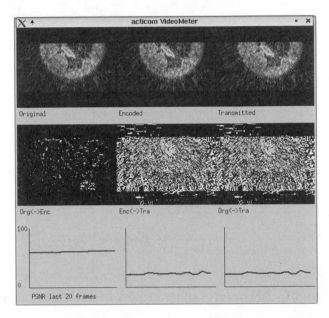

FIGURE 10.16 VideoMeter[4] for video quality evaluation.

TABLE 10.4 Average Header Size, Header Compression (Relative to 40-Byte IPv4 Header), Total Bandwidth Savings, and Average Video Frame PSNR Quality for H.26L Video over Error-Prone Link (BEP = 10^{-5}, QCIF, ROHC in Optimistic Mode).

Video Sequence	Quantization Scale	Avg. Header (Byte)	Header Compression (%)	Network Total Bandwidth Savings (%)	Avg. Video Frame PSNR (dB)
container	10	6.4	84.2	0.9	29.8
container	20	6.5	83.7	3.6	37.8
container	30	6.3	84.3	10.5	34.6
container	40	6.3	84.3	23.0	29.2
container	51	6.5	83.7	40.0	22.4
bridge close	30	6.2	84.5	8.6	18.3
carphone	30	6.4	83.9	5.5	33.1
claire	30	6.2	84.6	14.7	37.9
foreman	30	6.4	83.9	6.0	27.6
grandma	30	6.2	84.5	12.0	35.0
highway	30	6.3	84.3	10.2	34.1
mother and daughter	30	6.2	84.6	10.5	38.9
news	30	6.6	83.6	7.1	35.2
salesman	30	6.2	84.5	8.8	33.8
silent	30	6.3	84.3	7.2	32.4

This effect of lower quality in the decoded file (despite higher encoded video quality) would be even slightly worse without header compression.

Generally, we observe that ROHC reduces the total required bandwidth about 10% for intermediate quality video. For low-quality video streams, the savings are significantly larger. We also note that the bandwidth savings will be significantly larger for layered (scalable) encoded video, which is attractive for wireless environments because it can flexibly adapt to the wireless link conditions as well as the wide range of display and processing capabilities of the wireless devices. With layered encoded video, each encoding layer is considered a separate stream (flow) and has its own headers.[48] This tends to reduce the average ratio of frame sizes (packet pay loads) to headers.

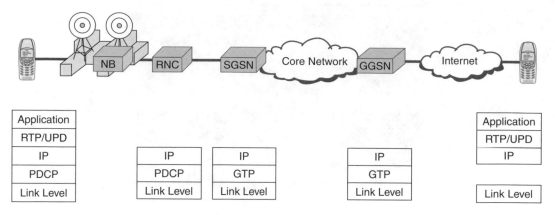

FIGURE 10.17   Location of ROHC in UTMS transmission chain. ROHC is provided by the packet data convergence protocol (PDCP) between the mobile end-system and RNC.

10.3.13   Deployment of Header Compression in Cellular Networks

The location of header compression in the networking protocol stack depends on the communication system. Generally, the packet header information has to be fully available for the IP routing. Header compression is therefore designed for a single link (between two adjacent IP interfaces). Furthermore, bandwidth is generally plentiful in the wired Internet compared with wireless links. Therefore, header compression is needed only for the wireless link. In the existing and upcoming cellular systems, header compression is placed in different network components. The goal of 3GPP is an All-IP network. Higher releases of the UMTS standard push the IP endpoint to the radio network subsystem. With every release, the topology or related protocols change. Next, we give the location of header compression for different generations or releases of cellular networks.

In 2.5G networks, header compression is performed by means of the Subnetwork Dependent Convergence Protocol (SNDCP).[49] The protocol is located at the mobile end-system and the Supporting GPRS Serving Node (SGSN) as part of the GSM recommendations. SNDCP allows transparent data transfer between the mobile end-system and the SGSN. For 2.5G networks, only RFC 1144[1] and RFC 2507[17] have been standardized. Because these compression schemes are not within our focus, we refer you to reference 30 for a more detailed implementation discussion of header compression within the GPRS subsystem.

For the next-higher generation, 3GPP specifies that header compression mechanisms are provided by means of the Packet Data Convergence Protocol (PDCP) specified in reference 50. In Release 99, only RFC 2507[17] is recommended for PDCP. Higher releases such as R4 and R5 introduce the usage of ROHC. For these releases, header compression is placed in the mobile end-system and the Radio Network Controller (RNC) as specified in 3G TS 25.322. The placement in these two entities was chosen to achieve both transparency and spectral efficiency.

In Figure 10.17, the location of the ROHC header compression mechanism as an integral part of PDCP is illustrated as specified by the 3GPP in reference 50. The figure gives an example of IP communication between two nodes. Without loss of generality, we assume that one node is connected directly to the Internet. The second node is assumed to be a wireless terminal connected to the UMTS Terrestrial Radio Access Network (UTRAN). Using the PDCP protocol, it communicates with the RNC via Node B. As stated earlier, the ROHC is placed in the PDCP. As a result of this design, the ROHC communication starts and ends at the RNC. This implicates that IP routing information is fully available at this point (looking toward the backbone) and bandwidth efficiency is achieved at the radio access network (looking toward the mobile end system). The SGSN and the GGSN make use of the GPRS Tunneling Protocol (GTP) to transport IP packets between the RNC and the Internet.

A mobile end system can be connected to several Node Bs while an IP session is ongoing. The radio network subsystem (RNS) takes care of the mobility and hands over the mobile end systems from one Node B to another with possible changes of the RNC. An RNC with several Node Bs is referred to as the serving radio network subsystem (SRNS). Whereas the compressor/decompressor pair in the mobile end system remains the same, new compressor/decompressor pairs have to be established in the RNC if an SRNS relocation takes place. This results in lost contexts for the RNCs. In dependency of the chosen ROHC mode (unicast, optimistic, or reliable) the performance degrades with every handover that requires SRNS relocation. Clearly, for the unicast mode the performance degradation is the highest, because the lost context will be detected only after a given timeout. ROHC, in contrast to other header compression schemes, is designed to be robust, which, in turn, helps in such situations of SRNS relocation. On the other hand, UMTS cells will be smaller than installed GSM cells, and thus handovers will occur more frequently. Therefore, the 3GPP specifies in reference 7 the possibility to forward the decompression context from the old RNC to the serving RNC (SRNC), where the new decompressor is located. In this case, the performance degradation does not occur at the expense of higher signaling overhead. But this seems reasonable because backbone capacity is not scarce, compared with the wireless bandwidth.

10.4   Conclusion

In this chapter, we gave an overview of the basic mechanisms for protocol header compression on wireless links. We examined the large protocol header overhead when streaming audio and video, the key motivation for using header compression. We traced the evolution of header compression mechanisms from the early proposals for compressing TCP/IP headers to the mechanims designed for the RTP/UDP/IP protocol stack typically used for streaming. We discussed the challenges of header compression on wireless links with large and varying bit-error rates and large round-trip times. In addition, we reviewed a number of refinements to header compression that have been developed to address these challenges. These developments have culminated in the ROHC framework. We gave an overview of the compression mechanisms used in ROHC and then evaluated the compression performance of ROHC for packet voice communication and video streaming over wireless links. We found that in typical scenarios, ROHC cuts the bandwidth required for voice service by a factor of nearly two and at the same time improves the voice quality (expecially for larger residual bit errors that are not corrected by physical or link-layer techniques). For video streaming with single-layer (nonscalable) encoded video, we found that ROHC reduces the bandwidth requirement by about 10% for intermediate quality video. For lower-quality video, the bandwidth reductions are significantly larger (up to about 40%). Similarly, our results indicate that ROHC will achieve significant reductions of the required bandwidth for streaming layered (scalable) encoded video (which requires a unique flow for each encoding layer). Finally, we discussed the deployment of ROHC in the evolving wireless systems, which push the IP endpoint closer and closer to the wireless terminal. With the current standards, ROHC is operating between the wireless terminal and its radio network controller.

Acknowledgments

We are grateful to Acticom GmbH for the ROHC implementation and technical support; Stephan Rein of TU Berlin, Germany, (visiting Arizona State Univerity in spring/summer of 2003); Michele Rossi of the University of Ferrara, Italy; Andrea Carpene of the University of Milano, Italy; and Thomas Sikora of the TU, Berlin, Germany, for many insightful discussions and comments on drafts of this chapter.

This work was supported in part by the National Science Foundation under grant No. Career ANI-0133252 and grant No. ANI-0136774.

TABLE 10.5 Acronyms and Abbreviations

2G	Second-generation mobile communication systems	NACK	Negative acknowledgement
2.5G	Extension of 2G with packet switched services	NC	No context (state)
3G	Third-generation mobile communication systems	O-Mode	Optimistic mode
3GPP	Third-Generation Partnership Project	PDA	Personal data assistant
ACK	Acknowledgement	PDCP	Packet data convergence protocol
BEP	Bit-error probability	PEP	Packet error probability
CH	Compressed header	PPP	Point-to-point protocol
CID	Context identifier	PSNR	Peak signal-to-noise ratio
CRC	Cyclic redundancy check	QCIF	Quarter common intermediate format
CRTP	Compressed real time protocol (RFC 2508)	QoS	Quality of service
CSRC	Contributing source	RFC	Request for comment
CTCP	Compressed transport control protocol (RFC 1144)	RNC	Radio network controller
ECRTP	Enhanced compressed RTP	ROCCO	Robust checksum-based compression
ESP	Encapsulating security payload	ROHC	Robust header compression (RFC 3095)
FO	First order (state)	RTP	Real time protocol (RFC)
G.711	Standard for pulse code modulation of voice frequencies	SC	Static context (state)
GGSN	Gateway GPRS support node	SGSN	Supporting GPRS serving node
GoP	Group of pictures	SigComp	Signaling compression
GPRS	General packet radio service	SIP	Session initiation protocol
GRE	Generic routing encapsulaton (RFC 2784)	SNDCP	Sub-network dependent convergence protocol
GSM	Global System for Mobile Communications	SO	Second order (state)
GTP	GPRS tunneling protocol	sSNR	Segmental signal-to-noise ratio
H.26L	Video coding standard, also known as H.264, MPEG4 Annex 10	TCP	Transport control protocol
		UDP	User datagram protocol
HTTP	Hypertext Transfer Protocol	UDVM	Universal decompressor virtual machine
IEEE	Institute of Electrical and Electronics Engineers	U-Mode	Unidirectional mode
IP	Internet protocol	UMTS	Universal Mobile Telecommunications System
IPHC	IP header compression (RFC 2507)	UTRAN	Radio interface of UMTS
IR	Initialization and refresh (state)	VJHC	Van Jacobson header compression (RFC 1144)
LAN	Local area network	W-LSB	Windowed least significant bit
LPC	Linear predictive coding	WWW	World Wide Web
LZW	Lempel Ziv Welch (data compression algorithm)		

References

1. Jacobson, V., Compressing TCP/IP headers for low-speed serial links, RFC 1144, Feb. 1990.
2. Perkins, S.J. and M.W., Mutka, Dependency removal for transport protocol header compression over noisy channels, in *Proc. of IEEE International Conference on Communications*, 2, Montreal, Canada, June 1997, pp. 1025–1029.
3. Köpsel, A., NetMeter tool for evaluation of ROHC compressions, acticom GmbH, Berlin, Germany, 2002.
4. Seeling, P., Fitzek, F.H.P., and Reisslein, M., Videometer, *IEEE Network Magazine*, 17, 1, Jan. 2003, p. 5.
5. Fitzek, F.H.P., Köpsel, A,. Wolisz, A., Reisslein, M., and Krishnam, M.A., Providing application-level QoS in 3G/4G wireless systems: a comprehensive framework based on multi-rate CDMA, in *IEEE International Conference on Third Generation Wireless Communications*, June 2001, pp. 344–349.
6. Bormann, C., Burmeister, C., Degermark, M., Fukushima, H., Hannu, H., Jonsson, L-E., Hakenberg, R., Koren, T., Le, K., Liu, Z., Martensson, A., Miyazaki, A., Svanbro, K., Wiebke, T., Yoshimura, T., and Zheng, H., RObust Header Compression: ROHC: Framework and four profiles: RTP, UDP, ESP, and uncompressed, RFC 3095, July 2001.
7. Third Generation Partnership Project, Radio access bearer support enhancements, 3GPP, Tech. Rep., 2002.

8. Postel, J., User Datagram Protocol, RFC 768, Aug. 1980.
9. Schulzrinne, H., Casner, S., Frederick, R., and Jacobson, V., RTP: a transport protocol for real time applications, RFC 1889, Jan. 1996.
10. Reisslein, M., Lassetter, J., Ratnam, S., Lotfallah, O., Fitzek, F., and Panchanathan, S., Traffic and quality characterization of scalable encoded video: a large-scale trace-based study, Arizona State University, Dept. of Electrical Eng., Tech. Rep., 2002, http://trace.eas.asu.edu.
11. Fitzek, F.H.P., Seeling, P., and Reisslein, M., H.26L Pre-Standard Evaluation, acticom GmbH, Tech. Rep., Nov. 2002, http://www.acticom.de, http://trace.eas.asu.edu.
12. JVT, JM/TML Software Encoder/decoder, http://bs.hhi.de/~suehring/tml/, 2003, H.26L Software Coordination by Carsten Suehring.
13. Wiegand, T., H.26L Test Model Long-Term Number 9 (TML-9) draft0, ITU-T Study Group 16, Dec. 2001.
14. Calveras, A., Arnau, M., and Paradells, J., A controlled overhead for TCP/IP header compression algorithm over wireless links, in *Proc. of The 11th International Conference on Wireless Communications (Wireless'99)*, Calgary, Canada, 1999.
15. Calveras, A., Arnau, M., and Paradells, J., An improvement of TCP/IP header compression algorithm for wireless links, in *Proc. of Third World Multiconference on Systemics, Cybernetics and Informatics (SCI'99) and the Fifth International Conference on Information Systems Analysis and Synthesis (ISAS'99)*, 4, Orlando, FL, July/Aug. 1999, pp. 39–46.
16. Calveras, A. and Paradells, J., TCP/IP over wireless links: performance evaluation, in *Proc. of 48th IEEE Vehicular Technology Conference VTC'98*, 3, Ottawa, Canada, May 1998, pp. 1755–1759.
17. Degermark, M., Nordgren, B., and Pink, S., IP header compression, RFC 2507, Feb. 1999.
18. Rossi, M., Giovanardi, A., Zorzi, M., and Mazzini, G., Improved header compression for TCP/IP over wireless links, *Electronics Letters*, 36, 23, Nov. 2000, pp. 1958–1960.
19. Rossi, M., Giovanardi, A., Zorzi, M., and Mazzini, G., TCP/IP header compression: proposal and performance investigation on a WCDMA air interface, in *Proc. of the 12th IEEE International Symposium on Personal, Indoor and Mobile Radio Communications*, 1, Sep. 2001, pp. A-78–A-82.
20. Degermak, M., Engan, M., Nordgren, B., and Pink, S., Low-loss TCP/IP header compression for wireless networks, in *Proceedings of ACM MobiCom'96*, 3, New York, Oct. 1997, pp. 375–387.
21. Perkins, C. and Crowcroft, J., Effects of interleaving on RTP header compression, in *Proceedings of IEEE Infocom 2000*, Tel Aviv, Israel, 2000, pp. 111–117.
22. Lim, J. and Stern, H., IPv6 header compression algorithm supporting mobility in wireless networks, in *Proceedings of the Southeastcon 2000*, 2000, pp. 535–540.
23. Welch, T.A., A technique for high performance data compression, *IEEE Computer*, 6, 17, June 1984, pp. 8–19.
24. Ziv, J. and Lempel, A., A universal algorithm for sequential data compression, *IEEE Transactions on Information Theory*, 23, May 1977, pp. 337–343.
25. Casner, S. and Jacobson, V., Compressing IP/UDP/RTP headers for low-speed serial links, RFC 2508, Feb. 1999.
26. Degermark, M., Hannu, H., Jonsson, L., and Svanbro, K., Evaluation of CRTP performance over cellular radio links, *IEEE Personal Communications*, 7, 4, 2000, pp. 20–25.
27. Svanbro, K., Hannu, H. Jonsson, L.-E., and Degermark, M., Wireless real-time IP services enabled by header compression, in *Proceedings of the IEEE Vehicular Technology Conference (VTC)*, 2, Tokyo, Japan, 2000, pp. 1150–1154.
28. Cellatoglu, A., Fabri, S., Worral, S., Sadka, A., and Kondoz, A., Robust header compression for real-time services in cellular networks, in *Proceedings of the IEE 3G 2001*, London, GB, Mar. 2001, pp. 124–128.
29. Chen, W.-T., Chuang, D.-W., and Hsiao, H.-C., Enhancing CRTP by retransmission for wireless networks, in *Proceedings of the Tenth International Conference on Computer Communications and Networks*, 2001, pp. 426–431.

30. West, M.A., Conroy, L.W., Hancock, R.E., Price, R., and Surtees, A.H., IP header and signalling compression for 3G systems, in *Proc. of 3G Mobile Communication Technologies*, May 2002, pp. 102–106.

31. Khiem, L., Clanton, C., Zhigang, L., and Haihong, Z., Efficient and robust header compression for real-time services, in *Proceedings of the IEEE Wireless Communications and Networking Conference (WCNC)*, 2, Chicago, IL, 2000, pp. 924–928.

32. Kostic, Z., Xiaoxin, Q., and Chang, L., Impact of TCP/IP header compression on the performance of a cellular system, in *Proceedings of the Wireless Communications and Networking Conference*, 1, Chicago, IL, 2000, pp. 281–286.

33. Liao, H., Zhang, Q., Zhu, W., and Zhang, Y.-Q., A robust TCP/IP header compression scheme for wireless networks, in *Proceedings of the IEEE International Conference on 3G Wireless and Beyond*, San Francisco, June 2001.

34. Boggia, G., Camarda, P., and Squeo, V.G., ROHC+: a new header compression scheme for TCP streams in 3G wireless systems, in *Proceedings of the IEEE International Conference on Communications (ICC)*, 5, 2002, pp. 3271–3278.

35. Jiao, C., Schwiebert, L., and Richard, G., Adaptive header compression for wireless networks, in *Proceedings of the 26th Annual IEEE Conference on Local Computer Networks*, Nov. 2001, pp. 377–378.

36. Pelletier, G., Zhang, Q., Jonsson, L.-E., Liao, H., and West, M., RObust Header Compression (ROHC): TCP/IP profile (ROHC-TCP), Internet draft, Work in progress, Nov. 2002.

37. Price, R., Hancock, R., McCann, S., West, M. A., Surtees, A., Ollis, P., Zhang, Q., Liao, H., Zhu, W., and Zhang, Y.-Q., TCP/IP compression for ROHC, Internet draft, Work in progress, Nov. 2001.

38. Lilley, J., Yang, J., Balakrishnan, H., and Seshan, S., A unified header compression framework for low-bandwidth links, in *Proceedings of ACM MobiCom*, 2000, pp. 131–142.

39. Farinacci, D., Li, T., Hanks, S., Meyer, D., and Traina, P., GRE: Generic routing encapsulation, RFC 2784, Mar. 2000.

40. Dommety, G., Key and sequence number extensions to GRE, RFC 2890, Sep. 2000.

41. Larzon, L.-A., Degermark, M., Pink, S., and Fairhurst, G., The UDP-Lite Protocol, Internet draft, Work in progress, Dec. 2002, draft-ietf-tsvwg-udp-lite-01.txt.

42. Wang, Y. and Zhu, Q., Error control and concealment for video communication: a review, *Proceedings of the IEEE*, 86, 5, May 1998, pp. 974–997.

43. Simpson, W., The Point-to-Point Protocol (PPP), RFC 1661, July 1994.

44. Price, R., Bormann, C., Christoffersson, J., Hannu, H., Liu, Z,. and Rosenberg, J., SigComp: signaling compression, RFC 3261, Jan. 2003.

45. Rosenberg, J., Schulzrinne, H., Camarillo, G., Johnston, A., Peterson, J., Sparks, R., Handley, M., and Schooler, E., SIP: session initiation protocol, RFC 3261, June 2000.

46. Wang, B., Schwefel, H., Chua, K., Kutka, R., and Schmidt, C., On implementation and improvement of robust header compression in UMTS, in *Proceedings of the 13th IEEE International Symposium on Personal, Indoor and Mobile Radio Communications*, 2002, pp. 1151–1155.

47. Rein, S., Performance measurements of voice quality over error-prone wireless networks using Robust Header Compression, Master's thesis, Communication Systems Group, Technical University of Berlin, Mar. 2003.

48. Ekmekci, S. and Sikora, T., Unbalanced quantized multiple description video transmission using path diversity, in *IS&T/SPIE's Electronic Imaging 2003*, 2003, Santa Clara, CA.

49. Shah, M.J., IP header compression in the SGSN, in *Proc. of IEEE SoutheastCon*, 2002, pp. 158–161.

50. Third Generation Partnership Project, Packet Data Convergence Protocol (PDCP) Specification, 3GPP, Tech. Rep., 2002.

11

Video Streaming in Wireless Internet

Frank H.P. Fitzek
Acticom GmbH

Patrick Seeling
Arizona State University

Martin Reisslein
Arizona State University

11.1 Introduction

Video services are becoming an integral part of future communication systems. Especially for the upcoming third-generation (3G) wireless networks such as Universal Mobile Telecommunications System (UMTS), video may very well turn out to be the key that achieves the required return of investment. Whereas previous generations of wireless communication systems were primarily designed and used for voice services, next-generation systems have to support a broad range of applications in a wide variety of settings. Novel wireless applications, such as telematics and fleet management, have introduced wireless networking to the enterprise domain. At the same time, the private-market sector is booming with the availability of low-priced wireless equipment. The early market stages were characterized by the needs of early adopters, mostly for professional use. As the market matures from the early adopters to normal users, new services will be demanded. These demands will likely converge toward the demands that exist for wired telecommunications services. These demands, also referred to as "the usual suspects," consist of a variety of different services, including Internet access for browsing, chatting, and gaming. In addition, entertainment services, such as television, cable television, and pay-per-view movies, are also in demand. With the availability of wireless services, the location is no longer of importance to private users. Thus, the demands of mobile users, connected over wireless networks, will approach this mixture of services. With the omnipresence of wireless services, the usage schemes will become independent from location and connection type. One example for such a wireless service is mobile gaming. Private users who are waiting at an airport, for instance, are able to join Internet-based multiplayer games to bridge time gaps. Among the various entertainment services, mobile video will likely account for a large portion of the entertainment services, as are cinemas, video rentals, and television now. This application scenario covers wireless entertainment broadcasting and video on demand. A second, professional application of video in the wireless domain is telemedicine, where remote specialists are enabled to respond to emergencies. The wide area of video services in wireless environments and the expectations for wireless communication systems call for an understanding of the basic principles of wireless video streaming.

Generally, the video delivered to wireless users is either (1) *live* video, such as the live coverage of a sporting event, concert, or conference, or (2) *prerecorded* (stored) video, such as a television show, entertainment movie, or instructional video. Some videos fall in both categories. For instance, in a typical distance-learning system, the lecture video is available to distance learners live (i.e., while the lecture is ongoing) and also as stored video (i.e., distance learners can request the lecture video later in the day or week from a video server).

In general, there are two ways to deliver video over a packet-switched network (including packet-oriented wireless networks): *file download* or *streaming*. With file download, the entire video is down-loaded to the user's terminal before the playback commences. The video file is downloaded with a conventional reliable transport protocol, such as Transmission Control Protocol (TCP). The advantage of file download is that it is relatively simple and ensures a high video quality. This is because losses on the wireless links are remedied by the reliable transport protocol and the playout does not commence until the video file is downloaded completely and without errors. The drawback of file download is the large response time, typically referred to as *startup delay*. The startup delay is the time from when the user requests the video until playback commences. Especially for large video files and small bandwidth wireless links, the startup delay can be very large.

With video streaming, on the other hand, playback commences before the entire file is downloaded to the user's terminal. Typically, only a small part of the video ranging from a few video frames to several hundreds or thousands of frames (corresponding to video playback durations on the order of hundreds of milliseconds to several seconds or minutes) are downloaded before the streaming commences. The remaining part of the video is transmitted to the user while the video playback is in progress. One of the key trade-offs in video streaming is between the startup delay and the video quality. That is, the smaller the amount of the video that is downloaded before streaming commences, the more the continuous video playback relies on the timely delivery of the remaining video over the unreliable wireless links. The errors on the wireless links may compromise the quality of the delivered video in that only basic low-quality (and low bit rate) video frames are delivered or some video frames are skipped entirely. Thus, video streaming gives the user shorter startup delays at the expense of reduced video quality. The challenge of video streaming lies in keeping the quality degradation to a level that is hardly noticeable or tolerable while utilizing the wireless resources efficiently (i.e., supporting as many simultaneous streams as possible).

Note that file download and streaming with some startup delay are suitable only for prerecorded video. The delivery of live video, on the other hand, requires an extreme form of video streaming with essentially no preplayback download. Another consideration in the delivery of prerecorded video is user interaction (i.e., videocassette recorder [VCR] functions such as fast forward, pause, and rewind). Some of these interactions may result in a new startup delay in video streaming.

In this chapter, we introduce video streaming in wireless environments. We first give an introduction to the world of digital video and outline the fundamental need for compression. Next, we introduce the basics of video compression. We show how video compression is achieved by exploiting various types of redundancies in the raw (uncompressed) video stream. We do this by following one sample video sequence on its way from the raw video to the compressed video stream. This sample video sequence,

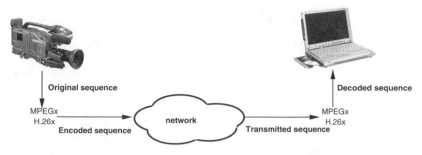

FIGURE 11.1 Generic wireless video streaming system.

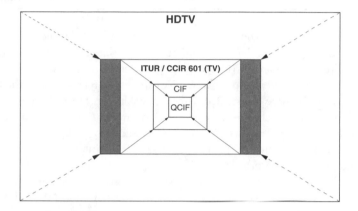

FIGURE 11.2 Illustration of image formats.

which is called *Highway*, is publicly available.[4] Next, we introduce the characteristics of the wireless communication channel and study the impact of the wireless link errors on the video quality. Finally, we discuss the various protocols and adaption techniques for streaming the video content over the wireless channel. The model of a transmission chain of a general wireless communication system for video streaming is given in Figure 11.1.

At the sender side, the video source is passed to the video (source) encoder. The compressed video stream is passed to the transport process, which, in turn, passes the stream plus some overhead information to the channel coder and modulation part for transmission over the wireless link. At the receiver side, the process is reversed.

We note that the abbreviations used throughout this chapter can be found in Table 11.3 (at end of chapter).

11.2 Basics of Video Compression

Video compression is undergoing constant changes as new coding/decoding (codec) systems are developed and introduced to the market. Nevertheless, the internationally standardized video compression schemes (see Section 11.2.5), such as the H.26*x* and MPEG-*n* standards, are based on a common set of fundamental encoding principles. In this section we give an overview of these encoding principles. For more details on video compression, refer to references 5 to 11.

11.2.1 Digital Video

The sizes of the pictures in the current video formats are illustrated in Figure 11.2. Note that the ITU-R/CCIR 601 format (i.e., the common TV image format) and the Common Intermediate Format (CIF) and Quarter Common Intermediate Format (QCIF) have the same ratio of width to height. In contrast, the High-Definition Television (HDTV) image format has a larger width to height ratio (i.e., is perceived

Table 11.1   Characteristics of Video Formats

Standard	QCIF ITU-T H.261		CIF ITU-T H.261		TV ITU-R/CCIR-601		HDTV ITU-R 709-3	
Format	PAL (25 Hz)	NTSC (30 Hz)	PAL (25 Hz)	NTSC (30 Hz)	PAL (25 Hz)	NTSC (30 Hz)	PAL (25 Hz)	NTSC (30 Hz)
Sub-sampling	4:2:0		4:2:0		4:2:2		4:2:2	
Columns (Y)	176		352		720		1920	
Rows (Y)	144		288		576	480	1080	
Columns (U,V)	88		176		360	360	960	
Rows (U,V)	72		144		576	480	1080	
Frame size (byte)	38016		152064		1244160	1036800	4147200	
Data Rate (Mbit/s)	7.6	9.1	30.4	36.5	248.8	298.6	829.4	995.3

as "wider"). Each individual image is composed of picture elements (usually referred to as pixels or pels). The specific width and height (in pixels) of the different formats are summarized in Table 11.1. Today, typical formats for wireless video are QCIF (176×144 pixel) and CIF (352×288 pixel).

A single pixel can be represented in two different color spaces: RGB and YUV. In the RGB color space, the pixel is composed of the three colors: Red, Green, and Blue. In the YUV color space, the pixel is represented by its luminance (Y), chrominance (U), and saturation (V). The U and V components are often referred to as the chrominance values. The RGB color space is typically used for computer displays, whereas the YUV color space is common for television sets. Television sets convert the incoming YUV signal to RGB while displaying the picture. The YUV color representation was necessary to broadcast color television signals while allowing the old black-and-white sets to function without modifications. Because the luminance information is located in different frequency bands, the old tuners are capable of tuning in on the Y signal alone. The conversion between these two color spaces is defined by a fixed conversion matrix. Most common video compression schemes take the YUV color space as input, and we therefore focus on this color space for the remainder of this section.

In digital video, the (analog) values of the three components Y, U, and V (or R, G, and B) are quantized to 8-bit representations — that is, there are 3 bytes representing each pixel. The human eye is far more sensitive to changes in luminance than to the two chrominance components. It is therefore common to reduce the information that is stored per picture by *chrominance subsampling*. In the unsampled YUV format, also referred to as YUV 4:4:4 format, every pixel is represented by 3 bytes, as just noted. With subsampling, the ratio of chrominance to luminance bytes is reduced. More specifically, subsampling represents a group of typically four pixels by their four luminance components (bytes) and one set of two chrominance values. Each of these two chrominance values is typically obtained by averaging the corresponding chrominance values in the group. In case the four pixels are grouped as a block of 2×2 pixels, the format is YUV 4:2:0. If the grouped pixels are forming a line of 4×1, the format is referred to as YUV 4:1:1. These two most common YUV sampling formats are illustrated in Figure 11.3 and Figure 11.4.

Thus, the size of one YUV frame with 4:2:0 (or 4:1:1) chrominance subsampling in the QCIF format (176-pixel columns by 144-pixel rows for the luminance) is shown in Equation 11.1:

$$176 \cdot 144 \cdot \left(8\ bit + \frac{2 \cdot 8\ bit}{4} \right) = 304{,}128\ bit = 38{,}016\ byte. \tag{11.1}$$

The frame sizes and data rates for the different video formats and frame rates are summarized in Table 11.1. Note that although chrominance subsampling reduces the bit rate significantly, the subsampled video is commonly referred to as uncompressed (raw) video, as it is the input to the video encoding (compression). For television applications, YUV 4:2:2 is used. This format samples the chrominance for every second luminance value. Several formats for the sequence of sampling and storage exist.

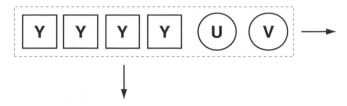

FIGURE 11.3 YUV 4:1:1 subsampling.

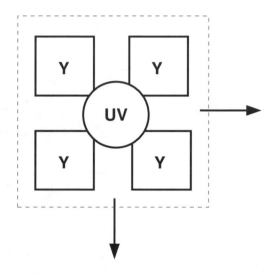

FIGURE 11.4 YUV 4:2:0 subsampling.

The different frame rates of 25 frames per second (fps) in Phase Alternation by Line (PAL)-compliant video and 30 fps in National Television Standards Committee (NTSC)-compliant video are due to the different frequencies in the power supplies in Europe/Asia and the U.S. Given the enormous bit rates of the raw video streams and the limited bandwidths provided by wireless links, it is clear that some form of compression is required to make wireless video viable.

In general, digital video is not processed, stored, compressed, and transmitted on a per-pixel basis but in a hierarchy,[10] as illustrated in Figure 11.5.

At the top of this hierarchy is the *video sequence*, which is divided into *Groups of Pictures (GoPs)*. Each GoP in turn consists of multiple *video frames*. A single frame is divided into *slices*. Each slice consists of several *macroblocks (MBs)*, each typically consisting of 4×4 *blocks*. Each block typically consists of 4×4 *pixels*.

Because of the large bit rates, digital video is almost always encoded (compressed) before transmission over a packet-oriented network. The limited bandwidths of wireless links make compression especially important. Video compression generally exploits three types of redundancies.[10] On a per-frame basis (i.e., single picture), neighboring pixels tend to be correlated and thus have *spatial* redundancy.[11] Intraframe encoding is used to reduce the spatial redundancy in a given frame. In addition, consecutive frames have similarities and therefore *temporal* redundancy. These temporal redundancies are reduced by interframe coding techniques. The result of the reduction of these two redundancies is a stream of codewords (symbols) that has some redundancy at the symbol level. The redundancy between these symbols is reduced by variable-length coding before the binary code is passed on to the output channel. The elimination of these redundancies is explained in the following three subsections.

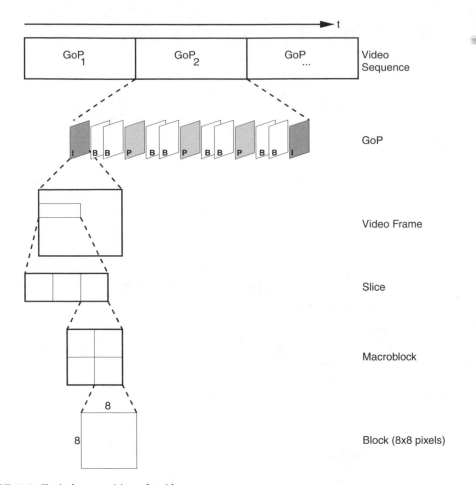

FIGURE 11.5 Typical composition of a video sequence.

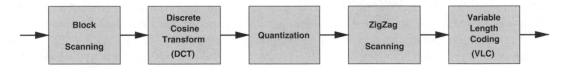

FIGURE 11.6 DCT coding concept.

11.2.2 Intraframe Coding

The intracoding (compression) of an individual video frame resembles still-picture encoding. It is commonly based on the Discrete Cosine Transformation (DCT). Wavelet-based transformation schemes have also emerged. Studies indicate that in the field of video encoding, the wavelet-based approach does not improve the quality of the transformed video significantly.[12] Essentially all internationally standardized video compression schemes, however, are based on the DCT, and we will therefore focus on the DCT in our discussion.

The intraframe coding proceeds by partitioning the frame into blocks, also referred to as block scanning. The size of these blocks today is typically 8×8 pixels (previously, also 4×4 and 16×16 were used). The DCT is then applied to the individual blocks. The resulting DCT coefficients are quantized and zig-zag-scanned according to their importance to the image quality. An overview of these steps is given in Figure 11.6.

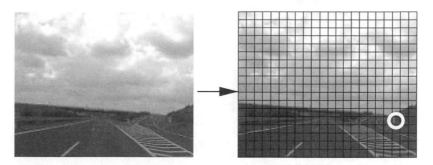

FIGURE 11.7 Video frame subdivision into blocks (QCIF format into 22 × 18 blocks of 8 × 8 pixels each).

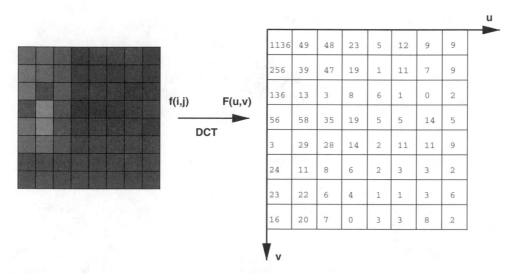

FIGURE 11.8 An 8 × 8 block of luminance values (visual representation and numerical values) and the resulting DCT transform coefficients (decimal places truncated).

11.2.2.1 Block Scanning

To reduce the computational power required for the DCT, the original frame is subdivided into blocks, because efficient algorithms exist for a block-based DCT.[13] The utilization of block shapes for encoding is one of the limitations for DCT-based compression systems. The typical object shapes in natural pictures are irregular and thus cannot be fitted into rectangular blocks, as illustrated in Figure 11.7. To increase the compression efficiency, different block sizes can be utilized at the cost of increased complexity.[14]

11.2.2.2 Discrete Cosine Transformation (DCT)

The DCT is used to convert a block of pixels (e.g., for the luminance component 8 × 8 pixels, represented by 8 bits each, for a total of 256 bits) into a block of transform coefficients. The transform coefficients represent the spatial frequency components of the original block. An example for this transformation of the block marked in Figure 11.7 is illustrated in Figure 11.8.

This transformation is lossless; it merely changes the representation of the block of pixels, or more precisely the block of luminance (chrominance) values. A two-dimensional DCT for an $N \times N$ block of pixels can be described as two consecutive one-dimensional DCTs (i.e., horizontal and vertical). With $f(i,j)$ denoting the pixel values and $F(u,v)$ denoting the transform coefficients, we have Equation 11.2:

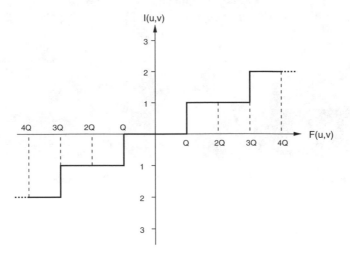

FIGURE 11.9   Illustration of quantization, $T = Q$.

$$F(u,v) = \frac{2}{N} \cdot C(u) \cdot C(v) \cdot \sum_{i=0}^{N-1} \sum_{j=0}^{N-1} f(i,j) \cos\left(\frac{(2i+1)u\pi}{2N}\right) \cos\left(\frac{(2j+1)v\pi}{2N}\right) \qquad (11.2)$$

where
$$C(x) = \begin{cases} \dfrac{1}{\sqrt{2}}, x = 0 \\ 1, otherwise \end{cases}$$

The lowest-order coefficient is usually referred to as the *DC* component, whereas the other components are referred to as *AC* components.

11.2.2.3   Quantization

In a typical video frame, the energy is concentrated in low-frequency coefficients. That is, a few coefficients with u and v close to zero have a high significance for the representation of the original block. On the other hand, most higher-frequency coefficients (i.e., $F(u,v)$s for larger u and v) are small. To compress this spatial frequency representation of the block, a quantization of the coefficients is performed. Two factors determine the amount of compression and the loss of information in this quantization:

1. Coefficients $F(u,v)$ with an absolute value smaller than the *quantizer threshold T* are set to zero (i.e., they are considered to be in the so-called dead zone).
2. Coefficients $F(u,v)$ with an absolute value larger than or equal to the quantizer threshold T are divided by twice the *quantizer step size Q* and rounded to the nearest integer.

In summary, the quantized DCT coefficients $I(u,v)$ are given by Equation 11.4:

$$I(u,v) = \begin{cases} 0 \, for \, |F(u,v)| < T \\ \left[\dfrac{F(u,v)}{2Q}\right] for \, |F(u,v)| \geq T \end{cases} \qquad (11.3)$$

A quantizer with $T = Q$, as typically used in practice, is illustrated in Figure 11.9. Figure 11.10 continues the example from Figure 11.8 and shows the quantized values for $T = Q = 16$. As illustrated here, typically many DCT coefficients are zero after quantization.[9] The larger the step size, the larger the compression

36	2	2	1	0	0	0	0
8	1	1	1	0	0	0	0
4	0	0	0	0	0	0	0
2	2	1	1	0	0	0	0
0	1	1	0	0	0	0	0
1	0	0	0	0	0	0	0
1	1	0	0	0	0	0	0
1	1	0	0	0	0	0	0

36	2	2	1	0	0	0	0
8	1	1	1	0	0	0	0
4	0	0	0	0	0	0	0
2	2	1	1	0	0	0	0
0	1	1	0	0	0	0	0
1	0	0	0	0	0	0	0
1	1	0	0	0	0	0	0
1	1	0	0	0	0	0	0

36,2,8,4,1...

FIGURE 11.10 Quantized DCT coefficients ($q = 16$) and zig-zag scanning pattern.

FIGURE 11.11 Small quantization parameter ($q = 16$, *PSNR* = 44.9579 dB, frame size = 4640 bytes).

FIGURE 11.12 Medium quantization parameter ($q = 35$, *PSNR* = 33.0344 dB, frame size = 135 bytes).

gain, as well as the loss of information.[15] The trade-off between compression and decodable image quality is controlled by setting the quantizer step size (and quantizer threshold).[10] This trade-off between image quality and compression (frame size in bytes after quantization) is illustrated for one frame from the *Highway* test sequence[4] in Figure 11.11, Figure 11.12, and Figure 11.13.

FIGURE 11.13   Large quantization parameter (q = 40, *PSNR* = 30.2765 dB, frame size = 79 bytes).

The frame was encoded with a fixed quantization scale and subsequently decoded. Encoders employ optimized static quantizers for each DCT coefficient which are stored in a quantization matrix. The quantizers are typically optimized for the perceived quality of the encoded video given the importance of the individual coefficient. The quantization scale factor q is then multiplied with the individual quantizers of the quantization matrix to determine the actually used quantizer Q for each DCT coefficient. This process is typically performed using matrix multiplication. Notice that the quality of the video frame visibly decreases as q increases. As can be seen from the frame sizes, the quality loss is also reflected in the amount of data needed. We also report the *Peak Signal-to-Noise Ratio (PSNR)* for these encodings. The PSNR is a commonly used objective quality metric; see Section 11.3.1 for details. As illustrated here, a smaller PSNR value corresponds to a worse image quality.

Figure 11.11, Figure 11.12, and Figure 11.13 represent the encoding result without rate control. Rate control is applied during the encoding process to adjust the resulting video frame sizes to the bandwidth available. The quantization is adjusted in a closed-loop process (i.e., the result of the quantization is measured for its size and as required encoded again with a different quantizer step size) to apply a compression in dependence of video content and the resulting frame size. The result is a constant bit rate (CBR) video stream but with varying quantization and thus quality. The opposite of CBR is variable bit rate (VBR) encoding. Here the quantization process remains constant; thus, it is referred to as open-loop encoding (i.e., the result of the quantization process is no longer subject to change in order to meet bandwidth requirements). To achieve a constant quality, VBR encoding has to be used.[16,17]

11.2.2.4   Zig-Zag Scanning

The coefficient values obtained from the quantization are scanned by starting with the DC component and then continuing to the higher-frequency components in a zig-zag fashion, as illustrated in Figure 11.10. The zig-zag scanning facilitates the subsequent variable-length encoding by encountering the most likely nonzero elements first. Once all nonzero coefficients are scanned, the obtained sequence of values is further encoded to reduce codeword redundancy; see Section 11.2.4. The scanning can be stopped before collecting all quantized nonzero coefficients to achieve further (lossy) compression.

11.2.3   Interframe Coding: Motion Estimation and Compensation

Video encoders commonly use interframe coding to reduce the temporal redundancy between successive frames (images). The basic idea of interframe coding is that the content of a given current video frame is typically similar to a past video frame. The past frame is used as a reference frame to predict the content of the current frame. This prediction is typically performed on a macroblock or block basis.[18] For a given

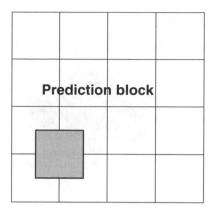

FIGURE 11.14 Reference (past) frame.

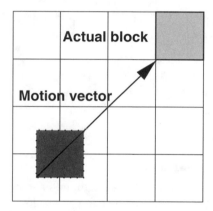

FIGURE 11.15 Predicted (current) frame.

actual block in the current frame, a *Block-Matching Algorithm (BMA)* searches for the most similar prediction block in the past frame, as illustrated in Figure 11.14 and Figure 11.15. The goal of the search is to determine the *motion vector* (i.e., the displacement of the most similar prediction block from the actual block).

This search, which is also referred to as motion estimation, is performed over a specific range of n pixels around the location of the block in the current frame. Several matching (similarity) criteria, such as the cross-correlation function, mean squared error, or mean absolute error, can be applied. To find the best match by full search, $(2n+1)2$ comparisons are required. Several fast-motion estimation schemes such as the *three-step search*[19] or the *hierarchical block-matching algorithm*[20] have evolved to reduce the processing. Once the motion vector is determined, the difference between the prediction block and the actual block is encoded using the intraframe coding techniques discussed in the preceding section. These differences may be due to lighting conditions, angles, and other factors that slightly change the content of blocks. These differences are typically small and allow for efficient encoding with the intraframe coding techniques (and Variable-Length Coding [VLC] Techniques). The quantizer step size can be set independently for the coding of these differences. The encoding of these differences accounts for the differences between the prediction block and the actual block and is referred to as *motion compensation*. The intercoded frame is represented by (1) the motion vectors (motion estimation) and (2) the encoded error or difference between the current frame and a previously transmitted reference frame (motion compensation).

f10 frames   **forward prediction**   **f**   **backward prediction**   **f+10 frames**

FIGURE 11.16 Changing video-frame content.

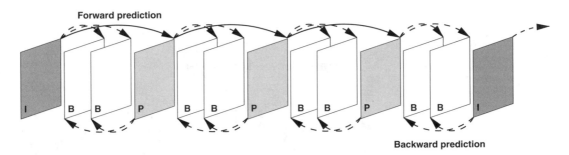

FIGURE 11.17 Typical MPEG GoPs (frames 1 to 12).

11.2.3.1 Concept of I, P, and B Frames

Blocks in video frames often reveal parts of the background or scene that were not visible before the actual frame.[9] Motion vectors of these areas can therefore not be found by referencing previous frames, but only by considering also future frames, as illustrated in Figure 11.16. For this reason, interframe coding often considers prediction from both past reference frames as well as future reference frames.

There are three basic methods for encoding the original pictures in the temporal domain: I (Intracoded), P (Intercoded), and B (Bi-directional), as introduced in the MPEG-1 standard.[8] These encoding methods are applied on the frame or block level, depending on the codec. The intercoded frames use motion estimation relying on the previous inter- or intracoded frame. The bi-directional-encoded frames rely on a previous as well as a following intra- or intercoded frame. Intracoded frames or blocks are not relying on other video frames and thus are important to stop error propagation. The sequence of frames between two intracoded frames is referred to as a GoP. The relationship among these various encoding types and how frames rely on each other in a typical MPEG frame sequence is illustrated in Figure 11.17.

11.2.4 Variable-Length Coding (VLC)

The purpose of VLC is to reduce the statistical redundancy in the sequence of code words obtained from zig-zag scanning an intracoded block (or block of differences for a predicted block). Short code words are assigned to values with high probabilities. Longer code words are assigned to less probable outcomes of the quantization. The mapping between these original values and the code symbols is done within the VLC. The mapping has to be known by both the sender and the receiver. As shown before, the quantization and zig-zag scanning result in a large number of zeros. These values are encoded using run-level coding. This encoding transmits only the number of zeros instead of the individual zeros. In addition, when no other values are trailing the coefficients with zeros (and this is the most likely case), an end-of-block (EOB) code word is inserted into the resulting bitstream. Huffman coding[21] and arithmetic coding[22,21]

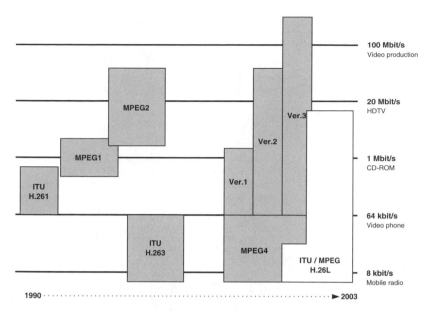

FIGURE 11.18   Video coding standards of ITU-T and ISO/MPEG.

and their respective derivatives are used to implement VLC. Huffman coding is fairly simple to implement but achieves lower compression ratios. On the other hand, arithmetic coding schemes are computationally more demanding but achieve better compression. Because processing power is abundant in many of today's systems, newer codecs mostly apply arithmetic coding.[10,23]

11.2.5   ISO/MPEG and ITU-T Standards for Video Encoding

We close this section by giving a quick overview of video standards. Despite a large variety of video coding and decoding systems (e.g., the proprietary Real-Media codec etc.), standardization on an international level is performed by two major bodies: International Telecommunications Union-Telecommunication Standardization Sector (ITU-T) and International Standardization Organization/Moving Pictures Experts Group (ISO/MPEG). The early H.261 codec of the ITU-T was focused on delivering video over Integrated Services Digital Network (ISDN) networks with a fixed bit rate of $n \times 64$ kbit/s, where n denotes the number of multiplexed ISDN lines. From this starting point, codecs were developed for different purposes, such as the storage of digital media or delivery over packet-oriented networks. The latest codec development H.26L (or MPEG-4 Annex 10, H.264/AVC) is still under way (at the time of this writing) and is being worked on by a Joint Video Team (JVT) of the ITU-T and the ISO/MPEG bodies. The evolving standards achieved better quality (in terms of PSNR; see Section 11.3.1) with lower bit rates and thus better rate-distortion performance. Figure 11.18 sketches an overview of the video standards development to date.

11.3   Traffic and Quality Characteristics of Encoded Video

Having outlined the encoding (compression) of video in the preceding section, we now turn our attention to the output of the video encoder. In particular, we study the traffic characteristics of the encoded video stream as well as the video quality. The video traffic is governed by the sizes (in bytes or bits) of the individual video frames and the frame periods (display times of the individual video frames). For NTSC video without any skipped frames, the frame period is the reciprocal of the fixed frame rate (i.e., the frame period is 1/30 fps = 33.33 msec). For networking research purposes, the traffic of encoded video

TABLE 11.2 Frame Statistics for the *Highway* Sequence (QCIF, $Q = 16$)

Aggregation Level	Attribute	Value
Video Sequence	Total size (byte)	9047512
	Total # of frames	1999
	Playing time (sec)	79.96
Frame	Minimum size (byte)	3347
	Maximum size (byte)	9178
	Mean size (byte)	4526.02
	Frame size variance	1261651.93
	Frame coefficient of variation	0.25
	Mean frame bit rate (bis/sec)	905203.80
	Peak frame bit rate (bit/sec)	2936960.00
	Peak-to-mean ratio	3.24
	Mean I-frame size	7861.53
	Mean P-frame size	4526.02
	Mean B-frame size	4036.88
GoP	Number of GoPs	166
	Minimum GoP size (byte)	7283
	Maximum GoP size (byte)	84719
	Mean GoP size (byte)	54205.58
	Mean GoP bit rate (bit/sec)	903426.33
	Peak GoP bit rate (bit/sec)	1411983.3
	GoP peak to mean	1.56
	GoP variance	20190776.03
	GoP coefficient of variation	0.08

is often recorded in *video traces*, which give the frame number (index), type (e.g., I, P, or B), playout time (when the frame is decoded and displayed on screen), and the frame size (length), as illustrated by the following excerpt of a video trace of an H.26L encoding:[24]

```
Frame  No.     Frametype     Time  [ms]     Length  [byte]
0              I             0.0             7969
1              P             120.0           5542
...            ...           ...             ...
```

Video traces are available for a few existing video standards.[25–28] The traces are the basis for video-traffic studies and video-traffic modeling. The traces are also used to drive simulations of video-streaming protocols and mechanisms (parsers that read the traces into commonly used discrete event simulator packages [e.g., NS,[29] Omnet++,[30] or Ptolemy[31]] are available).[32]

Continuing the example used throughout this chapter, Table 11.2 gives elementary statistics for the video frame sizes and bit rates of a QCIF *Highway* encoding. The statistical analysis of the video traffic also typically includes metrics capturing the correlations in the traffic and the long-range dependence (self-similarity) properties of the traffic, which are beyond the scope of this chapter.

Figure 11.19 gives a plot of the frame sizes averaged over GoPs as a function of time. Notice that the frame size hovers around 5500 bytes, with occasional peaks reaching up to 7000 and 8000 bytes.

This behavior is governed by the video content, as illustrated by three figures. The highlighted regions represent different content dynamics. The regions highlighted by eclipses have monotone content dynamics and result in roughly constant frame sizes. In contrast, the street sign and the bridge increase the content dynamics and result in larger frame sizes. Note that the fraction of the dynamic content in the frame influences the frame sizes. The street sign is smaller than the bridge that crosses the entire screen and results in a smaller frame-size peak than the bridge. Video frame sizes relate to the content of the

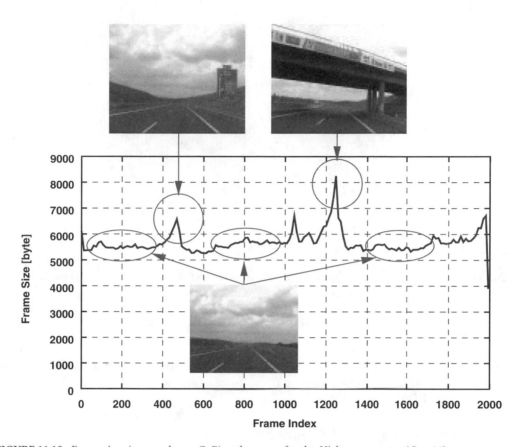

FIGURE 11.19 Frame sizes (averaged over GoP) and content for the *Highway* sequence ($Q = 16$).

video, different encoding methods, and encoder settings. These dependencies cause the frame sizes to have long-range dependencies and complicate the modeling of video traffic.

Note that the traffic variabilities in the *Highway* sequence considered here are relatively small (a peak-to-mean ratio of 3.24 and a coefficient of variation of 0.25), due to the overall fairly monotone content. Many typical entertainment videos are significantly more variable with frame size peak-to-mean ratios of 10 or more and coefficients of variation close to 1. Accommodating this highly variable (bursty) traffic in a packet-switched network is very challenging. In wireless video streaming, the second major challenge is to overcome the highly variable wireless link errors; see Section 11.4.

11.3.1 Objective Video Quality Measurements

To estimate the video quality, an approach is needed that compares the reconstructed frame at the receiver side with the original frame. The PSNR is the most commonly utilized metric. Other algorithms for objective video-frame quality assessment exist but are not proven to achieve better results than the PSNR.[33] The PSNR represents the objective video quality for each video frame by a single number. Because the human visual systems is more sensitive to the luminance component than the chrominance components, the PSNR is typically evaluated only for the Y (luminance) component. For a video frame composed of $N \cdot M$ pixels, the mean squared error (MSE) and the PSNR in decibels are computed as

$$MSE = \frac{\sum_{\forall i,j}[f(i,j) - \hat{f}(i,j)]^2}{N \cdot M} \qquad (11.5)$$

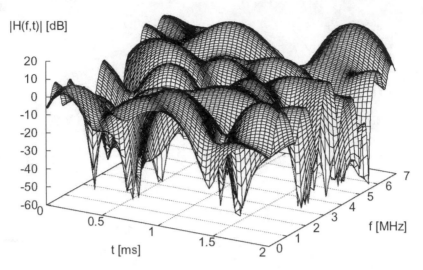

FIGURE 11.20 Typical signal-quality fluctuations on a wireless channel as a function of time and frequency.[1]

$$PSNR = 20 \cdot \log_{10}\left(\frac{255}{\sqrt{MSE}}\right) \qquad (11.6)$$

where $f(i,j)$ represents the luminance value of pixel (i,j) in the original source frame and $\hat{f}(i,j)$ the corresponding luminance value in the decoded frame. Some recent video traces include both the frame sizes as well as the PSNR frame qualities.[34]

11.4   Wireless-Channel Characteristics

Before we have a closer look at the protocols and mechanisms used for video streaming, we give a short introduction to the relevant characteristics of the wireless channel. A basic understanding of the impairments of the wireless channel is important, as the video streaming mechanisms have to compensate for (1) the errors on the wireless channels (discussed in this section) and (2) the burstiness of the video traffic (discussed in Section 11.3). Wireless channels are typically time varying and frequency selective, as illustrated in Figure 11.20 and Figure 11.21. These channel fluctuations are the result of a combination of attenuation, multipath fading, and shadowing[35] (see Section 11.4.1). Reflections result in signals being transmitted over multiple paths between sender and receiver (see Section 11.4.2).

11.4.1   Free-Space Propagation

The wireless signal is attenuated on its path from the transmitter to the sender (i.e., the power level [strength] of the received signal is lower than the transmit power level). The free-space propagation model is used to predict the received signal strength at the receiver under the assumption that only one line-of-sight path between sender and receiver exists. In such a scenario, the power level of the received signal P_r depends on the transmitted power P_s, the distance d between sender and receiver, the antenna gain of sender and receiver, G_t and G_r, the wavelength λ, and the path loss L. The so-called free-space equation[35,36] gives the received signal power as

$$P_r(d) = \frac{P_s G_t G_r \lambda^2}{(4\pi d)^2 L} \qquad (11.7)$$

Note that the received power decreases with the square of the distance.

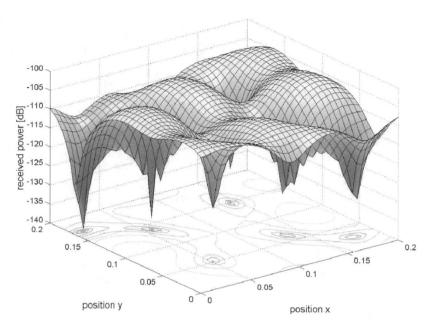

FIGURE 11.21 Typical signal-quality fluctuations on a wireless channel as a function of time and location.[2]

11.4.2 Multipath Fading

Considering a mobile and wireless communication between one sender and one receiver, the sender-side signal arrives at the receiver's antenna over multiple paths. Assuming an omni-directional antenna at the sender side, the signal is transmitted in all directions. Therefore, the sender's signal typically arrives over a direct path and multiple indirect paths. The direct path is called the *Line-of-Sight (LoS)* path, whereas all indirect paths are called the *Non-Line-of-Sight (NLoS)* path. NLoS paths are caused by reflection, diffraction, and scattering of the signal off mountains, buildings, and moving obstacles in the wireless environment. *Multipath propagation* refers to the situation where multiple copies of the unique sender-side signals arrive at the receiver. Because of the reflections and diffractions, the received multipath signals differ in amplitude, phase, and delay. The signals arriving at different times at the receiver typically reduce the received signal strength. Because symbols of the same signal interfere with each other, this type of interference is called *Intersymbol Interference (ISI)*. The level of interference depends on delay spread of a multipath signal. The delay spread is the time duration between the first incoming signal (LoS) and the last incoming signal with a significant power level. A typical example for multipath communication is illustrated in Figure 11.22 with one LoS path and two NLoS paths.

Suppose the sender transmits a signal at some time instance, and after τ_1 the LoS signal arrives at the receiver. The NLoS signals arrive after τ_2 and τ_3 (with $\tau_3 > \tau_2$) at the receiver. The delay spread in this example is $\tau_3 - \tau_1$. Because of the propagation in free space and the possible reflections, the received signals differ in their amplitudes. In real wireless systems, the delay spread and the number of paths strongly depend on the environmental setting. Figure 11.23, Figure 11.24, and Figure 11.25 give typical delay spreads for the settings *Rural Area*, *Hilly Terrain*, and *Bad Urban*[3] according to the European Cooperation in the field of Scientific and Technical Research (COST) 207.

We illustrate the impact of multipath propagation with the following example. Suppose two different signals, *SN* and *SH*, with low and high rates are transmitted by the sender over the channel. Whereas the symbol of the low rate signal, *SN*, has 400 time units, the symbol length of the high-data channel is only 100 time units. Both signals are given in Figure 11.26, where the low-data rate signal is on the left side and the high-data rate is on the right side.

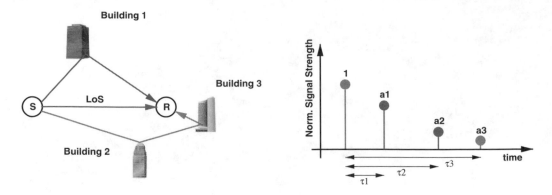

FIGURE 11.22   Example of multipath propagation for three paths.

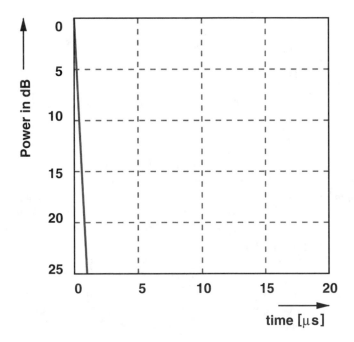

FIGURE 11.23   Delay spread for *Rural Area* (COST 207).[3]

   The characteristics of the multipath channel are given in Figure 11.27. The signal is received with three copies at the receiver side. After the LoS signal, two NLoS signals are received with a delay of 75 and 90 time units. Attenuation causes the signal to be received with a lower amplitude of 70%, 60%, and 20%. The three copies of the signals are given in Figure 11.28. Each of the copies is delayed and attenuated as described by the channel characteristics. If we have now a closer look at the received signals as illustrated in Figure 11.29, we note that these are distorted compared with the original sequences. Furthermore, it is easier to recognize and restore the low-data rate signal than the high-data rate signal. By this simple example, we see that our brain or an equalizer have to work harder to recognize high-data signals. Thus, there is a limit for single carrier systems using higher-data rates. (The effects of the ISI are mitigated in practice by using equalizers, the RAKE sender/receiver, and directional antennas.) We close this brief intuitive discussion by noting that multipath propagation can severely degrade system performance. Multipath propagation in conjunction with the Doppler-Effect due to the velocity of the mobile terminal causes changes in the received power of about 30 to 40 dB compared with the mean received power.[37]

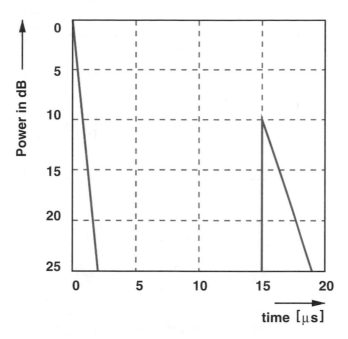

FIGURE 11.24 Delay spread for *Hilly Terrain* (COST 207).[3]

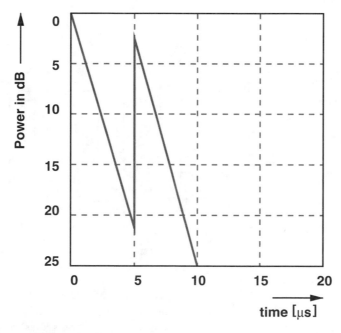

FIGURE 11.25 Delay spread for *Bad Urban* (COST 207).[3]

11.4.3 Shadowing Fading

Shadow fading is caused by obstructions, such as buildings and vegetation. These obstructions may block the signal from the wireless mobile terminal. Typically, the fading effect due to shadowing has a relatively long time scale (on the order of hundreds of msec to seconds), depending, of course, to a large degree on the velocity of the mobile terminal.

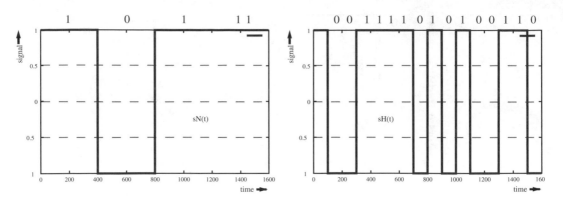

FIGURE 11.26   Low- and high-data rate signal.

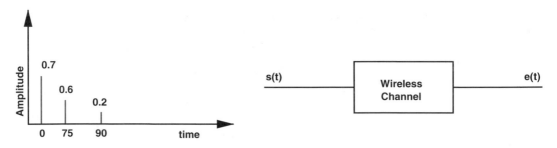

FIGURE 11.27   Channel characteristics.

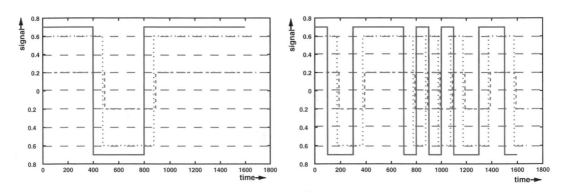

FIGURE 11.28   LoS and NLoS copies of the original signal.

The combined fading effects typically degrade the signal power over a long time period, resulting in *bursty* or correlated errors on the wireless link.[38,39] These correlated link errors, in turn, result in bursty packet drops (i.e., several consecutive packets are dropped on the wireless links). A popular model for this effect from the perspective of the higher protocol layers is the Gilbert-Elliot two-state Markov chain model.[40,41] The two states of this Markov chain represent the "bad" channel state, where packets are dropped with high probability, and the "good" channel state, where the packet-drop probability is small. The underlying Markov chain captures the effect of bursty errors in that a bad channel in a given time slot is more likely followed by a bad channel than a good channel in the next time slot, and vice versa. Higher-order Markov models with more states provide a refined model.[42]

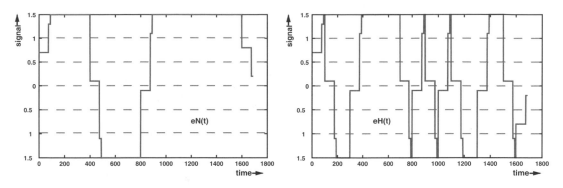

FIGURE 11.29 Received signal.

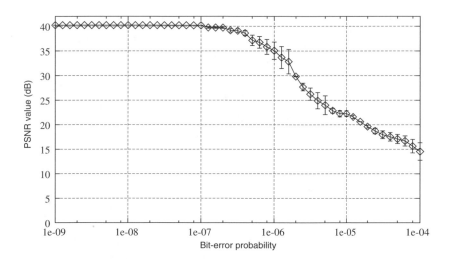

FIGURE 11.30 Average video frame PSNR as a function of bit error rate, III GoP.

11.5 Impact of Wireless-Channel Errors on Video Quality

Having discussed both the variability of the video traffic and the wireless channel, we now briefly illustrate their combined effects on the video quality over an error-prone channel. We simulate this by inflicting random bit errors on the encoded video. (Note that we do not explicitly introduce bursty errors, which would render the outcome differently. As decoders differ in their error resilience, sometimes bit errors have the same effect as a complete frame being lost on the channel. Therefore, we assume the outcome of both bursty errors and decoder-dependent frame drops is similar.) In this experiment, no error corrections or retransmissions were applied. Two different encoding schemes were used to show the differences of quality and bit rates. The illustrated results were obtained with an early version of the forthcoming H.26L standard[23] reference encoder. Figure 11.30 and Figure 11.31 illustrate the average video frame PSNR at the prospective receiver as a function of the mean bit-error probability.

Figure 11.32 and Figure 11.33 illustrate the respective encoded video traffic on the GoP-level. The first encoding scheme shown in the previous two figures consists only of intracoded frames (i.e., the GoP length is only one frame, and the frame type pattern is thus IIIIIIIIII…). With the lack of differential encoding, the video traffic generated is seemingly high. The objective video quality drops as the error probability increases. The second GoP is using a different encoding scheme and applies advanced features of H.26L (i.e., the GoP length is 12 frames, with a frame type pattern of IBBPBBSPBBPBBI…). The SP-frame

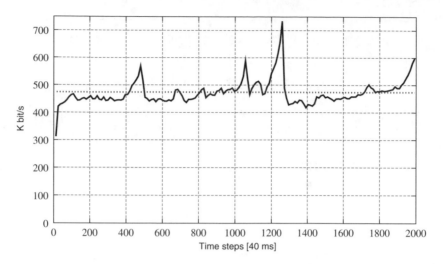

FIGURE 11.31 Video traffic (bit rate) as a function of time, III GoP, average bit rate = 471.8 kbit/sec.

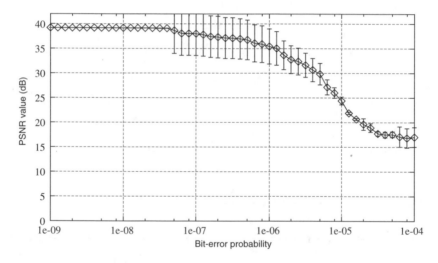

FIGURE 11.32 Average video frame PSNR as a function of bit-error rate, IBBPB GoP.

types are introduced in H.26L to add stream switching error resilience to the encoded video. The video traffic that is generated by the encoding is visibly lower than for the mere intracoded sequence. The objective video quality obtained from the decoding process, however, is similar to if not better than the intracoded sequence.

The comparison of Figure 11.32 and Figure 11.33 shows a very similar traffic characteristic. This is surprising because differentially encoded sequences should be more sensitive to errors. This result is explained as follows. As shown in Section 11.2.3.1, different frame types have certain dependencies. For successful decoding of frames that are differentially encoded, the referenced frames have to be present at the time of decoding. If referenced frames are missing, successful decoding cannot be achieved for their dependent frames. This effect is also referred to as error propagation. Thus, the impact of these referenced frames is higher than the impact of frames that are depending on them. Video sequences that are only intracoded should thus be less error sensitive than sequences with differential encoding (in the experiment shown, the SP frames within the encoded video stream are reducing the effect of error propagation). In addition, a second effect has to be regarded. The frame sizes of intracoded sequences are higher than those of sequences that also feature differential intercoding. With a given bit-error

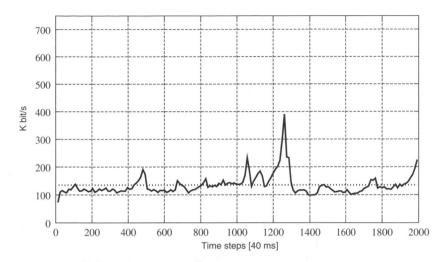

FIGURE 11.33 Bit rate as a function of time, IBBPB, average bit rate 135.2 kbit/sec.

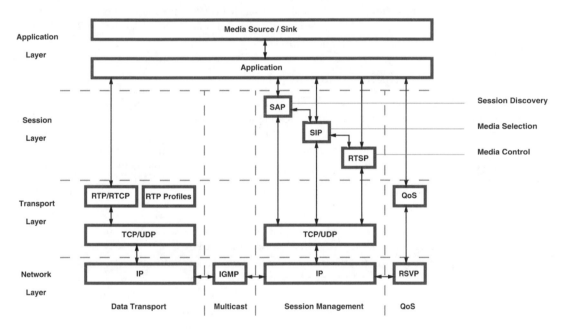

FIGURE 11.34 IP-based streaming protocol suite.

probability, the larger intracoded frames are more likely to be subject to transmission errors. These effects are visible for the intra- and intercoded GoP scheme.

This comparison yields the result that differential encoding of video data is not introducing a severe change in the quality at the receiver side. Additionally, the network benefits from a reduction of the bandwidth needed for the transmission of the encoded video sequence.

To compensate for errors on the network, three general methods of error handling exist. Forward error correction uses additional parity information to allow for correction of transmission errors. As errors in wireless transmissions tend to occur in bursts, the parity information may be lost with the original information. The second method is the adaption of retransmission schemes. Retransmissions, however, have to be handled carefully, especially in the domain of real-time streaming. Third, error-concealment techniques can be applied on the receiver side of the video stream. We will show how these techniques are applied as we examine the protocols and streaming of video in the following sections.

11.6   Internet Protocol Stack for Video Streaming

In this section, we move up the networking protocol stack and consider the protocols commonly used for wireless video streaming at the network (IP) layer and higher. The key enabling protocols for multimedia streaming over IP networks are the Real-Time Protocol (RTP) in combination with the Real-Time Control Protocol (RTCP). These run on top of the User Datagram Protocol (UDP) for data transport. At the same time, the Session Announcement Protocol (SAP), Session Initiation Protocol (SIP), and the Real-Time Streaming Protocol (RTSP) are used for session management, as illustrated in Figure 11.34.

The SAP, in conjunction with the SIP or RTSP protocols, initiate the streaming of a video. SAP announces both multicast and unicast sessions to a group of users. SIP initiates multimedia session in a client/server manner. An open issue is how the client retrieves the destination address. Possible solutions are that the address is well known or is provided by SAP. RTSP is simply a "remote control" used to control unicast streams in a server/client manner. SIP, SAP, and RTSP use the Session Description Protocol (SDP) to describe the media content and run over either the TCP or UDP protocol. (Note that SDP is missing in Figure 11.34; it is not a real protocol but is instead a language, such as HTTP.)

After establishing a session, the application is able to start the data exchange using the RTP protocol. RTP is used for data transmission, whereas quality-of-service (QoS) information between sender and receiver is exchanged using RTCP. The underlying IP network may independently provide QoS and multicasting (multicasting is a part of IP, not RTP).

11.6.1   Session Description Protocol (SDP)

SDP is used to describe a multimedia session. The SDP message contains a textual coding that describes the session; more specifically, it gives (1) the transport protocol used to convey the data, (2) a type field to distinguish the media (video, audio, etc.), and (3) the media format (MPEG-4, GSM, etc.). Furthermore, the SDP message may contain the duration of the session, security information (encryption keys), and the session name in addition to the subject information (e.g., Arielle © Disney). The following is an example of the textual coding of an SDP message:

```
v=0 // Version number
o=mjh 2890844526 2890842807 IN IP4 192.16.6.202 // Originator
s=Wireless Internet Demonstrator // session title
i=A seminar on Internet multimedia //session information
u=http://www.acticom.de //URL for more information
e=fitzek@acticom.de (Frank Fitzek) //  Email address to contact
c=IN IP4 224.2.17.12/127 // connection information
a=recvonly // attribute, this telling session is receive only
m=audio 1789 RTP/AVP 0 // PCM audio using RTP on port 1789
m=application 1990 udp wb // "wb" application on port 1990
a=orient:portrait // "wb" in portrait mode
m=video 2003 RTP/AVP 31 // H.261 video using RTP on port 2003
```

SDP messages can be carried in any protocol, including HTTP, SIP, RTSP, and SAP. Originally SDP was designed for the support of multicast sessions. The information relating to the multicast session was conveyed using SAP. More recently, SDP is also used in combination with SIP and RTSP.

11.6.2   Session Announcement Protocol (SAP)

The SAP[43] is used for advertising multicast sessions. In brief, SAP discovers ongoing multicast sessions and seeks out the relevant information to set up a session. (In case of a unicast session, the setup information might be exchanged or known by the participants.) Once all the information required for initiating a session is known, SIP is used to initiate the session.

11.6.3   Session Initiation Protocol (SIP)

Signaling protocols are needed to create sessions between two or more entities. For this purpose, the H.323 and SIP protocols have been standardized by two different standardization committees. H.323 was standardized by the ITU. The Internet Engineering Task Force (IETF) proposed the SIP specified in RFC 3261.[44] In contrast to other signaling protocols, SIP is text-based, like SDP.

SIP, a client/server-oriented protocol, is able to create, modify, and terminate sessions with one or multiple participants. Multiparty conferencing is enabled through IP multicast or a mesh of unicast connections. Clients generate requests and transmit them to an SIP proxy. The proxy, in turn, typically contacts an SIP registrar to obtain the user's current IP address. Users register with the SIP registrar whenever they start up an SIP application on a device (PDA, laptop, etc.). This allows the SIP registrar to keep track of the user's current IP address. With SIP it is thus possible to reach users who are on the move, making SIP very relevant for wireless streaming.

Using the INVITE request, a connection is set up. To release an existing connection, a BYE request is used. Besides these two requests, further requests are OPTIONS, STATUS, ACK, CANCEL, and REGISTER. SIP reuses HTTP header fields to ease an integration of SIP servers with Web servers. In the SIP terminology, the client is called a *user agent*. A host can simultaneously operate as client and as server.

The call identifiers used in SIP include the current IP addresses of the users wishing to communicate and the type of media encoding used (e.g., MPEG-4 in the case of video).

11.6.4   Real-Time Streaming Protocol (RTSP)

RTSP[45] may be thought of as a "remote control" for media streaming. More specifically, it is used to implement interactive features known from the VCR, such as pause and fast forward. RTSP has many additional functionalities (see reference 45 for details) and has been adopted by RealNetworks.

RTSP exchanges RTSP messages over an underlying transport protocol, such as TCP or UDP. The RTSP messages are ASCII text and very similar to HTTP messages. RTSP uses out-of-band signaling to control the streaming.

11.6.5   Real-Time Protocol (RTP)

The RTP[46] is a transport mechanism for real-time data. It consists of two components: RTP and RTCP (the RTP Control Protocol). Both RTP and RTCP typically run over UDP but can use any other packet-oriented transport protocol. In a videoconference, audio and video streams are separated and transmitted over different RTP sessions using different UDP port addresses.

To send multimedia content with RTP, the host packetizes the media, adds content-dependent header fields and the RTP header, and then passes the message to the underlying protocol layers. The content-dependent header field informs the receiver about the video codec used and the parameters used (as explained shortly in an example). The content-dependent header field has variable length, depending on the specific content and encoding used. On the other hand, the RTP header illustrated in Figure 11.35 has a fixed structure and is always 12 bytes long. The components that comprise the RTP header are in more detail:

- *VERSION (V):* The version number of the RTP protocol, represented by 2 bits. Currently, version number 2 is used.
- *PADDING (P):* This flag indicates that padding is used (if set to one). Padding might be used for encryption.
- *EXTENSION (X):* This flag indicates whether a content-dependent header is used, which is placed between the RTP header and the payload.

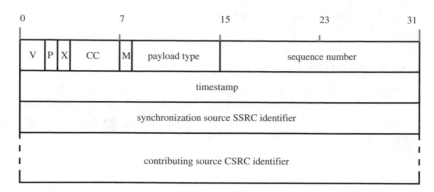

FIGURE 11.35 RTP header format.

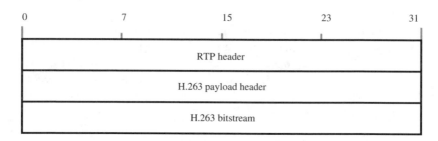

FIGURE 11.36 RTP payload header format and location.

- *PAYLOAD TYPE (PT):* This field identifies the format of the RTP payload.
- *SEQUENCE NUMBER:* The initial sequence number is chosen randomly and then in increments by one for each RTP packet. A random starting number was chosen to complicate plain-text attacks. By means of the sequence number, lost packets can be detected or disordered packets can be detected and replaced.
- *TIMESTAMP:* Four octets are used to reflect the sampling time. This information is important to ensure the playout at the receiver side is correct.

The CSRC counter field (CC), the marker field (M), and the source identifiers (SSRC) are beyond the scope of this discussion. See reference 46 for details.

Next, we take a closer look at the content-dependent header used to identify, among other things, the video codec used.

11.6.5.1 RTP Payload Format Specification Example

If the extension bit (X) is set in the RTP header, a content-dependent header is placed between the RTP header and the RTP payload. By means of an H.263 example, we illustrate the usage of such a header. As illustrated in Figure 11.36, the H.263 payload header is placed between the RTP header and the H.263 bit stream.

RFC 2190 specifies the payload format for encapsulating an H.263 bit stream in RTP. H.263 is an advancement of the H.261 video encoding scheme. Three different modes (Mode A, Mode B, and Mode C) are defined for the H.263 payload header. An RTP packet is forced to use only one of the three modes for H.263 video streams. The choice depends on the desired network packet size and H.263 encoding options used; see reference 47 for details. Mode A supports fragmentation at the level of group of block (GOB) boundaries, whereas Mode B and Mode C support more fine-grained fragmentation at macroblock boundaries. The modes have an impact on total packet size. In this example, we will refer only to Mode A. Mode A uses the payload header illustrated in Figure 11.37:

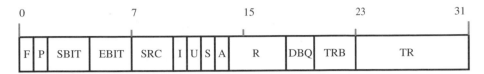

FIGURE 11.37 H.263 payload encapsulation header.

- *F flag:* The different modes (A, B, and C) are indicated by this flag. In case the flag is not set, Mode A is used; otherwise, flag P determines which mode is used.
- *P flag:* H.263 defines the usage of PB Frames. If this flag is set, PB frames are used. Additionally, if F is set and P is not set, Mode B is used; Mode C is used if P is set.
- *SBIT field and EBIT field:* Specifies the number of most/least significant bits that should be ignored in the first/last data byte.
- *SRC field:* Specifies the video format used. The values of the PTYPE field (bit 6 up to 8) of H.263 are used here.
- *I flag:* This flag is set in case intercoded video is used.
- *U flag:* This flag indicates whether the encoder used the unrestricted motion vector, an advanced motion compensation mechanism; see reference 47 for details.
- *S flag:* Using bit 11 of the PTYPE field.
- *A flag:* Using bit 12 of the PTYPE field.
- *R field:* Four bits are reserved and set to zero.
- *DBQ field:* If PB frames are not used, this field contains only zeros. Otherwise, the values of the DBQUANT field defined by H.263 are copied here.
- *TRQ field:* Temporal reference for B frames. This field contains only zeros if PB frames are not used.
- *TR field:* Temporal reference for P frames. This field contains only zeros if PB frames are not used.

As illustrated by this example, the content-dependent header (payload header) contains information that the receiver uses to decode and display the video.

11.6.5.2 Video-Related RFCs

Besides RFC 2190, multiple RTP packetization schemes for multimedia applications are given in various RFCs. We give a short overview of each of the related RFCs in the following section. For more detailed information, refer to the individual RFCs.

The RTP packetization scheme for the CellB video encoding is described in RFC 2029. The Cell image compression algorithm supports the variable bit-rate video coding,[48] CellB, which is derived from CellA and has been optimized for network-based video applications.

A description of how to packetize an H.261 video stream for RTP transmission is given in RFC 2032. The ITU-T Recommendation H.261 specifies the encodings used by ITU-T–compliant videoconference codecs. H.261 was originally specified for fixed data rate ISDN circuits with multiples of 64 kbit/sec, but H.261 can also be used over packet-switched networks, such as the Internet and wireless packet networks.

Whereas the H263 Standard from 1996 is referred to as H.263, the updated version is called H.263+. One of the major improvements of H263+ is the introduction of bit-stream scalability. Temporal, signal-to-noise ratio, and spatial scalability are used in H.263+. Numerous coding options were changed to improve coding performance. RFC 2429 considers the changes in the encoder.

RFC 2435 (which replaces RFC 2035) gives the RTP payload format for JPEG-compressed encoded video streams. The Joint Photographic Experts Group (JPEG) standard defines still-image compression. By combining a set of still images, it is considered *motion JPEG*. The packet format was optimized for real-time environments assuming that changes in the codec parameter are rare.

RFC 2250 (the revision of RFC 2038) gives the packetization of MPEG-1/MPEG-2 audio and video for RTP. In the RFC, two different approaches are presented. One approach focuses on the compatibility with other RTP-encapsulated media streams, whereas a second approach targets maximum interoperability with MPEG-system environments.

RFC 2431 specifies the RTP payload format for encapsulating ITU Recommendation BT.656-3 (digital-television equipment operating on the 525-line or 625-line standards) video streams in RTP. RTP packet lengths depend on the number of scan lines. To rebuild the video frame, each RTP packet contains information about fragmentation, decoding, and positioning information.

RFC 3016 specifies the RTP payload format for MPEG-4-based audio and video streams. This proposal for packet encapsulating allows the direct mapping of MPEG-4 streams onto RTP packets.

11.6.6   Real-Time Control Protocol (RTCP)

The companion control protocol for RTP is RTCP. It was introduced together with RTP in RFC 1889. Sender and receiver exchange RTCP packets to exchange QoS information periodically.

Five types of messages exist:

1. Sender Reports (SR)
2. Receiver Reports (RR)
3. Source Descriptions (SDES)
4. Application-Specific Information (APP)
5. Session Termination Packets (BYE)

Each report type serves a different function. The SR report is sent by any host that generated RTP packets. The SR includes the amount of data that was sent so far, as well as some timing information for the synchronization process. Hosts that receive RTP streams generate the Receiver Report. This report includes information about the loss rate and the delay jitter of the RTP packets received so far. Also included is the last timestamp and delay since the last SR was received. This allows the sender to estimate the delay and jitter between sender and receiver. The rate of the RTCP packets is adjusted to reflect the number of users per multicast group.

In general, RTCP provides the following services:

1. *QoS monitoring and congestion control:* This is the primary function of RTCP. RTCP provides feedback to an application about the quality of data distribution. The control information is useful to the senders, receivers, and third-party monitors. The sender can adjust its transmission based on the receiver report feedback. The receivers can determine whether congestion is local, regional, or global. Network managers can evaluate the network performance for multicast distribution.
2. *Source identification:* In RTP data packets, sources are identified by randomly generated 32-bit identifiers. These identifiers are not convenient for human users. RTCP SDES (source description) packets contain text information, called canonical names, as globally unique identifiers of the session participants. It may include the user's name, telephone number, e-mail address, and other information.
3. *Intermedia synchronization:* RTCP sender reports contain an indication of real time and the corresponding RTP timestamp. This can be used in intermedia synchronization like lip synchronization in video.
4. *Control information scaling:* RTCP packets are sent periodically among participants. When the number of participants increases, it is necessary to balance between getting up-to-date control information and limiting the control traffic. In order to scale up to large multicast groups, RTCP has to prevent the control traffic from overwhelming network resources. RTP limits the control traffic to at most 5% of the overall session traffic. This is enforced by adjusting the RTCP generating rate according to the number of participants.

11.7   Video Streaming Mechanisms

The combination of the stringent QoS requirements of encoded video and the unreliability of wireless links make the real-time streaming over wireless links a very challenging problem. For uninterrupted video playback in a real-time streaming scenario, the client has to decode and display a new video frame periodically (typically every 33 msec). This imposes tight timing constraints on the transmission of the video frames, once the playback at the wireless client has commenced. If a video frame is not completely delivered in time, the client loses a part of or all the frame. Generally, a small probability of frame loss (playback starvation) on the order of 10^{-5} to 10^{-2} is required for good perceived video quality. In addition, the video frame sizes (in bytes) of the more efficient VBR video encodings are highly variable, typically with peak-to-mean ratios of 4:10 as noted in Section 11.3; see also reference 25. Wireless links, on the other hand, are typically highly error prone, as discussed in Section 11.4. They introduce a significant number of bit errors that may render a transmitted packet undecodable. The tight timing constraints of real-time video streaming, however, allow only for limited retransmissions.[49] Moreover, the wireless-link errors are typically time varying and bursty. An error burst (which may persist for hundreds of milliseconds) may make the transmission to the affected client temporarily impossible. All these properties and requirements make real-time streaming of video over wireless networks a challenging problem. (Note that real-time video streaming is a significant challenge even for wired networks.[50]) In this section, we give an overview of the different strategies that can be used to overcome the challenges of video streaming.

We organize our discussion according to the Internet protocol stack. First we discuss mechanisms that are used at the channel coder level. We outline the basic concept of Forward Error Correction (FEC), a general technique to combat the bit errors on wireless links, and then examine some adaptive FEC techniques that have been specifically designed for wireless video streaming. Next, we move up to the link layer. We describe the basic ideas behind the concept of Automatic Repeat Request (ARQ) and then explore a number of ARQ schemes that have been tailored for video streaming. Strategies for video streaming often combine mechanisms that work at different levels of the protocol stack in *cross-layer designs*. We will discuss a few of the strategies that combine ARQ and FEC mechanisms; these strategies are often termed *hybrid ARQ*. Moving up to the network and transport layers, we will explore the issue of video streaming over the UDP versus streaming over the TCP. Finally, we move up to the application layer — that is, the video application and, in particular, the video (source) coder. We discuss techniques that adapt the video encoding in response to variations in the wireless channel. We will also describe scalable coding techniques that generate encoded video streams that can be adapted to the channel conditions at the lower protocol layers.

11.7.1 Forward Error Control Mechanisms

Shannon's channel coding theorem states that if the channel capacity is larger than the data rate, a coding scheme can be found to achieve small error probabilities. The basic idea behind FEC is to add redundancy to each information (payload) packet at the transmitter. At the receiver, this redundancy is used to detect and correct errors within the information packet.

The binary (n,k) *Bose-Chaudhuri-Hocquenghem (BCH)* code is a common FEC scheme based on block coding.[51] This code adds redundancy bits to payload bits to form code words and can correct a certain number of bit errors; see reference 51 for details. An important subclass of nonbinary BCH codes consists of the *Reed Solomon (RS)* codes. An RS code groups the bits into symbols and thus achieves good burst error-suppression capability.

An advantage of FEC is constant throughput with fixed (deterministic) delays independent of errors on the wireless channel. To achieve error-free (or close to error-free) communication, however, FEC schemes must be implemented for the worst-case channel characteristics. This results in unnecessary overhead on the typically highly variable wireless links.[52] In particular, when the channel is currently good, the FEC dimensioned for the worst-case conditions results in inefficient utilization of the wireless link. In addition, complex hardware structures may be required to implement the powerful, long codes required to combat the worst-case error patterns. Also note that by adding redundancy bits the latency of all packets increases by a constant value. This additional delay may not be acceptable for streaming with very tight timing constraints.

To overcome the outlined drawbacks of FEC, *adaptive* FEC was introduced. Adaptive FEC adds redundancy as a function of the current wireless channel characteristics. Adaptive FEC for wireless communication has been studied extensively over the past 5 years or so. A number of adaptive FEC techniques have been specifically designed and evaluated for video streaming. The scheme in reference 53, for instance, estimates the long-term fading on the wireless channel and adapts the FEC to proactively protect the packets from loss.

Generally, adaptive FEC is an important component of an error-control strategy and is often used in conjunction with the ARQ mechanisms we discuss next to form hybrid error-control schemes. Note that adaptive FEC may increase the hardware complexity and possibly the signaling overhead. Some recent approaches adjust the transmit power level to ensure the successful transmission of video packets; see, for instance, references 54 and 55.

11.7.2   ARQ Mechanisms

The basic idea of ARQ is to detect erroneous packets and to retransmit these packets until they are correctly received. Errors are typically detected by applying an FEC code that has good error-detection capabilities (and may have weak error-correction performance). A wide variety of ARQ mechanisms have been studied over the years. The studied approaches differ in complexity and efficiency of the retransmission process. The simplest ARQ protocol is called *Send and Wait*. The basic idea behind this approach is to send a new packet only if the previous packet is transmitted successfully and has been acknowledged. In case an acknowledgment is missing, the packet is retransmitted until it is successfully received. Improved ARQ protocols are *Go-back-N* and *Selective Repeat*. ARQ achieves reasonable throughput levels if the error probability on the wireless link is not very large. More recently, ARQ techniques that use the features of modern Code Division Multiple Access (CDMA) systems have been developed. The Simultaneous MAC Packet Transmission (SMPT) scheme,[56] for instance, retransmits dropped packets on additional (parallel) CDMA codes. Roughly speaking, the scheme ramps up the number of used CDMA codes when losses occur. This avoids excessive delays and stabilizes the throughput at the link layer, thus reducing the probability of frame drop. Note that in a similar manner, the scheme in reference 57 ramps up the transmit power.

11.7.3   Hybrid ARQ Techniques

The area of hybrid ARQ techniques that combine FEC and ARQ for more effective video streaming has been very active for the past couple of years. We discuss here a few of the major lines of work. One of the first studies to examine hybrid ARQ for wireless video is reference 49. Many refined hybrid adaptive ARQ schemes for streaming low-bit rate video over wireless links have been developed; see, for instance, reference 58.

Some of the schemes are designed for particular wireless networks. WLANs based on 802.11, for instance, are considered in references 59 and 60. Some schemes exploit the features of the encoded video for increasing the efficiency of the streaming. The scheme in reference 61 exploits the fact that the different frame types of the MPEG-encoded video have typically different sizes. The streams are coordinated such that the I frames of one video stream do not coincide with the I frames of another video stream. One fundamental aspect of this approach is the central coordination of the ongoing video streams. Central coordination generally allows for more efficient utilization of the wireless resources but requires a central entity. In a cellular wireless system, for instance, the coordination of the streaming in the forward (downlink) direction in the base station can significantly improve efficiency. The scheme in reference 62 exploits this central coordination. In addition, the scheme exploits buffers in the wireless clients and the fact that streamed video is typically stored (prerecorded) for the prefetching of video data. During periods of good wireless-channel conditions and low activity of the video traffic, the scheme transmits future video frames to the client buffer. The prefetched reserve of video frames allows the client to continue uninterrupted playback during periods of adverse wireless-channel conditions or high-bit rare video segments. Along the same lines, the scheme in reference 63 uses the receiver buffer to smooth an individual video stream (i.e., without exploiting central coordination).

11.7.4 Transport Layer: TCP or UDP?

Generally, it is widely accepted that video streaming applications run over the UDP. UDP provides a connectionless and unreliable datagram service with a minimum of overhead. In addition, UDP allows streaming applications to transmit as fast as they desire. UDP is thus generally well suited for video streaming; a few refinements to improve its performance in conjunction with error-resilient video coders have been developed (see, for instance, references 64 and 65).

The drawback of UDP is that it does not enforce any congestion control. Thus, the excessive use of UDP in the Internet can result in congestion collapse.[66] TCP provides a reliable connection between two terminals on top of an unreliable communication network, which may include wireless links. TCP enforces (1) congestion control to avoid excessive congestion in the network and (2) flow control to avoid overwhelming the receiver with data. TCP's congestion control uses an addittive-increase-multiplicative-decrease congestion window mechanism, which provides fair bandwidth allocation to all ongoing flows. One key limitation of using TCP over wireless links is that, by design, TCP interprets lost packets as a sign of congestion, causing TCP to throttle the transmission rate. When wireless links are involved, however, packets are typically dropped due to errors on the wireless link and not because of congestion. Thus, cutting back the transmission rate may not be the best strategy and could result in poor performance of TCP transmissions over wireless links. This issue has been studied in some detail in the general context of wireless data transmissions, and a number of remedies have been explored (see, for instance, reference 67). In the context of video streaming, this issue has received relatively less interest so far (see, for instance, reference 68).

Another limitation of video streaming with TCP is that TCP's congestion window may limit the transmission rate to below the video bit rate. This is an issue in both streaming over wired networks and wireless networks and has also received only limited attention so far; see, for example, reference 69.

A final consideration is that UDP supports multicasting whereas TCP does not. As discussed in some detail in Section 11.7.6, multicast may be one way to efficiently utilize the scarce wireless bandwidth for video streaming.

11.7.5 Adaptation in the Video Coding

The adaptation schemes at the application layer — that is, the level of the video (source) coder — fall into two broad categories:

1. Adaptive encoding on-the-fly
2. Scalable offline encoding

With on-the-fly adaptive encoding, the video coder changes the encoding parameters (such as the quantization scale and the bit rate or error resilience produced in the encoding) to adapt to the fluctuations on the wireless link. This technique is suitable for transmissions of live video, as well as for re-encoding or transcoding stored video for the transmission over the wireless medium. Scalable offline encoding techniques, on the other hand, produce an encoded bit stream that can be scaled (adapted) to the current conditions on the wireless link without requiring any on-the-fly encoding or reencoding. Typically, scalable encoding techniques produce a base layer and one or several enhancement layers. The base layer provides basic video quality. Adding the enhancement layers (when the wireless-link conditions allow for it) improves the video quality.

In this section, we first discuss streaming mechanisms that incorporate on-the-fly adaptive encoding. We then outline the various scalable encoding techniques and discuss streaming mechanisms for scalable video.

11.7.5.1 Streaming Mechanisms with Adaptive Encoding

The approaches for adaptive encoding may be broadly classified into approaches that adjust the transmission bit rate to adapt to the wireless channel and approaches that adapt the error resilience of the encoding. The goal of both approaches is to minimize the distortion of the video delivered over the wireless link.

An example of the rate adaptation approach is the scheme developed in reference 70. In this scheme, the acknowledgments for successfully received packets from the receiver are used in conjunction with a wireless-channel model to predict the future wireless-channel conditions. The predicted channel conditions and the playout deadlines of the video frames are translated into bit-rate constraints for the video encoder. A variety of similar approaches has been explored; see, for instance, references 71 and 72.

An approach that adapts the error resilience for transmission over the wireless link is developed in reference 73. This approach is tailored for transcoding (i.e., it takes an encoded video stream that is delivered over a wired network as input). Depending on the current conditions on the wireless link, the transcoder injects error-resilience information into the encoded bit stream. Both temporal resilience, which prevents bit errors from propagating to future frames, and spatial concealment, which limits the loss of synchronization when decoding the variable-length codes (see also reference 74), are added.

11.7.5.2 Streaming Mechanisms for Scalable Encoded Video

With scalable encoding, the encoder produces typically multiple layers. The base layer provides a basic quality (e.g., low spatial or temporal resolution video), and adding enhancement layers improves the video quality (e.g., increases spatial resolution or frame rate). A variety of scalable encoding techniques have been developed, which we will introduce in this section. We will then discuss mechanisms for streaming scalable encoded video over wireless links. These approaches control the transmission of the layers so as to compensate for the variations of the wireless channel. Aside from adapting to the wireless channel, scalable encoding is a convenient way to adapt to the wide variety of video-capable hardware in wireless scenarios (e.g., PDAs, laptops). Each of these devices has different constraints due to processing power, viewing size, and so on. Scalable encoding can satisfy these different constraints with one encoding of the video. We briefly note that an alternative to scalable encoding is to encode the video into different versions, each with a different quality level, bit rate, and spatial and temporal resolution. The advantage of having different versions is that it does not require the more sophisticated scalable encoders and does not incur the extra overhead due to the scalable encoding. The drawback is that the multiple versions take up more space on servers and possibly need to be streamed all together (simulcast) over the wired network to be able to choose the appropriate version at any given time over the wireless hop. Transcoding, as already noted above, is another alternative to scalable encoding. Transcoding can be used to adapt to the wireless-link conditions as in reference 73 or to adapt to different desired video formats.[75] This transcoding approach requires typically a high-performance intermediate node.

Having given an overview of the general issues around scalable video encoding, we now introduce the different approaches to scalable encoding. Refer to reference 76 for more details.

11.7.5.2.1 *Data Partitioning*

Although not explicitly a scalable encoding technique, data partitioning divides the bitstream of non-scalable video standards such as MPEG-2[7] into two parts. The base layer contains critical data such as motion vectors and low-order DCT coefficients, whereas the enhancement layer contains, for example, the higher-order DCT coefficients.[5] The *priority breakpoint* determines where to stop in the quantization and scanning process for the base-layer coefficients to be further encoded,[10] as shown in Figure 11.38. The remaining coefficients are then encoded by resuming the zig-zag scanning pattern at the breakpoint and are stored in the enhancement layer.

11.7.5.2.2 *Temporal*

Temporal scalability reduces the number of frames in the base layer. The removed frames are encoded into the enhancement layer and reference the frames of the base layer. Different patterns of combination of frames in base and enhancement layer exist.[6] In Figure 11.39, an enhancement layer consisting of all B frames is given, as already used in video trace evaluations.[34] No other frames depend on the successful decoding of B frames. If the enhancement layer is not decodable, the decoding of the other frame types is not affected. Nevertheless, because the number of frames that are reconstructed changes, the rate of fps is to be adjusted accordingly for viewing and quality evaluation methods (e.g., the last successfully

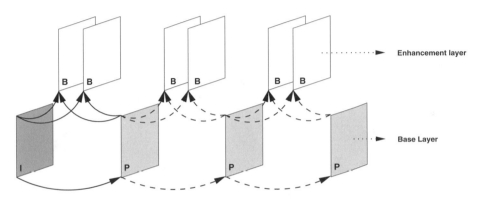

Base Layer

36	2	2	1	0	0	0	0
8	1	1	1	0	0	0	0
4	0	0	0	0	0	0	0
2	2	1	1	0	0	0	0
0	1	1	0	0	0	0	0
1	0	0	0	0	0	0	0
1	1	0	0	0	0	0	0
1	1	0	0	0	0	0	0

Enhancement Layer

FIGURE 11.38 Data partitioning by priority breakpoint setting.

FIGURE 11.39 Temporal scalability with all B frames in enhancement layer.

decoded frame is displayed for a longer period, also called *freezing*). This adjustment is inflicting the loss in viewable video quality.

11.7.5.2.3 Spatial Scalability

Scalability in the spatial domain is applying different resolutions to the base and the enhancement layer. If, for example, the original sequence is in the CIF format (352×288), the base layer is downsampled into the QCIF format (176×144) before the encoding. Spatial scalability is therefore also known as *pyramid coding*. In addition to the application of different resolutions, different GoP structures are used in the two layers. The GoP pattern in the enhancement layer is referencing the frames in the base layer. An exemplary layout of the resulting dependencies is illustrated in Figure 11.40.

The content of the enhancement layer is the difference between the layers, as well as the frame-based reference of previous and following frames of the same layer. A study of the traffic and quality characteristics of temporal and spatial scalable encoded video is given in reference 34.

11.7.5.2.4 Signal-to-Noise Ratio (SNR) Scalability

SNR scalability provides two (or more) different video layers of the same resolution but with different qualities. The base layer is coded by itself and provides a basic quality in terms of the (P)SNR (see Section 11.3.1). The enhancement layer is encoded to provide additional quality when added back to the base layer. The encoding is performed in two consecutive steps: first the base layer is encoded with a low-quality setting, then the difference between the decoded base layer and the input video is encoded with higher-quality settings in a second step,[76] as illustrated in Figure 11.41.

At the receiver side, the base quality is obtained simply by decoding the base layer. For enhanced quality, the enhanced layer is decoded and the result is added to the base layer. There is no explicit need

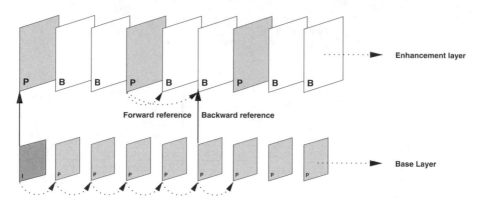

FIGURE 11.40 Example of spatial scalability and cross-layer references.

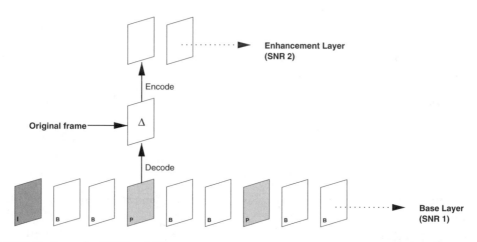

FIGURE 11.41 Example of SNR scalability.

for both layers to be encoded by the same video compression standard, though for ease of use it is advisable to do so.

11.7.5.2.5 Object Scalability

Another scalability feature is possible within video standards that support the composition of video frames out of several different objects, such as MPEG-4.[6] The base layer contains only the information that could not be fitted or identified as video objects. The enhancement layers are made up of the respective information for the video objects, such as shape and texture. The example shown in Figure 11.42 presents a case where the background (landscape) and an object (car) were separated. In this case, the background is encoded independently from the object.

11.7.5.2.6 Fine-Grain Scalability

Fine-Grain Scalability (FGS) is a relatively new form of scalable video encoding.[77] With FGS the video is encoded into a base layer and one enhancement layer. This enhancement layer has the special property that it can be cut at any bit rate and all the bits that are transmitted contribute toward improving the decoded video quality. FGS thus removes the restriction of conventional layered encoding where an enhancement layer must be completely received for successful decoding. The flexibility of FGS makes it attractive for video streaming, but this flexibility comes at the expense of reduced coding efficiency. Efforts are currently under way to improve the coding efficiency while maintaining the FGS flexibility.

FIGURE 11.42 Example of object-based scalability.

11.7.5.2.7 *Multiple Descriptions*

Multiple Description (MD) coding is yet another form of scalable coding of images and videos. With MD coding, an image or video is coded into multiple descriptions (which can be thought of as layers or streams). These descriptions have no explicit hierarchy (i.e., there is no notion of a base description and enhancement descriptions). Instead, any of the descriptions can be combined and decoded.[78–80] The more the encoded descriptions that are available at the decoder, the higher the decoded image or video quality.

Having given a brief introduction to scalable video coding, we now discuss streaming mechanisms for scalable video in wireless networks. General frameworks for streaming scalable video in wireless networks have been proposed in references 81 to 83. A number of schemes for efficiently streaming the layers of scalable encoded video have been developed; see, for instance, reference 84. These approaches focus on the packet scheduling over the wireless links. A number of other approaches incorporate multiple priorities for transmission (e.g., reference 85) or hybrid ARQ/FEC and power control (e.g., reference 86) for the efficient transmission of scalable video over wireless links. The emergence of the FGS coding technique has also prompted studies of the traffic characteristics of FGS encoded video[87] and transport schemes for wireless networks.[88,89]

Yet another direction of research develops specialized scalable encoding schemes for wireless video streaming. See, for instance, references 90 to 93.

In this context, a number of studies explore the joint optimization of the video (source) coding and the (channel) coding for the transmission over the wireless links. See, for example, references 94 to 96.

11.7.6 Wireless Multicast of Video Streams

Multicast is gaining a lot of interest in the wireless sector because it allows introducing video streaming services in a multicast fashion at a favorable price. Considering the scare bandwidth and the price paid for the frequency bands for 3G networks, the introduction of video services using unicast connection would be very hard. So, it is likely that the first video services in 3G networks will be based on multicast.

UMTS networks distinguish between two kinds of channels: dedicated channels and common channels. The dedicated channel is a point-to-point channel with power control. Using the dedicated channel, each wireless terminal has a unique channel. The common channel is a point-to-multipoint channel without power control. Multiple wireless terminals may connect to the common channel. Therefore, multicast can be applied in two different fashions, such as a mesh of unicast connection over dedicated channels or using the common channel. It depends on the distance of the user and the base station which solution is chosen to achieve high-bandwidth efficiency. Generally, for a large number of wireless terminals, the common channel seems to be preferable. On top of these channels, the IP protocol stack shown in Figure 11.34 is used.

TABLE 11.3 Acronyms and Abbreviations

3G	Third-generation mobile communication systems
AC	Alternating current
APP	Aplication specific information
ARQ	Automatic repeat request
BCH	Bose-Chaudhuri-Hocquenghem
BER	Bit-error rate
BMA	Block-matching algorithm
BYE	Session termination packet
CBR	Constant Bit Rate
CC	CSRC counter field
CDMA	Code Division Multiple Access
CIF	Common Interchange Format
CW	Congestion window
DC	Direct current
DCT	Discrete cosine transform
DLL	Data-link layer
EOB	End of block
FEC	Forward error correction
FGS	Fine Grain Scalability
GOB	Group of Blocks
GoP	Group of pictures
GSM	Global system for mobile communication
H.26x	Video standards of the ITU: H.261, H.263, H.263+, H.26L
HDTV	High-definition television
HTTP	Hypertext Transfer Protocol
IETF	Internet Engineering Task Force
IP	Internet Protocol
ISDN	Integrated Services Digital Network
ISI	Intersymbol interference
ISO	International Standardization Organization
ITU	International Telecommunication Union
JPEG	Joint Photographic Experts Group
JVT	Joint Video Team
LoS	Line of sight
LPDU	Link-layer packet data unit
M	Marker Field
MB	Macroblock
MD	Multiple Description coding
MPEG	Moving Picture Experts Group
MSE	Mean squared error
NLos	Non-line of sight
NTSC	National Television System Committee
PDA	Personal data assistant
PSNR	Peak signal-to-noise ratio
QCIF	Quarter CIF
QoS	Quality of service
RFC	Request for Comments
RGB	Color space format consisting of the values of red, green, and blue
RR	Receiver Report
RS	Reed-Solomon
RSVP	Resource Reservation Protocol
RTCP	Real-Time Control Protocol
RTO	Retransmission timeout
RTP	Real-Time Protocol
RTSP	Real-Time Streaming Protocol
RTT	Round-trip time
SAP	Session Announcement Protocol

TABLE 11.3 Acronyms and Abbreviations (continued)

SDES	Source description
SDP	Session Description Protocol
SIP	Session Initiation Protocol
SMPT	Simultaneous MAC Packet Transmission
SNR	Signal-to-noise ratio
SR	Sender Report
TCP	Transmission Control Protocol
TV	Television
UDP	User Datagram Protocol
UMTS	Universal Mobile Telecommunications System
VBR	Variable Bit Rate
VLC	Variable-length coding
YUV	Color space format consisting of the values for luminance, chrominance, and saturation

11.8 Conclusion

In this chapter, we gave an overview of wireless video streaming. We presented an intuitive introduction to the basic principles of video encoding. Next, we reviewed the properties of the wireless channel and how these properties affect video transmissions. In addition, we provided an overview of the Internet protocols involved in wireless video streaming. Finally, we discussed the currently available strategies for video streaming over wireless links. Note in closing that this is an active area of research, with many fundamental advances to be expected over the next few years. One trend in this ongoing research is to incorporate the application's or user's perspective more into the streaming.[97,98]

Acknowledgments

This work was supported in part by the National Science Foundation through grants Career ANI-0133252 and ANI-0136774.

References

1. Gross, J., Fitzek, F., Wolisz, A., Chen, B., and Gruenheid, R., Framework for combined optimization of dl and physical layer in mobile ofdm systems, in *Proceedings of the 6th International OFDM-Workshop 2001*, Hamburg, Germany, Sep. 2001, pp. 32-1–32-5.

2. Kravcenko, V., Boche, H., Fitzek, F., and Wolisz, A., No need for signaling: investigation of capacity and quality of service for multi-code CDMA systems using the WBE++ approach, in *Proceedings of the 4th IEEE Conference on Mobile and Wireless Communications Networks (MWCN 2002)*, Stockholm, Sweden, Sep. 2002, pp. 110–114.

3. Commission of the European Communities, Digital Land Mobile Radio Communications — COST 207, Office for Official Publications of the European Communities, Luxembourg, Tech. Rep., 1989, final report.

4. Video traces for network performance evaluation, http://trace.eas.asu.edu/.

5. Sikora, T., MPEG digital video coding standards, in *Digital Electronics Consumer Handbook*, McGraw Hill, New York, 1997.

6. Pereiera, F. and Touradj, E., *The MPEG-4 Book*, Prentice Hall PTR, Upper Saddle River, NJ, 2002.

7. MPEG-2, Generic coding of moving pictures and associated ausio information, ISO/IEC 13818-2, 1994, draft International Standard.

8. MPEG-1, Coding of moving pictures and associated audio for digital storage media at up to 1.5 mbps, ISO/IEC 11172, 1993.

9. Riley, M. and Richardson, I., *Digital Video Communications*, Artech House, Norwood, MA, 1997.
10. Ghanbari, M., *Video coding — an introduction to standard codecs*, Institution of Electrical Engineers, 1999.
11. Netravali, A.N. and Haskell, B.G., *Digital Pictures: Representation, Compression, and Standards*, Plenum Press, New York, 1995, ch. 3.
12. Xiong, Z., Ramchandran, K., Orchard, M.T., and Zhang, Y.-Q., A comparative study of DCT- and wavelet-based image coding, *IEEE Transactions on Circuits and Systems for Video Technology*, 9, 5, Aug. 1999.
13. Chen, W., Smith, C., and Fralick, S., A fast computational algorithm for the discrete cosine transform, *IEEE Transcript on Communications*, Sep. 1997, pp. 1004–1009.
14. Dinstein, I., Rose, K., and Heimann, A., Variable block-size transform image coder, *IEEE Transcript on Communications*, Nov. 1990, pp. 2073–2078.
15. Turaga, D. and Chen, T., Fundamentals of video compression: H.263 as an example, in *Compressed Video over Networks*, Reibman, A.R. and Sun, M.-T. , Eds. Marcel Dekker, New York, 2001, pp. 3–34.
16. Lakshman, T., Ortega, A., and Reibman, A., VBR video: tradeoffs and potentials, *Proceedings of the IEEE*, 86, 5, May 1998, pp. 952–973.
17. Ortega, A., Variable bit rate video coding, in *Compressed Video over Networks*, Sun, M.-T. and Reibman, A., Eds., Marcel Dekker, New York, 2001, pp. 343–383.
18. Shi, Y. and Sun, H., *Image and Video Compression for Multimedia Engineering*, CRC Press, Boca Raton, FL, 2000.
19. Koga, T. et. al., Motion compensated interframe coding for video conferencing, New Orleans: National Telecommunication Conference, 1981, pp. G5.3.1–G5.3.5.
20. Bierling, M., Displacement estimation by hierarchical block matching, *Visual Communications and Image Processing*, 1001, 1988, pp. 942–951.
21. Lelewer, D. and Hirschberg, D., Data compression, *ACM Computing Surveys (CSUR)*, 19, 3, 1987, pp. 261–296.
22. Howard, P. and Vitter, J., Analysis of arithmetic coding for data compression, *Information Processing and Management*, 28, 6, 1992, pp. 749–764.
23. Wiegand, T., H.26L test model long-term number 9 (TML-9) draft0, ITU-T Study Group 16, Dec. 2001.
24. Fitzek, F., Seeling, P., and Reisslein, M., H.26l pre-standard evaluation, acticom GmbH, Tech. Rep., Nov. 2002, http://www.acticom.de/.
25. Fitzek, F. and Reisslein, M., MPEG-4 and H.263 video traces for network performance evaluation, *IEEE Network*, 15, 6, Nov./Dec. 2001, pp. 40–54.
26. Krunz, M., Sass, R., and Hughes, H., Statistical characteristics and multiplexing of MPEG streams, in *Proceedings of IEEE Infocom'95*, Apr. 1995, pp. 455–462.
27. Rose, O., Statistical properties of MPEG video traffic and their impact on traffic modelling in ATM systems, University of Wuerzburg, Institute of Computer Science, Tech. Rep. 101, Feb. 1995.
28. Feng, W.-C., *Buffering Techniques for Delivery of Compressed Video in Video-on-Demand Systems*, Kluwer Academic Publisher, New York, 1997.
29. The network simulator - ns-2, http://www.isi.edu/nsnam/ns/index.html.
30. Varga, A., Omnet++, *IEEE Network Interactive*, 16, 4, 2002, http://www.hit.bme.hu/phd/vargaa/omnetpp.
31. Buck, J., Ha, E.L.S., and Messerschmitt, D., Ptolemy: a framework for simulating and prototyping heterogeneous systems, *International Journal of Computer Simulation*, 4, Apr. 1994, pp. 155–182, http://ptolemy.eecs.berkeley.edu/index.htm.
32. Fitzek, F., Seeling, P., and Reisslein, M., Using network simulators with video traces, Arizona State University, Dept. of Electrical Eng, Tech. Rep., Mar. 2003, http://trace.eas.asu.edu/.
33. Marie, A et. al., Video quality experts group: current results and future directions, 4067, Perth, Australia, 2000, pp. 742–453.

34. Reisslein, M., Lassetter, J., Ratnam, S., Lotfallah, O., Fitzek, F., and Panchanathan, S., Traffic and quality characterization of scalable encoded video: a large-scale trace-based study, Tech. Rep. Arizona State University, Dept. of Electrical Engineering, Aug. 2003, http://trace.eas.asu.edu.

35. Rappaport, T., Seidel, S., and Takamizawa, K., Statistical channel impulse response models for factory and open plan building radio communication system design, *IEEE Transactions on Communications*, 39, 5, May 1991, pp. 794–807.

36. Gibson, J. D., *Mobile Communications Handbook*, 2, IEEE Press, New York, NY, 1999.

37. Aldinger, M., Die simulation des mobilfunk-kanals auf einem digitalrechner, in *FREQUENZ*, 36, 4/5, 1982, pp. 145–152.

38. Kittel, L., Analoge und diskrete Kanalmodelle für die Signalübertragung beim beweglichen Funk, in *FREQUENZ*, 36, 4/5, 1982, pp. 152–160.

39. Zorzi, M. and Rao, R., Error-constrained error control for wireless channels, *IEEE Personal Communications*, Dec. 1997, pp. 27-33.

40. Rao, R.R., Higher layer perspectives on modeling the wireless channel, in *Proceedings of IEEE ITW*, Killarney, Ireland, June 1998, pp. 137–138.

41. Zorzi, M., Rao, R.R., and Milstein, L.B., On the accuracy of a first-order Markovian model for data block transmission on fading channels, in *Proceedings of IEEE International Conference on Universal Personal Communications*, Nov. 1995, pp. 211–215.

42. Elwalid, A., Heyman, D., Lakshman, T., Mitra, D., and Weiss, A., Fundamental bounds and approximations for ATM multiplexers with application to video teleconferencing, *IEEE Journal on Selected Areas in Communications*, 13, 6, Aug. 1995, pp. 1004–1016.

43. Handley, M., Perkins, C., and Whelan, E., SAP: session announcement protocol, RFC 2974, Oct. 2000.

44. Rosenberg, J. et. al., RFC 3261: SIP: session initiation protocol, Feb. 1999.

45. Schulzrinne, H., Rao, A., and Lanphier, R., Real time streaming protocol (RTSP), RFC 2326, Apr. 1998.

46. Audio-Video Transport Working Group, Schulzrinne, H., Casner, S., Frederick, R., and Jacobson, V., RTP: a transport protocol for real-time applications, RFC 1889, Jan. 1996.

47. Zhu, C., RTP payload format for H.263 video streams, RFC 2190, Sep. 1997.

48. Speer, M. and Hoffman, D., RTP payload format of Sun's CellB video encoding, RFC 2029, Oct. 1996.

49. Liu, H. and Zarki, M.E., Performance of H.263 video transmission over wireless networks using hybrid ARQ, *IEEE Journal on Selected Areas in Communications*, 15, 9, Dec. 1997, pp. 1775–1786.

50. Karlsson, G., Asynchronous transfer of video, *IEEE Communications Magazine*, 34, 8, Feb. 1996, pp. 106–113.

51. Lin, S., *Error Control Coding: Fundamentals and Applications*, Prentice Hall, Upper Saddle River, NJ, 1983.

52. Liu, H., Ma, H., Zarki, M.E., and Gupta, S., Error control schemes for networks: an overview, *Mobile Networks and Applications*, 2, 1997, pp. 167–182.

53. Kumwilaisak, W., Kim, J., and Kuo, C., Reliable wireless video transmission via fading channel estimation and adaptation, in *Proceedings of IEEE WCNC*, Chicago, Sep. 2000, pp. 185–190.

54. Kim, I.-M. and Kim, H.-M., Efficient power management schemes for video service in CDMA systems, *IEEE Electronics Letters*, 36, 13, June 2000, pp. 1149–1150.

55. Kim, I.-M. and Kim, H.-M., An optimum power management scheme for wireless video service in CDMA systems, *IEEE Transactions on Wireless Communications*, 2, 1, Jan. 2003, pp. 81–91.

56. Fitzek, F.H.P., Reisslein, M., and Wolisz, A., Uncoordinated real-time video transmission in wireless multicode CDMA systems: an SMPT-based approach, *IEEE Wireless Communications*, 9, 5, Oct. 2002, pp. 100–110.

57. Hwang, S.-H., Kim, B., and Kim, Y.-S., A hybrid ARQ scheme with power ramping, in *Proceedings of the 54th IEEE Vehicular Technology Conference*, 3, 2001, pp. 1579–1583.

58. Qiao, D. and Shin, K.G., A two-step adaptive error recovery scheme for video transmission over wireless networks, in *Proceedings of IEEE Infocom*, Tel Aviv, Israel, Mar. 2000.

59. Majumdar, A., Sachs, D., Kozintsev, I.V., Ramchandran, K., and Yeung, M., Multicast and unicast real-time video streaming over wireless LANs, *IEEE Transactions on Circuits and Systems for Video Technology*, 12, 6, June 2002, pp. 524–534.

60. Manghavamanid, R., Demirhan, M., Raikumar, R., and Raychaudhuri, D., Size-based scheduling for delay-sensitive variable bit rate traffic over wireless channels, Real-time and Multimedia Systems Laboratory. Carnegie Mellon University, Tech. Rep., Jan. 2003.

61. Chang, P.-R. and Lin, C.-F., Wireless ATM-based multicode CDMA transport architecture for MPEG-2 video transmission, *Proceedings of the IEEE*, 87, 10, Oct. 1999, pp. 1807–1824.

62. Fitzek, F. and Reisslein, M., A prefetching protocol for continuous media streaming in wireless environments, *IEEE Journal on Selected Areas in Communications*, 19, 10, Oct. 2001, pp. 2015–2028.

63. Iskander, C. and Mathiopoulos, P.T., Rate-adaptive transmission of H.263 video for multicode DS/CDMA cellular systems in multipath fading, in *Proceedings of the IEEE Vehicular Technology Conference*, spring 2002, pp. 473–477.

64. Singh, A., Konrad, A., and Joseph, A.D., Performance evaluation of UDP Lite for cellular video, in *Proceedings of the 11th International Workshop on Network and Operating System Support for Digital Audio and Video (NOSSDAV)*, Port Jefferson, NY, June 2001.

65. Zheng, H. and Boyce, J. An improved UDP protocol for video transmission over internet-to-wireless networks, *IEEE Transactions on Multimedia*, 3, 3, Sep. 2001, pp. 356–365.

66. Floyd, S. and Fall, K., Promoting the use of end-to-end congestion control in the internet, *IEEE/ACM Transactions on Networking*, 7, 4, Aug. 1999, pp. 458–472.

67. Balakrishnan, H., Padmanabhan, V., Seshan, S., and Katz, R., A comparison of mechanisms for improving TCP performance over wireless links, *IEEE ACM Transactions on Networking*, Dec. 1997.

68. Fitzek, F.H.P., Supatrio, R., Wolisz, A., Krishnam, M., and Reisslein, M., Streaming video applications over TCP in CDMA based networks, in *Proceedings of the 2002 International Conference on Third Generation Wireless and Beyond (3Gwireless 2002)*, San Francisco, May 2002, pp. 755–760.

69. Krasic, C., Li, K., and Walpole, J., The case for streaming multimedia with TCP, in *Proceedings of the 8th International Workshop on Interactive Distributed Multimedia Systems (IDMS)*, Lancaster, UK, Sep. 2001.

70. Hsu, C., Ortega, A., and Khansari, M., Rate control for robust video transmission over burst-error wireless channels, *IEEE Journal on Selected Areas in Communications*, 17, 5, May 1999, pp. 756–773.

71. Hu, P.-C., Zhang, Z.-L., and Kaveh, M., Channel condition ARQ rate control for real-time wireless video under buffer constraints, in *Proceedings of the IEEE International Conference on Image Processing*, 2000, pp. 124–127.

72. Tosun, A. and Feng, W., On improving quality of video for H.263 over wireless CDMA networks, in *Proceedings of the IEEE WCNC*, Chicago, Sep. 2000, pp. 1421–1426.

73. Reyes, G., Reibman, A.R., Chang, S.-F., and Chuang, J., Error-resilient transcoding for video over wireless channels, *IEEE Journal on Selected Areas in Communications*, 18, 6, June 2000, pp. 1063–1074.

74. Cote, G., Kossentini, F., and Wenger, S., Error resilience coding, in *Compressed Video over Networks*, Sun, M.-T. and Reibman, A.R., Eds. Marcel Dekker, New York, 2001, pp. 309–341.

75. Shanableh, T. and Ghanbari, M., Heterogeneous video transcoding to lower spatio-temporal resolutions and different encoding formats, *IEEE Transactions on Multimedia*, 2, 2, June 2000, pp. 101–110.

76. Ghanbari, M., Layered coding, in *Compressed Video over Networks*, Reibman, A.R. and Sun, M.-T., Eds. Marcel Dekker, New York, 2001, pp. 251–308.

77. Li, W., Overview of fine granularity scalability in MPEG-4 video standard, *IEEE Transactions on Circuits and Systems for Video Technology*, 11, 3, Mar. 2001, pp. 301–317.

78. Goyal, V., Multiple description coding: compression meets the network, *IEEE Signal Processing Magazine*, 18, Sep. 2001, pp. 74–91.

79. Gao, L., Karam, L., Reisslein, M., and Abousleman, G., Error-resilient image coding and transmission over wireless channels, in *Proceedings of the IEEE International Symposium on Circuits and Systems (ISCAS)*, Scottsdale, AZ, May 2002, pp. 629–632.

80. Ekmekci, S. and Sikora, T., Unbalanced quantized multiple description video transmission using path diversity, in *IS&T/SPIE's Electronic Imaging 2003*, 2003, Santa Clara, CA.

81. Naghshineh, M. and LeMair,M., End-to-end QoS provisioning in multimedia wireless/mobile networks using an adaptive framework, *IEEE Communications Magazine*, 35, 11, Nov. 1997, pp. 72–81.

82. Fankhauser, G., Dasen, M., Weiler, N., Plattner, B., and Stiller, B., The WaveVideo system and network architecture: design and implementation, Computer Engineering and Networks Laboratory (TIK), ETH Zurich, Tech. Rep., June 1998.

83. Wu, D., Hou, Y.T., and Zhang, Y.-Q., Scalable video coding and transport over broad-band wireless networks, *Proceedings of the IEEE*, 89, 1, Jan. 2001, pp. 6–20.

84. Jiang, Z. and Kleinrock, L., A packet selection algorithm for adaptive transmission of smoothed video over a wireless channel, *IEEE Transactions on Parallel and Distributed Systems*, 60, 4, Apr. 2000, pp. 494–509.

85. Gharavi, H. and Alamouti, S.M., Multipriority video transmission for third-generation wireless communication systems, *Proceedings of the IEEE*, , 10, Oct. 1999, pp. 1751–1763.

86. Zhao, S., Xiong, Z., and Wang, X., Joint error control and power allocation for video transmission over CDMA networks with multiuser detection, *IEEE Transactions on Circuits and Systems for Video Technology*, 12, 6, June 2002, pp. 425–437.

87. De Cuetos, P., Reisslein, M., and Ross, K.W., Analysis of a large library of MPEG-4 FGS rate-distortion traces for streaming video, Institut Eurecom, Tech. Rep., Dec. 2002, traces available at http://trace.eas.asu.edu/.

88. Van der Schaar, M. and Radha, H., Adaptive motion-compensation fine-granular-scalability (AMC-FGS) for wireless video, *IEEE Transactions on Circuits and Systems for Video Technology*, 12, 6, June 2002, pp. 360–371.

89. Stockhammer, T., Jenkac, H., and Weiss, C., Feedback and error protection strategies for wireless progressive video transmission, *IEEE Transactions on Circuits and Systems for Video Technology*, 12, 6, June 2002, pp. 465–482.

90. Horn, U., Girod, B., and Belzer, B., Scalable video coding for multimedia applications and robust transmission over wireless channels, in *7th International Workshop on Packet Video*, Mar. 1996, http://citeseer.nj.nec.com/horn96scalable.html.

91. Steinbach, E., Färber, N., and Girod, B., Standard compatible extension of h.263 for robust video transmission in mobile environments, *IEEE Transactions on Circuits and Systems for Video Technology*, 7, Dec. 1997, pp. 872–881.

92. Yu, Y. and Chen, C., SNR scalable transcoding for video over wireless channels, in *Proceedings of the IEEE WCNC*, Chicago, Sep. 2000, pp. 1398–1402.

93. Kondi, L., Batalama, S., Pados, D., and Katsaggelos, A., Joint source-channel coding for scalable video over DS-CDMA multipath fading channels, in *Proceedings of the 2001 International Conference on Image Processing*, 2001, pp. 994–997.

94. Srinivasan, M. and Chellappa, R., Adaptive source-channel subband video coding for wireless channels, *IEEE Journal for Selected Areas in Communication*, 16, Dec. 1998, pp. 1830–1839.

95. Chang, P.-R., Spread spectrum CDMA systems for subband image transmission, *IEEE Transactions on Vehicular Technology*, 46, Feb. 1997, pp. 80–95.

96. Iun, E. and Khandani, H.-M., Combined source-channel coding forthe transmission of still images over a code division multiple access (CDMA) channel, *IEEE Communication Letters*, 2, June 1998, pp. 168–170.

97. Lu, X., Morando, R.O., and ElZarki, M., Understanding video quality and its use in feedback control, in *Proceedings of the International Packet Video Workshop 2002*, Pittsburgh, 2002.

98. Meng, X., Yang, H., and Lu, S., Application-oriented multimedia scheduling over lossy wireless links, in *Proceedings of the 11th International Conference on Computer Communications and Networks*, Miami, Oct. 2002, pp. 256–261.

12

Integration of WLANs and Cellular Data Networks for Next-Generation Mobile Internet Access

Apostolis K. Salkintzis
Motorola, Inc.

12.1 Introduction

The ongoing Wireless Local Area Network (WLAN) standardization and research and development activities worldwide, which target bit rates higher than 100 Mbps, combined with the successful deployment of WLANs in numerous hotspots worldwide, confirm the fact that WLAN technology will play a key role in next-generation mobile Internet access. Cellular network operators have recognized this fact and are striving to exploit WLAN technology and integrate it into their cellular data networks. For this reason, there is a strong need for interworking mechanisms between WLANs and cellular data networks.

In this chapter, we focus on these interworking mechanisms, which effectively combine WLANs and cellular data networks into integrated, heterogeneous wireless data networks capable of ubiquitous data services and very high data rates in hotspot locations. We discuss the general aspects of such integrated WLAN/cellular data networks, and we examine the generic interworking architectures that have been proposed in the technical literature. In addition, we review the standardization activities in the area of WLAN/cellular data network integration. Subsequently, we propose and explain two different interworking architectures, which feature different coupling mechanisms. We illustrate the operational characteristics of these mechanisms and then compare them by summarizing their advantages and drawbacks. Finally, we concentrate on the specific architectures proposed for interworking between WLANs and Third-Generation Partnership Project (3GPP) systems. By the term 3GPP systems, we refer to the cellular telecommunication systems standardized by 3GPP (http://www.3gpp.org) — that is, the evolved Global System for Mobile Communications (GSM) systems such as the well-known General Packet Radio Service (GPRS) and the Universal Mobile Telecommunications System (UMTS). For the purposes of our discussion, it suffices to focus only on the wireless data capabilities of these systems and, in particular, on their *packet-switched* wireless data capabilities. We assume that the fundamental aspects of these systems as well as of WLANs are known; background material for GPRS and WLANs can be found in Chapter 3 and Chapter 2, respectively.

12.2   Integrated WLANs and Cellular Data Networks

In today's world economy, the ability to communicate on the move is becoming less of a luxury and more of a necessity. Second-generation (2G) cellular systems have enabled a high level of mobility, with wired-equivalent quality, for voice services and low-speed data (<9.6 kbps) services. This is done via the implementation of global standards based on digital technology, such as GSM and CDMAOne, with roaming agreements among operators acting as the glue that binds disparate networks together into one ubiquitous system from the end user's perspective.

Although 2G technology is adequate in meeting the voice communication needs of the typical cellular subscriber, its data communication capabilities are cumbersome and limited. The low bandwidth and complexity associated with these services have discouraged the average consumer from investing in wireless data. In contrast, wired data service, which can offer high bandwidth and always-on connectivity, has grown in popularity due to its availability and affordability. To compete with this technology, third-generation (3G) cellular systems promise competitive data rates, at speeds of up to 300 kbps initially and increasing up to 2 Mbps, with the same always-on connectivity of wired technology.

However, due to the delay of deployment of 3G cellular networks and the large investments made for new spectrum in which to offer 3G services, cellular operators are looking for ways to augment their offerings with "3G-like" services in efforts to generate new revenue stream in today's environment. 2.5G cellular data technology, in particular, the GPRS (see Chapter 3 and reference 2), which provides wireless data services at speeds of up to approximately 100 kbps, is gaining support as a wide-area data solution but has limited potential because it cannot support the high data rates required in business and multimedia applications. Therefore, because 2.5G cellular data technology is not sufficient to meet the market needs, and because 3G cellular data technology faces deployment delays and cost concerns, mobile network operators are turning to WLAN technology. This interest in WLAN technology also arises from the recent evolution and successful deployment of WLAN systems worldwide, as well as from the very high data rates (in excess of 100 Mbps) that future WLAN developments promise.

As mentioned in Chapter 2, WLAN systems are expected to be widely deployed in public locations, such as hotels and coffee shops, as well as in enterprises and homes. Specifically, WLANs will provide wireless data coverage by means of *hotspot* deployments. To generate revenue in the WLAN space with cellular customers, it is commonly believed that operators must provide a seamless user experience between the cellular and the WLAN access networks. This calls for interworking mechanisms between WLANs and cellular data networks capable to provide integrated authentication, integrated billing, roaming, terminal mobility, and service mobility.

A cellular data network can provide relatively low-speed (up to 100 kbps per user) data service but over a large coverage area. On the other hand, the WLAN provides high-speed data service (up to 11 Mbps with 802.11b and 54 Mbps with 802.11a) but over a geographically small area. An integrated

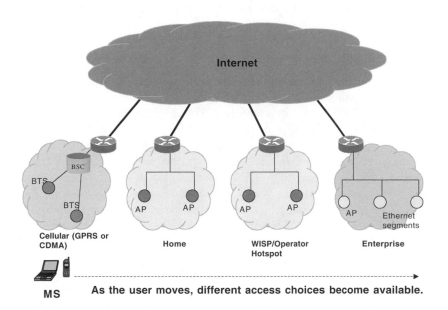

FIGURE 12.1 Multiple access options in an integrated data environment.

network combines the strengths of each, resulting in a wide-area system capable of providing users with ubiquitous data service, which ranges from low speed to high speed in strategic locations.

12.2.1   Roaming

Figure 12.1 illustrates an integrated environment in which a subscriber has multiple access network options. We will discuss two example configurations concerning the integration of the WLAN and cellular networks. These configurations vary in the area of ownership/management of the WLAN network. In the first example, the cellular operator owns and manages the WLAN network. In the second, a Wireless Internet Service Provider (WISP) or enterprise is the owner.

Consider a cellular data customer waiting in an airport where the cellular operator has deployed a WLAN. While waiting, the customer can avail of the high bandwidth and wireless connectivity of the WLAN to gain access to the data services provided by the operator. By augmenting their cellular data systems with WLANs, operators are able to enhance their data service capabilities with high-speed data connectivity in strategic locations such as airports and hotels. In doing this, the cellular operators are able to gain a competitive advantage by offering their current customer base the ability to roam into these hotspots.

In operator-owned WLANs, the cellular operators have the advantage (over the WISP) of an established customer base to which they can market such capabilities. Additionally, operators have authentication and billing mechanisms in place for their users, which they can leverage in the WLAN space.

Although the mechanisms may be slightly different if the cellular operators do not own the WLAN, the same user experience may be achieved. A multilateral roaming agreement between WISPs and cellular operators may allow a cellular customer to use a WISP-operated WLAN (this is further discussed in Section 12.8). The billing and authentication services would continue to be provided by the cellular operator. The WISP may partake in revenue sharing with the cellular operator, based on the roaming agreements between the two parties.

In the case of an enterprise WLAN, the enterprise may choose its own authentication and billing mechanisms. Most enterprises do not have any billing systems for data services and have only limited authentication mechanisms. However, cellular and wireline operators have considerable interest in providing their customers with service through the enterprise WLAN system. Such services would be facilitated by mechanisms similar to those required for a WISP-owned WLAN.

12.2.2 Session Mobility

Session mobility may be seen as an evolutionary step from roaming in this integrated environment. *Session* is defined here as a flow of IP packets between the end user and an external entity; for example, a File Transfer Protocol (FTP) session or a Hypertext Transfer Protocol (HTTP) session. Consider, for instance, a mobile device that is capable of connecting to the data network through WLANs and cellular. This could be, for example, a laptop with an integrated WLAN-GPRS card or a personal digital assistant (PDA) attached to a dual-access card. The end user is connected to the data network and is in a session flow through one access network, say, the WLAN. As the user moves out of the coverage of the WLAN system, the end device detects the failing WLAN coverage and seamlessly switches the flow to a GPRS network. The end-to-end session remains unaffected. Typically, no user intervention would be required to perform the switchover from WLAN to GPRS. Moreover, the user would not observe this handover. When the user moves back into the coverage of a WLAN system, the flow would be handed back to the WLAN network.

The mobility function is distinct from the roaming in the following way. Mobility requires no user intervention and preserves any IP-based session during handovers between cellular data and the WLAN. With roaming, the switchover between the WLAN and the cellular data requires explicit user intervention and in most cases would result in teardown of any existing sessions.

12.2.3 Enhanced Mobile Applications

Given the possibility of cellular customers being connected to two different access networks, say, a WLAN and GPRS, a number of enhanced applications can be enabled. The applications could take advantage of the fact that the end user is always connected through a low-speed cellular network and sometimes connected through a high-speed WLAN. An example is a mobile e-mail application, which schedules the delivery of attachment/large files, for example, when the mobile is connected to the WLAN and delivers only a synopsis of the e-mails when the mobile is connected to the GPRS network.

12.3 Generic Interworking Architectures

Several approaches have been proposed for interworking between WLANs and cellular data networks. These approaches are discussed below in detail. Without any loss of generality, we will specifically discuss the interworking between WLAN and GPRS. Nevertheless, the illustrated interworking principles can readily be generalized to any type of cellular data network.

The European Telecommunications Standardization Institute (ETSI) specifies in reference 1 two generic approaches for interworking: the so-called *loose coupling* and the *tight coupling*. With loose coupling, the WLAN is deployed as an access network *complementary* to the GPRS network. In this case, the WLAN utilizes the subscriber databases in the GPRS network but features no data interfaces to the GPRS core network. Considering the simplified GPRS reference diagram displayed in Figure 12.2, we may argue that the loose coupling between the GPRS and the WLAN is carried out at the G_i reference point. This means that with loose coupling the WLAN bypasses the GPRS network and provides direct data access to the external Packet Data Networks (PDNs). On the other hand, with tight coupling the WLAN is connected to the GPRS core network in the same manner as any other radio access network (RAN), such as GPRS Enhanced RAN (GERAN) and UMTS Terrestrial RAN (UTRAN). In this case, the WLAN data traffic goes through the GPRS core network before reaching the external PDNs. As shown in Figure 12.2, with tight coupling the WLAN is connected to either G_b or $I_{u\text{-}ps}$ reference points. A detailed description of the GPRS reference diagram, including the reference points and the functional nodes, can be found in reference 2.

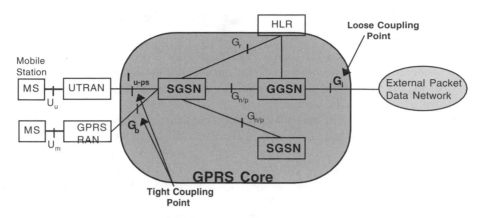

FIGURE 12.2 GPRS reference diagram showing the WLAN coupling points.

Reference 1 also describes a specific tight-coupling architecture for interworking between HiperLAN/2^3 and GPRS networks. In this architecture, the HiperLAN/2 RAN is connected to the standard $I_{u\text{-}ps}$ interface. The loose-coupling and tight-coupling approaches are further discussed in subsequent sections, where we propose more specific architectures and describe them in detail.

Currently, the short-term trend is to follow the loose-coupling approach and use (U)SIM-based authentication and billing. With this approach, subscribers can reuse their Subscriber Identity Module (SIM) card or their User Services Identity Module (USIM) to access a set of wireless data services over a WLAN. (Typically, GSM subscribers are equipped with an SIM card and UMTS subscribers with a USIM card.) However, as explained later, this approach features limited session mobility capabilities as compared with tight coupling. Section 12.8 discusses the particular aspects of the loose-coupling architecture proposed for the interworking of WLAN and 3GPP systems.

Notable references to other interworking architectures include references 6 to 8, which describe five different architectures for implementing handover between GPRS and IEEE 802.11[9] WLAN networks. However, all these architectures mainly refer to very high-level concepts and do not discuss any operational details. On the contrary, in subsequent sections of this chapter we propose detailed architectures and thoroughly discuss their key functional aspects.

12.4 Standardization Activities

Apart from the standardization activities in ETSI, which are summarized in the Technical Report (TR) 101 957[1] discussed earlier, other standardization bodies have been involved with the integration of WLANs and cellular telecommunication networks. Most of these activities are promoted by cellular operators, who want to benefit from the rapidly evolved WLAN technology and offer advanced, high-speed data services to their subscribers with one subscription, one bill, one set of services, and so forth. The goal of standardization activities is to define standard interworking interfaces and assure interworking across multivendor equipments and across several types of WLANs and cellular networks.

Several WLAN standardization bodies, in particular, ETSI Broadband Radio Access Networks (BRAN), IEEE 802.11, IEEE 802.15, and Multimedia Mobile Access Communication (MMAC), have agreed to set up a joint Wireless Interworking Group (WIG) to deal with the interworking between WLANs and cellular networks. This activity is being driven primarily from Europe by ETSI BRAN; the first WIG meeting took place in September 2002.

The most intense standardization activities have taken place in the 3GPP, a standardization body that maintains and evolves the GSM and UMTS specifications. 3GPP has approved a "WLAN/Cellular Interworking" work item, which aims to specify one or more techniques for interworking between WLANs and 3GPP networks (Section 12.8 is devoted to the interworking between WLANs and 3GPP systems). This standardization work was started in 2002, and its first phase is scheduled for completion late in

2003. In the context of this work, several interworking requirements have been specified and categorized into the six interworking scenarios discussed next:[4]

Scenario 1 — Common billing and customer care: The simplest form of interworking, this provides only a common bill and customer care to the subscriber but otherwise features no real interworking between the WLAN and the 3GPP network. For this reason, this scenario does not require any particular standardization activities.

Scenario 2 — 3GPP-based access control and charging: This scenario requires Authentication, Authorization and Accounting (AAA) for subscribers in the WLAN to be based on the same AAA procedures utilized in the 3GPP system. For example, subscribers in a WLAN can use their (U)SIM card for authentication, as they normally do in a 3GPP environment. Also, authorization is provided by the 3GPP system itself based on subscription data. This scenario basically enables "IP connectivity via WLAN" for 3GPP WLAN subscribers. In other words, the user is merely provided with a connection from the WLAN *directly* to the Internet or an intranet. Apart from the IP connectivity over the WLAN, it is not necessary to provide a specific set of services. You can find a more detailed discussion on this scenario in Section 12.8.2.

Scenario 3 — Access to 3GPP packet-switched services: The goal of this scenario is to extend the access to 3GPP packet switch (PS)–based services to subscribers in a WLAN environment. For example, if an operator maintains a Wireless Application Protocol (WAP) gateway for providing WAP and Multimedia Messaging Service (MMS) services to subscribers, under the interworking scenario 3, these WAP and MMS services should be accessible to subscribers also in a WLAN environment. As explained later, to gain access to the 3GPP PS-based services the user data traffic is tunneled between the WLAN and the 3GPP infrastructure. In addition, an *IP Service Selection* scheme is used for selecting the PS-based service to connect to. In general, with scenario 3 the user can have access to services such as IP Multimedia Services, location-based services, instant messaging, presence-based services, MMS, and so forth. However, although the user is offered access to the same PS-based services over several access networks (e.g., UTRAN and WLAN), no service continuity across these access networks is required in scenario 3. You can find more detailed discussion on this scenario in Section 12.8.3.

Scenario 4 — Access to 3GPP PS-based services with service continuity: The goal of this scenario is to allow access to PS-based services as required in scenario 3 and, in addition, to maintain service continuity across the 3GPP and WLAN access systems. For example, a user starting a WAP session from the 3GPP radio network should be able to continue this session after moving to a WLAN system, and vice versa. Although service continuity is required, the service continuity requirements are not very strict. This means that, under the interworking scenario 4, some services may not be able to continue after a vertical handover to or from the WLAN (e.g., due to varying capabilities and characteristics of access technologies). A typical example could be a PS-based service that requires tight delay performance, which cannot be met in a WLAN system. In this case, the service would most likely be terminated after the user moves to a WLAN. Also, under scenario 4, change in service quality is possible as a consequence of vertical handovers.

Scenario 5 — Access to 3GPP PS-based services with seamless service continuity: This scenario is one step ahead of scenario 4. Its goal is to provide *seamless* service continuity between the 3GPP and the WLAN radio access technologies. That is, PS-based services should be utilized across the 3GPP and WLAN radio access technologies in a seamless manner (i.e., without the user noticing any significant differences).

Scenario 6 — Access to 3GPP circuit switch (CS)-based services with seamless mobility: The goal of this scenario is to allow access to 3GPP CS-based services (e.g., normal CS voice calls) from the WLAN system. Seamless mobility for these services should be provided.

Table 12.1 summarizes the various interworking scenarios and indicates the operational capabilities of each one. Note that as we move from scenario 1 to scenario 6, the WLAN and cellular system become more tightly integrated; therefore, more and more demanding interworking requirements appear.

TABLE 12.1 Some WLAN/3GPP Interworking Scenarios and Their Characteristics

	Scenario 1	Scenario 2	Scenario 3	Scenario 4	Scenario 5	Scenario 6
Common billing	X	X	X	X	X	X
Common customer care	X	X	X	X	X	X
3GPP-based access control		X	X	X	X	X
3GPP-based access charging		X	X	X	X	X
Access to 3GPP PS-based services			X	X	X	X
Access to 3GPP PS-based services with service continuity				X	X	X
Access to 3GPP PS-based services with seamless service continuity					X	X
Access to 3GPP CS-based services with seamless mobility						X

After reviewing the most important standardization activities for interworking between WLANs and cellular data networks, we present in the following sections two architectures for WLAN-GPRS integration.

12.5   A Tight-Coupling Architecture

In this section, we propose a tight-coupling architecture that can fulfill the requirements of scenarios 1 to 4. Depending on the WLAN technology and, in particular, on whether the WLAN can support Quality of Service (QoS) equivalent to the GPRS Release 1999 QoS (as specified in reference 11), the proposed architecture can also satisfy the requirements of scenario 5.

The proposed architecture follows the principles of the aforementioned tight-coupling approach, as described in reference 1. However, in contrast to reference 1, the proposal specifies a tight-coupling architecture for interworking between 802.11 WLAN (not HiperLAN/2) and GPRS networks. In addition, the proposal assumes that the 802.11 WLAN is connected to the standard G_b interface (not I_{u-ps}), which is already deployed in live GPRS networks. It is noted that G_b is specified from GPRS Release 1997 onward, whereas I_{u-ps} is specified from GPRS Release 1999 onward. A discussion of the different GPRS releases can be found in reference 12.

In general, the proposed tight-coupling architecture provides a novel solution for interworking between 802.11 WLANs and GPRS. (In the rest of this section, the term *WLAN* refers to IEEE 802.11 WLAN, unless otherwise specified.) This architecture features many benefits, such as the following:

- Seamless service continuation across WLAN and GPRS. The users are able to maintain their data sessions as they move from WLAN to GPRS, and vice versa. For services with tight QoS requirements, seamless service continuation is subject to WLAN QoS capabilities.
- Reuse of GPRS AAA.
- Reuse of GPRS infrastructure (e.g., core network resources, subscriber databases, billing systems) and protection of cellular operator's investment.
- Support of Lawful Interception for WLAN subscribers (see reference 13).
- Increased security because GPRS authentication and ciphering can be applied on top of WLAN ciphering.
- Common provisioning and customer care.
- Access to core GPRS services such as Short Messaging Service (SMS), location-based Services, MMS, etc.

12.5.1   System Description

Figure 12.3 illustrates the proposed system architecture. A WLAN network is deployed with one or more off-the-shelf access points (APs), which are connected by means of a Distribution System (DS). In our system, the DS is a LAN, typically compliant with IEEE 802.3. The WLAN is deployed in an infrastructure

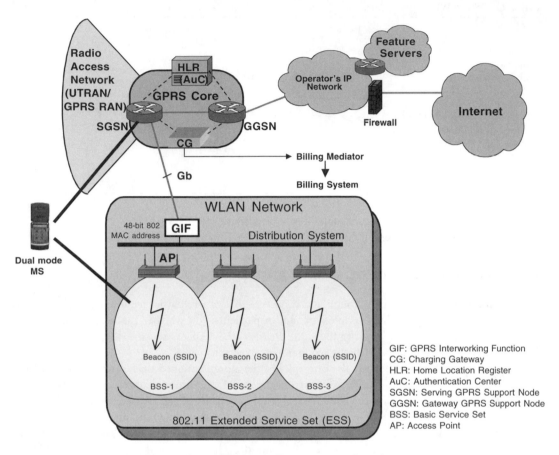

FIGURE 12.3 WLAN-GPRS integration with tight coupling: system configuration.

configuration — that is, APs behave like base stations and mobiles exchange data only with APs. The service area of one sole AP is termed as Basic Service Set (BSS).[9] Each WLAN is typically composed of many BSSs, which together form an Extended Service Set (ESS).[9]

The WLAN network is deployed as an alternative RAN and connects to the GPRS core network through the standard Gb interface. From the core network point of view, the WLAN is considered to be any other GPRS Routing Area (RA) in the system. In other words, the GPRS core network does not really identify the difference between an RA with a WLAN radio technology and an RA with GPRS radio technology.

The key functional element in the system is the *GPRS Interworking Function (GIF)*, which is connected to DS and to a Serving GPRS Support Node (SGSN) via the standard Gb interface. The main function of GIF is to provide a standardized interface to the GPRS core network and to virtually hide the WLAN particularities. The GIF is the function that makes the SGSN consider the WLAN to be a typical RA (composed of only one cell).

As discussed below, the existing GPRS protocols in the mobile are fully reused. In particular, the Logical Link Control (LLC),[14] Sub-Network Dependent Convergence Protocol (SNDCP), GPRS Mobility Management (GMM), and Session Management (SM) are used both in a standard GPRS cell and in a WLAN area. Therefore, the WLAN merely provides a new radio transport for these protocols.

When a mobile station (MS) is outside the WLAN area, its WLAN interface is in passive scan mode — that is, it scans a specific frequency band and searches for a beacon signal. (The terms *mobile station* and *user equipment [UE]* are used interchangeably in this chapter.) When a beacon is received, the Service Set Identifier (SSID) may be checked and compared against a preconfigured SSID. The SSID serves as a WLAN identifier and can help mobiles attach to the "correct" WLAN. For example, an operator could

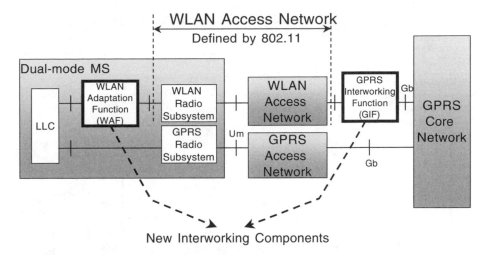

FIGURE 12.4 Tight coupling over Gb: reference diagram.

use a unique SSID and request that subscribers configure their mobiles so as to consider only this SSID as valid. As noted in Section 12.8.1, a WLAN can generally support more than one SSID.

When an MS detects a valid SSID, it performs the typical authentication and association procedures as specified in reference 9. It then enables its WLAN interface, and further GPRS signaling is carried over this interface.

MSs are dual mode — that is, they support both GPRS access and WLAN access in a seamless fashion. Seamless mobility is achieved by means of the RA Update (RAU) procedure, which is the core mobility management procedure in GPRS (see reference 2). Typically, when a mobile enters a WLAN area, an RAU procedure takes place and subsequent GPRS signaling and user data transmission is carried over the WLAN interface. Similarly, when the mobile exits a WLAN area, another RAU procedure takes place and the GPRS interface is enabled and used to carry further data and signaling traffic. From the core network point of view, handover between WLAN and GPRS is considered as handover between two individual cells.

It is important to point out that in an 802.11 WLAN, mobile terminals use 48-bit IEEE 802 addresses as Medium Access Control (MAC) addresses, which are hard coded in their network interface cards. Such addresses are also used for addressing in DSs; therefore, terminals attached to DS (e.g., GIF) and WLAN terminals share the same address space. In the configuration shown in Figure 12.3, MSs in the WLAN send uplink GPRS traffic to the MAC address of GIF, and, similarly, downlink GPRS traffic is sent from GIF to the MAC addresses of MSs. Below we explain how MSs discover the MAC address of GIF as well as the identity of the RA that corresponds to the WLAN.

The reference diagram of the proposed architecture is illustrated in Figure 12.4. The MS has two radio subsystems — one for GPRS access and another for WLAN access. The WLAN Adaptation Function (WAF) identifies when the WLAN radio subsystem is enabled (i.e., when the MS associates with a valid AP) and informs the LLC layer, which subsequently redirects signaling and data traffic to the WLAN. Note that all standard GPRS protocols operating ontop of LLC (SNDCP, GMM, SM) function as usual and do not identify which radio subsystem is used. WAF is a key component in the MS and is discussed below in greater detail.

12.5.2 Protocol Architecture

The protocol architecture is illustrated in Figure 12.5. As shown, the MS supports two radio subsystems (or interfaces) for transporting GPRS signaling and user data. The first interface is implemented with the GPRS-specific Radio Link Control (RLC)/MAC and physical layers, whereas the second is implemented with the 802.11-specific MAC and physical layers. These two interfaces provide two alternative

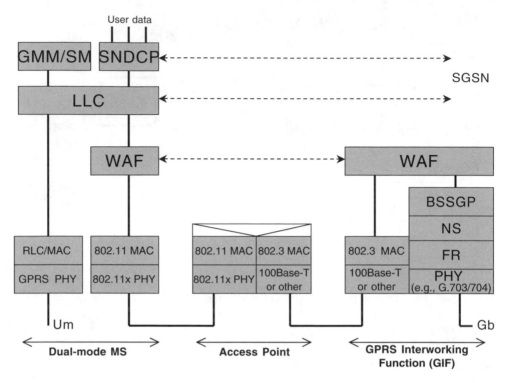

FIGURE 12.5 Tight coupling over Gb: protocol architecture.

means for transporting of LLC Packet Data Units (PDUs). Typically, when the MS is outside a WLAN area, LLC PDUs are transmitted over the GPRS interface (Um). However, when the mobile enters a WLAN area, LLC PDUs are transmitted over the WLAN interface. This switching is performed with the aid of WAF and could be completely transparent to the user and to upper GPRS layers.

As shown in Figure 12.5, WAF operates in both MS and GIF. It provides an adaptation function for interworking between LLC and 802.11 MAC (in the mobile) and between 802.3 MAC and Base Station Subsystem GPRS Protocol (BSSGP) (in the GIF). The signaling exchanged between the two WAF peers is described in the next section. Apart from WAF, GIF implements also the GPRS protocols specified on the G_b interface — that is, Frame Relay (FR), Network Service (NS), and BSSGP.[15]

The AP implements the 802.11 and 802.3 protocols and a simple interworking function that provides bridging between them. Such APs are already available in the market.

12.5.3 WLAN Adaptation Function

The main component in the proposed tight-coupling system is the WAF, which is implemented in every dual-mode MS and in the GIF and supports the appropriate interworking functions. With the aid of WAF, it becomes feasible to transport GPRS signaling and data over 802.11 WLANs.

WAF provides the following functions:

- It signals the activation of WLAN interface when the mobile enters a WLAN area. It also signals the change of RA to GMM when a mobile enters a WLAN area and gets associated with an AP.
- It supports the GIF/RA Identity (RAI) discovery procedure (discussed below), which is initiated by MSs in order to discover the MAC address of GIF and the RAI of the WLAN.
- It supports the paging procedure on G_b, used when the SGSN needs to page an MS. During this procedure, WAF sends an appropriate signaling message to the MS in order to alert it and respond to the page.

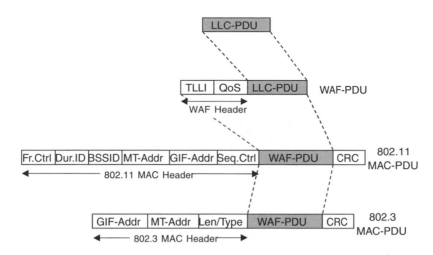

FIGURE 12.6 Encapsulation scheme.

- It transfers uplink LLC PDUs from the MS to the GIF by using the transport services provided by the 802.11 MAC. It also transfers downlink LLC PDUs from the GIF to mobiles.
- It supports QoS by implementing transmission scheduling in GIF and in the MS.
- It transfers the Temporary Logical Link Identifier (TLLI) and QoS information in the WAF header. The TLLI is a temporary MS identifier used by LLC layer for addressing purposes (see reference 14 for further details).

The encapsulation scheme used in the uplink direction as well as the format of a WAF PDU are illustrated in Figure 12.6. Each LLC PDU is encapsulated into a WAF PDU, which includes TLLI and QoS in the header. TLLI is used by GIF to update an internal mapping table that correlates TLLIs with 802 MAC addresses. In order to support the paging procedure, GIF also needs to correlate the International Mobile Subscriber Identities (IMSIs) with 802 MAC addresses. The correlation between TLLIs and 802 MAC addresses is used for forwarding downlink LLC PDUs received on G_b interface to the correct mobile on the WLAN. Note that the SGSN uses TLLI on G_b as address information, whereas the WLAN utilizes 802 MAC addresses. In the uplink direction, QoS contains the following attributes: (1) Peak Throughput, (2) Radio Priority, and (3) RLC mode (see references 14 and 15). These QoS attributes are primarily used for scheduling in the MS and GIF. In the downlink direction, the QoS may be empty because there is no need to transfer any QoS parameters to the mobile.

The 802.11 MAC header and 802.3 MAC header shown in Figure 12.6 are the standard headers specified by 802.11 and 802.3 standards, respectively.

12.5.4 GIF/RAI Discovery Procedure

The GIF/RAI discovery is a key procedure carried out immediately after an MS enters an 802.11 WLAN area and gets associated with an AP. The WAF in the MS initiates this procedure for the following reasons:

- To discover the 802 MAC address of GIF. All uplink LLC PDUs are subsequently transmitted to this MAC address.
- To discover the RAI that corresponds to the WLAN network.
- To send the MS's IMSI value to GIF. This value is subsequently used by GIF to support the GPRS paging procedure. By learning the IMSI of a particular MS, GIF can forward subsequent paging messages from the SGSN.

Figure 12.7 illustrates the signaling flow during the GIF/RAI procedure. This procedure is initiated after the 802.11 MAC layer is enabled — that is, after the mobile gets associated with a particular AP.

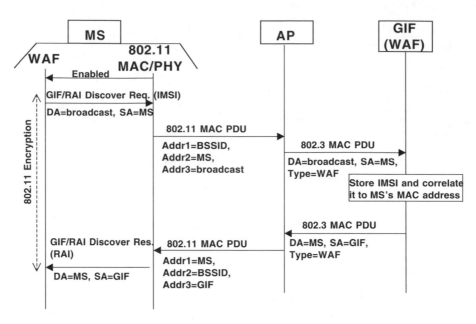

FIGURE 12.7   Signaling flow during a GIF/RAI discovery procedure.

The WAF layer in MS sends a request to 802.11 MAC to transmit a PDU with a source address (SA) equal to the MS's MAC address and a destination address equal to *broadcast*. This PDU is a *GIF/RAI Discover Request* message that includes the IMSI value of the MS. The 802.11 MAC layer transmits an 802.11 MAC PDU with the appropriate address information (designated as Addr1, Addr2, Addr3). Note that this PDU is directed to the AP with the identity Basic Service Set Identification (*BSSID*). The AP relays this message to the DS and is finally received by GIF, which associates the IMSI with the MS's 802 MAC address (designated as the MS). Subsequently, the WAF in GIF responds with a *GIF/RAI Discover Response* that includes the RAI of the WLAN. The MS receives this response, stores the GIF address and the RAI, and notifies the GMM layer that the current GPRS RA has changed. In response, the GMM layer initiates the normal GPRS RAU procedure and notifies the SGSN that the MS has changed RA. After that, normal GPRS data and signaling is performed over the WLAN.

In conclusion, with the proposed tight-coupling system, we have shown that, with the aid of WAF, MSs can seamlessly move between WLAN and GPRS radio access and use the normal GPRS procedures for mobility management.

12.6   A Loose-Coupling Architecture

As mentioned previously, loose coupling is another approach that provides interworking between GPRS and WLAN at the G_i interface (see Figure 12.2). In this section, we propose a specific loosely coupled interworking architecture and explain its key aspects.

Figure 12.8 illustrates the proposed loosely coupled architecture. As you can see, the WLAN is coupled with the GPRS network in the operator's IP network. Note that, in contrast to tight coupling, the WLAN data traffic does not pass through the GPRS core network but goes directly to the operator's IP network (or Internet). In this architecture, SIM-based authentication may be supported in both the GPRS and the WLAN networks to gain access to the operator's services. This architecture also supports integrated billing, via the Billing Mediator, into a common Billing System. The WLAN network may be owned by a third party, with roaming and mobility enabled via a dedicated connection between the operator and the WLAN or over an existing public network, such as the Internet (although only one method of roaming is required; both connections are shown in Figure 12.8 for completeness).

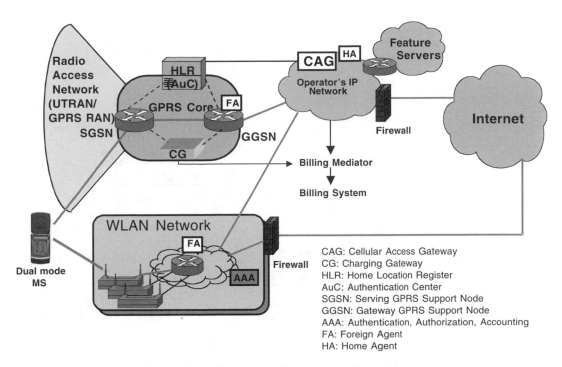

FIGURE 12.8 WLAN-GPRS integration with loose coupling: system configuration.

Loose coupling utilizes standard Internet Engineering Task Force (IETF)-based protocols for authentication, accounting, and mobility. It is therefore not necessary to introduce cellular technology into the WLAN, as is the case with the tight coupling. Roaming can be enabled across all types of WLAN implementations, regardless of who owns the WLAN network, solely via roaming agreements. The following sections describe the key aspects of security, billing, and mobility of this architecture in more detail.

12.6.1 Authentication

An authentication similar to GPRS may occur within the WLAN, depending on the particular implementation. If the GPRS operator owns the WLAN, it is most likely that the operator will want to reuse SIM-based authentication (or USIM-based for UMTS subscribers) within the WLAN environment. Similarly, for a subscriber to access services provided by a GPRS operator over any WLAN access network, regardless of whether the WLAN is owned by a GPRS operator, (U)SIM-based authentication may be used. The architecture proposed here, and illustrated in Figure 12.9, supports (U)SIM-based authentication, in much the same way as in GPRS. The Cellular Access Gateway (CAG) acts as an authenticator for WLAN users.

The authentication procedure shown in Figure 12.9 is based on the deployment of IEEE 802.1X[16] with 802.11.[9] In this architecture, the CAG provides the AAA server functionality in the cellular operator's IP core. The CAG interworks with the Home Location Register (HLR) to obtain the authentication credentials used in creating the authentication challenge to the MS and validating the response to the challenge. To do this, the CAG must interact with the HLR in a way similar to how SGSN interacts with HLR in the standard GPRS authentication procedure. The Extensible Authentication Protocol (EAP)[17] is used in the WLAN to perform the authentication of the MS, passing the subscriber identity, SIM-based authentication data, and encrypted session key(s) to be used for encryption for the life of the session. If it is undesirable to use SIM-based authentication in the WLAN system, standard username and password procedures may be used.

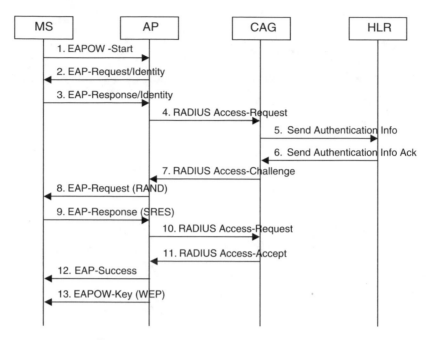

FIGURE 12.9 SIM-based authentication over a WLAN.

The signaling diagram in Figure 12.9 is conforming to the principles in 802.1x and to GSM/AKA authentication as specified in reference 19. The authentication flow starts right after the MS has associated with an AP. At first, the MS sends an EAP-over-WLAN (EAPOW) Start message to trigger the initiation of 802.1x authentication. In steps 2 and 3, the identity of the MS is obtained with the standard EAP-Request/Response messages defined in reference 17. Next, the AP initiates a RADIUS[18] dialogue with the CAG by sending an Access-Request message that contains the identity reported by the MS. In our case (we consider SIM-based authentication), this identity typically includes the IMSI value stored in the SIM card. CAG uses IMSI and possibly other information included in the identity (e.g., a domain name) to derive the address of the HLR that holds subscription data for that particular MS. In steps 5 and 6, the CAG retrieves one or more authentication vectors from the HLR. These could be either UMTS authentication vectors (if the MS is equipped with a USIM) or GSM authentication vectors (see reference 2 for their differences). In both cases, a random challenge (RAND) and an expected response (XRES) is included in every authentication vector. In steps 7 and 8, the RAND is sent to the MS, which runs the authentication algorithm implemented in the (U)SIM card and generates a challenge response value (SRES). In steps 9 and 10, SRES is transferred to CAG and compared against the corresponding XRES value received from the HLR. If these values match, a RADIUS Access-Accept is generated in step 11 (otherwise, a RADIUS Access-Reject is generated). This instructs AP to authorize the 802.1x port and to allow subsequent data packets from the MS. Note that the RADIUS Access-Accept message may also include authorization attributes, such as packet filters, which are used for controlling the user's access rights in the specific WLAN environment. In step 12, the AP transmits a standard EAP-Success message and subsequently an EAPOW-Key message (defined in reference 16) for configuring the session key in the MS.

Note that the authentication and authorization in the above procedure is controlled by the MS's home environment (i.e., the home GPRS network). The AP in the visited WLAN implements 802.1x and RADIUS but relies on the HLR in the home environment to authenticate the user. As shown in Figure 12.10, which illustrates the protocol architecture for the aforementioned authentication, the MS is ultimately authenticated by HLR by means of either GSM AKA or UMTS AKA mechanisms.

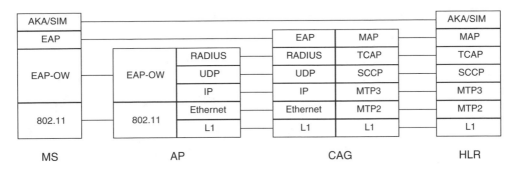

FIGURE 12.10 Loosely coupled WLAN control plane for authentication.

12.6.2 Encryption

A commonly believed weakness of the 802.11 WLAN standard is in its encryption technology. Wired Equivalent Privacy (WEP) has been shown to be a relatively inefficient encryption scheme if used as-is as the only form of encryption. With the use of EAP, WEP may be enhanced by the use of a unique session key for each user of the WLAN. Typically, a WEP-encrypted channel may be compromised in a matter of hours due to its weak encryption algorithm and the standard practice of a universal key, shared by all users of the WLAN network. By implementing 802.1x, the key sharing is no longer necessary, because a new key is derived on a per-user, per-session basis. Even if the session duration exceeds the amount of time for a breech to occur, the attacker may use the key to decrypt data only of the session to which the key belongs and only for the remaining duration of that session, and all other sessions remain secure. To further enhance security, the WLAN system may support the more advanced encryption schemes specified in reference 20.

12.6.3 Billing

Integrated billing is achieved via the Billing Mediator function in the loosely coupled architecture of Figure 12.8. The AP in the WLAN network may report accounting statistics to the CAG via standard AAA procedures (e.g., RADIUS Accounting), which will subsequently report these statistics to the Billing Mediator. Similarly, the GPRS core (i.e., SGSN/GGSN) will report accounting statistics pertaining to GPRS usage. The functionality of the Billing Mediator is to convert accounting statistics from both the GPRS and the WLAN access networks into a format native to the particular Billing System used by the operator. The Billing System may be an existing GPRS system or a standard, IP-based system used primarily by wireline ISPs.

12.6.4 Session Mobility

Contrary to the tight-coupling approach, which uses GPRS mobility management for session mobility, in the loose-coupling approach Mobile IP (MIP) can be used to provide session mobility across GPRS and WLAN domains. The MIP framework[21] consists of an MIP client (the MS), a foreign agent (FA), and a home agent (HA). As shown in Figure 12.8, the FA in the GPRS network resides in the GGSN, whereas the FA in the WLAN can reside in an access router. On the other hand, the HA is located in the operator's IP network. When the MS moves from GPRS to WLAN, it performs an MIP registration via the FA that resides in the WLAN. The FA completes the registration with the HA by providing a care-of address (CoA) to the HA to be used as a forwarding address for packets destined to the MS. The FA then associates the CoA with that of the MS and acts as a "proxy" on behalf of the MS for the life of the registration. This way, the MS does not need to change its IP address when it moves to WLAN (the same holds true when the MS moves from WLAN to GPRS). A more detailed description of MIP operation can be found in reference 21.

TABLE 12.2 Loose vs. Tight Coupling: A Side-by-Side Comparison

Category	Tight Coupling	Loose Coupling
Authentication	Reuse GPRS authentication for WLAN user. Reuse GPRS ciphering key for WLAN encryption.	CAG to provide SIM-based authentication interworking. Only RADIUS-based authentication is an alternative.
Accounting	Reuse GPRS accounting.	Billing Mediator to provide common accounting.
WLAN-Cellular Mobility	SGSN is the call anchor, and Intra-SGSN handovers provide mobility.	Home agent is the call anchor, and Mobile IP handovers between GGSN and access router provide mobility. Home agent could be collocated at the GGSN, CAG, or somewhere in an external network.
Context Transfer	Fine-grained context information is available, e.g., QoS parameters, information about multiple flows, etc.	Limited Context Transfer possible between GGSN and WLAN through current draft proposals in IETF Seamoby working group.
System Engineering	Impact of high-speed WLAN network on existing GSN from bearer and signaling standpoint is an issue.	WLAN and GPRS networks can be engineered separately.
New Development	WLAN Terminal modifications for GPRS signaling modifications in WLAN network or modifications in SGSN	CAG for SIM-based authentication. Billing Mediator for Accounting.
Standardization	A new interface in the SGSN might be required, specifically for connecting to WLANs.	EAP-SIM and EAP-AKA is being pursued in IETF PPPext working group.
Target Usage	Applies primarily to WLAN networks owned by cellular operators. Has limited application when the WISP is different from cellular operator.	Applies more broadly.

12.7   Tight vs. Loose Coupling

As mentioned earlier, there are two generic approaches for WLAN-cellular integration: tight coupling and loose coupling. In Table 12.2 we show a brief comparison between these two coupling methods.

With tight coupling, the WLAN connects to the SGSN either via an already standardized interface (e.g., G_b or $I_{u\text{-}ps}$) or via a new interface, specified for optimal performance with WLANs. Tight coupling provides a firm coupling between WLAN and GPRS, and its main advantage is the enhanced mobility across the two domains, which is entirely based on GPRS mobility management protocols. In addition, tight coupling offers reuse of GPRS AAA and protects the operator's investment by reusing the GPRS core network resources, subscriber databases, billing systems, and so forth. Moreover, it can support GPRS core services such as SMS and Lawful Interception for WLAN subscribers.

Nevertheless, tight coupling is primarily tailored for WLANs owned by cellular operators and cannot easily support third-party WLANs. Also, some cost and capacity concerns are associated with the connection of a WLAN to an SGSN. For example, the throughput capacity of an SGSN could be sufficient for supporting thousands of low bit-rate GPRS terminals but would not be sufficient for supporting hundreds of high bit-rate WLAN terminals. More important, tight coupling cannot support legacy WLAN terminals, which do not implement the GPRS protocols. These are some of the reasons that account for the current trend toward loose coupling.

Loose coupling is mainly based on IETF protocols, which are already implemented in live WLANs. Consequently, it imposes minimal requirements on WLANs. However, it mandates the provisioning of new equipment by the cellular operator — mainly, specific AAA servers for interworking with WLANs (see Section 12.8.2). It also requires the implementation of MIP for supporting mobility across the two access networks. The typical high latency associated with MIP registrations is an issue and might not permit seamless session handovers for some demanding applications.

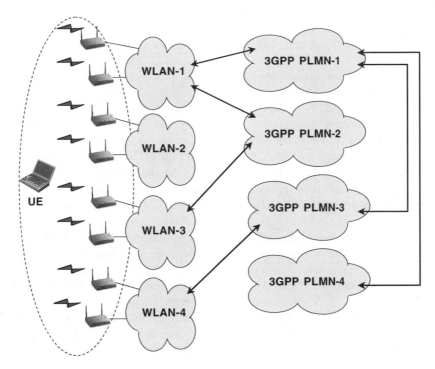

FIGURE 12.11 High-level WLAN/3GPP interworking model.

In general, the choice of the optimal interworking architecture should be determined by a number of factors. For example, if the wireless network is composed of a large number of WLAN operators and cellular operators, then the loosely coupled architecture would be the best choice. On the other hand, if the WLAN network is exclusively owned and operated by a cellular operator, then the tightly coupled architecture might become more attractive. Independent of the choice of the architecture, WLAN technology can and will play an important role in supplementing wide-area cellular data networks and enabling IP multimedia services in hotspots.

12.8 Interworking between WLANs and 3GPP Networks

After reviewing the generic approaches for interworking between WLANs and cellular telecommunication networks (i.e., the loose and tight coupling) and discussing the details of some proposed interworking techniques, we turn attention to the specific characteristics of the loose coupling architecture specified by 3GPP for the interworking of WLANs and 3GPP networks. As noted before, the 3GPP cellular systems have evolved out of GSM and are specified in phases, formally called *releases*, e.g., 3GPP Release 1999 (or UMTS phase 1), 3GPP Release 5, and 3GPP Release 6. All 3GPP specifications for each release can be found online at http://www.3gpp.org/ftp/specs; a detailed description of the various releases can also be found in reference 12.

12.8.1 WLAN/3GPP Interworking Model

The general WLAN/3GPP interworking environment we assume is illustrated in Figure 12.11. In this figure, the User Equipment (UE) receives signal simultaneously from four WLANs, namely, WLAN-1, WLAN-2, WLAN-3, and WLAN-4. (As noted before, the terms UE and MS are used interchangeably in this chapter.) As before, we assume that each WLAN is based on IEEE 802.11 technology and therefore each one is broadcasting a single SSID. Note that multiple SSIDs can be supported by an IEEE 802.11

WLAN, but only one is broadcast. The nonbroadcast SSIDs are provided in response to a probe request from a WLAN client (see probe request/response in reference 9).

The arrows in Figure 12.11 indicate direct roaming agreements. For example, WLAN-1 has a direct roaming agreement with 3GPP Public Land Mobile Network-1 (PLMN-1) and 3GPP PLMN-2, whereas WLAN-3 has a direct roaming agreement with only 3GPP PLMN-2. In addition, 3GPP PLMN-1 and 3GPP PLMN-3 have a direct roaming agreement between each other. Note that 3GPP PLMN-4 has a roaming agreement with only 3GPP PLMN-1 (i.e., it does not have a direct roaming agreement with any WLAN). Also, indirect roaming agreements are feasible (e.g., between WLAN-1 and 3GPP PLMN-4).

The UE shown in Figure 12.11 is assumed to be WLAN capable (i.e., it features a WLAN radio interface) and to have a 3GPP subscription — that is, it is equipped with a (U)SIM card. As noted before, one of the primary requirements is that the UE must be authenticated by its 3GPP home PLMN using its (U)SIM credentials and algorithms, and then (provided the authentication is successful) it must use the WLAN for obtaining IP connectivity. To satisfy such a requirement, several challenges need to be addressed. For instance, the UE must choose a WLAN that is 3GPP interworking capable and also has a direct or indirect roaming agreement with its 3GPP home PLMN. Assume, for example, that the 3GPP home PLMN of the UE in Figure 12.11 is 3GPP PLMN-1. It is clear, then, that the UE must choose WLAN-1 because it is the only one with a roaming agreement with 3GPP PLMN-1. On the other hand, if the home PLMN of UE is 3GPP PLMN-3, then the WLAN-4 should ideally be chosen and, finally, if the home PLMN of UE is 3GPP PLMN-4, then WLAN-1 should again be chosen because it is the only one that provides indirect interworking with 3GPP PLMN-4 (via 3GPP PLMN-1). Such network selection problems are vital in the context of WLAN/3GPP interworking and are discussed in more detail in Section 12.8.1.1.

The WLANs illustrated in Figure 12.11 may be owned by either of the following:

- A 3GPP system operator
- A public network operator (e.g., a public WLAN operator) who is not a 3GPP PLMN operator
- An entity providing WLAN access in a local area (i.e., building manager or owner or airport authority) but who is otherwise not a public network operator
- A business entity that may be providing a WLAN for its internal use that also wishes to allow interconnection, and possibly visitor use, for some or all of their WLANs

A WLAN may support several types of access control, such as *open access*, *Universal Access Mechanism (UAM)*,[29] and *IEEE 802.1x*.[16] Open access is useful for providing free access to some type of services, such as advertisements and local information. Open access uses the IEEE 802.11 open authentication, which is merely a null authentication scheme. Most IEEE 802.11 WLANs today support open authentication at the 802.11 MAC layer for allowing access to free services. The access to chargeable services is controlled by an access control scheme implemented "on-top" of IEEE 802.11, such as the UAM.[29] With UAM, the UE is associated with an AP and then assigned an IP address usually with Dynamic Host Configuration Protocol (DHCP). Access to chargeable services is blocked by a gateway, referred to as a Public Access Control (PAC) gateway. To get authenticated to the PAC gateway, the WLAN subscriber uses a Web browser to access a "welcome" page, which prompts for the user's credentials (usually, a username and password). The privacy of these credentials is ensured by running HTTP over a secure layer, such as Secure Sockets Layer (SSL) or Transport Layer Security (TLS). With 802.1x access control, the UE uses an EAP method (as opposed to the HTTP method in UAM) to get authenticated. One particular characteristic of 802.1x access control is that it is enforced right after the UE associates with an AP. Therefore, with 802.1x a user must always be authenticated before accessing the WLAN network. In this case, access to free services can be realized by using, for example, a well-known "guest" username and password. Further details on 802.1x are provided later on.

As mentioned earlier, one primary requirement of WLAN/3GPP interworking is to provide (U)SIM-based authentication, which implies that a WLAN capable to interwork with 3GPP systems must support 802.1x access control. However, a public WLAN is also required to support the legacy UAM and open access control schemes because these schemes are in wide use today and supported by all legacy WLAN clients. This calls for several access control schemes to be supported by a single WLAN. One way to make that feasible is to use at least two different SSIDs and select the access control scheme based on the SSID used by a UE in its association request. In particular, when a UE tries to associate to the broadcast SSID, then the open/UAM access control could be used (in order to support legacy clients), and when a UE tries to associate to a private (nonbroadcast) SSID, the 802.1x access control could be used. This way, both legacy WLAN clients as well as new 3GPP WLAN clients could be supported by a single WLAN. As discussed in the next subsection, a predefined "well-known" private SSID could be used by 3GPP UEs, which needs to be supported by all WLANs that interwork with 3GPP systems.

12.8.1.1 Network Selection

In the environment shown in Figure 12.11, the UE needs to perform the following selection procedures:

1. Select a WLAN that provides interworking with 3GPP PLMNs, i.e., it has a direct roaming agreement with at least one 3GPP PLMN. For example, the UE in Figure 12.11 must not select WLAN-2 because it does not support interworking with 3GPP networks.
2. In addition to selecting a suitable WLAN, the UE has to select one of the PLMNs that interwork with this WLAN. For example, if the preferred PLMN of the UE in Figure 12.11 is 3GPP PLMN-2, then after associating with WLAN-1 the UE has to specify that it prefers to use 3GPP PLMN-2 and not 3GPP PLMN-1. It is assumed that the UE maintains a list of preferred PLMNs, typically in its smart card.

To deal with the above network-selection problem, several solutions have been proposed. (Note that this problem covers the selection of the WLAN *and* the selection of the 3GPP visited and home PLMN.) For example, in its paper Nokia[26] proposes WLANs that interwork with 3GPP PLMNs to broadcast an SSID in the format WLAN_NAME:3GPP:MCC. From this SSID, the UE knows that the WLAN provides interworking with 3GPP PLMNs in the country with Mobile Country Code <MCC>. Then it determines whether its preferred PLMN in this country is supported by sending a probe request with SSID=3GPP:Operator_Brand:*PLMN_ID*, where *PLMN_ID* is the two- or three-digit Mobile Network Code (MNC) code of its preferred PLMN. Because this solution requires changes to the broadcasted SSID, it is not compatible with legacy WLANs.

Cisco and Intel[27] in their paper do not propose the use of SSIDs for network selection. To support discovery of meta-level data about a WLAN, they propose to encode such data as Extensible Markup Language (XML) and deliver it to a smart client via an EAP- or HTTP-based method. Such data could include network capabilities and provide information such as (1) the MCC/MNC of the interworking PLMN; (2) the URL of a Web page for branding and help information to assist subscribers of roaming partners; (3) the ability of the WLAN to support QoS; (4) detailed charging rates, terms, and so forth. According to this proposal, a smart client would first associate using the broadcast SSID, recover information stored in the "welcome page" of the WLAN, and then optionally trigger EAP authentication by reassociating with a private SSID. (This private SSID is different for every preferred PLMN and is part of the data included in the XML file.) The UE's Network Address Identifier (NAI), which is included in EAP signaling, is used by the WLAN for taking AAA routing decisions and thus for using the preferred PLMN.

RIM and Motorola[28] in their paper propose that the UE identify whether a WLAN is 3GPP interworking capable by transmitting a probe request including a predefined "well-known" SSID. If the WLAN is 3GPP interworking capable, the UE is associated by using this SSID and receives XML metadata through a new EAP method.

What was finally decided by 3GPP is to specify a well-known 3GPP-specific SSID, which should be supported by all WLANs interworking with 3GPP systems. This SSID may either be the broadcast SSID or a private SSID. After a UE confirms that a WLAN supports the 3GPP-specific SSID, it performs an association by using this SSID, which results in an 802.1x access control procedure to initiate. To indicate its preferred 3GPP visited PLMN, as well as its 3GPP home PLMN, the UE uses an appropriately encoded NAI value. This is further explained in Section 12.8.2.2.

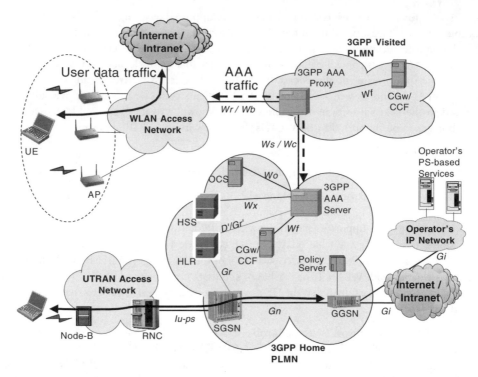

FIGURE 12.12   WLAN/3GPP interworking architecture for scenario 2 (roaming case).

12.8.2   Interworking Architecture for Scenario 2

Figure 12.12 illustrates the proposed WLAN/3GPP interworking architecture that meets the requirements of scenario 2, for the general case when the WLAN is not directly connected to the UE's 3GPP home Public Land Mobile Network (PLMN) (roaming case). The architecture for the nonroaming case can be straightforwardly derived — that is, the 3GPP AAA server is directly connected to the WLAN over the *Wr/Wb* interface, and the *Ws/Wc* interface is not used. The 3GPP home PLMN illustrated in this figure supports UTRAN UEs that access the 3GPP core network over the typical UTRAN access network and, in addition, supports WLAN UEs. Note that the user data traffic of UTRAN UEs and WLAN UEs is routed completely differently. In particular, the user data traffic of WLAN UEs is routed by the WLAN itself toward the Internet or an intranet, whereas the user data traffic of UTRAN UEs is routed through the 3GPP packet-switched core network, encompassing the SGSN and the GGSN elements. UTRAN UEs can typically have access to the Internet, a corporate intranet, or the 3GPP operator's PS-based services, such as WAP or MMS.

Note that in scenario 2 only AAA signaling is exchanged between the WLAN and the 3GPP home PLMN (via the 3GPP visited PLMN), for authenticating, authorizing, and charging purposes. The so-called *3GPP AAA server* is a new functional component that must be incorporated in a 3GPP PLMN in order to support interworking with WLANs. The 3GPP AAA server in the 3GPP home PLMN terminates all the AAA signaling with the WLAN and interfaces with internal 3GPP components, such as the HLR, the Charging Gateway/Charging Collection Function (CGw/CCF), the Home Subscriber Server (HSS), and the Online Charging System (OCS). Note that both the HLR and the HSS are basically subscription databases, used by the 3GPP AAA server for acquiring subscription information for particular WLAN UEs. Typically, if an HSS is available, the HLR need not be used.

The 3GPP AAA server can also route AAA signaling to or from another 3GPP PLMN, in which case it functions as a proxy and is referred to as *3GPP AAA Proxy*. In Figure 12.12, the 3GPP AAA server in the 3GPP visited PLMN takes the role of a proxy that routes AAA signaling to or from the 3GPP AAA server in the UE's home PLMN. The 3GPP AAA Proxy identifies the IP address of the 3GPP AAA server by using the NAI identity sent by the UE. There is also a WLAN AAA proxy (not shown in Figure 12.12), which resides in the WLAN and routes the AAA messages to the appropriate 3GPP AAA proxy/server based on the NAI identity sent by the UE. Note that NAI is used for two routing decisions: one at the WLAN AAA proxy and another at the 3GPP AAA proxy.

As noted earlier, the user data traffic of WLAN UEs is routed through the WLAN to an external intranet or Internet. Although this traffic is routed by the WLAN, the 3GPP PLMN (both home and visited) can apply a specific policy to this traffic. For example, during the authorization phase the 3GPP AAA server may define a set of restrictions to police user data traffic, such as "do not allow FTP traffic" or "allow HTTP traffic only to or from a specific IP address."

12.8.2.1 Reference Points

The reference points (also referred to as *interfaces*) illustrated in the architecture diagram of Figure 12.12 are briefly discussed below. A more thorough discussion can be found in reference 5. Note that we discuss only the interfaces relevant to WLAN/3GPP interworking; a brief description of the 3GPP internal interfaces, such as G_r, $I_{u\text{-}ps}$, G_i, and G_n, can be found in reference 2.

Wr/Wb: This interface carries AAA signaling between the WLAN and the 3GPP visited or home PLMN in a *secure manner*. The protocol across this interface is primarily *Diameter*,[23,24] which is used for providing AAA functions and for carrying the EAP messages exchanged between the UE and the 3GPP AAA server. However, because legacy WLANs need to be supported, the Radius[18] protocol must also be supported across *Wr/Wb*, and this calls for RADIUS-Diameter interworking functions across the legacy WLAN and the 3GPP AAA proxy or server. Diameter is the preferred AAA protocol because it can provide enhanced functionality compared with Radius. Note that the term *Wr* is used to specifically refer to the interface that carries authentication and authorization information, whereas *Wb* is used to refer to the interface that carries accounting information. *Wr* and *Wb* corresponds to the *Ls* and *Lp* interfaces specified by WIG.[30]

Ws/Wc: This interface provides the same functionality as *Wr/Wb* but runs between a 3GPP AAA proxy and a 3GPP AAA server. Because *Ws/Wc* is across two 3GPP networks, it does not have to support Radius (i.e., it supports only Diameter). Again, the term *Ws* is used to refer to the interface that carries authentication and authorization information, whereas *Wc* is used to refer to the interface that carries accounting information.

Wf: This interface is located between the 3GPP AAA server and the 3GPP Charging Gateway (CGw)/ Charging Collection Function (CCF). The prime purpose of the protocols crossing this interface is to transport/forward charging information toward 3GPP operator's CGw/CCF located in the visited or home PLMN. The application protocol on this interface is Diameter based. The *Wf* interface corresponds to the *La* interface specified by WIG.[30]

Wo: This interface is used by a 3GPP AAA server to communicate with the 3GPP OCS and exchange online charging information so as to perform credit control for the online-charged subscribers. The application protocol across this interface is again Diameter based.

Wx: This interface is located between the 3GPP AAA server and the HSS and is used primarily for accessing the WLAN subscription profiles of the users, retrieval of authentication vectors, etc. As noted before, if *Wx* is implemented, then the interface to the HLR is not required. The protocol across *Wx* is either Mobile Application Part (MAP) or Diameter based.

D′/Gr′: This optional interface is used for exchanging subscription information between the 3GPP AAA server and the HLR by means of the MAP protocol. It is typically used when the enhanced functionality provided by the HSS is not available. Note that this interface is based either on the *D* interface, which runs between the 3GPP Mobile Switching Center (MSC) and the HLR, or on the *Gr* interface, which runs between the SGSN and the HLR (see Figure 12.12). In other words, *D′/ Gr′* normally implements a subset of the *D* functionality (denoted as *D′*) or a subset of the *Gr* functionality (denoted as *Gr′*).

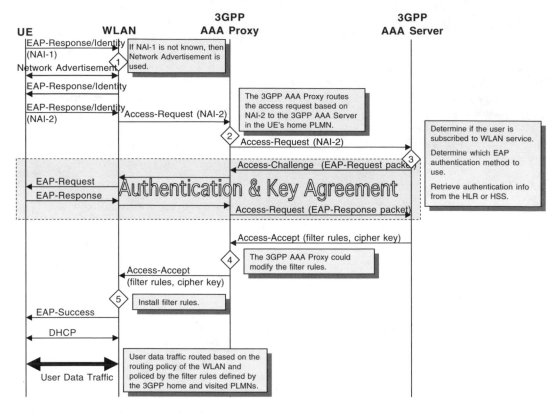

FIGURE 12.13   Generic AAA signaling for interworking with scenario 2 (roaming case).

12.8.2.2   AAA Signaling for Scenario 2

Figure 12.13 illustrates a typical message sequence diagram and some associated comments applicable to scenario 2 interworking. The figure corresponds to the roaming case shown in Figure 12.12. Initially, the UE sends its identity to the WLAN within an EAP-Response/Identity message. Note that the WLAN in Figure 12.12 encompasses an AP and a WLAN AAA proxy. The UE's identity, NAI-1, is formatted as an NAI and has the form *username@realm*, where the realm is formatted as an Internet domain name as specified in RFC 1035. As discussed before, in this case, a possible realm value could be *homeMNC.homeMCC.WLAN.3gppnetwork.org*, which points to the home PLMN of the UE. The WLAN identifies whether it can route AAA messages to this PLMN (e.g., by trying a DNS query to resolve the realm into an IP address). If the resolution is successful, a AAA access request is sent to the identified 3GPP AAA server over the Wr interface. Otherwise, the WLAN sends network advertisement data to the UE, which includes the 3GPP PLMNs it can interwork with. A proposed way to send this advertisement data is to use a new EAP method, called 3GPP-Info,[28] which carries an XML structure populated with a list of PLMN identities. Work is currently under way in IETF to specify a suitable set of EAP extensions for supporting the network advertisement. On the basis of the network advertisement data, as well as on its roaming preferences, the UE selects a preferred 3GPP visited PLMN and forms a second NAI (NAI-2) corresponding to this PLMN. NAI-2 is used by both the WLAN and the 3GPP AAA proxy for taking routing decisions and finally reaching the 3GPP AAA server in the UE's home PLMN. Therefore, NAI-2 needs to encode both the preferred 3GPP visited PLMN and the UE's 3GPP home PLMN. For this purpose, it could be encoded as *homeMNC.home MCC//username@visitedMNC.visitedMCC.WLAN.3gppnework.org*.

At point 3 in Figure 12.12, the 3GPP AAA server interacts with either the HLR or the HSS to retrieve the WLAN subscription profile of the UE and authentication information. Subsequently, the authentication and key agreement phase is initiated, which is based either on EAP-AKA or EAP-SIM methods. Message flows for this phase are detailed in reference 5 and in the EAP-AKA[19] and EAP-SIM[22] Internet drafts.

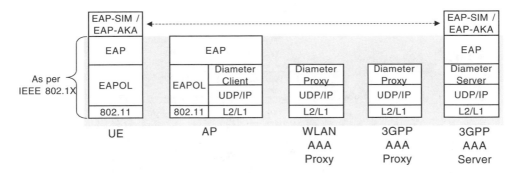

FIGURE 12.14 AAA protocol architecture (roaming case).

After the successful authentication and key agreement, the 3GPP AAA server transmits a AAA access accept message, which contains the cipher key to be used for data encryption on the radio interface and possibly some filter rules for policing user data traffic. These filter rules might be modified (e.g., more rules could be added) by the 3GPP AAA proxy based on the policy of the 3GPP visited PLMN. Note that after the EAP-Success message, which effectively terminates the authentication and authorization phase, the UE typically uses DHCP to get an IP address from within the address space of the WLAN. This may be either a public or a private IP address.

A simplified AAA protocol architecture that applies to the interworking scenario 2 (and scenario 3, as explained in the next section) is illustrated in Figure 12.14. Note that the method used to authenticate the UE and possibly the 3GPP home PLMN (EAP-AKA supports mutual authentication) needs to be implemented only by the UE and the 3GPP AAA server in UE's home PLMN. The WLAN needs to support only the generic EAP protocol and the EAP-over-LAN (EAPOL) specified in IEEE 802.1X.[16]

12.8.3   Interworking Architecture for Scenario 3

For satisfying the key requirement of scenario 3 — that is, access to 3GPP PS-based services — the user data traffic needs to be routed to the 3GPP system. A network architecture that fulfills this requirement is illustrated in Figure 12.15, which is basically an extension of the architecture of scenario 2 (see Figure 12.12). Note that Figure 12.15 also corresponds to a roaming case. As compared with Figure 12.12, note that although the AAA traffic follows the same route, the user data traffic is routed to the UE's 3GPP home PLMN to a new component called the *Packet Data Gateway (PDG)*. The user data traffic is also routed through a data gateway, referred to as the *Wireless Access Gateway (WAG)*, which in the case of roaming is in the 3GPP visited PLMN. As discussed below, this routing is enforced by establishing appropriate tunnels. The PDG functions much like a GGSN in a 3GPP PS network. It routes the user data traffic between the UE and an external PDN, which is selected based on the 3GPP PS-based service requested by the UE (WAP, MMS, IP Multimedia, etc). The requested 3GPP PS-based service is identified by the UE through a *WLAN-Access Point Name (W-APN)* that is included in the AAA signaling messages (see below for details). The PDG may also perform address translation, enforce policy, and generate charging records. The WAG functions mainly as a route policy element — that is, it ensures that user data traffic from authorized UEs is routed to the appropriate PDGs, located either in the same PLMN or in another PLMN (as shown in Figure 12.15).

It is easy to note from Figure 12.15 that WLAN UEs can access to the same services (or PDNs) used by the UTRAN UEs. In other words, the 3GPP subscribers can maintain access to the same set of services as they roam across WLAN- and 3GPP-specific radio networks.

Keep in mind that user data traffic need not always be routed to the UE's home PLMN as shown in Figure 12.15. Alternatively, user data could be routed to the PDG in the 3GPP visited PLMN in order to have access to the 3GPP PS-based services offered by the visited PLMN.

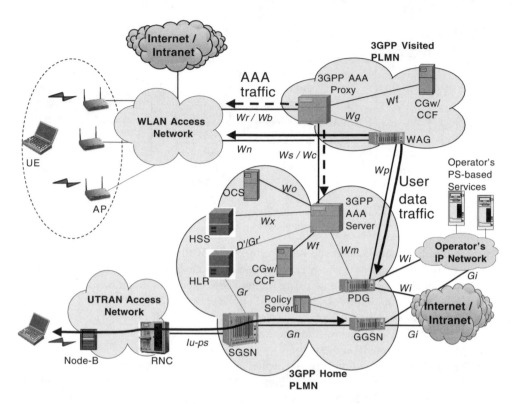

FIGURE 12.15 WLAN/3GPP interworking architecture for scenario 3 (roaming case).

12.8.3.1 Reference Points

As shown in Figure 12.15, several additional interfaces, namely, *Wn*, *Wm*, *Wi*, *Wg*, and *Wp*, are introduced for satisfying the requirements of scenario 3:

Wn: This interface is used for transporting tunneled user data between the WLAN and the WAG. WLANs that need to provide scenario 3 interworking with 3GPP networks have to deploy a *Wn*-compliant interface.

Wm: This interface is located between the 3GPP AAA server and the PDG and is used to enable the 3GPP AAA server to retrieve tunneling attributes and UE's IP configuration parameters from or via the PDG. The AAA protocol across this interface is Diameter based.

Wi: This interface is similar to the G_i reference point provided by the PS domain (see Figure 12.15). Interworking with packet data networks (PDN-1, PDN-2, etc.) is provided via the *Wi* interface based on IP. A PDN may be an external public or private packet data network or an intraoperator packet data network.

Wg: This interface is mainly used by the 3GPP AAA proxy to deliver routing policy enforcement information to the WAG. This information is used by the WAG to securely identify the user data traffic of a particular UE and apply the required routing policy.

Wp: This interface transports tunneled user data traffic between the WAG and the PDG. The *Wp* is not only used in the roaming case shown in Figure 12.15, but is also used in nonroaming scenarios. Even in the nonroaming case, user data traffic goes through the WAG in the home PLMN to the PDG also in the home PLMN. One candidate protocol across *Wp* is the GPRS Tunneling Protocol (GTP), which is used on the *Gn* interface between the SGSN and GGSN.

In addition to these interfaces, an interface from the PDG to the policy server could be implemented for supporting service-based local policy. An interface with similar functionality has already been defined for the GGSN (the so-called *Go* interface), and it is specified in 3GPP TS 29.208.

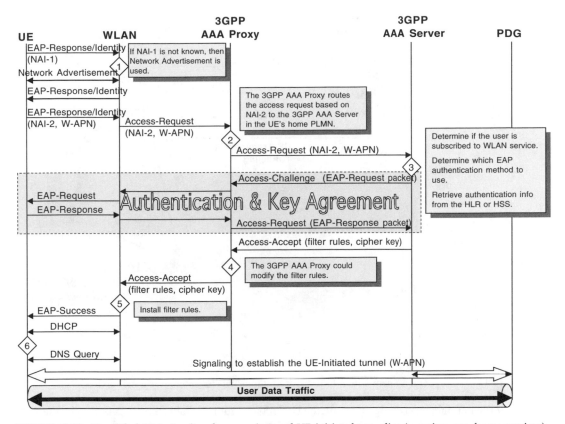

FIGURE 12.16 Simplified AAA signaling for scenario 3 and UE-initiated tunneling (roaming case, home services).

12.8.3.2 AAA Signaling for Scenario 3

The key requirement of scenario 3 is the routing enforcement of user data traffic between the UE and the PDG. This routing enforcement is applied through establishment of the appropriate tunnel. As discussed below, this tunnel may be either *UE-initiated* or *UE-transparent*. Note that although there are many tunneling options that could be considered, for the sake of brevity, we restrict our discussion here to only a few.

A particular characteristic of scenario 3 is that the UE conveys a W-APN to the 3GPP AAA server in order to explicitly indicate that access to a 3GPP PS-based service is requested. The W-APN is an identifier formatted as an Internet domain name that designates the PS-based service requested by the UE and possibly the operator through which this service is requested. In effect, the W-APN identifies the PDN that provides the requested PS-based service.

12.8.3.2.1 UE-Initiated Tunneling

The UE-initiated tunneling corresponds to the case when the tunnel establishment is initiated by the UE itself, and hence the tunnel originates at the UE and terminates at the PDG. (Another possible alternative is for the tunnel to terminate at the WAG in the visited PLMN.) Figure 12.16 illustrates a typical signaling flow for this case. Note that up to point 6 the signaling flow is the same as the one shown in Figure 12.14. After the EAP-Success message, the UE uses DHCP to obtain an IP address and then initiates a tunnel establishment with the PDG. To discover the IP address of the PDG that provides access to the requested PS-based service (or PDN), the UE performs a Domain Name System (DNS) query to resolve the requested W-APN into an IP address. For example, an W-APN encoded as *MMS.homeMNC.homeMCC. WLAN. 3gppoperator.org* would be resolved to the IP address of the PDG that provides access to the MMS service in the PLMN with the identity *<homeMCC, homeMNC>*. The W-APN is typically used by the 3GPP AAA server to select the appropriate PDG. As noted in Figure 12.16, the W-APN value may also

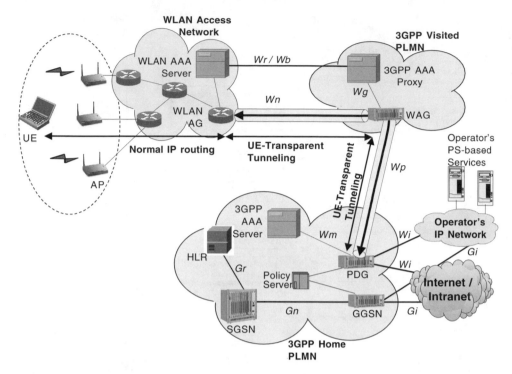

FIGURE 12.17   Routing enforcement with UE-Transparent tunneling (roaming case, home services).

be part of the parameters exchanged during the UE-initiated tunnel establishment. It is used by the PDG for service authorization, policy enforcement, and for routing the traffic on the UE-Initiated tunnel to and from the designated PDN. Before the tunnel is accepted by the PDG, tunnel authorization information is retrieved from the 3GPP AAA server.

The UE-initiated tunnel establishment may or may not request user intervention. The WLAN is expected to filter the user data traffic and allow only traffic to and from the PDG to pass through. One possible method for tunnel establishment is Mobile IP.[21]

12.8.3.2.2   UE-Transparent Tunneling

UE-transparent tunneling is used when the tunnel is established without any UE intervention. An example of such a configuration is depicted in Figure 12.17. In this case, the UE is roaming and uses PS-based services in the home PLMN. Two UE-transparent tunnels are established: one from a WLAN access gateway (WLAN AG) to the WAG and another from the WAG to the PDG in the home PLMN. In this case, all UE outbound traffic is routed to the WLAN AG with normal IP routing and then enters the first tunnel to the WAG and finally reaches the PDG over the second tunnel. Note that the WLAN AG, the WAG, and the PDG need to make forwarding decisions based on previously installed information, which is communicated during the AAA phase. The PDG would also perform address translation if the UE's IP address is not assigned from the address space of the 3GPP home PLMN. Several other alternative configurations can be realized for UE-transparent tunneling. For example, the tunnel from the WLAN AG to the WAG could be an *aggregate* tunnel (or site-to-site tunnel) that aggregates *all* user data traffic to and from the 3GPP visited PLMN. Also, the first tunnel could originate directly from an AP instead of the WLAN AG.

A simplified signaling flow that corresponds to the UE-transparent tunneling is depicted in Figure 12.18. After the UE is correctly authenticated, the 3GPP AAA server determines (e.g., from the UE's subscription) that UE-transparent tunneling should be used. Hence, it negotiates with the PDG the tunnel attributes that should be used across the *Wp* interface, and it informs the PDG about the W-APN requested by the UE. Note that the requested W-APN needs to be transferred from the UE to the 3GPP

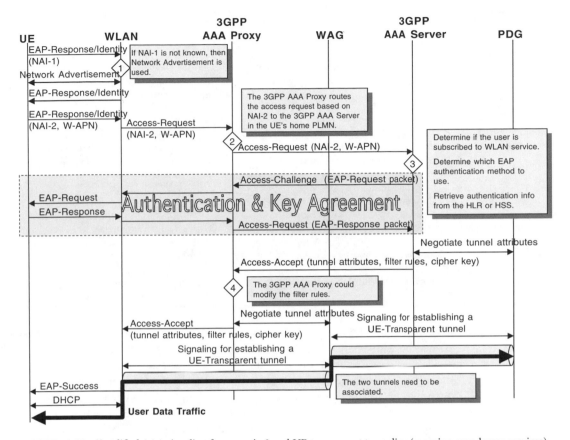

FIGURE 12.18 Simplified AAA signaling for scenario 3 and UE-transparent tunneling (roaming case, home services).

AAA server embedded in AAA messages. After the 3GPP AAA proxy receives the Access Accept (point 4 in Figure 12.18), it communicates to the WAG the tunnel attributes (the IP address of PDG, the tunnel type, etc.). In addition, the 3GPP AAA proxy decides what type of tunnel should be used across the *Wn* interface and forwards the corresponding tunnel attributes to the WLAN. After the two UE-transparent tunnels are successfully established, the UE receives the EAP-Success message and subsequently uses DHCP to receive an IP address. The DCHP packets could either be tunneled to the PDG, in which case the UE is assigned an IP address from the address space of the 3GPP home PLMN, or be processed by the WLAN, in which case the UE is assigned an IP address from the address space of the WLAN.

Table 12.3 provides a brief list of the advantages and disadvantages of UE-transparent and UE-initiated tunneling options. In general, a 3GPP PLMN could support both options or one of them based on the 3GPP operator preferences. The main advantage of UE-initiated tunneling is that is can support multiple simultaneous tunnels from the UE to one or more PDGs.

12.9   Concluding Remarks

With no doubt, the recent evolution and successful deployment of WLAN systems worldwide has fueled the need for interworking mechanisms between WLANs and cellular data networks, such as GPRS/UMTS. In response to this need, several forums and standardization bodies worldwide have initiated numerous activities for exploiting WLAN technology and integrating this technology into cellular data networks. In this chapter, we reviewed a number of generic interworking approaches (loose and tight coupling) and extensively discussed the technical details for some proposed interworking techniques. Although loose coupling is the short-term market preference, we have seen that tight coupling features significant

TABLE 12.3 Some Advantages and Disadvantages of UE-Transparent and UE-Initiated Tunneling

	UE-Transparent Tunneling	UE-Initiated Tunneling
UE intervention	Not required.	Required.
Simultaneous access to many PDNs	Because the tunnel is established as part of WLAN session setup, the UE can have access to only one PDN.	More than one tunnel can be established, each one associated with a different PDN. Hence, access to more than one PDN is feasible, e.g., access to the Internet and to a corporate intranet.
Mobility impact	The UE mobility may have a great impact on the tunnel. In particular, as the UE moves between APs, the UE-transparent tunnel might need to be reestablished. Note that the UE-transparent tunnel might not originate at an AP but from some other device in the WLAN, e.g., a WLAN access gateway as shown in Figure 12.17.	Tunnel reestablishment if not required as the UE hands over from one AP to another. So, UE mobility has minimal impact on the UE-initiated tunnel(s).
Air-interface impact	No additional air-interface overhead.	Overhead on the air interface is increased due to the encapsulation performed by the UE.
Security	More secure because the UE is not involved with the tunnel establishment.	Less secure.
UE processing	Demands no additional UE-processing capabilities.	Additional UE-processing capabilities required for performing tunnel management.

advantages (such as extended mobility support across diverse radio access technologies) and is better suited to WLANs owned by cellular operators. For the deployment of tight-coupling techniques, the development of interworking gateways is necessary in the border between the WLAN and the cellular data network. Tight coupling is also promising in scenarios in which the WLAN clients are dual-mode devices (e.g., UMTS and WLAN). In such situations, the WLAN clients have the UMTS radio protocol stack implemented, and this stack could also be exploited in WLAN radio environments to provide mobility and session management. Note, however, that the UMTS radio protocol stack is not directly operational on top of the WLAN physical and MAC layers, and the appropriate adaptation functions need to be developed (e.g., for supporting the UMTS logical channels on top of the WLAN MAC layer).

In this chapter, we also discussed the loose-coupling architecture standardized by 3GPP for interworking between WLAN and 3GPP systems. In this context, we focused specifically on scenarios 2 and 3 because they feature the highest market interest. At the time of this writing, the WLAN/3GPP interworking architecture has reached a stable state; however, it is still evolving and has not been approved yet. The architecture standardized by 3GPP will enable 3GPP subscribers to benefit from high throughput IP connectivity in strategic hotspot areas (airports, hotels, etc.) and maintain access to the same PS 3GPP services across WLAN, UTRAN, or GSM Enhanced RAN (GERAN) radio technologies. This consistent service availability across hybrid radio technologies (or service roaming) is important for maintaining a common user experience. However, session mobility (i.e., the continuation of IP sessions when changing radio technology) is also important as the user becomes more mobile and frequently moves among different radio access technologies. Session mobility is a requirement for scenarios 4, 5, and 6 (see Section 12.4) and will be an evolutional step after the service roaming of scenario 3. After the technical challenges (e.g., vertical handovers) of session mobility are efficiently addressed, cellular data users will be able to maintain the same multimedia sessions with the same quality of service as they move across various radio access technologies. This is the technical subject and the vision of many R&D projects carried out worldwide.

Acknowledgments

The author wishes to thank Chad Fors and Rajesh Pazhyannur of Motorola for their contributions to this chapter.

References

1. ETSI, Requirements and architectures for interworking between HIPERLAN/3 and 3rd generation cellular systems, ETSI TR 101 957, Aug. 2001.
2. 3GPP TS 23.060 v5.6.0, General Packet Radio Service (GPRS); service description; stage 2 (Release 5), June 2003.
3. ETSI, Broadband RANs (BRAN); HIPERLAN Type 2, system overview, ETSI TR 101 683, Feb. 2000.
4. 3GPP TR 22.934 v6.1.0, Feasibility study on 3GPP system to WLAN interworking (Release 6), Dec. 2002.
5. Draft 3GPP TS 23.234 v1.12.0, 3GPP system to WLAN interworking: system description (Release 6), July 2003. Latest draft available at http://www.3gpp.org/ftp/Specs/Latest-drafts/.
6. Pahlavan, K. et al., Handoff in hybrid mobile data networks, *IEEE Personal Communication*, Apr. 2000.
7. Krishnamurthy, P. et al., Handoff in 3G non-homogeneous mobile data networks, *European Microwave Week*, Oct. 1998.
8. Krishnamurthy, P. et al., Scenarios for inter-tech mobility, *WiLU Technical Report*, Jan. 1998.
9. IEEE, Wireless LAN Medium Access Control (MAC) and Physical Layer (PHY) specifications, IEEE standard 802.11, edition 1999.
10. 3GPP, IP multimedia subsystem: stage 2, 3GPP TS 23.228 v5.5.0, June 2002.
11. 3GPP, QoS concept and architecture, 3GPP TS 23.107 v5.5.0, June 2002.
12. Salkintzis, A.K., *Network Architecture and Reference Model, Broadband Wireless Mobile — 3Gwireless and Beyond*, ch. 3, John Wiley & Sons., Hoboken, NJ, 2002.
13. 3GPP, 3G security; lawful interception architecture and functions, 3GPP TS 33.107 v3.5.0, Mar. 2002.
14. 3GPP, Logical Link Control Layer specification, 3GPP TS 04.64 v8.7.0, Dec. 2001.
15. 3GPP, BSS GPRS protocol, 3GPP TS 08.18 v8.10.0, May 2002.
16. IEEE, Port-based network access control, IEEE standard 802.1X, edition 2001.
17. Blunk, L. and Vollbrecht, J., PPP Extensible Authentication Protocol (EAP), IETF RFC 2284, Mar. 1998.
18. Rigney, C. et al., Remote Authentication Dial In User Services (RADIUS), IETF RFC 2138, Apr. 1997.
19. Arkko, J. and Haverinen, H., EAP AKA authentication, Internet draft, draft-arkko-pppext-eap-aka-11, Oct. 2003.
20. IEEE, Specification for enhanced security, IEEE standard 802.11i/D4, May 2003.
21. Perkins, C., IP mobility support for IPv4, IETF RFC 3220, Jan. 2002.
22. Haverinen, H., *EAP SIM authentication*, Internet draft, draft-haverinen-pppext-eap-sim-12, Oct. 2003.
23. Calhoun, P. et al., DIAMETER base protocol, IETF RFC 3588, Sept. 2003.
24. Calhoun, P., et al., Diameter network access server application, http://www.ietf.org/internet-drafts/draft-ietf-aaa-diameter-nasreq-11.txt, Feb. 2003, (work in progress).
25. 3GPP TS 29.002 v3.14.0, Mobile Application Part (MAP) specification (Release 1999), Sep. 2002.
26. 3GPP technical document S2-031430, PLMN selection for 802.11 type of WLAN, Nokia, Apr. 2003. Available at http://www.3gpp.org/ftp/tsg_sa/WG2_Arch/TSGS2 _31_Seoul/tdocs/S2-031430.zip.
27. 3GPP technical document S2-031259, Network selection, Cisco and Intel, Apr. 2003. Available at http://www.3gpp.org/ftp/tsg_sa/WG2_Arch/TSGS2_31_ Seoul/tdocs/S2-031259.zip.
28. 3GPP technical document S2-031860, WLAN Network selection, RIM and Motorola, May 2003. Available at http://www.3gpp.org/ftp/tsg_sa/WG2_Arch/TSGS2_32_San_Diego/tdocs/S2-031860.zip.
29. WiFi Alliance — Wireless ISP Roaming (WISPr), Best current practices for wireless ISP roaming, version 1.0, Feb. 2003. Available at http://www.wifi.org/OpenSection/downloads/WISPr_V1.0.pdf.
30. ETSI technical document ETSI/BRAN 34d014, WLAN Interworking Group (WIG) baseline document II, Sep. 2003.

Index